Sustainable Agriculture and Crop Yield

Sustainable Agriculture and Crop Yield

Edited by Thelma Bosso

SYRAWOOD
PUBLISHING HOUSE

New York

Published by Syrawood Publishing House,
750 Third Avenue, 9th Floor,
New York, NY 10017, USA
www.syrawoodpublishinghouse.com

Sustainable Agriculture and Crop Yield
Edited by Thelma Bosso

© 2017 Syrawood Publishing House

International Standard Book Number: 978-1-68286-413-5 (Hardback)

Cataloging-in-publication Data

Sustainable agriculture and crop yield / edited by Thelma Bosso.
p. cm.
Includes bibliographical references and index.
ISBN 978-1-68286-413-5
1. Sustainable agriculture. 2. Crop yields. 3. Fisheries. 4. Sustainable fisheries. I. Bosso, Thelma.
S494.5.S86 S87 2017
630--dc23

Printed in the United States of America.

TABLE OF CONTENTS

PREFACE

Sustainable agriculture is defined as the practice of farming that is practiced alongside a systematic understanding of ecosystem management. It is aimed at strategic management of resources over a long period of time. This book is a valuable compilation of topics, ranging from the basic to the most complex advancements in the field. It is compiled in such a manner that it will provide in-depth knowledge about the theory and practice of sustainable agriculture. This book explores themes that discuss resource utilization and efficiency, management of farms according to needs as well as plant breeding and sowing patterns. Experts in the field of agricultural soil sciences, horticulture and microbiology will find this book to be extremely useful. Students that require an in-depth study on sustainable agriculture and its relation to the environment can use this book to aid their research.

This book unites the global concepts and researches in an organized manner for a comprehensive understanding of the subject. It is a ripe text for all researchers, students, scientists or anyone else who is interested in acquiring a better knowledge of this dynamic field.

 I extend my sincere thanks to the contributors for such eloquent research chapters. Finally, I thank my family for being a source of support and help.

Editor

Reconciling Pesticide Reduction with Economic and Environmental Sustainability in Arable Farming

Martin Lechenet[1], Vincent Bretagnolle[2], Christian Bockstaller[3,4], François Boissinot[5], Marie-Sophie Petit[6], Sandrine Petit[1], Nicolas M. Munier-Jolain[1]*

1 Institut National de la Recherche Agronomique, Unité Mixte de Recherche 1347 Agroécologie, Dijon, Côte d'Or, France, 2 Centre d'Etudes Biologiques de Chizé - Centre National de Recherche Scientifique, Beauvoir sur Niort, Deux-Sèvres, France, 3 Institut National de la Recherche Agronomique, Unité de Recherche 1121 Agronomie et Environnement, Colmar, Haut-Rhin, France, 4 Université de Lorraine, Vandœuvre-lès-Nancy, Meurthe-et-Moselle, France, 5 Chambre d'Agriculture des Pays de la Loire, Angers, Maine-et-Loire, France, 6 Chambre Régionale d'Agriculture de Bourgogne, Quetigny, Côte d'Or, France

Abstract

Reducing pesticide use is one of the high-priority targets in the quest for a sustainable agriculture. Until now, most studies dealing with pesticide use reduction have compared a limited number of experimental prototypes. Here we assessed the sustainability of 48 arable cropping systems from two major agricultural regions of France, including conventional, integrated and organic systems, with a wide range of pesticide use intensities and management (crop rotation, soil tillage, cultivars, fertilization, etc.). We assessed cropping system sustainability using a set of economic, environmental and social indicators. We failed to detect any positive correlation between pesticide use intensity and both productivity (when organic farms were excluded) and profitability. In addition, there was no relationship between pesticide use and workload. We found that crop rotation diversity was higher in cropping systems with low pesticide use, which would support the important role of crop rotation diversity in integrated and organic strategies. In comparison to conventional systems, integrated strategies showed a decrease in the use of both pesticides and nitrogen fertilizers, they consumed less energy and were frequently more energy efficient. Integrated systems therefore appeared as the best compromise in sustainability trade-offs. Our results could be used to re-design current cropping systems, by promoting diversified crop rotations and the combination of a wide range of available techniques contributing to pest management.

Editor: Raul Narciso Carvalho Guedes, Federal University of Viçosa, Brazil

Funding: Funding for the study was provided by the French National Research Agency ANR (STRA-08-02 Advherb project) and from Région Bourgogne. The Burgundy farms network was developed in the framework of the Réseau Mixte Technologique "Systèmes de Culture Innovants" and the project "Plus d'Agronomie, Moins d'intrants" initiated by Région Bourgogne. The long term experiment at Dijon-Epoisses was partly funded by the European Network of Excellence ENDURE. The funders had no role in study design, data collection and analysis, decision to publish, or preparation of the manuscript.

Competing Interests: The authors have declared that no competing interests exist.

* E-mail: nicolas.munier-jolain@dijon.inra.fr

Introduction

Reconciling agricultural productivity with other components of sustainability remains one of the greatest challenges for agriculture [1]. A key issue will be to achieve substantial reductions in the level of pesticide use for environmental and health reasons [2,3]. Agriculture in temperate climates is widely dominated by conventional intensive farming systems, with highly specialized crop productions and a heavy reliance on pesticides and mineral fertilizers [4]. However, increasing environmental concerns about intensive farming practices has contributed to the emergence of innovative farming systems, such as organic and integrated farming, typically presented as alternative paths to reduce pesticide use as compared to current conventional systems [5,6,7]. Whether these systems better meet sustainability criteria has been a matter of debate [8,9]. Integrated farming, recently promoted in Europe through the 2009/128/EC European directive [10], is defined as a crop protection management based on Integrated Pest Management (IPM) principles, which emphasizes physical and biological regulation strategies to control pests while reducing the reliance on pesticides [11]. It can be regarded as an intermediate between conventional farming, with high levels of inputs, and organic farming, which prohibits the use of synthetic pesticides and fertilizers. Organic and integrated farming have in common the combined use of management approaches to replace, at least in part, synthetic inputs. However, unlike organic farming which is growing both in Europe (by 40 to 50% between 2003 and 2010 [12]) and in the US (by 270% between 2000 and 2008 [13]), integrated arable crop production is not expanding because it is perceived by farmers as a complex system which is difficult to implement, labour-consuming, and associated with reduced and unpredictable economic profitability [14,15]. As a consequence, the amount of pesticides sprayed has only decreased slightly in Europe (−3.6% from 2000 to 2007 [16]) and in the US (−7.5% from 2000 to 2007 [17]). Moreover, this decrease can be partly attributed to the substitution of older chemistry, applied at high dosage, by new products that are efficient at lower doses, which actually cannot be considered as a reduction of pesticide reliance. In France, the national action plan, ECOPHYTO 2018, which had set a target of a 50% decrease in pesticide use by the year 2018, is currently far from achieving this goal [18].

So far, assessments of cropping system sustainability have compared few – typically two or three – experimental prototypes that represent conventional, organic or integrated strategies

[19,20]. However, this approach fails to capture the diversity within each of these farming strategies. Given the diversity of crop management options within a conventional, an integrated or an organic strategy, which might lead to contrasted performances, the generic value of experimental results ignoring this variability may be argued. We assessed the sustainability of 48 cropping systems located in regions of intensive arable farming and covering a wide range of pesticide use levels and cultivation techniques such as crop rotations, from monoculture to highly diversified crop rotations, soil tillage (e.g. inversion tillage, shallow tillage or direct drilling), fertilization (mineral or organic fertilizers), or weed management (e.g. only based on herbicide use, including mechanical weeding). More details about the cropping system sample are available in the online SI section (Dataset S1). All the studied cropping systems were followed for between three and 12 years, between 1999 and 2012. Eight cropping systems were organic, 30 were based on integrated farming and 10 were conventional (Figure 1). Using eight sustainability indicators to evaluate the performance of the study systems, our aims were: (i) to identify possible conflicts between the reduction of pesticide reliance and other components of sustainability; and, (ii) to assess the potential of organic and integrated strategies for improving agricultural sustainability.

As the performance of a cropping system depends not only on the combination of management options it implements, but also on the local production situation [21] (including biophysical and socio-economic local aspects), we standardized the indicators of performance and pesticide use, using a ratio of the performances of the cropping systems over those of a local reference system. This enabled us to focus solely on the effects of the management strategies on sustainability indicators. The local references were cropping systems selected as representative of the most widespread crops and practices within each production situation. Pesticide use

was measured as the Treatment Frequency Index (TFI), which is a commonly used indicator in Europe to estimate the cropping system dependence on pesticides [22]. In our sample, organic cropping systems did not use any pesticides (synthetic or natural) so their relative TFI, expressed as a ratio of the local reference TFI, was zero. Integrated cropping systems displayed TFI values that were on average half (−47%) of the local references (Table S1).

Results

Table S1 presents the mean and standard deviation for each performance indicator according to the management strategy (organic, integrated and conventional). The second tab of Dataset S1 provides performance details for each cropping system of the sample.

Productivity and energy efficiency

Given the primary role of agriculture remains to produce food and other goods, we used an indicator of productivity, expressed as the total yearly amount of energy produced by a cropping system, whatever the crops cultivated (Figure 2a). The productivity of organic cropping systems was below that of their local reference (Figure 3b), ranging from −22% to −76%. For non-organic cropping system, productivity was uncorrelated to relative TFI (Figure 2a and Table 1), with some cropping systems that had a low reliance on pesticides even exceeding the productivity of the local reference. Cropping system productivity may strongly depend on crop type, especially if the whole above-ground biomass is harvested or not. Crops other than grain crops were frequently grown in integrated farming, as they are typically associated with low pesticide requirements and can contribute to weed control in subsequent crops [23]. They typically consist of

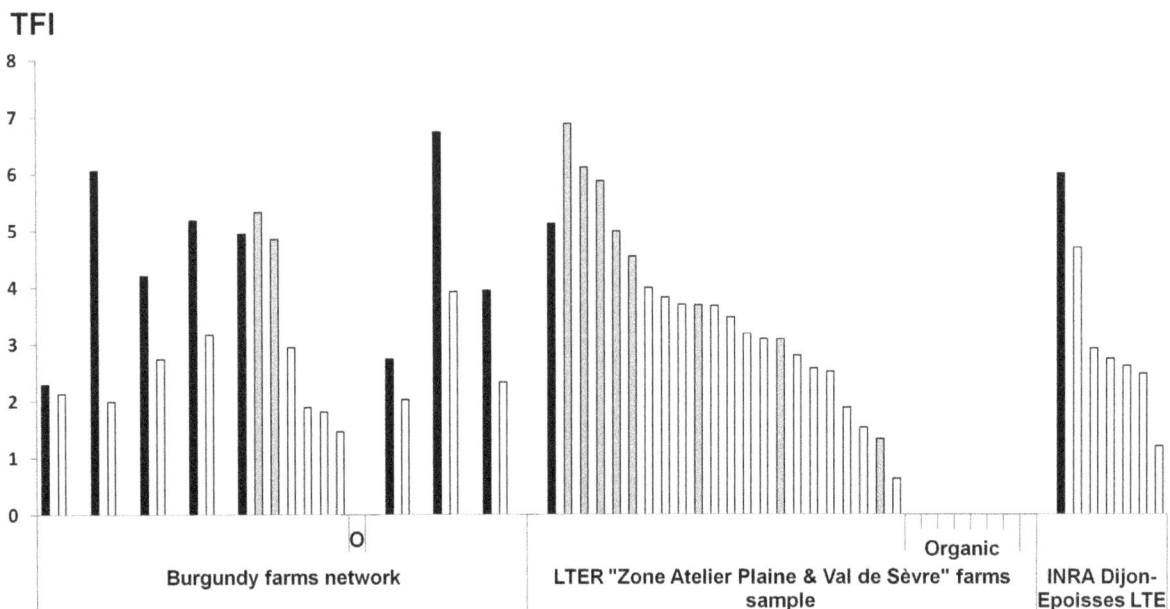

Figure 1. Distribution of the Treatment Frequency Index (TFI) for the studied arable cropping systems. Average TFI for each cropping system composing the study sample. At each site, black bars correspond to the local reference, grey bars to conventional cropping systems and white bars to integrated cropping systems. The sample also includes eight organic cropping systems with TFI = 0 and labelled with "O" or "Organic". Details about the cropping systems are available in Dataset S1.

forage crops, dedicated to livestock feeding with limited energy efficiency, or of crops used for non-food applications. However, distinguishing cropping systems based on grain crops or on crops in which all above-ground biomass is harvested did not change the observed pattern. In systems with grain crops only, productivity was not correlated with relative TFI (Table 1), suggesting that a reduction in pesticide use intensity may not be necessarily translated into a decrease in productivity. The second indicator of energy productivity we used was the energy efficiency of cropping systems, resulting from a ratio between energy output and energy input. It evaluated the ability of a cropping system to convert energy inputs into outputs. Organic cropping systems were significantly less energy efficient than other systems (Figures 2b and 3c, Table 2). Despite their energy consumption being lower (Table 2), notably due to their low reliance on nitrogen fertilizers, it was not sufficient to offset their limited productivity. Energy consumption was negatively correlated with relative TFI in integrated and conventional systems which cultivated only grain crops (Table 1). Energy efficiency was also negatively correlated with relative TFI in these systems, although the relationship was weak and only marginally significant ($r_s = -0.35$, $P = 0.07$). The systems with the highest energy efficiency, whether they included crops with all above-ground biomass harvested or not, were mostly integrated systems (Figures 2b and 3c).

Environmental impact

The environmental impact of cropping systems and their reliance on external inputs were assessed with the indicator I-Pest [24] and with estimates of fuel and nitrogen fertilizer consumption. I-Pest is a predictive indicator that assesses the environmental impacts of pesticide use as the risk of contamination of the air, and surface and ground waters (see Figure S1). As the organic cropping systems composing the sample did not used synthetic or natural pesticide, their cumulated I-Pest was 0. As expected for the rest of the sample, cumulated I-Pest was strongly and positively correlated to relative TFI (Figure 2c and Table 1). Fuel and nitrogen fertilizers together amounted to more than 60% of the total energy inputs for all tested cropping systems. Organic systems consumed more fuel than the rest of the sample (Table 2), with their average consumption exceeding the local references by 17% (Figure 3d). Organic cropping systems had, nonetheless, a lower reliance on N fertilization than the rest of the sample (Figure 3g, Table 2), in line with their lower yield targets and the frequent occurrence of crops with low N requirements used in organic rotations. No relation was detected between fuel consumption and relative TFI in non-organic systems (Figure 2d, Table 1), but a positive correlation was clearly visible between relative TFI and the amount of nitrogen fertilizers applied (Figure 2e).

Economic sustainability and workload

Economic sustainability was assessed by considering (i) the profitability, i.e. the average semi-net margin over a range of ten real price scenarios for agricultural products, fuel and fertilisers, and (ii) the sensitivity of this profitability in a context of price volatility, i.e. the relative standard deviation of the semi-net margin. The range of price scenarios used for the calculations was set to reflect the variability of the economic context over the last decade. Profitability, when averaged over the ten price scenarios was not correlated with relative TFI for integrated and conventional systems (Figure 2f and Table 1), and no significant difference appeared with organic systems (Mann-Whitney test, $P > 0.9$). It suggests that low pesticide use would not necessarily result in lower economic return. The strong variability observed within each class (Figure 3f), most notably for integrated cropping systems,

confirmed that strategies to reduce pesticide use could even lead to an increase in profitability. As integrated cropping systems were, in contrast to organic systems, evaluated with a conventional price reference, the most profitable integrated systems were able to efficiently reduce their production costs. No relation was detected between the sensitivity to price volatility and relative TFI in conventional and integrated systems (Figure 2g, Table 1). Sensitivity to price volatility was significantly lower in organic cropping systems than in other systems (Table 2), most probably because: (i) they were based on more diversified crop rotations, which spread risks and buffered semi-net margin at the farming system scale; and, (ii) their crop rotations typically included crops with low N demand, that had reduced reliance on N inputs, whose price is directly related to the volatile price of fossil fuels.

The issue of social sustainability was addressed using the 'workload' indicator, which gives emphasis to the potential for bottlenecks where available workforce is a limiting factor at the farm scale (Figure 2h). Workload was calculated for each technical operation but excluded time devoted to transport and crop monitoring. Workload was found not correlated with relative TFI in non-organic cropping systems (Table 1), and no significant difference was found with the organic group (Mann-Whitney test, $P > 0.1$), so that reducing pesticide use does not necessarily imply an increased workload. Indeed, in integrated systems, labour requirements ranged from low to high relative values (Fig 2h). The level of workload was, however, related to the type of fertilization, with cropping systems having organic fertilization requiring an average of 13% greater working time, as compared to mineral fertilizer-based cropping systems (Table 2).

Crop diversity

Diversification of crop rotations is often presented as an efficient management tool for controlling pests and to improve agricultural sustainability [25,26]. We used a crop sequence indicator, Isc [27], which estimates the consistency of the crop sequence with regard to the potential of input reduction, by addressing effects of crop rotation on pathogens, pests, weeds, soil structure and nitrogen supply of preceding crops. Even if no significant correlation appeared between Isc and relative TFI (Table 3), organic and integrated cropping systems displayed significantly higher Isc values than conventional systems (Table 2). A negative correlation between Isc and productivity suggests that diversifying crop rotation may reduce cropping system productivity (Table 3), but the Spearman correlation test was no longer significant when organic cropping systems were excluded ($P = 0.07$). We did not detect any significant relationship between energy efficiency and crop diversification, whether organic cropping systems were included or not ($P = 0.44$). No correlation was observed between Isc and semi-net margin, but workload appeared to be lower for systems with higher Isc (Table 3). We found the expected negative correlation between Isc and N fertilization rates, and consequently between Isc and energy consumption (Table 3). We focused therefore more particularly on cropping systems including legume crops, which also displayed higher Isc values than the rest of the sample (Table 2). The role of legume in improving energy efficiency at the cropping system scale was clearly demonstrated by the correlation between the frequency of occurrence of legumes in the crop rotation and the energy efficiency ($r_s = 0.37$, $P < 0.05$). The sensitivity to price volatility was negatively correlated with the frequency of occurrence of legumes in the crop rotation ($r_s = -0.33$, $P = 0.02$), but positively correlated with the level of N fertilization ($r_s = 0.49$, $P = 5*10^{-4}$). Fostering exogenous N independence therefore appeared as an efficient way to limit income variability.

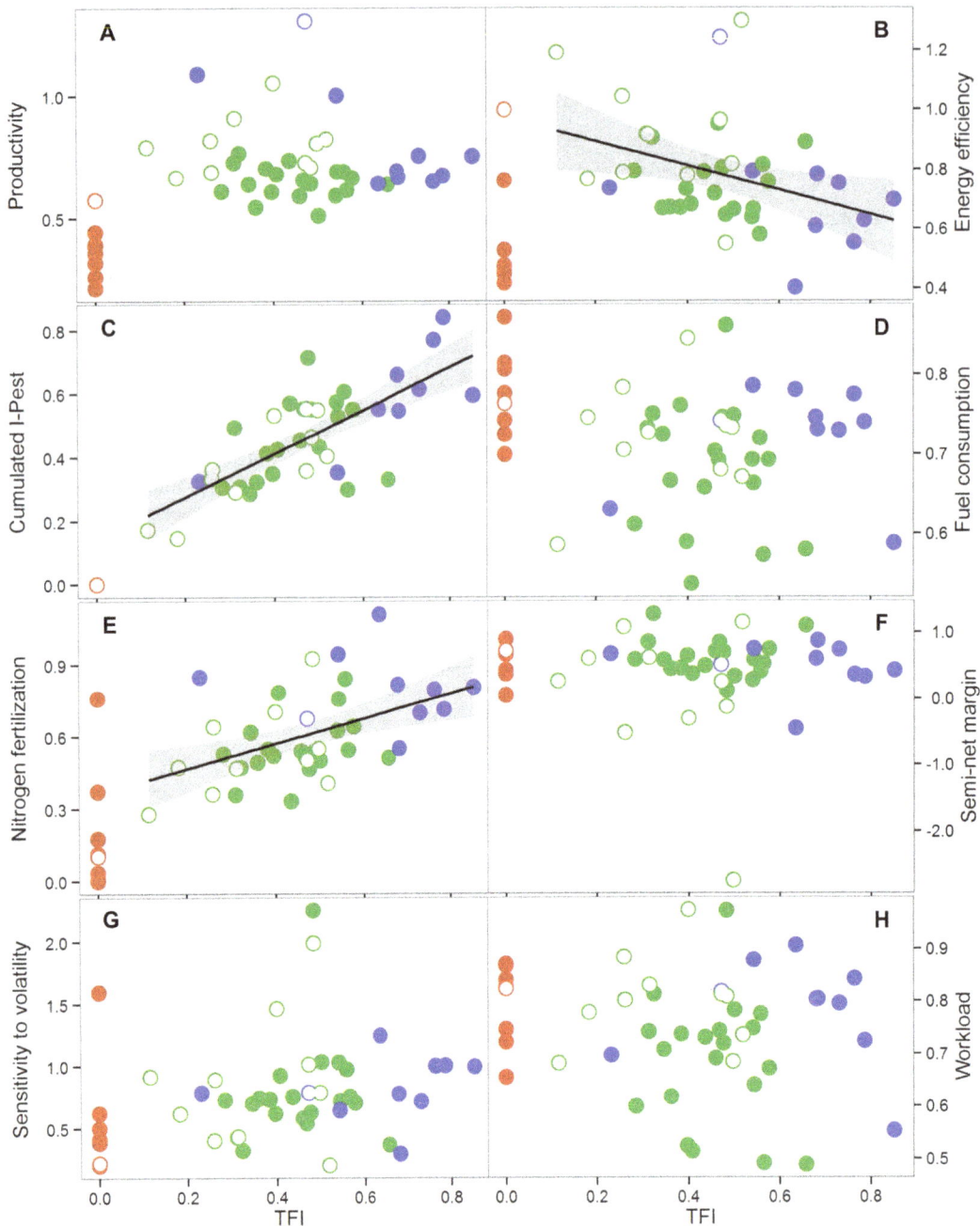

Figure 2. Relationship between sustainability indicators and relative TFI. Cropping system performances according to their relative TFI. Conventional, integrated and organic cropping systems are represented by blue, green and red symbols respectively. Filled symbols correspond to the cropping systems with grain crops only and empty symbols refer to the cropping systems including crops for which the whole above-ground biomass is harvested. Each sustainability indicators is expressed as the natural logarithm of the ratio between the cropping system and the local reference indicators. Linear regressions are represented with their standard error for cumulated I-Pest (Pearson correlation test: $r_p = 0.74$, $P = 5*10^{-8}$), nitrogen fertilization (Pearson correlation test: $r_p = 0.48$, $P = 0.002$), and energy efficiency (Pearson correlation test: $r_p = -0.38$, $P = 0.02$). Performance metric included: a) productivity, b) energy efficiency, c) cumulated I-Pest, d) fuel consumption, e) nitrogen fertilization, f) semi-net margin, g) sensitivity to price volatility, h) workload.

Discussion

This work was aimed at detecting cropping systems able to reconcile low pesticide use and other components of sustainability. Our original multiple dimensions approach, based on a precise description of management practices, was designed to compare and contrast numerous cropping systems from different produc-

tion situations. This approach, applied at the large-scale, was able to provide generic knowledge about potential trade-offs between the different issues of agricultural sustainability.

Sustainability of integrated and organic farming

Our results show that achieving a low level of pesticide use is possible without triggering negative side effect on any of the

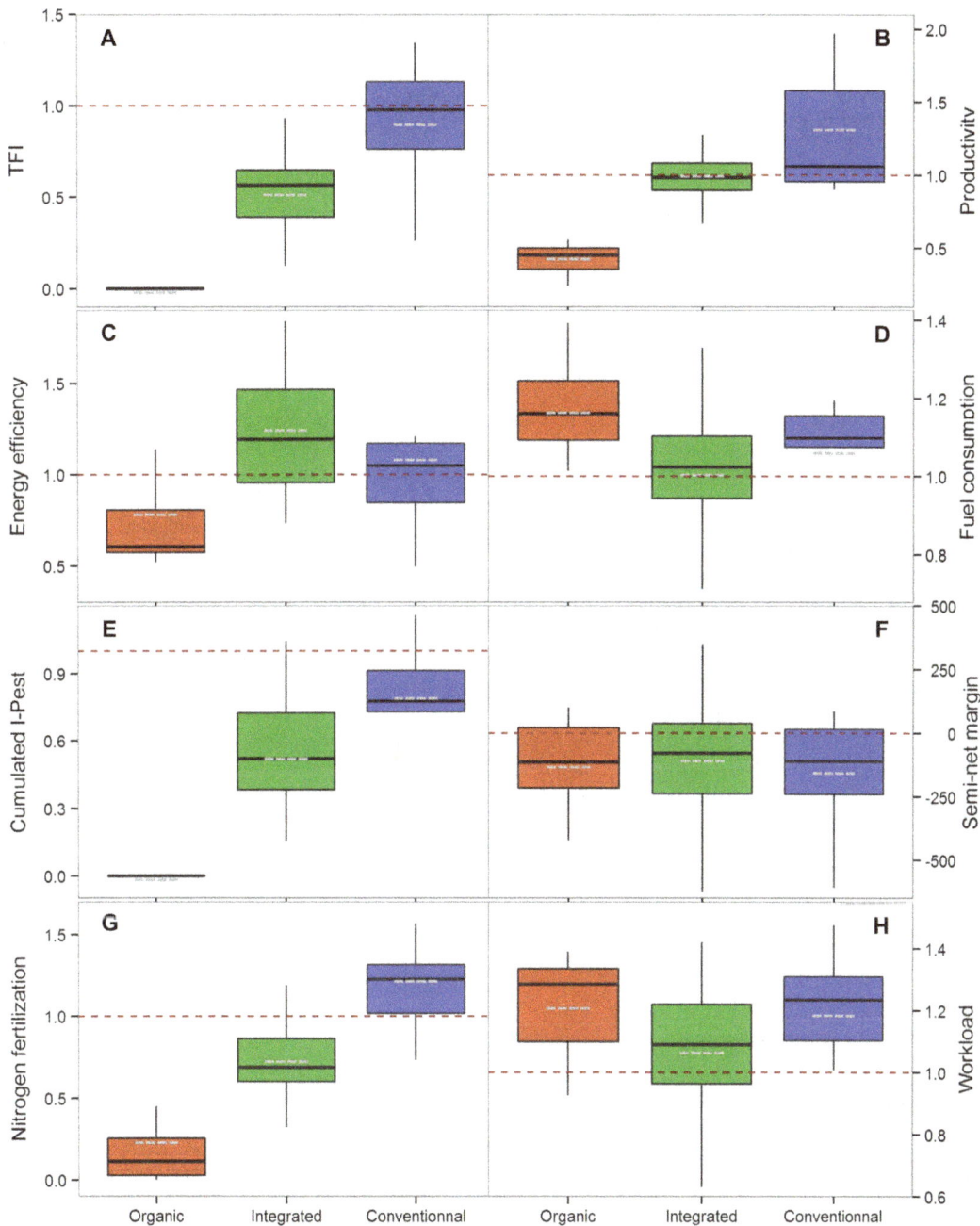

Figure 3. Cropping systems distribution according to sustainability indicators. Performance indicators are expressed as a ratio of the local reference indicator, except for semi-net margin, expressed as a difference with the local reference. Conventional, integrated and organic cropping systems are represented by blue, green and red box plots respectively. The horizontal black bars and grey dashed bars correspond to median and mean values respectively. The horizontal red dashed bar recalls the position of the local references. Outliers are not represented. Performance metrics included: a) Treatment Frequency Index, b) productivity (organic farming: one outlier, $v = 0.78$; integrated farming: two outliers, $v1 = 1.48$ and $v2 = 1.87$), c) energy efficiency (organic farming: one outlier, $v = 1.72$), d) fuel consumption (integrated farming: one outlier, $v = 1.37$; conventional farming: two outliers $v1 = 0.8$ and $v2 = 0.88$), e) cumulated I-Pest (conventional farming: two outliers $v1 = 0.38$ and $v2 = 0.42$), f) semi-net margin, g) nitrogen fertilization (organic farming: one outlier, $v = 1.13$; integrated farming: two outliers, $v1 = 1.32$ and $v2 = 1.52$), h) workload (conventional farming: one outlier $v = 0.74$).

components of cropping system sustainability we assessed in this study. Integrated cropping systems were not only associated with low pesticide use and low risks for contamination of air and water with pesticide residues; they also displayed lower energy consumption than more intensive cropping systems and are likely to improve energy efficiency without impact on productivity and profitability. Lower pesticide usage in arable cropping systems did not imply a heavier workload, another critical point conditioning strongly the adoption of an innovative strategy.

Organic farming prohibits the use of synthetic pesticides and fertilizers, and this approach is often associated with low nitrogen fertilization, as observed in our sample. In addition to the positive

Table 1. Rank correlation between TFI and sustainability indicators for integrated and conventional cropping systems.

Spearman correlation	Productivity	Productivity (grain crops only)	Energy consumption	Energy consumption (grain crops only)	N fertilization	Pesticides environmental impact
r_s	−0.17 (NS)	0.06 (NS)	0.30 (NS)	0.42	0.51	0.67
P-value	0.3	0.8	0.06	0.03	$9*10^{-4}$	$4*10^{-6}$
Spearman correlation	Fuel consumption	Energy efficiency	Semi-net margin	Sensitivity to prices volatility	Workload	Crop Sequence Indicator (Isc)
r_s	0.06 (NS)	−0.40	−0.05 (NS)	0.22 (NS)	$9*10^{-3}$ (NS)	−0.22
P-value	0.7	0.01	0.8	0.2	0.95	0.2

Spearman rank correlation tests ($\alpha = 0.05$). r_s is the Spearman correlation coefficient. Values of r_s followed by (NS) are not significant.

effects on environmental quality, numerous studies underlined other environmental benefits of organic farming such as effects on pollinator dynamics [28], on landscape floristic composition [29], as well as on soil microbial diversity [5,30]. Here we demonstrated that organic farming does not necessarily affect profitability and workload, and conversely, might strengthen farm financial stability in a variable and unpredictable economic context. Organic cropping systems were however less productive and less energy efficient than integrated systems in our sample. Although highly dependent on crops and production context, productivity in organic farming was already reported as lower than in conventional farming by other comparative studies [31]. The poor land use efficiency associated with organic farming is a key issue in the current land sharing – land sparing debate about the growing competition for land use [32], and notably urban sprawl [33] as well as the necessity to keep natural spaces undisturbed [34,35]. Both aspects – environmental benefits of organic farming and the limited productivity per unit of land – should therefore be considered by decision makers in their incentives for sustainable agriculture.

Crop diversification

Our results support the hypothesis that crop diversification may be an effective means to enhance cropping system performance. At the cropping system scale, crop diversification provides agronomic advantages, such as the regulation of pests, diseases and weeds [36,37,25]. In our sample, the most diversified cropping systems, which displayed the highest values of the crop sequence indicator, Isc, were indeed less dependent on pesticides. Their low environmental impact on water and air quality makes crop diversification an interesting potential pathway for reducing the damage caused by agriculture on natural resources (e.g., biodiversity [38]), as well as on human health (e.g. neurological degenerative disorders [39]). By mitigating the adverse effects of climate variability, crop diversification may also improve system resilience for productivity [40], with the increasing likelihood of extreme weather events requiring farm adaptation [41]. Economic market volatility is an additional source of variation and risk factor for farm economic stability. We found that crop diversification, particularly through the introduction of legumes in the crop rotation, is likely to limit dependence on inputs that have unstable prices. By allowing a decrease in the use of exogenous N fertilizer across a crop rotation, legume cultivation reduces production cost fluctuations and consequently makes the cropping system less sensitive to market volatility. Legumes come with a supplementary advantage [42] in the face of the considerable amount of fossil energy necessary to produce mineral N fertilizers, and we noted a substantial increase in energy efficiency for crop rotations where

Table 2. Significantly different groups for a given performance indicator.

Indicator	Test designation	P-value	Statistic W
Productivity	Difference of productivity between organic cropping systems and the rest of the sample	$2*10^{-8}$	318
Productivity	Difference of productivity between cropping systems including crops with the whole above-ground biomass harvested and the rest of the sample	$5*10^{-4}$	76
Energy efficiency	Difference of energy efficiency between organic cropping systems and the rest of the sample	0.005	258
Energy consumption	Difference of energy consumption between organic cropping systems and the rest of the sample	0.001	272
Fuel consumption	Difference of fuel consumption between organic cropping systems and the rest of the sample	0.02	72
N fertilization rate	Difference of N fertilization rate between organic cropping systems and the rest of the sample	$2*10^{-4}$	285
Sensitivity to price volatility	Difference of sensitivity to price volatility between organic cropping systems and the rest of the sample	0.01	250
Workload	Difference of workload between cropping systems based on organic fertilization and the rest of the sample	0.045	189
Crop sequence indicator	Difference of Isc between organic cropping systems and conventional cropping systems	0.01	69
Crop sequence indicator	Difference of Isc between integrated cropping systems and conventional cropping systems	0.001	255
Crop sequence indicator	Difference of Isc between cropping systems including legumes and the rest of the sample	$6*10^{-6}$	63

Mann-Whitney tests ($\alpha = 0.05$). All P-values are below 0.05, indicating that the differences between means of the sub-samples are significant for the corresponding indicators.

Table 3. Rank correlation between Crop Sequence Indicator Isc and sustainability indicators.

Spearman correlation	Productivity	Energy consumption	N fertilization	Pesticides environmental impact	Fuel consumption
r_s	−0.35	−0.42	−0.41	−0.39	−0.07 (NS)
P-value	0.01	0.003	0.004	0.006	0.7
Spearman correlation	**Energy efficiency**	**Semi-net margin**	**Sensitivity to price volatility**	**Workload**	
r_s	−0.05 (NS)	0.10 (NS)	−0.23 (NS)	−0.29	
P-value	0.7	0.5	0.1	0.04	

Spearman correlation tests ($\alpha = 0.05$). r_s is the Spearman correlation coefficient. Values of r_s followed by (NS) are not significant.

legume crops are more frequent. The most part of legumes introduced as diversification crops are however forage crops, and livestock production is commonly considered more energy consuming than plant production [43]. Conversely, the use of farmyard manure may contribute to reduce mineral fertilizers reliance for grain and forage crop production. The necessity of (i) integrating these situation-dependent parameters into energy balancing calculations, and, (ii) evaluating other environmental indicators [44] will be critical for the assessment of livestock production as a management option for enhancing agricultural sustainability.

A key agronomical advantage of crop diversification is related to the management of weed resistance to herbicides. Crop diversification is an efficient means to alternate herbicide modes of action and to introduce diversified measures of weed control, allowing changing selection pressure on weed communities and thus maintaining a sensitive weed population (i.e. maintaining high herbicide efficiency) [45].

Our results demonstrate a negative correlation between the Isc value and workload. We can nevertheless assume that diversifying crop rotations increases cropping system complexity and time devoted to field observations. Another aspect is that crop diversification may lead to a more evenly distributed workload over the seasons. Crops diversity implies a greater diversity in sowing and harvest periods, which are both times of peak labour that strongly influence task organisation and farmer decision making [15]. By reducing the amplitude of these peaks in labour, crop diversification could contribute to ensuring greater farmer decisional flexibility at the farm scale.

Beyond technical and organizational issues at the farm level, diversifying crop production as a component of an integrated strategy at regional or national scale would inevitably lead to important changes in production volumes, as well as markedly changing agricultural sectors within each production basin. It would definitely require an adaptation in the organisation of the whole agricultural sector and the development of new local markets. These economic and social lock-ins are rightly highlighted as the main limiting constraints hindering crop diversification [46]. However, by creating a particular economic sub-context, niche markets can be attractive and able to support innovation. Promoting such niche markets, for integrated farming development, would be the first step along an accelerating cycle of improvement based on mutually positive feed-backs between production and outlets.

Materials and Methods

In all cases, the field studies did not involve endangered or protected species.

For future permissions about the private farm network of Burgundy, please contact Marie-Sophie Petit (co-author of the research article, Chambre Régionale d'Agriculture de Bourgogne) and Sandrine Petit (co-author of the research article, INRA).

For future permissions about the private farms survey carried out on the LTER "Zone Atelier Plaine & Val de Sèvre", please contact Nicolas Munier-Jolain (corresponding author, INRA).

Study areas

The main objective of this study was to highlight potential conflicts between pesticide use and a set of sustainability indicators, so the cropping systems we consider were selected to maximize the contrast across the range of possible pesticide use intensities. The sample of cropping systems we used originates from:

(i) A long term experiment conducted since 2000 at the INRA Dijon-Epoisses farm in Bretenière (Burgundy, eastern France; 47°20′N, 5°2′E) in order to assess Integrated Weed Management-based cropping systems [15,26]. Seven cropping systems were tested between 2000 and 2012 including different combinations of technical levers likely to reduce pesticide reliance.

(ii) An experimental network (bringing together 14 cropping systems) monitored (1) by the local agricultural extension services and coordinated by the Chambre Régionale d'Agriculture de Bourgogne, and (2) by the INRA de Dijon. This network involved contrasting private farms of the Burgundy region, and was developed to test feasibility of innovative cropping systems with reduced pesticide use in a realistic context.

(iii) A survey of private farms carried out in 2010 on the LTER "Zone Atelier Plaine & Val de Sèvre" [47] located in the Poitou-Charentes region (450 km^2 study area in western France), and set up to explore a diversity of pesticide reliance, including organic farming, conventional intensive systems, and intermediate, IPM-based systems. Twenty nine varied cropping systems were surveyed in this area.

Cropping systems classification

Details of cropping systems, including crop sequences, performances, and detailed crop management operations are made available in the Dataset S1 and S2. Cropping systems were considered as conventional, integrated or organic according to the following rules. Cropping systems complying with the organic farming specifications were treated as 'organic'. Other systems were considered as 'integrated', when they were based either on diversified crop rotations including unusual alternative crops for the production situation (i.e. not present in the local reference crop rotation), or when crop management included at least one non-chemical management approach that contributed to the control of pests, diseases or weeds. These included for instance biocontrol, mechanical weeding and false seed bed techniques. Systems that were not classified as 'organic' or 'integrated' were classified as 'conventional'.

Local reference definition

For each of the 48 systems, a local cropping system reference was selected to reflect the most widespread crop rotation and associated technical management, as well as the typical agricultural performance in the production situation. Using this local cropping system reference made it possible to distinguish the effects of agronomic strategies from the effects of the production situation (soil, climate, economic and social context) when assessing the various components of sustainability of each cropping system. The Dijon-Epoisses experiment included a reference standard system that follows recommendations of local extension services [26], and which was used as the local reference. For each farm of the network across Burgundy, the local reference was defined as the cropping system implemented within the farm before the set-up of the alternative cropping system, even though crop management sequences were slightly updated according to expert appraisal to match with current standards (e.g. active ingredients allowed). For the Zone Atelier "Plaine et Val de Sèvre", local expert knowledge was used to select one system from the survey, with a standard crop rotation for the area, and a crop management representative of local practices. This system was then used as the local reference for all the remaining surveyed systems of the area.

Assessment of sustainability

The assessment of sustainability at the cropping system scale was based on a range of indicators covering economic, environmental and social issues. The Treatment Frequency Index (TFI) [22] estimates the number of registered doses applied, for each pesticide, per hectare and per crop season. Averaged over the cropping system, this indicator summarizes the level of dependence on pesticides, which should be distinguished from the environmental impact of pesticide use. This indicator is calculated for each pesticide application according to the following formula:

$$TFI = \frac{Application\ rate \times Treated\ surface\ area}{Registered\ dose \times Plot\ surface\ area}$$

The application rate and the registered dose were both expressed for a given commercial product (which possibly contains several active ingredients). The recommended application dose depends obviously on the treated crop and on the targeted pest. Here we defined the registered dose as the lowest application dose which is recommended for a given crop. The TFI for a given crop season was then calculated as the sum of the TFI for each pesticide application performed during this crop season. Productivity was evaluated as the amount of energy harvested yearly. This approach allowed the comparison of different crop rotations that included crops with different yielding potentials and different energy content. For each crop, yields were transformed into the energy metric using their Lower Heating Value (LHV) [48], which corresponds to the amount of energy released per unit of mass by the combustion of the harvested biomass. Energy consumption

was estimated from the conversion of inputs into energy according to the Dia'terre reference database [48]. Dia'terre is an assessment tool developed by the Agency for the Environment and Energy Management (ADEME) in the framework of the French Plan for Energy Performance (PPE) to evaluate a carbon-energy balance at the farm scale. The reference database used to design this assessment tool provides energy values for indirect energy consumption associated with the production of farming inputs. For instance, the calculation of the energy cost associated with the production of nitrogen fertilizers integrates the energy necessary from raw material production (e.g. Haber–Bosch process) through materials processing, manufacture and distribution. We used the reference energy cost provided by the carbon calculator Dia'terre to compute the energy balances of the cropping systems. In this way, the inputs necessary for crop production were converted into energy, using the energy cost of fertilizers, pesticides, seeds, water spread for irrigation, fuel consumed by the equipment and the amount of steel necessary to manufacture this equipment, i.e. energy cost of mechanization (see Dataset S3). The energy requirements for preparing farmyard manure are farm-specific and very difficult to quantify precisely. A simplification was consequently required: following previous studies based on energy balancing methods in crop production [49], the energy equivalent of farmyard manure was equated with that of the mineral fertilizers they substituted (using a substitution value related to the fertilizing efficiency of manure). Energy efficiency was computed from the ratio between productivity and energy consumption. For assessing the economic productivity, the gross product derived from the direct conversion of crops yields into economic values. The 'semi-net' margin was calculated as the gross product per hectare from which we subtracted the input costs (fertilizers, pesticides, seeds, fuel, water and mechanisation). This 'semi-net' margin assessed the system profitability without taking into account subsidies or incentives. The sensitivity to price volatility was defined as the relative standard deviation of the semi-net margin calculated over ten contrasting real price scenarios selected between 2000 and 2010, and thus measured the ability of a cropping system to generate a stable income in a variable economic context. The ten scenarios integrated the prices of crops but also the prices of volatile inputs such as fertilizers or fuel. Each price scenario was defined at a given moment between 2000 and 2010, and it therefore reflected the correlations between the prices of crop products and inputs. This approach notably made it possible to integrate better the effects coming from crop diversity (proportion of cereal crops, oil crops or protein crops) on cropping system profitability and economic stability. Fuel consumption and workload were estimated according to in-field cropping operations only, without considering fuel and time consumed for farm-to-field transports, or extra-workload dedicated to equipment maintenance or field observations. The size, fuel requirements and working output of the various equipment types were standardized for all cropping systems, and defined from a national database [50], consistent with the aim of evaluating management strategies, and of ignoring the potential effects of the equipment specifications (See Dataset S4 for the details of the equipment used for the calculations).

Pesticide environmental impact was expressed as cumulated I-Pest [24]. This indicator measures the risk associated with pesticide application for three compartments of the environment, namely the air, the surface water and the groundwater. This risk indicator, ranging from 0 to 1 (maximum risk), and calculated for each active substance application, is based on: (i) field inherent sensitivity to pesticide transfer toward these three compartments; (ii) characteristics of the active substance (e.g. ecotoxicity, mobility, half-life); and, (iii) information about the conditions of the spraying operation (e.g. amount of active substances employed, canopy cover at the date of treatment) in order to calculate three impact factors, one for each compartment. I-Pest index is obtained using fuzzy decision trees that allow the aggregation of these three impact factors into one synthetic indicator. The diagram presented in Figure S1 illustrates how this indicator of pesticide environmental impact was computed for each pesticide active substance that was sprayed within the field.

The crop sequence indicator Isc [27] is used as an additional indicator to quantify the agronomic effects of crop diversification. Isc ranges on a qualitative scale between 0 and 10 (best value) and is calculated as shown in the following equation:

$$Isc = kp \times kr \times kd$$

Isc is based on the assessment of the effects of the previous crop on the current crop (kp), with respect to the development of pathogens, pests and weeds, to soil structure and nitrogen supply. kp, ranging between 1 and 6, was assessed for 470 couples crop/previous crop. kp is corrected by two factors taking into account the crop frequency (kr ranging between 0.3 and 1.2) and the crop rotation whole diversity (kd ranging between 1.0 and 1.4). Isc yields respectively 0.5 for wheat monoculture, 3.3 for a rape/wheat rotation, 5.1 for a rape/wheat/barley rotation, and 7.6 for a maize/wheat/sunflower/spring barley rotation.

Computation of sustainability indicators at the cropping system level

As a first step each indicator was calculated for each cropping operation composing our database (Dataset S2). These values of indicators were summed over the crop season year, and then averaged across years and across plots, each plot being considered as a replicate of a given cropping system. Each indicator was therefore calculated at the cropping system level, integrating (i) the different crops composing the crop sequence, (ii) the variability of crop production related with the inter-annual climatic variability, and (iii) the possible variation in plot properties.

All sustainability indicators were expressed per hectare and per year. For distinguishing specifically the effects of the management strategy on cropping system sustainability from the effects of the production situation, each indicator computed for a given cropping system was then expressed as a ratio (or as a distance in the case of semi-net margin) between the system indicator and the local reference indicator. To increase the quality of the graphs drawing the relationship between sustainability indicators and pesticide use, values of assessment indicators were translated into natural logarithm (Figure 2), which reduced the visual effect of extreme values.

Statistical analyses

Spearman and Pearson correlations were estimated using the 'rcorr' correlation matrix function in the *Hmisc* package of R v2.15.0 [51]. The difference between the means of two sub-samples for a given indicator was tested with a non-parametric Mann-Whitney test ('wilcox.test' function with two samples) in the *stats* package of R v2.15.0.

References

1. Foley JA, Ramankutty N, Brauman KA, Cassidy ES, Gerber JS, et al. (2011) Solutions for a cultivated planet. Nature 478: 337–342.

Supporting Information

Figure S1 Simplified description of the assessment process of pesticide environmental impact in the I-Pest model.

Dataset S1 Cropping systems details. A.xlsx file describing the cropping systems of the studied sample (e.g. crop rotation, tillage and weed management strategies). This file also provides information about the local reference associated with the evaluation of each cropping system. The second tab provides the respective performances of each cropping system described in the first tab.

Dataset S2 Cropping operations database. A.xlsx file which provides the details of all cropping operations carried out in each cropping system: type of cropping operation, date (when recorded), application rates (for pesticides, fertilizers, seeds and irrigation) and proportion of the plot surface targeted.

Dataset S3 Energy balancing database. A.xlsx file with two sheets. The first sheet provides energy cost values for inputs: pesticides active substances, fuel, fertilizers, irrigation water and seeds. The second sheet includes the Lower Heating Values (LHV) for usual crops, that is to say the energy contained in one mass unit of crop harvested.

Dataset S4 Standard equipment characteristics. A.xlsx file describing the technical characteristics of the standard equipment we associated with each cropping operation. Details include the purchase price, the payback period and the maintenance cost to calculate the mechanization costs, but also the equipment size and weight, the working output, the fuel consumption rate and the energy cost value.

Table S1 Means and standard deviations for the range of performance indicators according to the management strategy. A.xlsx file summarizing and comparing the performances of organic, integrated and conventional cropping systems which compose the study sample. Significant difference between groups was tested with a Mann-Whitney test.

Acknowledgments

We thank A. Villard, C. Vivier and M. Geloen for their contribution to the experimental network in Burgundy, and D. Meunier, P. Farcy, P. Chamoy for technical assistance. We particularly want to thank D. Bohan for the precious advice on style and the language corrections.

Author Contributions

Conceived and designed the experiments: NMJ VB CB SP. Performed the experiments: ML FB. Analyzed the data: ML FB. Contributed reagents/materials/analysis tools: CB MSP. Wrote the paper: NMJ ML. Discussed the results and commented on the manuscript: NMJ ML VB CB SP MSP FB.

2. Pimentel D (1995) Amounts of pesticides reaching target pests: environmental impacts and ethics. Journal of Agricultural and Environmental Ethics 8: 17–29.

3. Richardson M (1998) Pesticides-friend or foe? Water science and technology 37: 19–25.

4. Tilman D, Cassman KG, Matson PA, Naylor R, Polasky S (2002) Agricultural sustainability and intensive production practices. Nature 418: 671–677.

5. Maeder P, Fliessbach A, Dubois D, Gunst L, Fried P, et al. (2002) Soil fertility and biodiversity in organic farming. Science 296: 1694–1697.

6. Holland JM, Frampton GK, Cilgi T, Wratten SD (1994) Arable acronyms analysed - a review of integrated arable farming systems research in Western Europe. Annals of applied biology 125: 399–438.

7. Ferron P, Deguine JP (2005) Crop protection, biological control, habitat management and integrated farming. A review. Agronomy for Sustainable Development 25: 17–24.

8. Trewavas A (2001) Urban myths of organic farming. Nature 410: 409–410.

9. Pimentel D, Hepperly P, Hanson J, Douds D, Seidel R (2005) Environmental, energetic, and economic comparisons of organic and conventional farming systems. Bioscience 55: 573.

10. Directive 2009/128/EC of the European Parliament and of the Council (2009) Official journal of the European Union. Available: http://eur-lex.europa.eu/LexUriServ/LexUriServ.do?uri = OJ:L:2009:309:0071:0086:en:PDF Accessed 2013 Sep 10.

11. Munier-Jolain N, Dongmo A (2010) Evaluation de la faisabilité technique de systèmes de Protection Intégrée en termes de fonctionnement d'exploitation et d'organisation du travail. Comment adapter les solutions aux conditions locales? Innovations Agronomiques 8: 57–67.

12. European commission (2010) Eurostat Agriculture online database. Available: http://epp.eurostat.ec.europa.eu/portal/page/portal/agriculture/farm_structure/database. Accessed 2013 Jul 19.

13. USDA Economic Research Service (2010) U.S. certified organic farmland acreage, livestock number, and farm operations. Available: http://www.ers.usda.gov/data-products/organic-production.aspx "\l ".UiWVOX8QO89. Accessed 2013 Jul 22.

14. Bastiaans L, Paolini R, Baumann DT (2008) Focus on ecological weed management: what is hindering adoption? Weed Research 48: 481–491.

15. Pardo G, Riravololona M, Munier-Jolain N (2010) Using a farming system model to evaluate cropping system prototypes: Are labour constraints and economic performances hampering the adoption of Integrated Weed Management? European Journal of Agronomy 33: 24–32.

16. Food and Agriculture Organization of the United Nations (FAO) (2013) FAOSTAT Resources: Pesticides Use. Available: http://faostat.fao.org/site/424/DesktopDefault.aspx?PageID = 424#ancor. Accessed 2013 Jul 25.

17. U.S. Environmental Protection Agency (EPA) (2011) Pesticides industry sales and usage: 2006 and 2007 market estimates. Washington, D.C.: U.S. Environmental Protection Agency. Available: www.epa.gov/opp00001/pestsales/07pestsales/market_estimates2007.pdf. Accessed 2013 Jul 25.

18. Ministère de l'Agriculture, de l'Agro-alimentaire et de la Forêt (2012) Note de suivi du plan Ecophyto 2018: tendances de 2008 à 2011 du recours aux produits phytopharmaceutiques. Available: http://agriculture.gouv.fr/IMG/pdf/121009_Note_de_suivi_2012_cle0a995a.pdf. Accessed 2013 Jul 26.

19. Reganold JP, Glover JD, Andrews PK, Hinman HR (2001) Sustainability of three apple production systems. Nature 410: 926–930.

20. Deike S, Pallutt B, Christen O (2008) Investigations on the energy efficiency of organic and integrated farming with specific emphasis on pesticide use intensity. European Journal of Agronomy 28: 461–470.

21. Aubertot JN, Robin MH (2013) Injury Profile SIMulator, a qualitative aggregate modelling framework to predict crop injury profile as a function of cropping practices, and the abiotic and biotic environment. I. Conceptual bases. PLoS one 8: e73202.

22. OECD (2001) Environmental Indicators for Agriculture, Volume 3: Methods and Results. Available: www.oecd.org/tad/sustainable-agriculture/40680869.pdf. Accessed 2014 Feb 26.

23. Meiss H, Mediene S, Waldhardt R, Caneill J, Bretagnolle V, et al. (2010) Perennial lucerne affects weed community trajectories in grain crop rotations. Weed Research 50: 331–340.

24. Van der Werf H, Zimmer C (1998) An indicator of pesticide environmental impact based on a fuzzy expert system. Chemosphere 36: 2225–2249.

25. Davis AS, Hill JD, Chase CA, Johanns AM, Liebman M (2012) Increasing cropping system diversity balances productivity, profitability and environmental health. PLoS one 7: e47149.

26. Chikowo R, Faloya V, Petit S, Munier-Jolain NM (2009) Integrated Weed Management systems allow reduced reliance on herbicides and long-term weed control. Agriculture, Ecosystems and Environment 132: 237–242.

27. Bockstaller C, Girardin P (2000) Using a crop sequence indicator to evaluate crop rotations. 3rd International Crop Science Congress 2000 ICSC, Hambourg, 17–22 August 2000, p. 195.

28. Andersson GKS, Rundlöf M, Smith HG (2012) Organic Farming Improves Pollination Success in Strawberries. PloS one 7: e31599.

29. Aavik T, Liira J (2010) Quantifying the effect of organic farming, field boundary type and landscape structure on the vegetation of field boundaries. Agriculture, Ecosystems and Environment 135: 178–186.

30. Li R, Khafipour E, Krause DO, Entz MH, de Kievit TR, et al. (2012) Pyrosequencing Reveals the Influence of Organic and Conventional Farming Systems on Bacterial Communities. PloS one 7: e51897.

31. Seufert V, Ramankutty N, Foley JA (2012) Comparing the yields of organic and conventional agriculture. Nature 485: 229–232.

32. Foley JA, Defries R, Asner GP, Barford C, Bonan G, et al. (2005) Global consequences of land use. Science 309: 570–574.

33. Theobald DM (2001) Land-use dynamics beyond the American urban fringe. Geographical Review 91: 544.

34. Phalan B, Onial M, Balmford A, Green R (2011) Reconciling food production and biodiversity conservation: land sharing and land sparing compared. Science 333: 1289–1291.

35. Hulme MF, Vickery JA, Green RE, Phalan B, Chamberlain DE, et al. (2013) Conserving the birds of Uganda's banana-coffee arc: land sparing and land sharing compared. PLoS One 8: e54597.

36. Altieri MA, Nicholls CI, Ponti L (2009) Crop diversification strategies for pest regulation in IPM systems. Integrated pest management Cambridge University Press, Cambridge, UK. pp. 116–130.

37. Krupinsky J, Bailey K, McMullen M, Gossen B, Turkington T (2002) Managing plant disease risk in diversified cropping systems. Agronomy Journal 94: 198–209.

38. Beketov MA, Kefford BJ, Schäfer RB, Liess M (2013) Pesticides reduce regional biodiversity of stream invertebrates. Proceedings of the National Academy of Sciences 110: 11039–11043.

39. Ascherio A, Chen H, Weisskopf MG, O'Reilly E, McCullough ML, et al. (2006) Pesticide exposure and risk for Parkinson's disease. Annals of Neurology 60: 197–203.

40. Di Falco S, Chavas JP (2008) Rainfall shocks, resilience, and the effects of crop biodiversity on agroecosystem productivity. Land Economics 84: 83–96.

41. Reidsma P, Ewert F, Lansink AO, Leemans R (2010) Adaptation to climate change and climate variability in European agriculture: The importance of farm level responses. European Journal of Agronomy 32: 91–102.

42. Nemecek T, von Richthofen JS, Dubois G, Casta P, Charles R, et al. (2008) Environmental impacts of introducing grain legumes into European crop rotations. European Journal of Agronomy 28: 380–393.

43. Pimentel D, Pimentel M (2003) Sustainability of meat-based and plant-based diets and the environment. The American Journal of Clinical Nutrition 78: 660S–663S.

44. Halberg N, van der Werf HMG, Basset-Mens C, Dalgaard R, de Boer IJM (2005) Environmental assessment tools for the evaluation and improvement of European livestock production systems. Livestock Production Science 96: 33–50.

45. Beckie HJ (2009) Herbicide Resistance in Weeds: Influence of Farm Practices. Prairie Soils and Crops 2:3.

46. Meynard JM, Messéan A, Charlier A, Charrier F, Farès M, et al. (2013) Freins et leviers à la diversification des cultures. Etudes au niveau des exploitations agricoles et des filières. Synthèse du rapport d'étude, INRA. Available: http://inra.dam.front.pad.brainsonic.com/ressources/afile/223799-6afe9-resource-etude-diversification-des-cultures-synthese.html. Accessed 2013 Mar 15.

47. Centre d'Etudes Biologiques de Chizé (2009) Zone Atelier «Plaine & Val de Sèvre». Available: http://www.zaplainevaldesevre.fr. Accessed 2013 Jun 5.

48. Agence de l'Environnement et de la Maîtrise de l'Energie (ADEME) (2011) Guide des valeurs Dia'terre. Version du référentiel 1.13. Available: http://www2.ademe.fr/servlet/KBaseShow?sort = -1&cid = 96&m = 3&catid = 24390. Accessed 2013 Jul 12.

49. Hülsbergen KJ, Feil B, Biermann S, Rathke GW, Kalk WD, et al. (2001) A method of energy balancing in crop production and its application in a long-term fertilizer trial. Agriculture, Ecosystems and Environment 86: 303–321.

50. Bureau de Coordination du Machinisme Agricole (BCMA) (2012) Simcoguide online decision tool. Available: http://simcoguide.pardessuslahaie.net/#accueil. Accessed 2013 Apr 12.

51. R Development Core Team (2012). R: A language and environment for statistical computing (R Foundation for Statistical Computing, Vienna, Austria).

2

Global Agricultural Land Resources – A High Resolution Suitability Evaluation and Its Perspectives until 2100 under Climate Change Conditions

Florian Zabel*, Birgitta Putzenlechner, Wolfram Mauser

Department of Geography, Ludwig Maximilians University, Munich, Germany

Abstract

Changing natural conditions determine the land's suitability for agriculture. The growing demand for food, feed, fiber and bioenergy increases pressure on land and causes trade-offs between different uses of land and ecosystem services. Accordingly, an inventory is required on the changing potentially suitable areas for agriculture under changing climate conditions. We applied a fuzzy logic approach to compute global agricultural suitability to grow the 16 most important food and energy crops according to the climatic, soil and topographic conditions at a spatial resolution of 30 arc seconds. We present our results for current climate conditions (1981–2010), considering today's irrigated areas and separately investigate the suitability of densely forested as well as protected areas, in order to investigate their potentials for agriculture. The impact of climate change under SRES A1B conditions, as simulated by the global climate model ECHAM5, on agricultural suitability is shown by comparing the time-period 2071–2100 with 1981–2010. Our results show that climate change will expand suitable cropland by additionally 5.6 million km^2, particularly in the Northern high latitudes (mainly in Canada, China and Russia). Most sensitive regions with decreasing suitability are found in the Global South, mainly in tropical regions, where also the suitability for multiple cropping decreases.

Editor: Juergen P. Kropp, Potsdam Institute for Climate Impact Research, Germany

Funding: This research was carried out within the framework of the GLUES (Global Assessment of Land Use Dynamics, Greenhouse Gas Emissions and Ecosystem Services) Project, which has been supported by the German Ministry of Education and Research (DMBF) program on sustainable land management (FKZ 01LL0901E). (http://modul-a.nachhaltiges-landmanagement.de/en/). The funders had no role in study design, data collection and analysis, decision to publish, or preparation of the manuscript.

Competing Interests: The authors have declared that no competing interests exist.

* Email: f.zabel@lmu.de

Introduction

Natural constraints are limiting the land's suitability for agriculture and cultivation practices. They consist of prevailing local climatic, soil and topographic conditions determining the available energy, water and nutrient supply for agricultural crops. Besides natural conditions, complex interactions of social, economic, political, and cultural aspects determine whether and how land is used for agriculture. Agricultural land has become one of the largest terrestrial biomes on the planet, occupying approx. 40% of the land surface [1]. Thereby, a variety of different land use types and intensities determine heterogeneously distributed patterns, including e.g. the choice of crop varieties, irrigation practices, fertilization, terracing and the level of technological input [2]. Thus, natural constraints are to a limited extent suspended by human actions [3].

The demand for agricultural products is expected to increase by 70–110% by 2050, driven by a projected world population of 9 billion people, increasing meat consumption and a growing use for bio-based materials and biofuel [4–15].

An increase in agricultural production can be accomplished by agricultural intensification and expansion, while considering social and environmental externalities and changing climate conditions

[5,16]. Bruinsma [16] concluded that additionally 1.2 million km^2 of converted land are projected to be necessary until 2030 and another 5% up to 2050 with most land expected to be transformed in South America and Sub Saharan Africa, while latest studies project an increase of cropland between 10-25% by 2050 compared to 2005 for different socio-economic and climate scenarios [17]. Nonetheless, the expansion of agricultural land into forested or protected areas must be viewed critically, in order to conserve valuable ecosystem services e.g. for regulating climate or conserving biodiversity [5–8].

Changing patterns of temperature and precipitation and man-made degradation affect the suitability of land for agricultural use. For example, 19-23 ha of suitable land are lost per minute due to soil erosion and desertification [18,19]. Additionally, the area of suitable land is decreasing due to urbanization, with an estimate of 1.5 million km^2 until 2030 [20,21].

When focusing on the natural potentials of land for agricultural use, suitability analyses give local evidence on todays and future availability and quality. Thus, they help answering questions for managing a transition towards a more environmentally efficient and sustainable land use and involve better information on the global scale impacts of land use decisions [1].

Table 1. List of investigated food, feed and energy crops.

Crop name
Barley (*hordeum vulgare*)
Cassava (*manihot esculenta*)
Groundnut (*arachis hypogaea*)
Maize (*zea mays*)
Millet (*pennisetum americanum*)
Oil palm (*elaeis guineensis*)
Potato (*solanum tuberosum*)
Rapeseed (*brassica napus*)
Paddy rice (*oryza sativa*)
Rye (*secale cereale*)
Sorghum (*sorghum bicolor*)
Soy (*glycine maximum*)
Sugarcane (*saccharum officinarum*)
Sunflower (*helianthus annus*)
Summer wheat (*triticum aestivum*)
Winter wheat (*triticum gestivum*)

The relationship between climate, soil, topography and agricultural suitability has long been recognized. As such, suitability analysis combine heterogeneous soil, terrain and climate information and determine whether specific crop requirements are fulfilled under the given local conditions and assumptions. A variety of regional suitability studies for specific crops exist [22–28], while only a few exist on a global scale and for a broad variety of crops [3,29–31].

In the meantime, global soil and topography data are available at high spatial resolution and global climate models have improved their capabilities and spatial resolution. Previous analysis showed that questions of scale play a major role in suitability analysis as coarse data affect the validity of results [32]. In this context, we present our results in modelling global crop-suitability using a fuzzy logic approach at a spatial resolution of 30 arc seconds. The results of this approach include the potentially suitable area for agriculture differentiated for 16 crops for rainfed and irrigated conditions, the start of the growing cycles and the number of crop cycles. We analyze global distribution of agricultural suitability and changes until 2100 considering the numbers of crop cycles.

Thereby, we identify changes, opportunities and challenges in global agriculture related to the expansion of agricultural land competing with protected and forested areas as ecosystem services.

Material and Methods

Local climate, soil and topography determine the natural suitability of land for agricultural use. Thereby, the climatic, soil and topographic requirements may vary over a wide range of different agricultural crops. This analysis investigates the suitability for the following 16 crops that are most important for the global economy, food security and biofuel issues (see Table 1).

We aggregated the world into 23 regions in order to regionally analyse the results (see Fig. 1). We applied a fuzzy-logic approach [33,34] in order to calculate the crops' suitability on the globe at a spatial resolution of 30 arc seconds ($0.00833°$, approx. 1 km^2 at the equator). The length of the growing cycle (*lgc*) and the '*membership functions*' that describe the crop-specific requirements for each of the crops during the growing period (Fig. 2) are derived from [35].

The membership functions representing climate constraints describe the degree of membership of each selected crop with regard to mean temperature and total precipitation during its

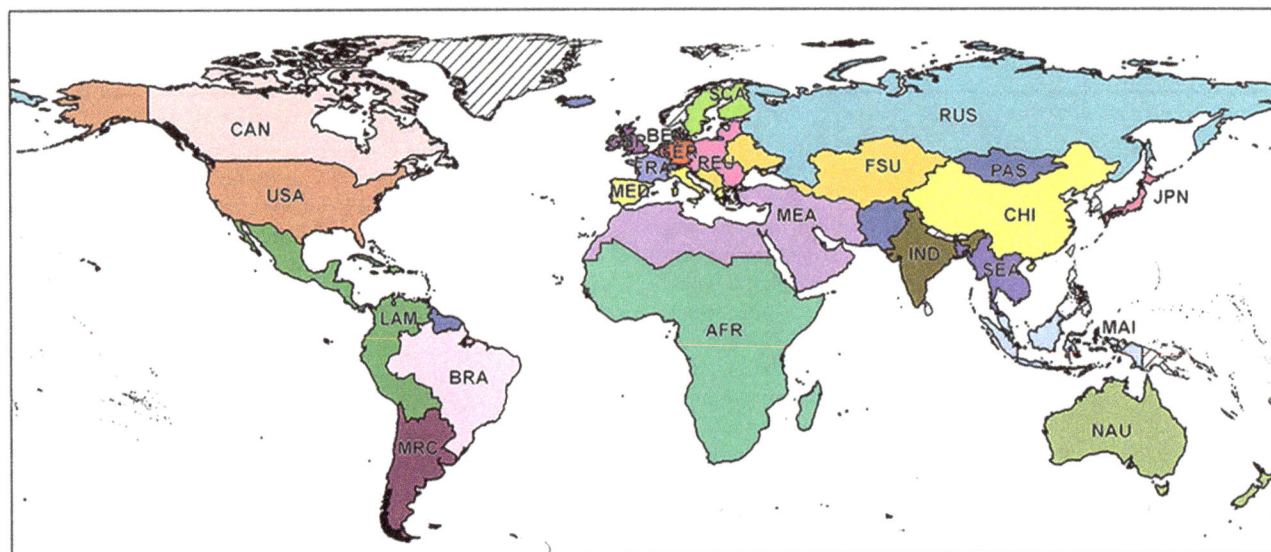

Figure 1. Map of the 23 world regions: AFR (Sub Saharan Africa), ANZ (Australia, New Zealand), BEN (Belgium, Netherlands, Luxemburg), BRA (Brazil), CAN (Canada), CHN (China), FRA (France), FSU (Rest of Former Soviet Union and Rest of Europe), GBR (Great Britain), GER (Germany), IND (India), JPN (Japan), LAM (Rest of Latin America), MAI (Malaysia, Indonesia), MEA (Middle East, North Africa), MED (Italy, Spain, Portugal, Greece, Malta, Cyprus), PAC (Paraguay, Argentina, Chile, Uruguay), ROW (Rest of the World), REU (Austria, Estonia, Latvia, Lithuania, Poland, Hungary, Slovakia, Slovenia, Czech Republic, Romania, Bulgaria), RUS (Russia), SCA (Finland, Denmark, Sweden), SEA (Cambodia, Laos, Thailand, Vietnam, Myanmar, Bangladesh), USA (United States of America).

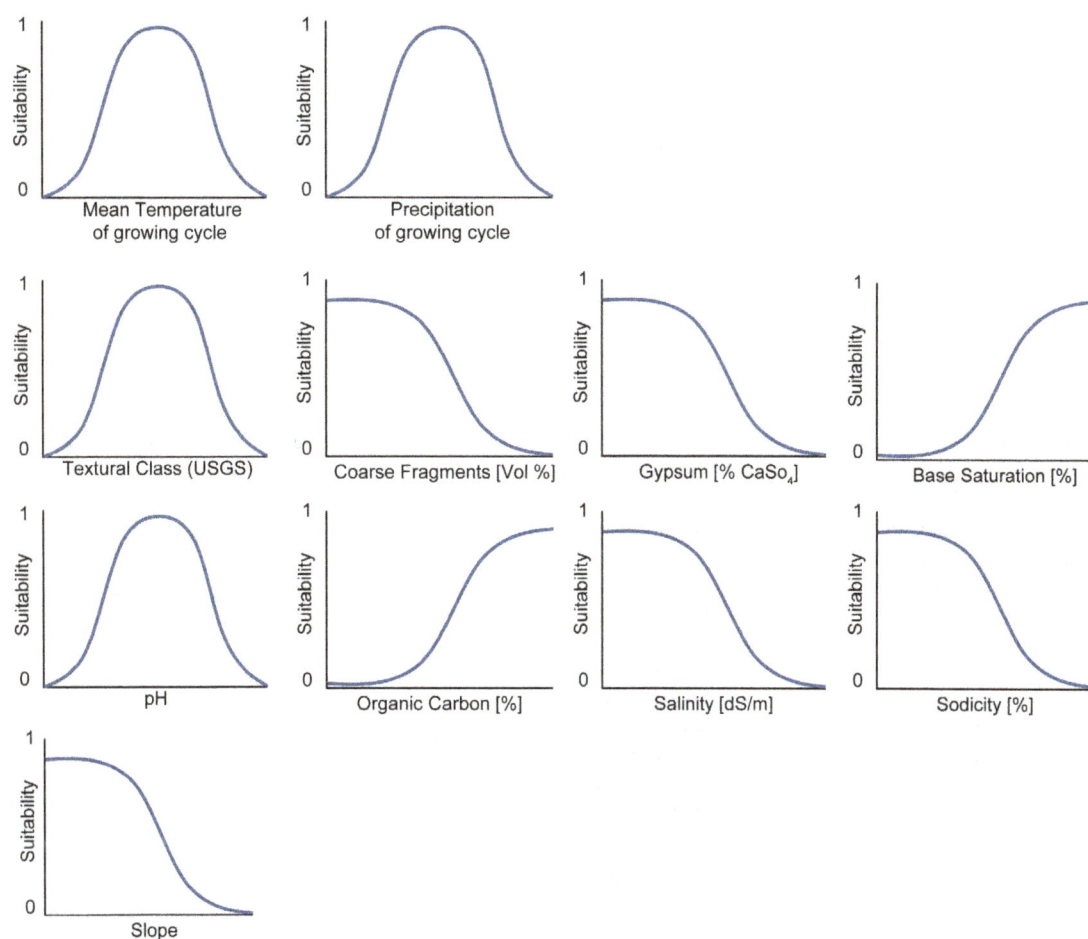

Figure 2. Membership functions for climatic, soil and topographic conditions.

respective growing cycle. Depending on the crop, membership functions have different curves according to [35]. Three shapes are in principle possible: 'more is better', 'less is better' and 'optimum'. For temperature e.g., the suitability is increasing from a minimum towards an optimal temperature and again decreasing until a maximum temperature is reached (Fig. 2). Eight soil parameters are considered: texture, proportion of coarse fragments and gypsum, base saturation, pH content, organic carbon content, salinity, and sodicity. Terrain is considered by the slope. The fuzzy-logic approach calculates *fuzzy values* based on the ecological rules (between 0 and 1), which determine the crops'

suitability on a specific location by the lowest membership value of all parameters.

An overview of the applied global datasets is given in Table 2. The climate data applied in this study are outputs from the global circulation model ECHAM5 of the Max-Planck Institute for Meteorology (MPI-M) [36,37]. It uses radiative forcing, sea surface temperature and sea ice concentrations from a 20th century/ SRES A1B scenario simulation. The 6-hourly dataset (temperature, precipitation) are converted to daily values for the climate period of 1981–2010 and 2071–2100. The daily data is spatially downscaled from its original resolution of 0.56° to 0.00833° (30

Table 2. Applied global datasets.

Parameter	Source	Detailed Description
Climate	ECHAM5	[37]
Soil	Harmonized World Soil Database (HWSD)	[41]
Topography	Space Shuttle Topography Mission (SRTM)	[39]
Crop-requirements	FAO Land Evaluation Part III: Crop Requirements	[35]
Irrigation	Global Map of Irrigation Areas (GMIA) v5.0	[44]
Protected Areas	International Union for Conservation of Nature (IUCN) Protected Areas	[45]
Forested Areas	GlobCover 2009	[46]

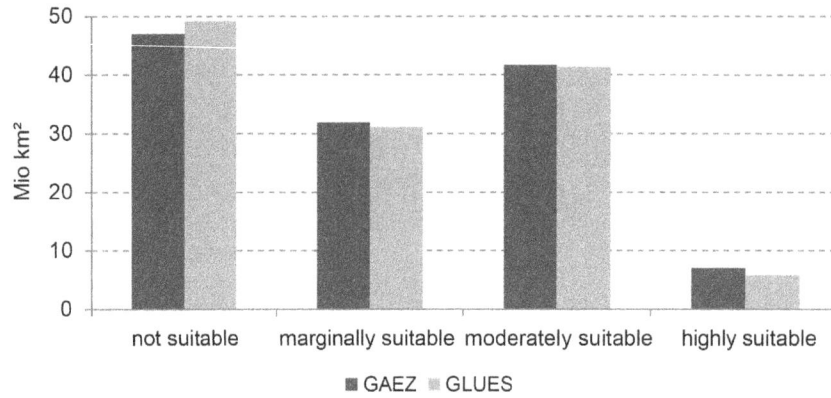

Figure 3. Global comparison of agriculturally suitable area between GAEZ (Baseline period 1961–1990) and GLUES (1961–1990).

arc seconds), based on an approach by [38], using sub-grid terrain information provided by the SRTM-dataset [39]. A bias-correction is executed during the downscaling procedure for temperature and precipitation based on monthly derived factors from the WorldClim dataset [40].

Mean temperature ($\bar{T}$) and total precipitation ($\bar{P}$) are calculated over the length of the growing cycle for each day of the year (doy) (see eq. 1 and eq. 2). Starting on the 1st of January ($doy = 1$), the growing cycle is shifted day by day until the 31st of December ($doy = 365$). The suitability value (S) is calculated for each doy as in eq. 3 for $\bar{T}$ and $\bar{P}$ according to the membership function (mf).

$$\bar{T}_{doy} = \overline{T_{doy}, \ldots, T_{doy+lgc}} \qquad (eq.1)$$

$$\bar{P}_{doy} = \sum_{doy}^{doy+lgc} P \qquad (eq.2)$$

$$S(\bar{T}_{doy}) = mf(\bar{T}_{doy}) \; ; \; S(\bar{P}_{doy}) = mf(\bar{P}_{doy}) \qquad (eq.3)$$

Since the natural suitability of crop growth is limited by the minimum value, the smaller value of the temperature and precipitation fuzzy value determines the climate suitability $S(C)$ which is calculated for each doy (eq. 4).

$$S(C_{doy}) = min\{S(\bar{T}_{doy}), S(\bar{P}_{doy})\} \qquad (eq.4)$$

Among all daily fuzzy values of $S(C)$ within the year, the maximum of $S(C)$ determines the climate suitability over the growing cycle and thus, the optimal start of the growing cycle (eq. 5) for cultivation of a single crop within the entire growing season.

$$S(C_{start\ of\ the\ growing\ cycle}) = max\{S(C_{doy\ 1}), \ldots, S(C_{doy\ 365})\} \quad (eq.5)$$

In order to allow for the calculation of multiple cropping, the fuzzy values for each possible combination of days for the start of the growing cycle are tested as to how often they would fit within one year. The number of multiple cropping is selected that generates the highest accumulated value. Multiple cropping and the start of the growing cycle(s) are obtained for single, double and triple cropping. Hereby, the start of the growing cycle(s) in the context of this paper describes an optimal time for cultivation of a crop to reach the maximum suitability within a year. Crop mixing is not considered. Regarding temporal demands for technical field work, we assume a break of two weeks between crop cycles.

Moreover, the following assumptions are made: At least 20 mm of precipitation are required within the first two weeks of the growing season in order to provide enough soil moisture for germination. No day within the growing period must be below 5°C and below 1°C for winter crops. Vernalisation requirements are considered separately from the growing period for winter

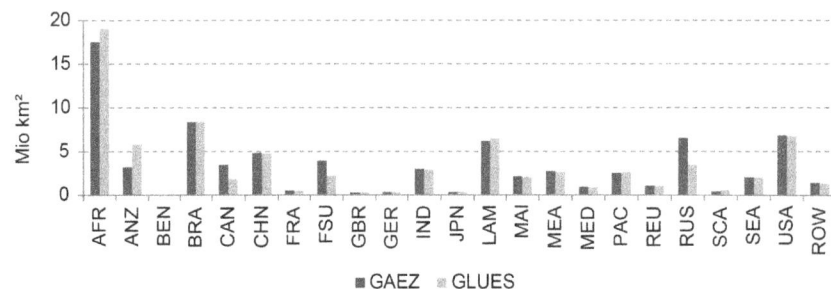

Figure 4. Comparison of total agriculturally suitable area of GAEZ (Baseline period 1961–1990) and GLUES (1961–1990) for different regions.

Table 3. Classified suitability considering rainfed conditions (1981–2010).

Not Suitable	Marginally Suitable	Moderately Suitable	Highly Suitable
49.8 million km^2	30.6 million km^2	41.3 million km^2	5.8 million km^2

crops: Vernalisation period starts 150 days before the start of the growing period. At least 20 days below 5°C must exist during the vernalisation period and there must not exist more than 3 days below −30°C. In order to consider permafrost conditions that exclude agricultural use, mean annual temperature must not be below 0°C. Mean daily incoming solar radiation must exceed 60 W/m^2 to provide enough energy for crop growth.

Thus, suitability values, number of crop cycles and the start of the growing cycle are calculated on each land surface pixel for both rainfed and irrigated conditions. For irrigated conditions, fuzzy values for precipitation are neglected during the calculation process. Due to a lack of global information on irrigation practices, we assume perennial irrigation on irrigated areas.

Besides climatic constraints, soil properties are limiting agricultural suitability. According to the membership functions (Fig. 2), the fuzzy values representing each of the soil properties are calculated. The minimum of the eight values represents the value of the soil suitability. Soil information was taken from the Harmonized World Soil Database (HWSD) [41], considering the topsoil (0–30 cm) of the dominant and all (up to 8) component soils at a spatial resolution of 30 arc seconds [42]. Within the calculation of soil suitability, fuzzy values of each of the component soils are calculated and weighted according to their share.

The suitability for crops to be cultivated is decreasing with increasing slope (see Fig. 2). The slope must not exceed 16% for the considered crops, except for oil palm and paddy rice. The slope was calculated and resampled to 30 arc seconds from Shuttle Radar Topography Mission (SRTM) data [39].

Across all climate, soil and topography fuzzy values, the lowest fuzzy value quantifies the crops' suitability at a certain location. The highest value across all crops determines the suitability for agriculture at a certain location.

This methodology does not allow for yield estimations, in which socio-economic and bio-physical aspects, which our approach does not consider, play an important role. However, this approach is well suited to draw conclusions about where areas are agriculturally suitable and how these areas may change with future climate conditions.

Results

The Earth surface consists of 510 million km^2 of which 149 million km^2 are land surface. Up to 60°S, excluding Antarctica, and considering a lack of input data, in total 127.5 million km^2 of land surface remain to be analyzed regarding their suitability for agriculture. We classified the results of the suitability analysis into

four categories: not suitable (0), marginally suitable (>0.0), moderately suitable (>0.33) and highly suitable (>0.75).

Comparison

Our results (further named GLUES in the Figures) highly correlate with existing studies, such as the GAEZ approach [29], when comparing the area of each of the four classified categories in each of the 23 World Regions ($R^2 = 0.99$).

The global aggregation of the classified areas and the regional distribution of not suitable and suitable areas show a high level of agreement (Fig. 3 and 4). Compared to the distribution of global cropland in the year 2000 [43], our approach identifies 95.5% of current cropland as suitable.

Rainfed

For the period 1981–2010, our suitability analysis shows that in total 77.7 million km^2 are potentially suitable for purely rainfed agricultural cultivation, while 49.8 million km^2 are not suitable for rainfed conditions (Table 3). Further, 30.6 million km^2 are marginally suitable, 41.3 million km^2 are moderately suitable and 5.8 million km^2 are highly suitable (Table 3).

Irrigation

Irrigated agriculture produces 40% of the world's food (FAO) on 3.1 million km^2 [44]. When considering irrigation, suitability is area weighted according to the fraction of rainfed and irrigated agricultural area (given by GMIA Version 5.0 [44]). Thereby, irrigation increases suitability on irrigated areas in global average by 0.13, adds 1.8 million km^2 of suitable land (Table 4) and allows for multiple cropping on 1.2 million km^2 (assuming sufficient water available for irrigation). Accordingly, huge areas e.g. in the Nile and Ganges delta are only becoming suitable due to irrigation. Overall, 79.6 million km^2 are suitable with spatially varying patterns (Fig. 5).

Figure 5 represents the global distribution of agricultural suitability as a result of local climate, soil and terrain conditions. In boreal regions, the growing season over all stages of phenology usually is too short for cultivation. The temperate zones seasonally have adequate temperatures and enough precipitation and often sufficient soil, while in subtropical regions, the annual distribution of precipitation strongly determines crop growth and soils often are alkaline. In inner tropics have adequate temperature and moisture throughout the year, but soil quality often restricts cultivation due to low organic content and acidity [3].

Table 4. Classified suitability considering rainfed and irrigated conditions (1981–2010).

Not Suitable	Marginally Suitable	Moderately Suitable	Highly Suitable
48.0 million km^2	31.8 million km^2	41.8 million km^2	5.9 million km^2

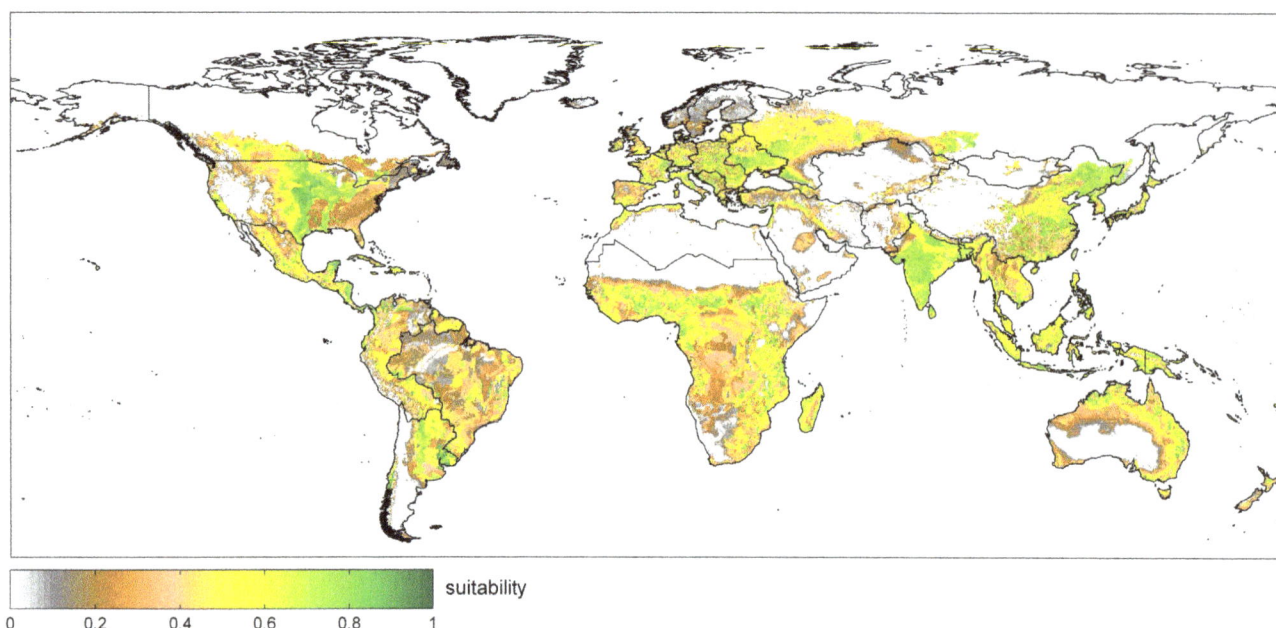

Figure 5. Agricultural suitability considering rainfed conditions and irrigated areas (1981–2010).

Protected Areas

Protected areas globally account for 8.3 million km^2. Information on actual protected areas is gathered from IUCN [45]. When excluding protected areas from the suitability calculation, 74.8 million km^2 remain suitable for cultivation. Thereby, protected areas are mainly situated in not suitable or marginally suitable areas (Table 5). Only 2% (0.2 million km^2) of the global protected area are located on land highly suitable for agriculture, 25% (2.1 million km^2) are on moderately suitable land, 30% (2.4 million km^2) on marginally suitable land while 43% (3.6 million km^2) are situated on unsuitable land. Overall, only 57% of global protected areas are suitable for agriculture.

Forested Areas

Dense forests are highly important to provide numerous ecosystem services. Densely forested areas account for 23.3 million km^2 according to GlobCover [46] and 23.5 million km^2 according to [47]. GlobCover defines forests as being dense when 75% of the pixel is forest [46]. Only 1.5 million km^2 or 6.2% of the global densely forested areas are currently protected.

4.9% (1.1 million km^2) of the densely forested areas (excluding forests within protected areas) are located in highly suitable land, 49.4% (11.1 million km^2) in moderately suitable land, 37.5% (8.4 million km^2) in marginally suitable land and only 8.2% (1.9 million km^2) are situated on unsuitable land. Overall, 92% of densely forested areas are potentially suitable for agriculture which indicates that global forests are subject to increasing societal stress.

Current Use of Suitable Land and Trade-Offs

When excluding both, protected areas and dense forests from the suitability calculation, 54.1 million km^2 remain suitable (Table 6). In comparison, currently used agricultural land (including pasture) today covers 49.1 million km^2, of which 15.5 million km^2 (status for 2011) are arable land (land under temporary and permanent crops; double-cropped areas are counted only once) [48]. Accordingly, 91% of all suitable land is already occupied by agriculture when today's protected and densely forested areas are preserved in the future. This illustrates that agricultural expansion is only possible by substituting other uses/covers of land which causes high social and ecological externalities. Figure 6 gives an overview of the current use/cover of suitable areas in the different regions of the world.

Figure 6 shows, that the current fraction of suitable area, which is not protected or dense forest is highly variable across regions. The most efficient use of current agriculturally suitable land is obvious in the USA, where only 2% of currently suitable land is not yet used or protected/dense forest.

In Africa, about 20% of the agriculturally suitable area is currently not used for agriculture or is statistically not recorded in the data of currently used agricultural land (Ramankutty et al., 2008). This shows the extraordinary potentials of Africa for future expansion of agricultural land. However, agricultural expansion would always take place at ecological costs (e.g. conversion of tropical rainforest, grassland and savannah). In Latin America large suitable areas are protected or covered with dense forest and

Table 5. Classified suitability for 1981–2010 considering rainfed and irrigated conditions, excluding protected areas.

Not Suitable	Marginally Suitable	Moderately Suitable	Highly Suitable
44.4 million km^2	29.4 million km^2	39.7 million km^2	5.7 million km^2

Table 6. Classified suitability for 1981–2010 considering rainfed and irrigated conditions, excluding protected and densely forested areas.

Not Suitable	Marginally Suitable	Moderately Suitable	Highly Suitable
42.6 million km^2	21.0 million km^2	28.6 million km^2	4.6 million km^2

Table 7. Classified suitability for 2071–2100 considering rainfed and irrigated conditions, excluding protected and densely forested areas.

Not Suitable	Marginally Suitable	Moderately Suitable	Highly Suitable
37.8 million km^2	24.8 million km^2	30.2 million km^2	3.9 million km^2

the current fraction of remaining suitable area is smaller than in Africa, India is the prototype of a country, which is already using very large parts of its suitable agricultural land - and by for using the largest proportion (58%) of current cropland. Australia and larger parts of Asia still have reasonable land resources left for future expansion (Fig 6).

Future Change

For the investigation of future agricultural suitability for the time-period 2071–2100 as determined by the simulated climate effects of the SRES A1B emission scenario, we assume no changes in irrigated areas, soil properties, terrain or any adaptations, such as crop breeding. As result, when again excluding protected and densely forested areas, the global area being highly suitable for agriculture decreases from 4.6 to 3.9 million km^2, while marginally and moderately suitable areas increase (Table 7). In total, agriculturally suitable areas increase by 4.8 million km^2 due to the selected climate change scenario. However, most of the additional area is only marginally suitable for agricultural use.

Without excluding any areas, the impact of climate change increases the potentially suitable areas on the globe by 5.6 million km^2. Marginally suitable areas increase by 4.2 million km^2, moderately suitable areas increase by 2.3 million km^2, while highly suitable areas decrease by 0.8 million km^2 (Fig. 7).

A more regional analysis shows that the world is divided into regions that receive additional suitable land and regions where land that used to be suitable turns into not suitable land (Fig. 8). Regions in the northern hemisphere, such as Canada (+2.1 million km^2 of suitable land), Russia (+3.1 million km^2) and China (+0.9 million km^2), benefit most.

On the global scale, suitability improves on 18.7 million km^2 and worsens on 22.2 million km^2. In total, the area with decreasing suitability is 3.5 million km^2 more than the area with increasing suitability (Fig. 9). The highest absolute net loss of suitable areas is found in Sub-Saharan Africa.

Thereby, the globally averaged suitability value (averaged over all suitable areas), decreases from 0.41 to 0.39. The greatest losses of suitability are simulated for France and the Mediterranean (Fig. 10). The changing suitability is mapped in Fig. 11.

Growing Cycle and Multiple Cropping

The seasonal development of temperature and precipitation determines the length of the growing season, the start of the growing cycle and the potential number of annual cropping. Thus, the option of multiple cropping represents an important measure for farmers to increase production. Figure 12 shows the spatial distribution of the start of the growing cycle for the time period 1981–2010, exemplarily for maize.

Changing climate does not only affect the suitability of land, but also the start and length of the growing cycle. As an example, the start of the growing cycle for maize in Germany shifts in average 23 days earlier in time, when comparing the period of 2071–2100 with 1981–2010. The shift of growing cycles again influences the possibility for multiple cropping. Today's maximal achievable multiple cropping according to the course of temperature and precipitation is shown in Fig. 13.

Our results suggest that climate change has huge impacts on the areas suitable for multiple cropping under the assumed climate scenario. Until 2100, 6.0 million km^2 are globally lost for triple cropping until 2100, while the area which is suitable for double cropping increases by 2.3 million km^2. Multiplying the area with

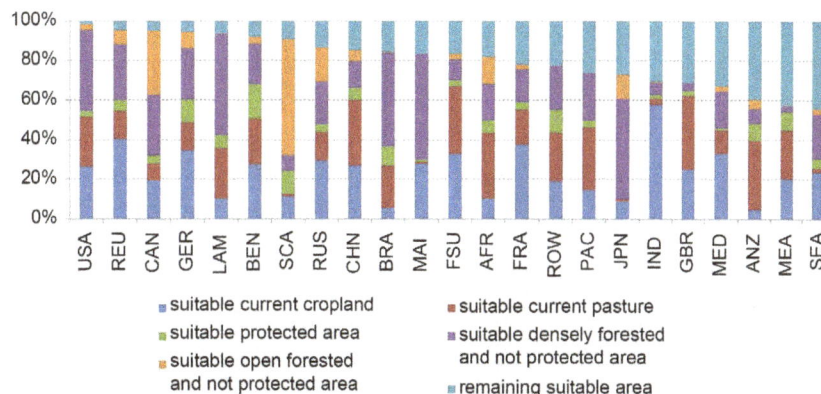

Figure 6. Current use of suitable areas (1981–2010), considering forest cover [46], **protected areas** [45] **and current pasture and cropland (Ramankutty et al., 2008).** If forested areas are agriculturally suitable and protected, they are attributed to 'suitable protected area'.

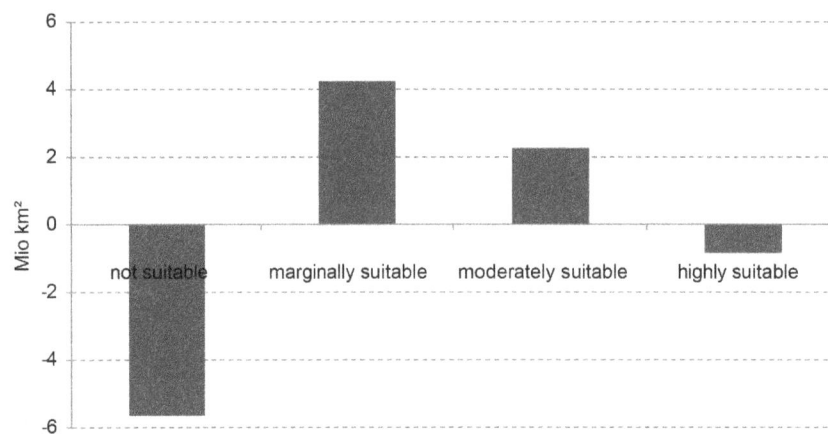

Figure 7. Global changes in agricultural suitability categories (million km²) between 1981–2010 and 2071–2100.

the number of cycles, this means a global decrease of 13.4 million km². Most of the increase in double cropping areas results from the transformation from triple to double cropping. Again, no change in irrigation is assumed in this calculation.

The largest decrease in multiple copping area can be found in Brazil (BRA) and in Sub-Saharan Africa (AFR), where areas suitable for triple cropping decrease by 1.7 (AFR) and 2.9 million km² (BRA) (Fig. 14), while the area for double cropping increases by 0.2 and 1.3 million km², respectively. In total, this means a decrease of multiple cropping area by 1.5 (AFR) and 1.6 million km² (BRA). This is equivalent to the amount of 4.7 and 6.1 million km² respectively, which are lost for agriculture, when multiplying the area with the number of possible crop cycles. This corresponds to 20.2 (AFR) and 28.8% (BRA) of today's potentially suitable area for multiple cropping. In the same manner, France (FRA) and the Mediterranean (MED) lose 24.1 (FRA) and 13.2% (MED) of their total equivalent area when considering the change of multiple cropping, which means a decrease by 93 (FRA) and 55% (MED) according to the multiple cropping area of 1981–2010. Regions where areas that potentially allow for more than one crop cycle increase due to climate change are CHI, IND, JPN, MEA, REU, RUS and USA, while the total area considerably increases mainly in the USA for both, double (0.35 million km²) and triple (0.12 million km²) cropping (Fig. 14).

Conclusions

The analyses of the present situation demonstrats that there is extraordinary potential e.g. for Sub Saharan Africa for future expansion of agricultural land without expanding into protected or forested areas. Further research is necessary to identify the environmental and social costs and consequences of agricultural expansion in these regions. Also further investigation is needed to give answers on how this land could be managed sustainable with benefit to local food systems and socio-economy.

Our results show at high spatial resolution how agricultural suitability may change until 2100 due to changing climate under the chosen scenario (SRES A1B), assuming no adaptation measurements by farmers. First, suitable areas increase especially in the northern regions such as Canada, China and Russia, where new land will be available for agricultural use. The increase in suitable areas mainly takes place in sparsely populated areas, which could imply a lack of labor for open up new agricultural land and prepare soils. Certainly, it will be related with high investment costs and it will take a long time to extend agriculture here. Secondly, global average suitability decreases under the chosen climate scenario. Especially the extend of highly suitable areas is reduced by the effect of climate change. Finally, suitable areas indirectly are reduced due to a substantial global reduction of the suitability for multiple cropping, especially in Sub Saharan Africa, and Brazil.

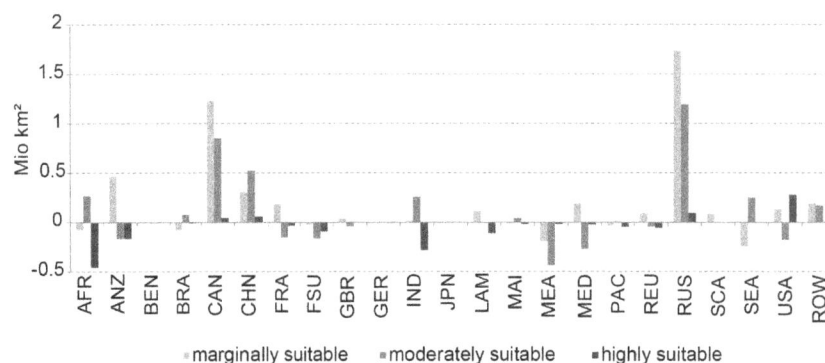

Figure 8. Regional change in agricultural suitability categories (million km²) between 1981–2010 and 2071–2100.

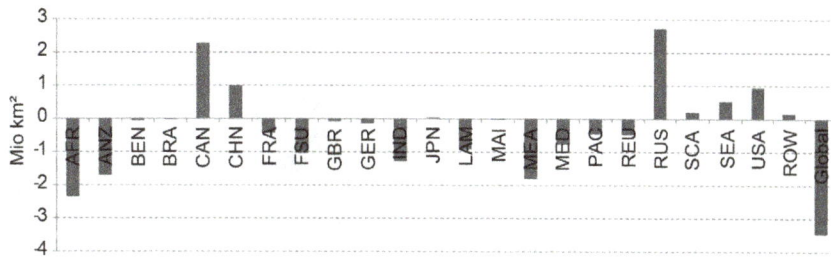

Figure 9. Regional change of agriculturally suitable area due to A1B climate change scenario between 1981–2010 and 2071–2100.

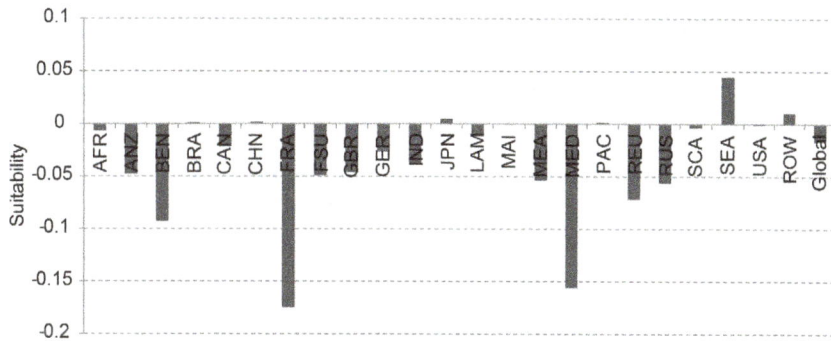

Figure 10. Regional changes in the average suitability between 1981–2010 and 2071–2100.

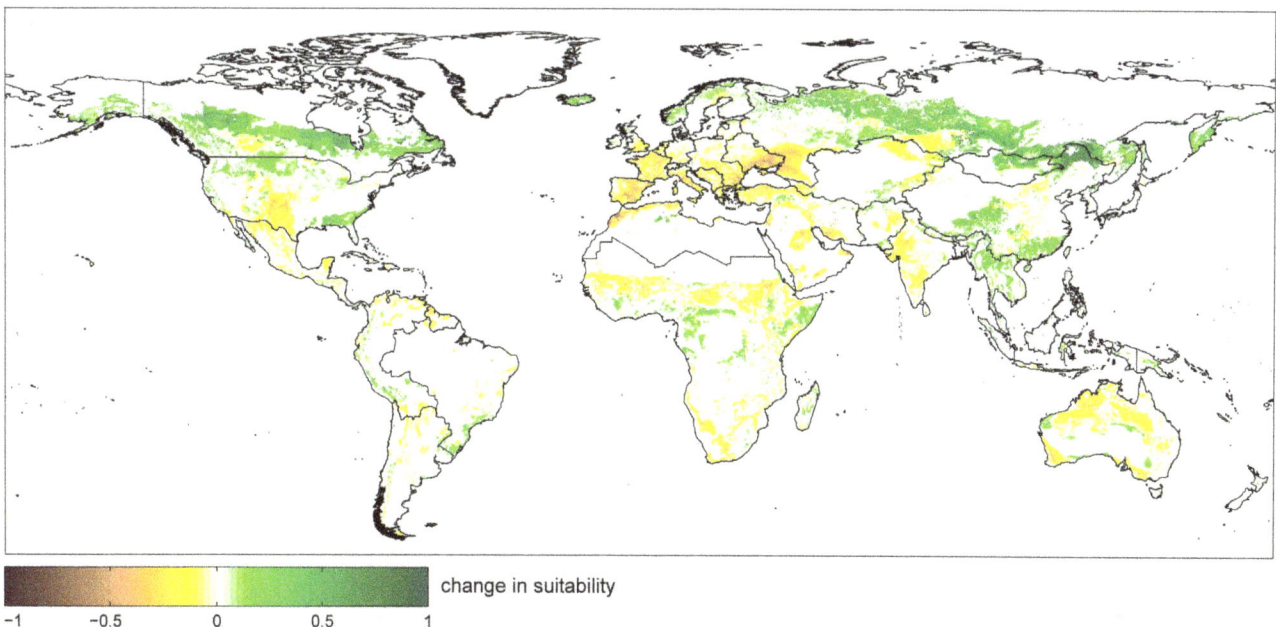

change in suitability

−1   −0.5   0   0.5   1

Figure 11. Change in agricultural suitability between 1981–2010 and 2071–2100. Green areas indicate an increase in suitability while brown areas show a decreasing suitability.

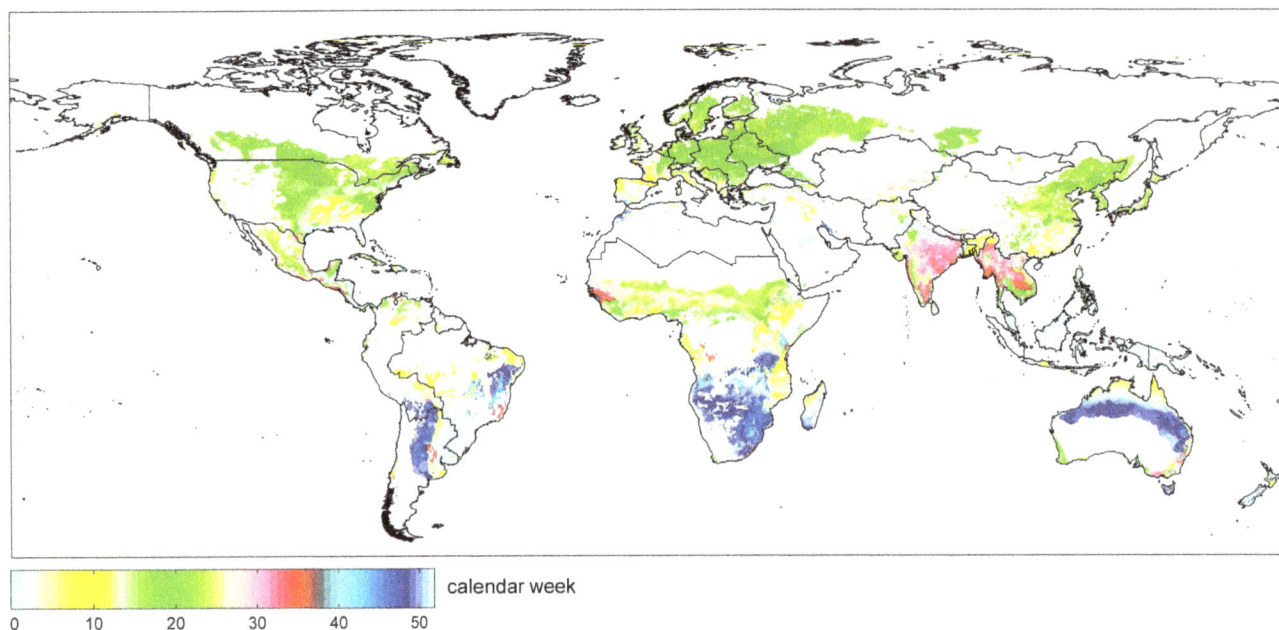

Figure 12. Start of the growing cycle for maize (1981–2010). The start of the growing cycle is illustrated for rainfed conditions and for irrigated conditions on predominantly irrigated areas (irrigated area > 50%). In case of multiple cropping, the map shows the start of the first growing cycle.

Overall, the Global North regionally increases suitability and the number of crop cycles, while the Global South and the Mediterranean area lose agriculturally suitable land without adaptations. This will decisively affect smallholder farmers as their options for adaptations through e.g. irrigation are limited.

Scientific knowledge on the geographical distribution has decisively being increased with the availability of global data sets, also based on remote sensing. The tensions between both limits of land expansion and intensification within the context of sustainable agricultural intensification stresses the ongoing debate on

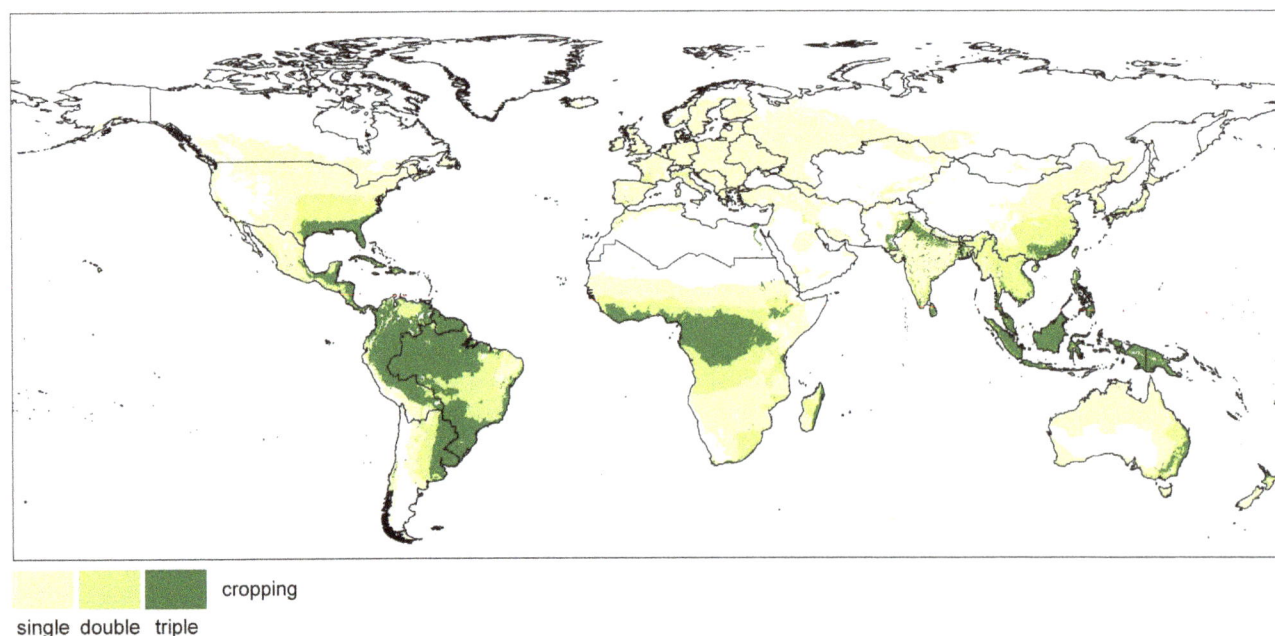

Figure 13. Suitable areas for single, double and triple cropping (1981–2010). Multiple cropping is illustrated for rainfed conditions and for irrigated conditions on predominantly irrigated areas (irrigated area > 50%).

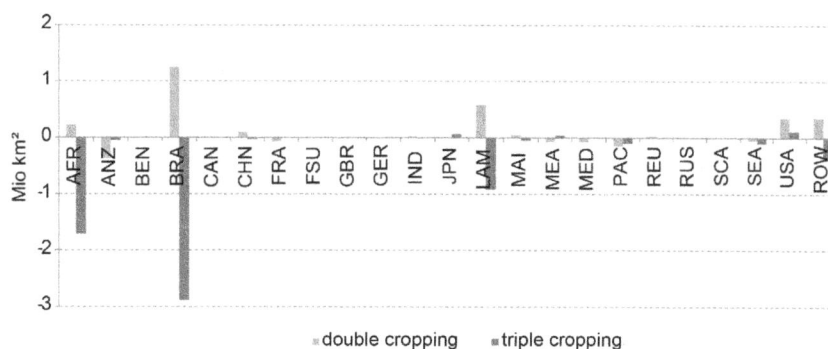

Figure 14. Change in area suitable for double and triple cropping (million km²) between 1981-2010 and 2071-2100.

global land management, considering the complex interplay and trade-offs between different uses of land and ecosystem services.

Acknowledgments

This research was carried out within the framework of the GLUES (Global Assessment of Land Use Dynamics, Greenhouse Gas Emissions and Ecosystem Services) Project. Thanks to all project members and to Jonas Maier who contributed to this study as a student assistant.

Author Contributions

Conceived and designed the experiments: FZ BP WM. Performed the experiments: FZ. Analyzed the data: FZ. Contributed reagents/materials/analysis tools: FZ. Contributed to the writing of the manuscript: FZ.

References

1. Foley JA, DeFries R, Asner GP, Barford C, Bonan G, et al. (2005) Global Consequences of Land Use. Science 309: 570–574.
2. Václavík T, Lautenbach S, Kuemmerle T and Seppelt R (2013) Mapping global land system archetypes. Global Environmental Change 23: 1637–1647.
3. Ramankutty N, Foley JA, Norman J, McSweeney K (2002) The global distribution of cultivable lands: current patterns and sensitivity to possible climate change. Global Ecology and Biogeography 11: 377–392.
4. Alexandratos N, Bruinsma J (2012) World agriculture towards 2030/2050: the 2012 revision. ESA Working Paper No 12-03. Rome: FAO.
5. Tilman D, Balzer C, Hill J, Befort BL (2011) Global food demand and the sustainable intensification of agriculture. Proceedings of the National Academy of Sciences 108: 20260–20264.
6. Foley JA, Ramankutty N, Brauman KA, Cassidy ES, Gerber JS, et al. (2011) Solutions for a cultivated planet. Nature 478: 337–342.
7. Godfray HCJ, Beddington JR, Crute IR, Haddad L, Lawrence D, et al. (2010) Food Security: The Challenge of Feeding 9 Billion People. Science 327: 812–818.
8. Pretty J, Sutherland WJ, Ashby J, Auburn J, Baulcombe D, et al. (2010) The top 100 questions of importance to the future of global agriculture. International Journal of Agricultural Sustainability 8: 219–236.
9. Gregory PJ and George TS (2011) Feeding nine billion: the challenge to sustainable crop production. Journal of Experimental Botany.
10. Ray DK, Mueller ND, West PC, Foley JA (2013) Yield Trends Are Insufficient to Double Global Crop Production by 2050. PLoS ONE 8: e66428.
11. Vermeulen S, Zougmoré R, Wollenberg E, Thornton P, Nelson G, et al. (2012) Climate change, agriculture and food security: a global partnership to link research and action for low-income agricultural producers and consumers. Current Opinion in Environmental Sustainability 4: 128–133.
12. Kastner T, Rivas MJI, Koch W, Nonhebel S (2012) Global changes in diets and the consequences for land requirements for food. Proceedings of the National Academy of Sciences 109: 6868–6872.
13. Cassidy ES, West PC, Gerber JS, Foley JA (2013) Redefining agricultural yields: from tonnes to people nourished per hectare. Environmental Research Letters 8: 034015.
14. Erb K-H, Haberl H, Krausmann F, Lauk C, Plutzar C, et al. (2009) Eating the Planet: Feeding and fuelling the world sustainably, fairly and humanely – a scoping study. Social Ecology Working Paper. Alpen-Adria Universität Klagenfurt.
15. Spiertz JHJ, Ewert F (2009) Crop production and resource use to meet the growing demand for food, feed and fuel: opportunities and constraints. NJAS - Wageningen Journal of Life Sciences 56: 281–300.
16. Bruinsma J (2011) The resources outlook: by how much do land, water and crop yields need to increase by 2050? In: P. Conforti , editor. Looking ahead in world food and agriculture: Perspectives to 2050. Rome: FAO.
17. Schmitz C, van Meijl H, Kyle P, Nelson GC, Fujimori S, et al. (2014) Land-use change trajectories up to 2050: insights from a global agro-economic model comparison. Agricultural Economics 45: 69–84.
18. Pimentel D, Harvey C, Resosudarmo P, Sinclair K, Kurz D, et al. (1995) Environmental and Economic Costs of Soil Erosion and Conservation Benefits. Science 267: 1117–1123.
19. UNCCD (2014) Desertification - The Invisible Frontline. Bonn, Germany: United Nations Convention to Combat Desertification.
20. Seto KC, Fragkias M, Güneralp B, Reilly MK (2011) A Meta-Analysis of Global Urban Land Expansion. PLoS ONE 6: e23777.
21. Avellan T, Meier J, Mauser W (2012) Are urban areas endangering the availability of rainfed crop suitable land? Remote Sensing Letters 3: 631–638.
22. Teka K, Haftu M (2012) Land Suitability Characterization for Crop and Fruit Production in Midlands of Tigray, Ethiopia. Momona Ethiopian Journal of Science 4: 12.
23. Kalogirou S (2002) Expert systems and GIS: an application of land suitability evaluation. Computers, Environment and Urban Systems 26: 89–112.
24. Baja S, Chapman DM, Dragovich D (2002) A Conceptual Model for Defining and Assessing Land Management Units Using a Fuzzy Modeling Approach in GIS Environment. Environmental Management 29: 647–661.
25. Braimoh AK, Vlek PLG, Stein A (2004) Land Evaluation for Maize Based on Fuzzy Set and Interpolation. Environmental Management 33: 226–238.
26. Kurtener D, Torbert HA, Krueger E (2008) Evaluation of Agricultural Land Suitability: Application of Fuzzy Indicators. In: O . Gervasi, B . Murgante, A . Laganûà, D . Taniar, Y . Mun and M . Gavrilova, editors. Computational Science and Its Applications – ICCSA 2008. Springer Berlin Heidelberg. pp. 475–490.
27. Nisar Ahamed TR, Gopal Rao K, Murthy JSR (2000) GIS-based fuzzy membership model for crop-land suitability analysis. Agricultural Systems 63: 75–95.
28. Van Ranst E, Tang H, Groenemam R, Sinthurahat S (1996) Application of fuzzy logic to land suitability for rubber production in peninsular Thailand. Geoderma 70: 1–19.
29. IIASA/FAO (2012) Global Agro-ecological Zones (GAEZ v3.0) - Model Documentation. IIASA, Laxenburg, Austria and FAO, Rome, Italy: IIASA, FAO.
30. Fischer G, Hizsnyik E, Prieler S, Wiberg D (2011) Scarcity and abundance of land resources: competing uses and the shrinking land resource base. SOLAW Background Thematic Report. FAO.
31. Lane A, Jarvis A (2007) Changes in Climate will modify the Geography of Crop Suitability: Agricultural Biodiversity can help with Adaptation. Journal of SAT Agricultural Research 4: 12.
32. Avellan T, Zabel F, Mauser W (2012) The influence of input data quality in determining areas suitable for crop growth at the global scale – a comparative analysis of two soil and climate datasets. Soil Use and Management 28: 249–265.
33. Burrough PA (1989) Fuzzy mathematical methods for soil survey and land evaluation. Journal of Soil Science 40: 477–492.
34. Burrough PA, Macmillan RA, van Deursen W (1992) Fuzzy classification methods for determining land suitability from soil profile observations and topography. Journal of Soil Science 43: 193–210.
35. Sys CO, van Ranst E, Debaveye J, Beernaert F (1993) Land evaluation: Part III Crop requirements. Agricultural Publications. Brussels: G.A.D.C.
36. Bengtsson L, Hodges KI, Keenlyside N (2009) Will Extratropical Storms Intensify in a Warmer Climate? Journal of Climate 22: 2276–2301.

37. Jungclaus JH, Keenlyside N, Botzet M, Haak H, Luo JJ, et al. (2006) Ocean Circulation and Tropical Variability in the Coupled Model ECHAM5/MPI-OM. Journal of Climate 19: 3952–3972.

38. Marke T, Mauser W, Pfeiffer A, Zängl G, Jacob D, et al. (2013) Application of a hydrometeorological model chain to investigate the effect of global boundaries and downscaling on simulated river discharge. Environ Earth Sci: 1–20.

39. Farr TG, Rosen PA, Caro E, Crippen R, Duren R, et al. (2007) The Shuttle Radar Topography Mission. Reviews of Geophysics 45: RG2004.

40. Hijmans RJ, Cameron SE, Parra JL, Jones PG, Jarvis A (2005) Very high resolution interpolated climate surfaces for global land areas. International Journal of Climatology 25: 1965–1978.

41. FAO/IIASA/ISRIC/ISSCAS/JRC (2012) Harmonized World Soil Database (version 1.2). FAO, Rome, Italy and IIASA, Laxenburg, Austria.

42. Avellan T, Zabel F, Putzenlechner B, Mauser W (2013) A Comparison of Using Dominant Soil and Weighted Average of the Component Soils in Determining Global Crop Growth Suitability. Environment and Pollution 2: 11.

43. Ramankutty N, Evan AT, Monfreda C, Foley JA (2008) Farming the planet: 1. Geographic distribution of global agricultural lands in the year 2000. Global Biogeochemical Cycles 22: GB1003.

44. Siebert S, Henrich V, Frenken K, Burke J (2013) Global Map of Irrigation Areas version 5. Rheinische Friedrich-Wilhelms-University, Bonn, Germany/Food and Agriculture Organization of the United Nations, Rome, Italy.

45. IUCN (2008) Guidelines for Applying Protected Area Management Categories In: N. Dudley, editor. Gland, Switzerland: IUCN.

46. Bontemps S, Defourny P, Bogaert Ev, Arino O, Kalogirou V (2009) GLOBCOVER 2009, Products Description and Validation Report. ESA, Univ. catholique de Louvain.

47. Hansen MC, Potapov PV, Moore R, Hancher M, Turubanova SA, et al. (2013) High-Resolution Global Maps of 21st-Century Forest Cover Change. Science 342: 850–853.

48. FAOSTAT (2014) FAOSTAT Land USE module. Retrieved 24 February 2014. Available at: http://faostat.fao.org/site/377/DesktopDefault.aspx?PageID=377#ancor.

3

The Dynamics of Latifundia Formation

Luis Fernando Chaves[1,2,3]*

1 Graduate School of Environmental Sciences, Hokkaido University, Sapporo, Japan, 2 Programa de Investigación en Enfermedades Tropicales (PIET), Escuela de Medicina Veterinaria, Universidad Nacional, Heredia, Costa Rica, 3 Institute of Tropical Medicine (NEKKEN), Nagasaki University, Nagasaki, Japan

Abstract

Land tenure inequity is a major social problem in developing nations worldwide. In societies, where land is a commodity, inequities in land tenure are associated with gaps in income distribution, poverty and biodiversity loss. A common pattern of land tenure inequities through the history of civilization has been the formation of latifundia [Zhuāngyuán in chinese], i.e., a pattern where land ownership is concentrated by a small fraction of the whole population. Here, we use simple Markov chain models to study the dynamics of latifundia formation in a heterogeneous landscape where land can transition between forest, agriculture and recovering land. We systematically study the likelihood of latifundia formation under the assumption of pre-capitalist trade, where trade is based on the average utility of land parcels belonging to each individual landowner during a discrete time step. By restricting land trade to that under recovery, we found the likelihood of latifundia formation to increase with the size of the system, i.e., the amount of land and individuals in the society. We found that an increase of the transition rate for land use changes, i.e., how quickly land use changes, promotes more equitable patterns of land ownership. Disease introduction in the system, which reduced land profitability for infected individual landowners, promoted the formation of latifundia, with an increased likelihood for latifundia formation when there were heterogeneities in the susceptibility to infection. Finally, our model suggests that land ownership reforms need to guarantee an equitative distribution of land among individuals in a society to avoid the formation of latifundia.

Editor: Rodrigo Huerta-Quintanilla, Cinvestav-Merida, Mexico

Funding: This work was funded by Japan Society for the Promotion of Science(JSPS). The funders had no role in study design, data collection and analysis, decision to publish, or preparation of the manuscript

Competing Interests: The author has declared that no competing interests exist.

* E-mail: lchaves@nagasaki-u.ac.jp

Introduction

The socialized nature of ecosystem transformation is a major theme of study within the broad field of environmental studies [1,2,3]. Over recent years the fields of ecology and epidemiology have become aware of the fundamental role humans play on ecosystem transformation [3] and the impacts of such transformations on biodiversity loss and disease emergence [4]. One socio-ecological phenomenon that has caught a great amount of attention is the dynamics of deforestation [3], and the transition between agricultural and forested land [5,6,7]. Transitions in land-use change have been modeled using Markov Chains [5,6]. Several models have shown that the proportion of forested land can be explained as a function of information flow, adaptive social learning and the rate of deforestation, which can be modulated by the utility of land according to its use, and the forest recovery rate, which is a biological attribute of the landscape [5,6].

Deforestation has also been long recognized as one of the major drivers for the emergence of infectious diseases affecting humans, with studies documenting the emergence of malaria [8,9], leishmaniasis [10,11] and Yellow fever [12] shortly after large scale land use changes. An additional insight from the study of the association between malaria emergence and deforestation was the correlation of malaria endemicity with the formation of latifundia, i.e., the accumulation of land tenure by a small number of landowners, a pattern observed both in the Agro-Pontino Romano for centuries [8], and Spain during the 1930s [13]. More specifically, it has been suggested, and documented, by the long historical records for the Roman Agro-Pontino [8], that defores-

tation and agricultural development led to ideal conditions for the development of mosquito vectors of malaria parasites [8], a fact biologically instantiated by ecological research over recent years [12]. The debilitating effects of malaria on farmers reduce their ability to harvest crops and lead to the sale or abandonment and adjudication of land by healthier and/or wealthier landowners that will underutilize land as latifundia, i.e., large states whose exploitation, because of the landmass size, require the labor of workers who do not own the land [8]. When land cover is primarily forested, land tenure can be redistributed for agricultural exploitation, and in turn result in a repeated cycle of agricultural exploitation, malaria transmission and latifundia formation [8]. The problem of latifundia formation has widespread consequences, for example, it can be at the basis of biodiversity loss in countries with extreme inequities in wealth, where latifundia and the lack of land property rights are among the major causes behind deforestation [14,15,16,17,18]. More generally, latifundia are also detrimental to society as demonstrated by the cliodinamical analysis of societies that declined or disappeared after they promoted the creation of latifundia, e.g., ancient Rome [19] or that switched from models of land tenure equity to latifundia, e.g., China at the end of the Tang dynasty [20].

Here, we present a model of latifundia formation that considers the dynamics of land use change, among the following land-use states: forest, agricultural land and land in recovery [5] and the pre-capitalist trade of land in recovery [21]. The pre-capitalist trade implies that land exploitation does not lead to the accumulation of capital, and that goods are traded by their "instantaneous" value [21]. First, we mathematically analyze the case for 2 landowners and then study the case for n>2 landowners through computer intensive simulations. In our model a finite and equal amount of land (which can be in any of the different land use "states") is divided among a fixed number of landowners which get different utilities from the land they own according to its state. In our model landowners only trade empty land in order to increase their profit. This null-model for latifundia formation successfully recreated patterns of latifundia formation or "land equity", i.e., a situation where a large proportion of the original landowners remained owning land, once land use and trade reached an equilibrium, i.e., when there were no changes after further model iterations. We then used this model, with parameters that favor "land equity" to test the influence of disease on latifundia formation. We found the assumption of a discounted land utility, i.e., that disease reduce the profits from land use, was a plausible driver of latifundia formation under conditions that would otherwise never lead to inequities in patterns of land ownership.

Materials and Methods

Data Patterns

Figure 1A shows the percent of land exploited as latifundia and Figure 1B the different degrees of malaria endemicity in Spain during the 1930s [13]. Data from the maps in Figure 1A and 1B were extracted using ARCGIS® from the maps in Beauchamp [13]. The association between latifundia, which can also be measured in terms of landmass, such as states larger than 25 ha., i.e., requiring external labor [13], and malaria endemicity, based on a cluster analysis (partition around medioids), for each province was studied using a multiple correspondence analysis between categories for the level of land exploited as latifundia and the dominant malaria endemicity level for each province (see Protocol S1 Appendix A for details). We also derived continuous endemicity indices employing principal components analysis and multidimensional scaling to better visualize the relationship between malaria transmission and land exploited as latifundia (see Protocol S1 Appendix B for details).

Figure 1C shows that epidemic malaria transmission was associated with large proportions of land being exploited as latifundia in Spain during the 1930s. This is supported by the proximity between the categories Int (intense malaria) and values above 40 (which indicate the proportion of land exploited as latifundia). Figure 1D shows how, in general, as malaria transmission intensity increased, the percent of land exploited as latifundia also increased, as well as, its variability. Data patterns indicate that a good null-model of latifundia formation should be able to present a wide variability in the likelihood of latifundia formation in a context with disease transmission (Figure S1).

Model Details

Basic land use model. Let's assume there is a landscape that is subdivided in **n**×**m** land parcels, where **n** individual landowners possess **m** of the parcels at the beginning of the dynamics. Each land parcel, k, belonging to landowner i, can transition between three possible states, S, at time t:

$$S_k^i(t) = \begin{pmatrix} F \\ A \\ E \end{pmatrix} \qquad (1)$$

where F denotes forested (native vegetation in a wider sense) land, A agricultural land and E stands for post agricultural land, e.g.,

land where agriculture is not profitable anymore. The transition from F to A is governed by a deforestation rate, r, from A to E by a degradation rate, η, and from E to A by a forest recovery rate, μ [5]. Assuming changes between land-use types are unidirectional, then the probabilities of staying in the same state is $1-r$ for F, $1-\eta$ for A and $1-\mu$ for E. Although the transition rates between different land uses can be modeled to change as function of global characteristics of the landscape [6] or information flux [7] we will fix these rates to ease the understanding of latifundia formation.

Basic Land Trade Model. To study the dynamics of land ownership and the mechanisms of latifundia formation, we can add a layer of complexity to the previous model by incorporating the trade between individual landowners. To incorporate the trade between individual landowners, we define a trade rate (t_{ji}) as the probability of individual i purchasing land from individual j. If we further assume that land trade is restricted to parcels in the E state, and that trade happens before land transition is realized, we can obtain the following equations for the amount of land belonging to an individual landowner i trading with $j \neq i$:

$$x_i(t) = (1-r)x_i(t-1) + (\mu)\left(\left(1 - \sum_j t_{ij}(t-1)\right)z_i(t-1) \right.$$
$$\left. + \sum_j t_{ji}(t-1)z_j(t-1) \right)$$

$$y_i(t) = (r)x_i(t-1) + (1-\eta)y_i(t-1) \qquad (2)$$

$$z_i(t) = (\eta)y_i(t-1) + (1-\mu)\left(\left(1 - \sum_j t_{ij}(t-1)\right)z_i(t-1) \right.$$
$$\left. + \sum_j t_{ji}(t-1)z_j(t-1) \right)$$

This model is illustrated in Figure 2, which shows how land changes status according to its use, and how land can be traded between owner i and j. Equation (2) allows the estimation of land in the different states belonging to a given individual i at a given time, i.e., $x_i(t)$, $y_i(t)$ and $z_i(t)$ respectively represent the amount of land that is forested, under agricultural production or in a post-agricultural stage at a given time t. t_{ij} the probability of individual i selling land to individual j. Finally, t_{ji}, μ, η and r parameters already defined in the text preceding equation (2).

Defining the trade rates. The trade rates will be a function of the purchasing power (P_i), which is the probability of acquiring land at a given time by a landowner i and the sale pressure (V_j) which is the probability of selling land by a landowner $j \neq i$. Let's further assume that purchasing ability, P_i, is a function of the utility, u, of each parcel, k, belonging to each individual, i, at time t:

$$u_k^i = \begin{cases} a & if & S=F \\ b & if & S=A \\ c & if & S=E \end{cases} \qquad (3)$$

where a indicates the forest value attributed to ecosystem services when the parcel is forested, b represents the monetary benefits by crop sales minus the cost of agricultural land management; c is the utility for a post-agricultural parcel. We define purchasing ability (or power), P_i, as the probability of a landowner to purchase land in relation to other landowners as defined by his/her assets in relation to those present in the whole population:

A

B

C

D

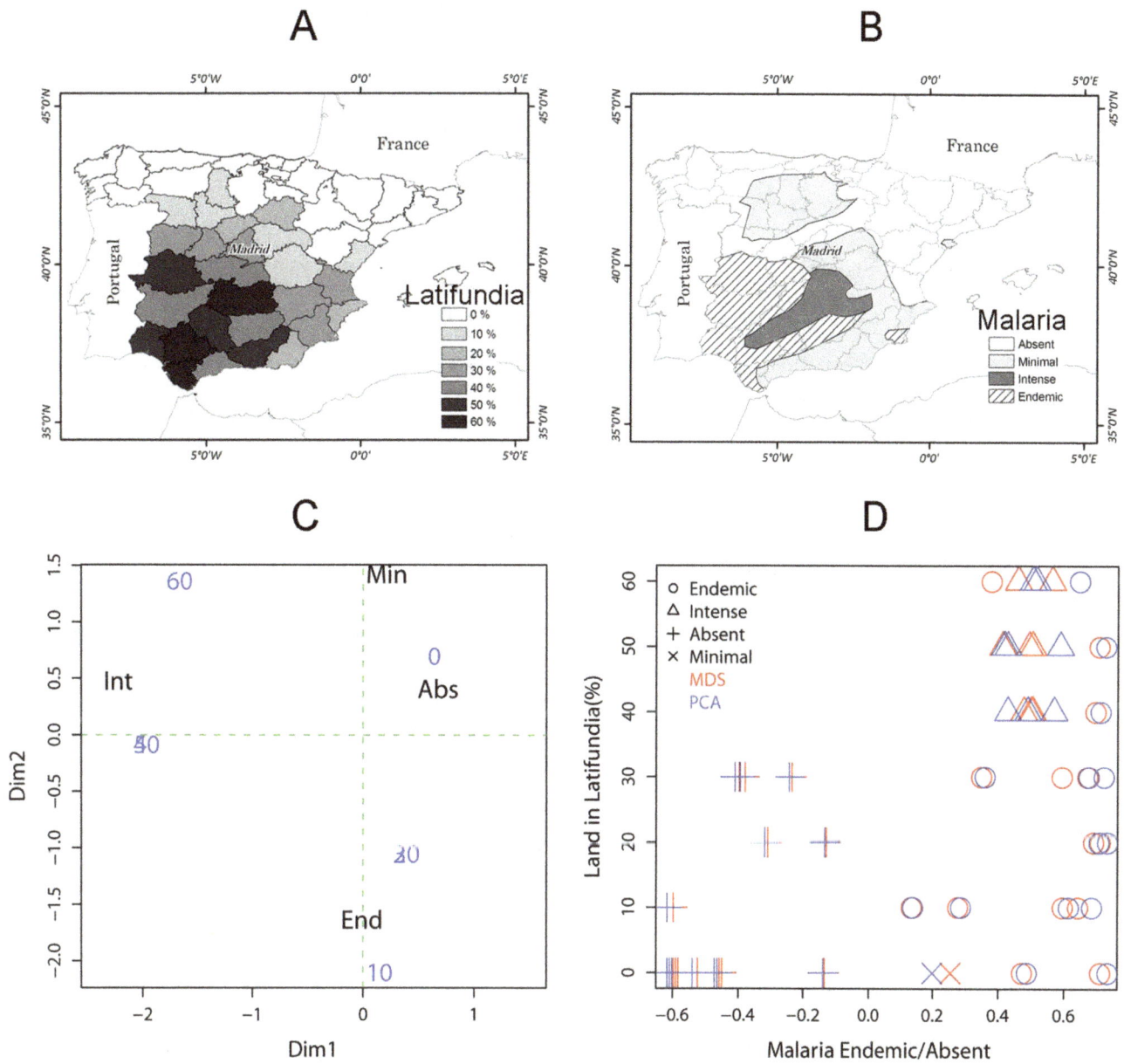

Figure 1. Latifundia, malaria and their association patterns in Spain during the 1930s. (**A**) Percent of land properties that were latifundia, i.e., land properties larger than 25 hectares (**B**) Malaria endemicity (Endemic>Intense>Minimal>Absent) (**C**) Correspondence analysis between malaria endemicity classes (Obtained from a partition around medoids cluster analysis, see Protocol S1 Appendix A for further details) and the percent of latifundia. Malaria categories are (End = endemic, Int = intense, Min = minimal and Abs = absent) (**D**) Percent of Land in Latifundia as function of malaria endemicity indices (the indices were based on the first component of a principal components analysis, PCA or Multidimensional Scaling, MDS, see legend for color and symbol explanation, see also Protocol S1 Appendix B for further details). (A) and (B) are re-drawn from Beauchamp [13].

$$P_i(t) = \frac{a * x_i(t-1) + b * y_i(t-1) + c * z_i(t-1)}{\sum_{j=1}^{n} \left(a * x_j(t-1) + b * y_j(t-1) + c * z_j(t-1) \right)} \quad (4)$$

For the sale pressure individuals in a population of landowners can use several criteria. First, we will consider when the sale pressure is the complement of the purchasing power. Under this scenario how likely a landowner is to sell his land is the complement of his/her purchasing power:

$$V_i(t) = 1 - P_i(t) \quad (5)$$

Under this assumption landowners with a higher purchasing power (P_i) are less likely to sell their land. This assumption is analogous to the one behind the territory size increase in the Colllins-Turchin model of geopolitics [22], where the more powerful (richer) state (landowner) is more likely to conquer (buy) land from the weaker (poorer) state (landowner). Two additional cases that we considered are presented in Protocol S1 Appendix C.

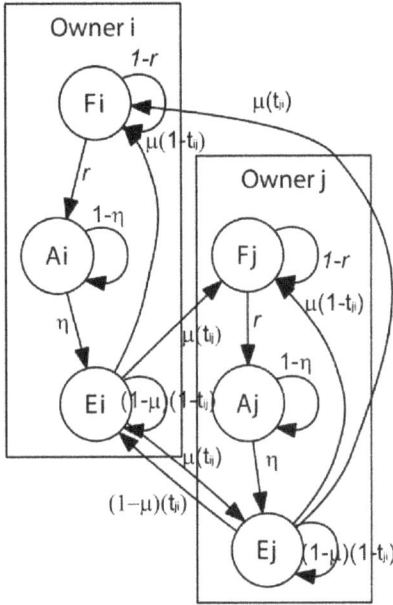

Figure 2. Model description. Graph of land use transition and trade between Owner i and Owner j, for a detailed explanation of the transitions see equations 1 & 2 in the methods section.

Model analysis in the case of two landowners i and j. As a first approach we can study the case of two landowners i and j presented in Figure 2. For trade to occur one landowner needs to sell land, and another needs to buy it. Let's assume, the trade from i to j, t_{ij} is defined by the purchase ability of j, P_j, and the sale pressure of i is the complement of the purchase power, i.e., $(1-P_i)$. Furthermore equation (5) implies that $P_i + P_j = 1$, therefore:

$$t_{ij} = (1 - P_i)^2 \qquad (6)$$

$$t_{ji} = P_i^2 \qquad (7)$$

which can be used to find a steady state solution (see Protocol S1 Appendix D) for the system of equations presented in (2):

$$\begin{bmatrix} \pi_{Fi} \\ \pi_{Ai} \\ \pi_{Ei} \\ \pi_{Fj} \\ \pi_{Aj} \\ \pi_{Ej} \end{bmatrix} = \begin{bmatrix} P_i \mu \eta / (r\eta + \mu(\eta + r)) \\ P_i r \mu / (r\eta + \mu(\eta + r)) \\ P_i r \eta / (r\eta + \mu(\eta + r)) \\ (1 - P_i)\mu\eta / (r\eta + \mu(\eta + r)) \\ (1 - P_i)r\mu / (r\eta + \mu(\eta + r)) \\ (1 - P_i)r\eta / (r\eta + \mu(\eta + r)) \end{bmatrix} \qquad (8)$$

From equation (8) we have that $P_i > 1 - P_i$ implies that land, independent of its use, will spend more time in the hands of landowner i, and if $P_i = 1$ all land belongs to landowner i. Only when $P_i = P_j = 0.5$ a land equity equilibrium can be expected.

Results of equation (8) assume that P_i and P_j are fixed through time. However, P_i and P_j can change through time because the total utility from the land of a given landowner can change through time following equation (4). When P_i follows the dynamics of equation (4), and V_i follows the dynamics of equation (5), we have that when $P_i(t = 0) > P_j(t = 0)$:

$$P_i(t \to \infty) = 1 \qquad (9)$$

meaning that when a landowner begins with an advantage in his/her utilities when compared with the other landowner he/she will become the latifundist as time goes on. The demonstration can be seen in the Protocol S1 Appendix D. This result also holds when V_i follows the assumptions presented in Protocol S1 Appendix C.

Model analysis for more than two owners. In reality, landscapes tend to be divided among more than two landowners. Although $P_i(t = 0) > 1/2$ is a general condition for latifundia formation that extends to cases with more than two landowners (i.e., $\mathbf{n} > 2$) we need to resort to computationally intensive simulations to understand the sensitivity of the dynamics to different system sizes (both increasing the number of landowners, $\mathbf{n}$, and land parcels per landowner, $\mathbf{m}$) and to different transition rates along the different land uses (S, see equation 1), as well as differences in the utilities of the different land types (u, equation 3). Thus, we performed a series of simulations to understand what conditions were the most likely to promote land equity and latifundia. In all simulations we generated the different initial conditions by assuming that for each individual landowner the amount of land in each state came from two uniform distributions whose mean was equal to $1/3$ of the amount of land per capita considered in the system, and the third state was the difference to complete $\mathbf{m}$, the number of land parcels per capita. We then iterated the equations presented in (3) until the model reached a steady state (i.e., the values of x_i, y_i and z_i kept constant through time for the $\mathbf{n}$ landowners) and then we computed whether the system reached an equilibrium of equity or converged into a latifundia (Figure 3). In the simulations land parcels were assumed to be a continuous variable, i.e., parcel fractions could be traded. To ease model implementation at each time step we computed a pool of the land on sale:

$$\bar{L}_j(t) = \sum_j (V_j) z_j(t - 1) \qquad (10)$$

which was then redistributed according to each individual purchasing power (P_i):

$$x_i(t) = (1 - r)x_i(t - 1) + (\mu)\big((1 - V_i(t - 1))z_i(t - 1) + P_i \bar{L}_j(t)\big)$$

$$y_i(t) = (r)x_i(t - 1) + (1 - \eta)y_i(t - 1) \qquad (11)$$

$$z_i(t) = (\eta)y_i(t - 1) + (1 - \mu)\big((1 - V_i(t - 1))z_i(t - 1) + P_i \bar{L}_j(t)\big)$$

To ease the interpretation of the model, we assumed that each simulation iteration corresponds to one year.

Incorporating disease dynamics. To test the hypothesis that disease transmission can promote or enhance latifundia formation [8] we coupled a Susceptible-Infected-Susceptible

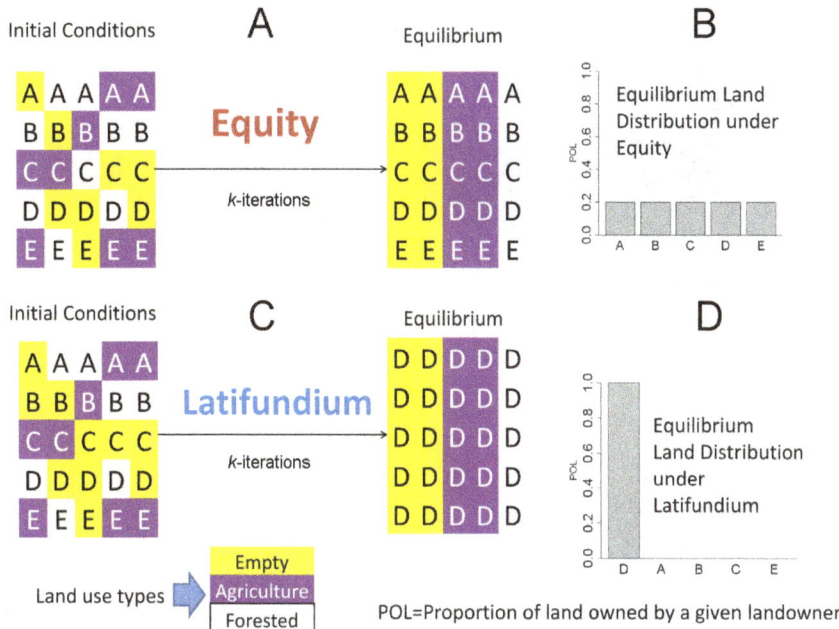

Figure 3. Model simulation scheme. At the beginning of the simulations all landowners have the same amount of land, however the state of the parcels is randomly assigned, after k iterations the model can converge to: (**A**) an equilibrium of equity, where all landowners have the same amount of land which is at equilibrium regarding the land use transitions, which implies (**B**) an uniform distribution in land ownership or (**C**) a Latifudium equilibrium, where all the land (in equilibrium regarding land use transitions) belongs to a single landowner, which implies (**D**) a skewed distribution with one (or a few) landowners accumulating land ownership. In (A) and (C) letters represent different landowners, and colors the land use, see figure legend for further details.

pathogen transmission model [23] to the land trade model. In this model we assumed that all individuals that were infected at a given time step will recover the next year and then we fixed the basic reproductive number (R_0) of the disease. We assumed disease transmission to be frequency dependent so that:

$$R_0 = \frac{\kappa}{n}\lambda \tag{12}$$

where κ was the number of infected individuals the previous year, n the total population size and λ the force of infection. Thus, depending on the variable number of κ we were able to estimate λ as follows:

$$\lambda = \frac{n}{\kappa}R_0 \tag{13}$$

This λ was used to assign an infected status to some of the landowners by the simple rejection simulation method, i.e., whenever a random value from a uniform distribution was more extreme than λ we identified an individual as infected. When an individual was infected the utilities of his land parcels were discounted by a parameter h, which represents the decreased productivity of a sick landowner. h defined in $(0,1)$ and equation (4) was rewritten as follows for an infected individual:

$$P_i(t) = \frac{h*(a*x_i(t-1)+b*y_i(t-1)+c*z_i(t-1))}{\sum_{j=1}^{(n-\kappa)}(a*x_j(t-1)+b*y_j(t-1)+c*z_j(t-1))+h*\sum_{j=1}^{\kappa}(a*x_j(t-1)+b*y_j(t-1)+c*z_j(t-1))} \tag{14}$$

And:

$$P_i(t) = \frac{a*x_i(t-1)+b*y_i(t-1)+c*z_i(t-1)}{\sum_{j=1}^{(n-\kappa)}(a*x_j(t-1)+b*y_j(t-1)+c*z_j(t-1))+h*\sum_{j=1}^{\kappa}(a*x_j(t-1)+b*y_j(t-1)+c*z_j(t-1))} \tag{15}$$

For an individual free of infection. In (14) and (15) κ is the number of infected individuals in the previous time step. In the models dealing with disease we employed equation (5) for the sale pressure.

In the land trade model with discounted utility for disease infected landowners we inquired whether differences in disease susceptibility increased the likelihood of latifundia formation by making some individuals insensitive to infection and comparing the results with simulations were such protection was absent in the population. To further understand the impacts of disease we proceeded with two kind of simulations, in one case we introduced the disease in a population where land equity was an initial condition (a state were no latifundia can be formed) and tested whether disease transmission was able to generate latifundia formation in the system. We also studied the impact of having a given proportion of the population protected from the disease on the dynamics of Latifundia formation. In all simulations we assumed the disease to either be endemically established ($R_0>1$) or epidemic ($R_0\approx1$).

Results

Simulations from our model (Figure 4) showed that a combination of utilities were empty land had the highest utility (denoted by c in equation 3), followed by the utility of agricultural land (b in equation 3) and forested land (a in equation 3), was the best combination of utilities able to produce a null model were

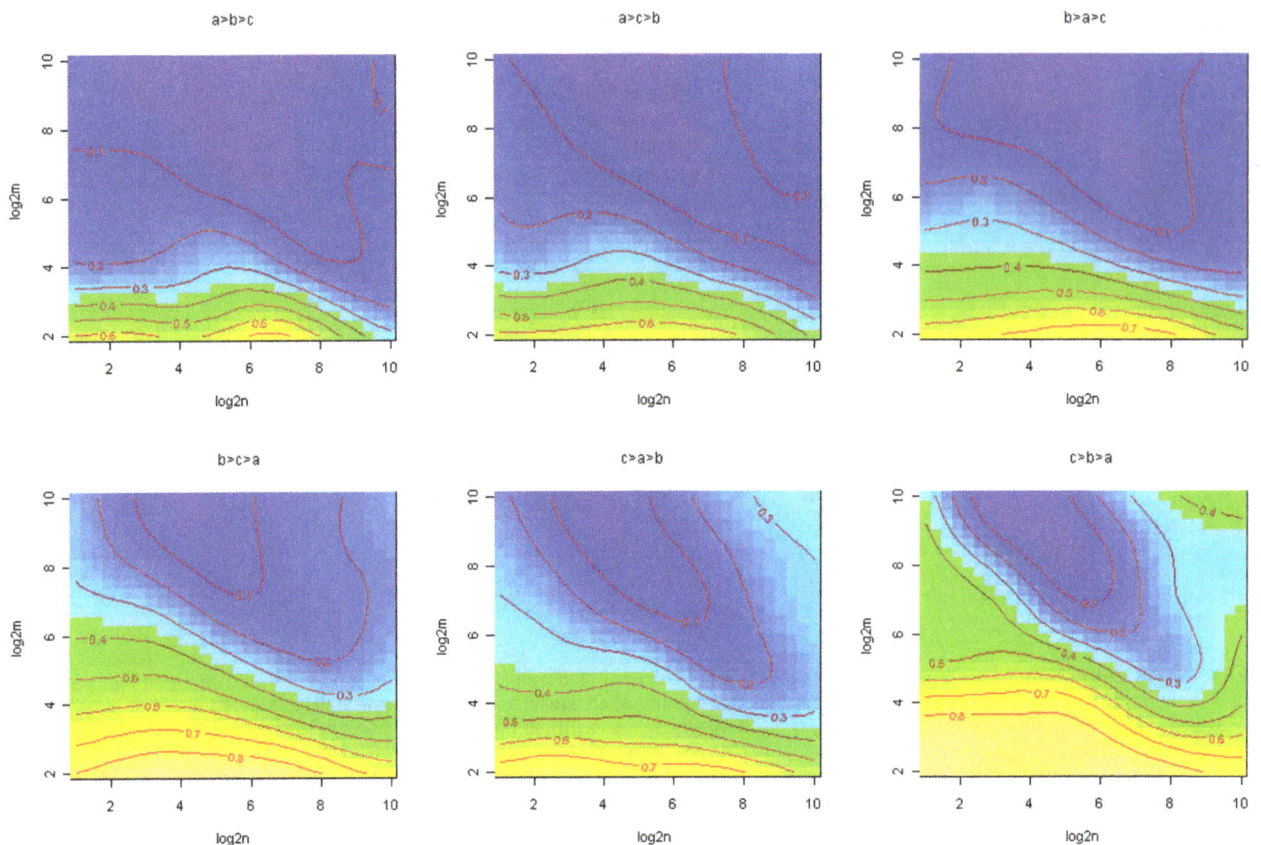

Figure 4. Impact of utility combinations on the likelihood of equity as equilibrium. In top of each panel the relations between the utilities is presented: a = forest, b = agriculture and c = empty (see equation 3 for further details). In each panel, the y axis represents the $\log_2$ of the number of parcels (**m**) per individual and the x axes present the $\log_2$ of the number individuals (**n**). Simulations were run 100 times for each combination of **n** and **m**, where the values of n and m were 2 to the power of the values in the x and y axes respectively. Contour lines give the probability of equity as equilibrium for a given parameter combination. Contour lines were obtained with a generalized additive model where the probability of latifundia formation was a smoothed function of the number of parcels and individuals in the model. In all the simulations run to draw this figure the transition rates across land use types were fixed equal to 0.5, i.e, $\mu = \eta = r = 0.5$. Regarding the utilities of each land type they were always 3, 2 and 1 for any sequence of $u_{k1} > u_{k2} > u_{k3}$, where $u_{k1} \in \{a,b,c\}$; $u_{k2} \in \{\{a,b,c\} - u_{k1}\}$ and $u_{k3} \in \{\{a,b,c\} - \{u_{k1} \cup u_{k2}\}\}$. In all panels blue corresponds to low probability of equity and green to high probability of equity.

latifundia was less likely to occur. As suggested by the data on malaria transmission and land in latifundia from Spain in the 1930s (Figure 1), the model where the utilities are ranked $c = 3 > b = 2 > a = 1$ is a good null model to test the effects of disease transmission on latifundia formation. This ranking of land use utilities can create scenarios where simulations from our model are less likely to generate latifundia.

We found that when sale pressure was defined as the complement of the purchasing power, latifundia were less common in the outputs from our model simulations. Figure S2 and S3 show, respectively, the outcomes for the case when landowners make decision for land sale based on the average or median landowner assets. These last two assumptions were more likely to promote the formation of latifundia under the conditions considered in our simulations.

Another important result shown by Figure 4 is that latifundia were more likely to emerge in a larger population of landowners, i.e., the proportion of land exploited as latifundia increased with the number of landowners used for model simulation. A detailed sensitivity analysis of this model showed that low rates of land use change or very different rates of land use change promoted the emergence of latifundia (Figure 5). By contrast, high and similar rates of land use change were associated with the emergence of

land tenure equity (Figure 5). In general similar results were observed when the number of parcels or landowners were slightly changed with respect to the values set for Figure 5 (see also, Figures S2, S3, S4, S5). Nevertheless, as population size increased there was an increasing trend in the probability of latifundia formation.

Figure 6 shows the impacts of disease transmission on the likelihood of latifundia formation. Figure 6A to 6D show the impacts of epidemic disease transmission, i.e., when the basic reproductive number of the disease, R_0, was close to 1. We observed that extremely high discount rates (very low values for the parameter h of equation 13) were able to promote the formation of latifundia (low ratios of initial to final number of landowners). Nevertheless, as the parameter h increased (or the discount rate decreased) it was observed that increasing the proportion of people unable to acquire the infection could lead to the formation of latifundia. By contrast, the scenario of endemic disease transmission, i.e., when the basic reproductive number of the disease was higher than 1 ($R_0 >> 1$) showed (Figure 6E to 6H) the extreme condition of $h = 0$ to invariably lead to the formation of latifundia, which was alleviated in the case of heterogeneous populations with individuals protected from transmission. Nevertheless, the final proportion of individual landowners was equal to

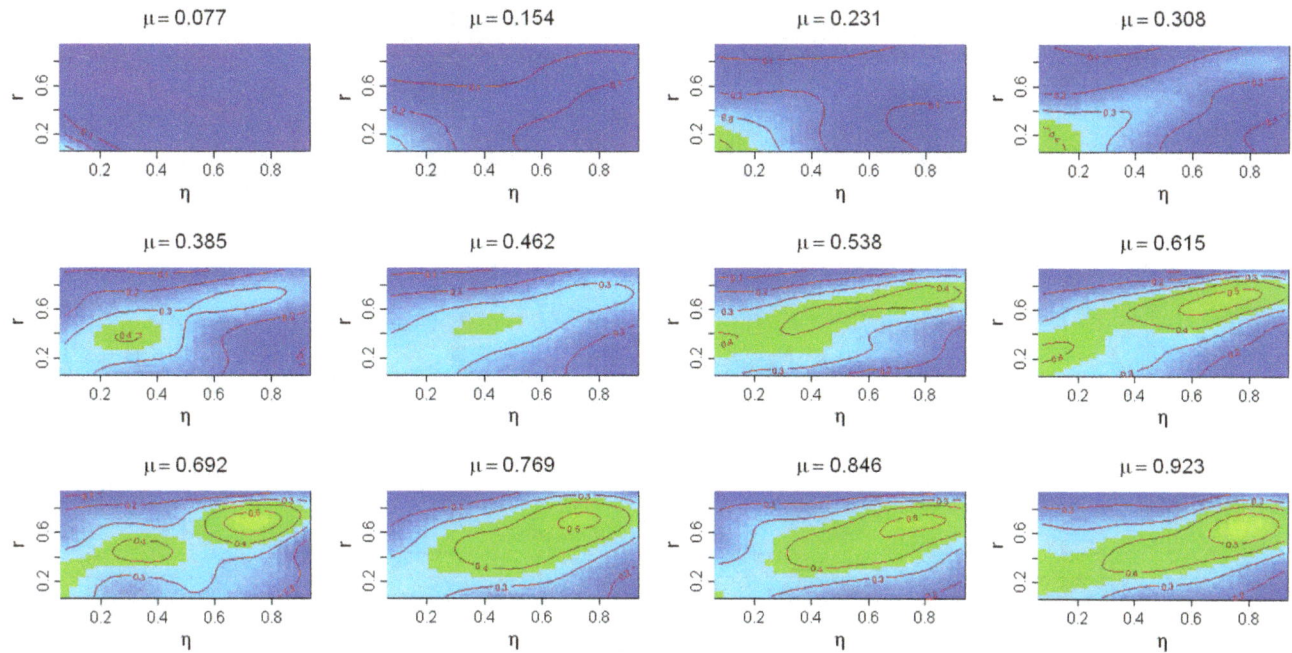

Figure 5. Sensitivity analysis to changes in the transition rates between land use. In the top of each panel the degraded land recovery into forest rate (μ) is presented, the x axis represents the agricultural land degradation rate (η) and y axis deforestation rate into agricultural land (r). Simulations were run 100 times for each combination of μ, η and r. Contour lines indicate the probability of equity for a given parameter combination. Contour lines were obtained with a generalized additive model where the probability of latifundia formation was a smoothed function of the rates considered in the x and y axis. In all the simulations run to draw this figure the number of parcels per individual (m) and the number of individuals in the population were fixed to 64, a quantity that we assume to reflect the structure of rural communities in ancient Rome [19,22] (see Figures S4, S5, S6, S7 for results with other values of m and n). Utilities were as follows: $c = 3$, $b = 2$, $a = 1$. In all panels blue corresponds to low probability of equity and green to high probability of equity.

the proportion of the population protected from disease transmission (Figure 6E to 6H). In both the epidemic and endemic scenarios, a decrease in the utility discount rate for sick landowners resulted in a decreased likelihood of latifundia formation (Figure 6). In synthesis, disease transmission might be a plausible driver for the formation of latifundia, especially when there are high discount rates in land use utility of sick landowners or when there are heterogeneities in susceptibility to infection among the landowners.

Discussion

Latifundia formation has been a cross cultural phenomenon. From records in the western world, dating back the "Historia Naturalis" of Pliny the Elder who foresaw latifundia as the cause of Roman decline [8], the decline of Tang China in the Oriental world [20], revolutionary movements seeking equity in land ownership in Latin America [24], and post-colonial struggles for land access in South Saharan Africa [25] latifundia formation remains an obstacle to social equity, and is linked with major environmental issues, especially deforestation and its associated loss of species diversity [14,15,18]. Therefore, understanding the processes underlying latifundia formation remains fundamental to propose land ownership policies that avoid their generation. Our model of land tenure dynamics successfully reproduced patterns of land ownership observed in Spain in the 1930s [13], where a variability in the degree of latifundia formation can emerge from slightly different initial conditions. The different analyses we performed show that our model is more likely to lead to the formation of latifundia as the number of landowners increases. This pattern is interesting as it could reflect some historical facts.

For example, China during the Tang dynasty and its immediate predecessors implemented the *juntien* land system which guaranteed equity in land tenure, since following the death of a landowner only 20% of his/her land will be inherited by his/her descendants, and the remaining 80% will be redistributed in the population according with the productivity of each individual able to work the land [20]. This system and other policies promoting socio-economic equity within classes have been suggested as pillars for the success of Tang dynasty as one of the most advanced societies ever [26]. Nevertheless, this system, that avoided the formation of latifundia, collapsed briefly after the An Lushan rebellion, which occurred at a time of demographic changes that significantly increased the size of the Chinese population [27]. This population expansion could have acted together with the fragmentation of land resulting from the juntien [20]. Thus, although our model did not explicitly consider population growth, our simulations show a likely outcome: latifundia might be more likely to emerge in larger populations. The second important inference from our model is the importance that land use transition speed could have on the formation of the latifundia. High and similar transition rates between different land uses also prevented the formation of latifundia, which is a pattern that could explain the absence of latifundia establishment after the colonization of new agricultural land. For example, in the North American Midwest farms have never reached the size that would make them latifundia [28] and this may be related to the similarity between the native prairie vegetation and the agricultural exploitation of crops like corn and wheat. At any rate, the transitions between different land uses are likely similar in the American Midwest.

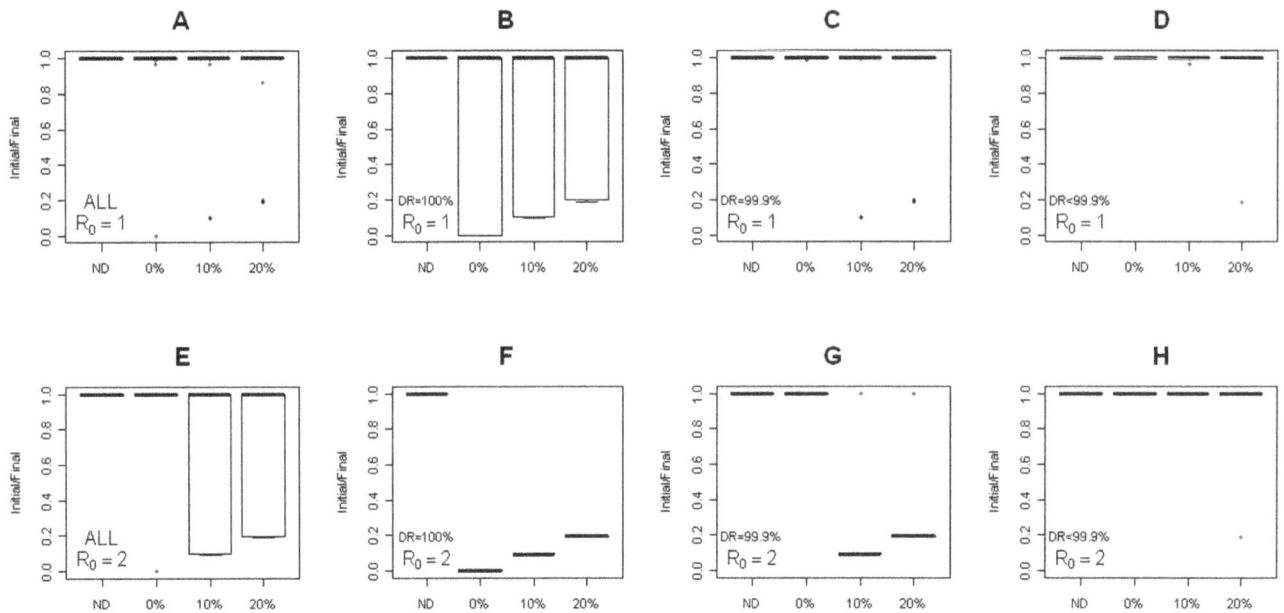

Figure 6. Ratio of initial to final landowners under disease transmission. Panels (A) to (D) show the patterns of latifundia formation (i.e., a low ratio of initial to final landowners) under epidemic conditions ($R_0 = 1$), panels (E) to (H) under endemic conditions ($R_0 = 2$). In Panels (A) and (E) there is no discount rate (DR = 0, h = 1 in equation 13), in panels (B) and (F) there is 100% discount rate (DR = 100, h = 0 in eq. 14), in panels (C) and (G) there a 99.9% DR (h = 0.01 in eq. 14) and panels (D) and (H) show the results for DR<99.9. In all panels the x represents the proportion of landowners that were protected from disease transmission (0, 10, 20) and the control simulation ran under conditions that do not lead to latifundia formation (i.e., perfect equity on land ownership and use, ND in the x axis). In all simulations, parameters for land use change were set up equal to 0.5, number of landowners (n) was 64 and parcels for landowner (m) was also 64. In all the simulations run to draw this figure the transition rates across land use types were fixed equal to 0.5, i.e, $\mu = \eta = r = 0.5$. Utilities were as follows: c = 3, b = 2, a = 1.

Our model does not account for other kind of historical phenomena which have created latifundia, for example, the unequal distribution of land following colonialism [24,25]. Nevertheless, our model can explain the maintenance of these unequal systems of land tenure. In fact, our model shows that small differences in land utilities can lead to the formation of latifundia even if a set of landowners begin with similar amount of land, and as we showed mathematically for the case of two landowners, when a landowner has an initial advantage, it is expected that he/she will eventually own all the land in the system. Therefore, the practical application of our model results is that land redistribution reforms need not only to ensure equity on the quantity of land a landless farmer receives, but also on the utility associated with the potential land uses.

Our model was also able to show the plausibility of disease as driver of latifundia formation, a hypothesis originally suggested by Angelo Celli in the 1930s [8]. Our simulations showed that two key elements of Celli's original hypothesis are, indeed, fundamental to the formation of latifundia: (i) the decreased utility of land exploitation by the reduced labor ability of sick landowners (which we modeled with the parameter h) and (ii) differences in the risk to acquire infections. In that sense our model can mechanistically confirm one of the main observations made by Celli [8], i.e., that people that were protected from disease transmission were more likely to either conserve and/or purchase land from people that were susceptible to disease transmission. Our model specifically showed that a high discount rate in the presence of disease can lead to formation of latifundia independently of the endemic or epidemic status of a disease, and that heterogeneities in disease susceptibility can further increase the likelihood of latifundia formation even if the discount rates in land utility are not 100% for sick landowners. From an applied perspective these results suggest

that inequities in the protection against a disease, for example the use and access to disease prevention devices, can promote further socio-economic inequities in societies where a disease is endemically persistent or has frequent epidemics. This phenomenon has been observed in malaria, where inequities in access to insecticides treated nets can feed positive feedback loops further increasing socio-economic differences within a host population [29,30].

Finally, in summary, our model was able to show: (i) that latifundia can emerge when there are initial patterns of inequity in land ownership, with landowners with the most resources likely acquiring land from landowners with less resources, (ii) that high transition rates in land use can hamper the formation of latifundia while slow transitions can enhance its formation, (iii) having the highest utility in traded land can also regulate the formation of latifundia and (iv) that both heterogeneities in disease susceptibility and high discount rates on land utility by sick landowners can also promote and enhance the formation of latifundia.

Supporting Information

Figure S1   Boxplot for malaria endemicity (index based on the 1st PC of land under different malaria transmission endemicity levels) as function of endemicity categories.

Figure S2   Exploring Alternative Rules for the Sale Pressure: Sale pressure based on the average of landowners assets.

Figure S3 Exploring Alternative Rules for the Sale Pressure: Sale pressure based on the median of landowners assets.

Figure S4 Sensitivity analysis to changes in the transition rates between land use and different _m_ (number of landparcels) and _n_ (number of landowners) _n_ = 64, _m_ = 32. For interpretation and other parameter values see legend of figure 5 in the main text.

Figure S5 Sensitivity analysis to changes in the transition rates between land use and different _m_ (number of landparcels) and _n_ (number of landowners) _n_ = 64, _m_ = 128. For interpretation and other parameter values see legend of figure 5 in the main text.

Figure S6 Sensitivity analysis to changes in the transition rates between land use and different _m_ (number of landparcels) and _n_ (number of landowners) _n_ = 32, _m_ = 32. For interpretation and other parameter values see legend of figure 5 in the main text.

Figure S7 Sensitivity analysis to changes in the transition rates between land use and different _m_ (number of

landparcels) and _n_ (number of landowners) _n_ = 128, _m_ = 128. For interpretation and other parameter values see legend of figure 5 in the main text.

Acknowledgments

Prof. Chuang Ting-Wu, Dr Gao Yan and Dr. Rebecca Levine offered valuable help to redraw the maps from Beauchamp and to extract quantitative information from the maps using ARCGIS®. Prof. Chuang Ting-Wu and Prof. Sun Yongkoon suggested valuable references about Land Ownership patterns in China during the Tang dynasty. Dr. François Feugier made valuable suggestions to develop the models presented here. Ass. Pro. A. Satake made comments on the ms. An anonymous reviewer is also thanked for his/her detailed comments to improve model presentation. LFC was a Japan Society for the Promotion of Science Postdoctoral Fellow during the realization of this study.

Author Contributions

Conceived and designed the experiments: LFC. Performed the experiments: LFC. Analyzed the data: LFC. Contributed reagents/materials/analysis tools: LFC. Wrote the paper: LFC.

References

1. Levins R (1995) Toward an integrated epidemiology. Trends Ecol Evol 10: 304.
2. Levins R, Lopez C (1999) Toward an ecosocial view of health. Int J Health Serv 29: 261–293.
3. Scheffer M (2009) Critical transitions in Nature and Society. Princeton: Princeton University Press.
4. Levins R, Awerbuch T, Brinkmann U, Eckardt I, Epstein P, et al. (1994) The emergence of new diseases. American Scientist 82: 52–60.
5. Satake A, Iwasa Y (2006) Coupled ecological and social dynamics in a forested landscape: the deviation of individual decisions from the social optimum. Ecological Research 21: 370–379.
6. Satake A, Rudel TK (2007) Modeling the forest transition: Forest scarcity and ecosystem service hypotheses. Ecological Applications 17: 2024–2036.
7. Rodrigues A, Koeppl H, Ohtsuki H, Satake A (2009) A game theoretical model of deforestation in human-environment relationships. Journal of Theoretical Biology 258: 127–134.
8. Celli A (1933) The history of malaria in the roman campagna from ancient times. London: John Bale, Sons & Danielsson, Ltd.
9. Sutter PS (2007) Nature's Agents or Agents of Empire? Entomological Workers and Environmental Change during the Construction of the Panama Canal. Isis 98: 724–754.
10. Chaves LF, Cohen JM, Pascual M, Wilson ML (2008) Social Exclusion Modifies Climate and Deforestation Impacts on a Vector-Borne Disease. PLoS Neglected Tropical Diseases 2: e176.
11. Saldaña A, Chaves LF, Rigg CA, Wald C, Smucker JE, et al. (2013) Clinical Cutaneous Leishmaniasis Rates Are Associated with Household _Lutzomyia gomezi_, _Lu. panamensis_, and _Lu. trapidoi_ Abundance in Trinidad de Las Minas, Western Panama. The American Journal of Tropical Medicine and Hygiene 88: 572–574.
12. Yasuoka J, Levins R (2007) Impact of deforestation and agricultural development on anopheline ecology and malaria epidemiology. American Journal of Tropical Medicine and Hygiene 76: 450–460.
13. Beauchamp C (1988) Fièvres d'hier, Paludisme d'aujourd'hui. Vie et mort d'une maladie. Annales ESC 43: 248–275.
14. Fearnside PM (1993) Deforestation in brazilian Amazonia - The effect of population and land-tenure. Ambio 22: 537–545.
15. Perfecto I, Vandermeer J (2008) Biodiversity conservation in tropical agroecosystems - A new conservation paradigm. Annals of the New York Academy of Sciences. Oxford: Blackwell Publishing. 173–200.
16. Rosero-Bixby L, Palloni A (1998) Population and deforestation in Costa Rica. Population and Environment 20: 149–185.
17. Rosero-Bixby L, Maldonado-Ulloa T, Bonilla-Carrion R (2002) Forests and population on the Osa Peninsula, Costa Rica. Revista De Biologia Tropical 50: 585–598.
18. Bonilla-Carrion R, Rosero-Bixby L (2004) Presión demográfica sobre los bosques y áreas protegidas, Costa Rica 2000. In: Rosero-Bixby L, editor. Costa Rica a la luz del censo del 2000. San José: Centro Centroamericano de Poblacion. 575–594.
19. Turchin P, Nefedov SA (2009) Secular Cycles. Princeton: Princeton University Press.
20. Xiong VC (1999) The land-tenure system of Tang China: a study of the equal-field system and the turfan documents. T'oung Pao 85: 328–390.
21. Marx K (1949) El Capital (Critica de la Economia Politica). Mexico: Fondo de Cultura Economica. 849 p.
22. Turchin P (2003) Hystorical Dynamics: why States Rise and Fall. PrincetonUSA: Princeton University Press. 245 p.
23. Mangel M (2006) The theoretical biologist's toolbox: quantitative methods for Ecology and Evolutionary Biology. Cambridge: Cambridge University Press. 375 p.
24. Wright AL, Wolford W (2003) To Inherit the Earth: The Landless Movement and the Struggle for a New Brazil. OaklandCA: Food First Books. 368 p.
25. Phombeah G (2005) Kenya: overthrowing colonial landlords. BBC Focus on Africa 16: 30–33.
26. Adshead SAM (2004) T'ang China: The Rise of the East in World History. Melbourne: Palgrave Macmillan. 256 p.
27. Pulleyblank EG (1955) The Background of the Rebellion of An Lu-Shan. London: Oxford University Press.
28. Lewontin RC, Levins R (2007) Biology Under the Influence: Dialectical Essays on Ecology, agriculture, and health. New York: Monthly Review Press.
29. Mathanga DP, Campbell CH, Taylor TE, Barlow R, Wilson ML (2006) Socially marketed insecticide-treated nets effectively reduce Plasmodium infection and anaemia among children in urban Malawi. Tropical Medicine & International Health 11: 1367–1374.
30. Honjo K, Chaves LF, Satake A, Kaneko A, Minakawa N (2013) When they don't bite, we smell money: understanding malaria bednet misuse. Parasitology 140: 580–586.

4

Atlantic Bluefin Tuna: A Novel Multistock Spatial Model for Assessing Population Biomass

Nathan G. Taylor[1]*, **Murdoch K. McAllister**[1], **Gareth L. Lawson**[2], **Tom Carruthers**[1], **Barbara A. Block**[3]

1 Fisheries Center, University of British Columbia, Vancouver, British Columbia, Canada, **2** Department of Biology, Woods Hole Oceanographic Institution, Woods Hole, Massachusetts, United States of America, **3** Hopkins Marine Station, Stanford University, Pacific Grove, California, United States of America

Abstract

Atlantic bluefin tuna (*Thunnus thynnus*) is considered to be overfished, but the status of its populations has been debated, partly because of uncertainties regarding the effects of mixing on fishing grounds. A better understanding of spatial structure and mixing may help fisheries managers to successfully rebuild populations to sustainable levels while maximizing catches. We formulate a new seasonally and spatially explicit fisheries model that is fitted to conventional and electronic tag data, historic catch-at-age reconstructions, and otolith microchemistry stock-composition data to improve the capacity to assess past, current, and future population sizes of Atlantic bluefin tuna. We apply the model to estimate spatial and temporal mixing of the eastern (Mediterranean) and western (Gulf of Mexico) populations, and to reconstruct abundances from 1950 to 2008. We show that western and eastern populations have been reduced to 17% and 33%, respectively, of 1950 spawning stock biomass levels. Overfishing to below the biomass that produces maximum sustainable yield occurred in the 1960s and the late 1990s for western and eastern populations, respectively. The model predicts that mixing depends on season, ontogeny, and location, and is highest in the western Atlantic. Assuming that future catches are zero, western and eastern populations are predicted to recover to levels at maximum sustainable yield by 2025 and 2015, respectively. However, the western population will not recover with catches of 1750 and 12,900 tonnes (the "rebuilding quotas") in the western and eastern Atlantic, respectively, with or without closures in the Gulf of Mexico. If future catches are double the rebuilding quotas, then rebuilding of both populations will be compromised. If fishing were to continue in the eastern Atlantic at the unregulated levels of 2007, both stocks would continue to decline. Since populations mix on North Atlantic foraging grounds, successful rebuilding policies will benefit from trans-Atlantic cooperation.

Editor: Stuart A. Sandin, University of California San Diego, United States of America

Funding: This work was supported by grants from the TAG A Giant Foundation, the Monterey Bay Aquarium Foundation, the Lenfest Ocean Program, Washington, DC, USA, the Canadian Fisheries and Oceans International Governance Strategies Fund and the National Oceanic and Atmospheric Administration (NOAA) of the United States. The funders had no role in study design, data collection and analysis, decision to publish, or preparation of the manuscript.

Competing Interests: The authors have declared that no competing interests exist.

* E-mail: nathan.taylor@dfo-mpo.gc.ca

Introduction

Atlantic bluefin tuna *(Thunnus thynnus)* is a large, endothermic, and highly migratory member of the tuna family, Scombridae. They can reach a mass of 650 kg and live to be over 32 years old [1]. Historically, its range has encompassed much of the North Atlantic, from the waters off Norway and the Faroe Islands to the South Atlantic and the west coast of Africa. Atlantic bluefin occurrences have been reported from Mauritania [2] and off South Africa [3]. In the western Atlantic Ocean, the species' historic range extended from Canada to Brazil, including the Gulf of Mexico and Caribbean Sea. In the twentieth century, the population appears to have disappeared from the southern part of its range and the North Sea [4].

Recent studies have shown that spatial population structure and movements are more complicated than previously thought. Conventional [5] and electronic tagging [6,7,8] studies, as well as genetic [9], organochlorine tracer [10], and otolith microchemistry [11] studies, indicate that three or more populations of Atlantic bluefin tuna exist [12]. Genetic studies indicate that at least two populations spawn in the Mediterranean Sea in summer months [12]. In the Gulf of Mexico, a smaller population spawns

in the spring months (April–June). Histological sampling of fisheries catches indicates half of the fish spawned in the Gulf of Mexico are sexually mature at age 12 [13]. This has been corroborated by electronic tagging data for which the mean age of individuals returning to the Gulf to spawn was 11.8 years [7]. In comparison, the age at maturity in the eastern Mediterranean population is considered to be 4 years [14]. However, fish tagged in the western Atlantic that return to breed in the western Mediterranean spawning areas enter at Gibraltar on average at ages 7–9 [7]. Site-directed fidelity has been observed [7,15] and is hypothesized to maintain genetic structuring [12].

Atlantic bluefin tuna populations are managed by the International Commission for the Conservation of Atlantic Tunas (ICCAT) as western and eastern populations, or stocks, separated by the 45° meridian. Both populations are considered to be overfished [13,15,16], and rebuilding policies in the western Atlantic do not appear to have been successful to date. Bycatch of bluefin tuna in areas closed to directed bluefin fishing, such as the Gulf of Mexico, remains problematic [17]. Illegal and underreported catches, due in part to widespread tuna ranching, have been a severe problem in the Mediterranean Sea, and scientists have had to adjust reported catches using Japanese import records for assessments [13]. For example, in 2006, the reported eastern Atlantic and Mediterranean catches were 31 kilotonnes (kt), but import records suggested that as much as 54 kt were caught [13].

The determination of whether a stock is overexploited requires the prediction of historical spawning stock biomass (SSB), which has proven difficult to determine for Atlantic bluefin tuna. A central reason is that the multiple Atlantic stocks are mixed on fishing grounds and demonstrate stock-specific movements. In much of the western Atlantic Ocean, biological markers [9–11] show that eastern and western Atlantic bluefin stocks co-occur. Tagging studies indicate that large-scale migrations of 7400 km or more routinely take place across the ICCAT stock boundary and between the western and eastern Atlantic and the Mediterranean Sea by bluefin tuna of all ages [6,7,18]. In addition, recent results from tagging, genetics, and microchemistry markers demonstrate stock-specific seasonal and/or ontogenetic movements [6,7,18,19]. Ontogeny and population origin influence which areas a bluefin tuna utilizes in the North Atlantic, so that in any given area, age or season, there can be different proportions of each population. Mixing of populations compromises the accuracy of the single-stock models that are currently used to determine SSB declines because some catches have been attributed to the incorrect stock of origin.

Because ICCAT does not routinely consider population mixing in assessments, it may not effectively understand or control fishing mortality on individual Atlantic bluefin tuna populations. Even though a mixed-stock assessment model exists [20,21], current ICCAT bluefin tuna assessments primarily use single-stock virtual population analysis (VPA) [22]. The single-stock VPA assumes that all bluefin tuna catches west of the 45° meridian are from the western spawning population, and that all fish to the east of this longitude are from the eastern population. Failure to accurately account for seasonal movements and ontogenetic distinctions, as well as western and eastern stock mixing, can therefore compromise the reliability of current and future population size estimates. In turn, projections of the effects of various policy actions are likely to be unreliable.

In this paper, we provide a new seasonally and spatially explicit fisheries model that incorporates population mixing in an effort to improve our capacity to assess Atlantic bluefin tuna population sizes. This is a multi-stock age structured tag integrated assessment model that we refer to as MAST. This population dynamics model runs on quarterly intervals, incorporates catch data from 1950 to 2008, and is fitted to (1) age-composition landings data from 1960 to 2008; (2) 29 stock-trend time-series derived from commercial and research catch-per-unit-effort (CPUE) series [13]; (3) ICCAT conventional ("spaghetti") tagging data [5]; (4) archival and pop-up satellite archival tag data [7]; and (5) published otolith microchemistry data [11]. The model assumes time-invariant gear selectivity and the reporting rates for conventional tags documented by Kurota et al. [23]. Using this model, we applied Bayesian integration using Markov Chain Monte Carlo Simulation (MCMC) to account for uncertainties in model parameters and predictions of spawning stock biomass and fishing mortality rates from 1950 to the present.

We evaluated the rebuilding efficacy of management scenarios that capture a range of alternatives previously considered for Atlantic bluefin tuna management. We examined five cases: near-complete fisheries closures that could have occurred under a Convention for the International Trade in Endangered Species (CITES) listing [15]; 2010 ICCAT quotas, with and without a Gulf of Mexico spawning area closure with catch redistributed; a scenario that assumed that actual eastern catches were double the 2010 ICCAT quotas; and, finally, a scenario of very high eastern

Atlantic and Mediterranean catch levels that occurred from the late 1990s to 2007.

Methods

Modeling Approach

The Multistock Age-Structured Tag-integrated assessment model (MAST) is a mixed-stock, seasonal, and spatially explicit statistical catch-at-age model that can be fitted to relative abundance indices, age proportions, and otolith microchemistry, as well as to conventional and electronic mark-recapture data. The model was written and fitted to data using the software AD Model Builder, which is freely available from www.admb-project.org (ADMB Project 2009). The model and statistical fitting procedure are described in detail in the online Text S1. We characterized parameter uncertainty using Markov Chain Monte Carlo simulation.

The MAST model consists of four major components:

1. Initialization of the model based on steady-state conditions (unfished numbers and biomass) given the model's parameters;

2. Updating the state variables (numbers and biomass at age in each area);

3. Relating the state variables to observations on relative abundance, age-composition information, and mark-recapture observations; and

4. Evaluating the probability of model parameters given the data.

We provide a description of each model component in the main text, and refer readers to the Model Description in the online Text S1 for further detail.

We defined five geographic areas for quantifying movement dynamics: the Gulf of Mexico, which we assume is the western-stock spawning area; the Gulf of St. Lawrence, which we assume contains primarily western-stock fish [11]; the western and eastern Atlantic Ocean, which we assume to be mixed-stock areas; and the Mediterranean Sea, which we assume is the eastern-stock spawning area. We used these areas because they are either mixed-stock areas in the eastern and western Atlantic Ocean basins that have historical importance at ICCAT, or because they appear to be nearly exclusively western-stock (the Gulf of Mexico and Gulf of St. Lawrence) or eastern-stock (Mediterranean Sea) areas. Figure S1 shows the model areas and the electronic tag geoposition data. By including the Gulf of St. Lawrence as a distinct area, additional tagging data and an additional CPUE abundance index can be used in modeling the population dynamics of the western stock [11]. MAST models western-stock fish as moving between the Gulf of Mexico, the Gulf of St. Lawrence, and the western and eastern Atlantic (corresponding to area indices 1–4). MAST models eastern-stock fish as moving between the Mediterranean Sea and the eastern and western Atlantic (area indices 3–5).

We parameterized movement matrices using gravity models for the model-fitting base-case. The probability of fish moving from one area to another is defined in terms of a movement matrix μ. Each movement matrix consists of rows representing area of origin and columns representing the destination area. Each row element of μ therefore represents the probability of fish moving from area j (rows) to area j' (columns); each row represents a probability vector v, where $v = (v_1, v_2, ..., v_n)$ and $\Sigma v = 1$. Here we estimated a single propensity of fish to stay in a given area—that is, "gravity" (the diagonal elements of μ)—which is assumed to capture the attractiveness of that area relative to the areas associated with the off-diagonal elements. These latter elements are given as (1-

$\mu_{ji})/(n-1)$, where n is the number of stock areas for stock i. An alternative to the gravity model is the bulk-transfer model, in which the full matrix of movement probabilities is estimated. Some biological detail is lost in using this gravity parameterization; the main advantage is that it substantially reduces the number of estimated parameters compared with the bulk-transfer case. We discuss the bulk-transfer parameterization and the sensitivity of the model in the online Text S1.

For Atlantic bluefin tuna, there is strong seasonal and ontogenetic dependence of movement rates, where fish of different ages use different habitats during the year for foraging or spawning [6,7,18]. To account for these phenomena, we modeled quarterly time-steps and two age-groups: 0–7 and 8+ We assumed that movement transitions to spawning areas (the Gulf of Mexico for the western stock, and the Mediterranean Sea for the eastern stock) during the spawning quarter were given by the maturity-at-age schedule.

The MAST model uses the management-oriented approach [28], meaning that the model is initialized using maximum sustainable yield *(MSY)* and the fishing rate that produces maximum sustainable yield *(F_{msy})*. Under this formulation, *MSY* and F_{msy} are the leading estimated parameters. Then, using estimated gear selectivity as well as input growth [1], mortality, and maturity parameters, we derived the recruitment compensation parameter κ [29], initial numbers, and initial biomass B_0. Maturity-at-age schedules were based on [30] for the western stock and [31] for the eastern stock. The model was parameterized with initial numbers-at-age in the spawning area and then run for 25 years to allow the model to equilibrate between areas (see Tables S1–S7).

Initial numbers-at-age for each stock (see Tables S1–S7) were updated in each time-step (i.e., the next quarter) according to natural and fishing mortality, as well as migration parameters. Age-zero recruits were predicted by a Beverton-Holt stock recruit function (Eq. 25, Table S5). Details of the state dynamics are described in the online Text S1.

Parameter Estimation

We estimated the parameters that define the model by fitting predicted observations (component 3) to observed data. The modeling procedure starts with initial parameter values (see Tables S1–S5) that define the state variables, then proceeds through the state dynamics (Tables S4 and S5), where state variables such as numbers-at-age for each stock are updated at each time-step. Ultimately, the model calculates the statistical objective function value, which represents the probability of the model given the data (Table S6). Parameters are estimated using a conventional nonlinear optimization procedure. AD Model Builder was used to implement the model.

For all data types except electronic tag data, we used conventional statistical likelihoods. We fitted relative abundance indices using Walters and Ludwig's [32] formulation (see Tables S5 and S6). We fitted otolith microchemistry using binomial likelihoods, and conventional mark-recapture data using negative binomial likelihoods (Tables S4 and S5). If the stock-of-origin for a given cohort was known through a cohort's area of visitation or marking, then that cohort's survival and movement dynamics were modeled according to the movement probability matrix for that stock. However, for 85% of cohorts (Table S10), stock of origin was unknown. In these cases, the likelihood was computed twice; that is, using movement probability matrices from western and eastern stocks (Eq. 1.38, Table S5), with likelihood weights given by the ratio of vulnerable numbers of stock i to total vulnerable numbers in that area at that time.

For electronic tag data, we used discrete, state-space likelihoods. We modeled the state of tags through discrete states at each model time-step [33] (see Eqs. 1 and 2 in Table S2). We assumed that the tag was attached to a live fish in area j; captured on a fishing vessel; attached to a fish that died of natural causes; or shed from the fish (Table S8). For electronic tags, equations describing state transitions are listed in Table S8, and parameters for electronic tag observation probabilities $p(y_t|s_t)$ are given in Table S9. When modeling tag tracks, capture probabilities represent the probability of obtaining a geoposition for a particular tag type. In the case of pop-up satellite archival tags, there are complete tag tracks (i.e., spatial positions at each quarter), so these observation probabilities are 1. For archival tags, however, not all tag tracks are complete; there can be missing geolocations at times between the last geolocation recorded by the tag and the location given by the vessel position at time of recapture. In these cases, we estimated a single observation probability parameter for archival tags (Table S9) that represents the proportion of time between the release of the tag and its recovery in which it was possible to determine the geoposition of the tag.

Data

We used catch data from the 2010 ICCAT CATDIS database (www.iccat.int). CATDIS is the official database that contains catches in 5×5 degree grid squares, by quarter and gear. We separated catches into four gear categories: longline, purse seine, bait boat, and other. For each catch record, an area was assigned according to Fig. S1. The input data for MAST consisted of total catches by fleet, area, and quarter from 1950 to 2008 (Fig. S2). To account for large catch underreporting from 1998 to 2007 in the Mediterranean Sea, we inflated catches reported in the Mediterranean. We used the same procedure as RUN 14 of the 2008 ICCAT stock assessment, where total eastern catches were assumed to be 50,000 metric tonnes (mt) from 1998 to 2006 and 60,000 mt in 2007 [13]. At the time of writing, catch data for 2008 and 2009 were not yet available, so we assumed that the total eastern and western catches in these years were the recommended quotas. This may be a reasonable assumption, since there is evidence that compliance has improved considerably with ICCAT's introduction of a vessel monitoring system in 2008 [13].

We aggregated conventional tag data into cohorts *(h)* for fish that had the same assigned age and were captured in the same quarter; these data are available in ICCAT's conventional tag database (www.iccat.int) and are summarized in [5]. (We used the version of the database updated in September 2009.) The data were filtered to remove incomplete records that were missing size or location data at either release or recapture. The filtered conventional tag data set consisted of 47,439 releases that were distilled into 1732 cohorts. Of these, 125 tag cohorts were assigned to the western stock and 142 to the eastern stock, and 1465 were unknown (Table S10). Details of how tagged fish (both electronic and conventional) were assigned to stocks and age-groups can be found in the online Text S1.

Between 1996 and 2008, a total of 968 bluefin tuna were electronically tagged with internally implanted archival tags and/ or externally attached pop-up satellite archival tags at tagging locations along the U.S. East Coast, in the Gulf of Mexico, in the Gulf of St. Lawrence, off Ireland, and in the Mediterranean Sea [6,7,18]. The daily geopositions of electronic tags were aggregated to quarterly area assignments. If a tag was reported being in more than one discrete stock area, it was assigned to the area where it spent the greatest proportion of time. Additional details of how satellite geopositions were determined are given in the online Text S1.

We used the commercial CPUE time-series and catch-at-age proportions from the ICCAT assessment document [13]. We used the catch-at-age data from the assessment to define catch-at-age proportions [13] from age 1 to 10+in the western Atlantic (areas 1–3) and eastern Atlantic (areas 4–5), which were based on western and eastern catch-at-age data from 1960–2007 and 1970–2007, respectively. Table S11 is a summary of which CPUE series we used, as well as the corresponding quarters and area for each.

We extracted otolith microchemistry stock-composition data from Rooker *et al.* [11], who divided their data into three age-groups: giant (age 10+), medium (age 5–9), and school (age 4 or younger). They had stock-composition samples for the Mediterranean, Gulf of Mexico, Gulf of St. Lawrence, Gulf of Maine, and the Mid-Atlantic Bight. We fitted the model (see below) to stock-composition ratios from the Gulf of Maine and the Mid-Atlantic Bight only because it was assumed that the Gulf of Mexico and Gulf of St. Lawrence areas were 100% western stock and the Mediterranean was 100% eastern stock. The stock-composition data used in the model are summarized in Table S12.

Uncertainty and Projections

We computed marginal posterior probability distributions for all estimated parameters using MCMC simulation with six chains. One value was sampled for every ten iterations, and we ran the MCMC until the multivariate posterior scale reduction factor [34] was below 1.05. We present fishing mortality rate reconstructions by area and gear type at the posterior mode; posterior samples of western and eastern bluefin tuna spawning stock biomasses; stock status relative to maximum sustainable yield; and stock composition in mixed-stock areas.

We ran the base-case model with a series of management options, including complete fisheries closures, spatial closures, and other quota options. We chose scenarios to reflect a broad range of possibilities in Atlantic bluefin tuna management. The first scenario (i) represents total closures, which might have occurred with listing under CITES. For the quota scenarios, we assumed future bluefin tuna catches west (W) and east (E) of the 45° meridian from 2010 to 2025 to be: (ii) 1750 mt W/12,900 mt E, with no Gulf of Mexico closure; (iii) 1750 mt W/12,900 mt E, assuming a Gulf of Mexico closure with catches redistributed to the western Atlantic; (iv) if eastern catches continued to be double the current quotas, that is, 1750 mt W/25,800 mt E; and (v) an eastern overfishing case of 1750 mt W/60,000 mt E. This final scenario was intended to capture what might have occurred if Atlantic bluefin tuna catches continued at 2007 levels.

In addition to parameter uncertainty, we examined the sensitivity of the base-case results to a suite of alternative model parameterizations and reporting-rate-prior distributions. The details of each sensitivity case and the corresponding effect of each to key stock status metrics are listed in Table S13, and the effect on conventional tag reporting rates is given in Table S14.

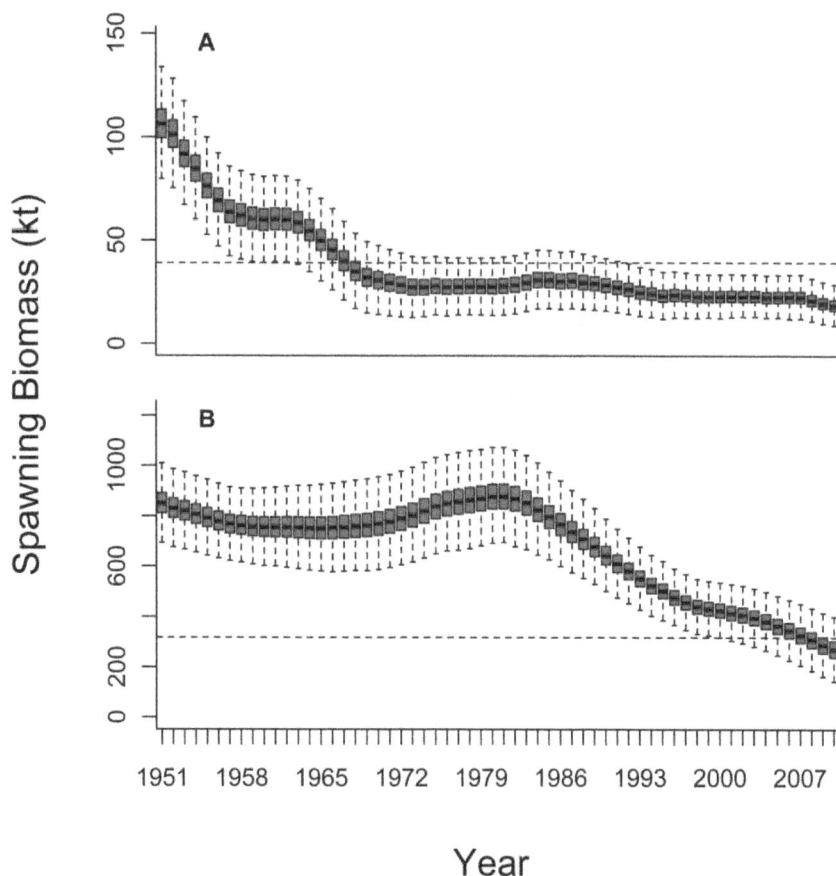

Figure 1. Box plots of posterior samples of the spawning stock biomass (kt) of (A) western and (B) eastern Atlantic bluefin tuna. The horizontal lines within the blue bars represent the posterior median values, the blue bars represent the interquartile values, and whiskers are 1.5 times the interquartile range. The dashed horizontal lines represent the spawning stock biomass that would produce maximum sustainable yield.

Results

Historical Abundance and Exploitation of Atlantic Bluefin Tuna

MAST estimates the initial stock size (B_0) of the western population to be 100–120 kt, and that of the eastern stock to be 800–900 kt (Figs. 1A and B). The ranges reflect the most credible interquartile ranges. The estimates of the maximum sustainable yield (MSY) from which these initial biomasses were calculated are 3.9 and 25 kt for the western and eastern stocks, respectively. Predictions of stock depletion rates relative to 1950 are 17% for the western stock and 33% for the eastern stock.

Furthermore, MAST indicates that the western bluefin tuna stock was subject to overfishing and was depleted to below the MSY stock biomass level (B_{msy}) relatively early in the fishery. Longline and purse-seine fishing in the Northwest Atlantic in the 1960s depleted the stock to levels below MSY before 1970 (Fig. 1A). The large annual Gulf of Mexico longline catches (approximately 3–4 kt) that occurred in the 1970s corresponded with high fishing mortality rates (Fig. 2B) on western-stock spawners, which further depleted the stock.

Observed declines in western Atlantic biomass have also been the result of a declining eastern population. The model predicts that the decline of the eastern stock to below B_{msy} has occurred as recently as the last 10 years (Fig. 1B), owing largely to substantial illegal and unreported catches in the east [13]. Concurrent with the depletion of eastern populations over the last 15 years, the model predicts a steady increase in the ratio of western to eastern fish in the western Atlantic Ocean (Fig. 3A).

The model points to serial, regional depletions as fishing effort has shifted spatially over time. Japanese longlining catches in the Gulf of Mexico in the 1970s were relatively small, but concentrated on a smaller number of western-stock spawners (Fig. 2A). In the western Atlantic, high fishing mortality rates occurred initially from longlining and purse seining, which removed as much as 20,000 mt annually in the 1960s off the coastal United States (Fig. 2A). The Norwegian purse-seine fisheries caught approximately 20,000 mt annually in the early 1960s, exerting mean fishing mortalities of up to 0.8 yr^{-1} until this fishery rapidly collapsed in 1963 [24]. These early fisheries occurred in mixed-stock areas of the western and eastern Atlantic,

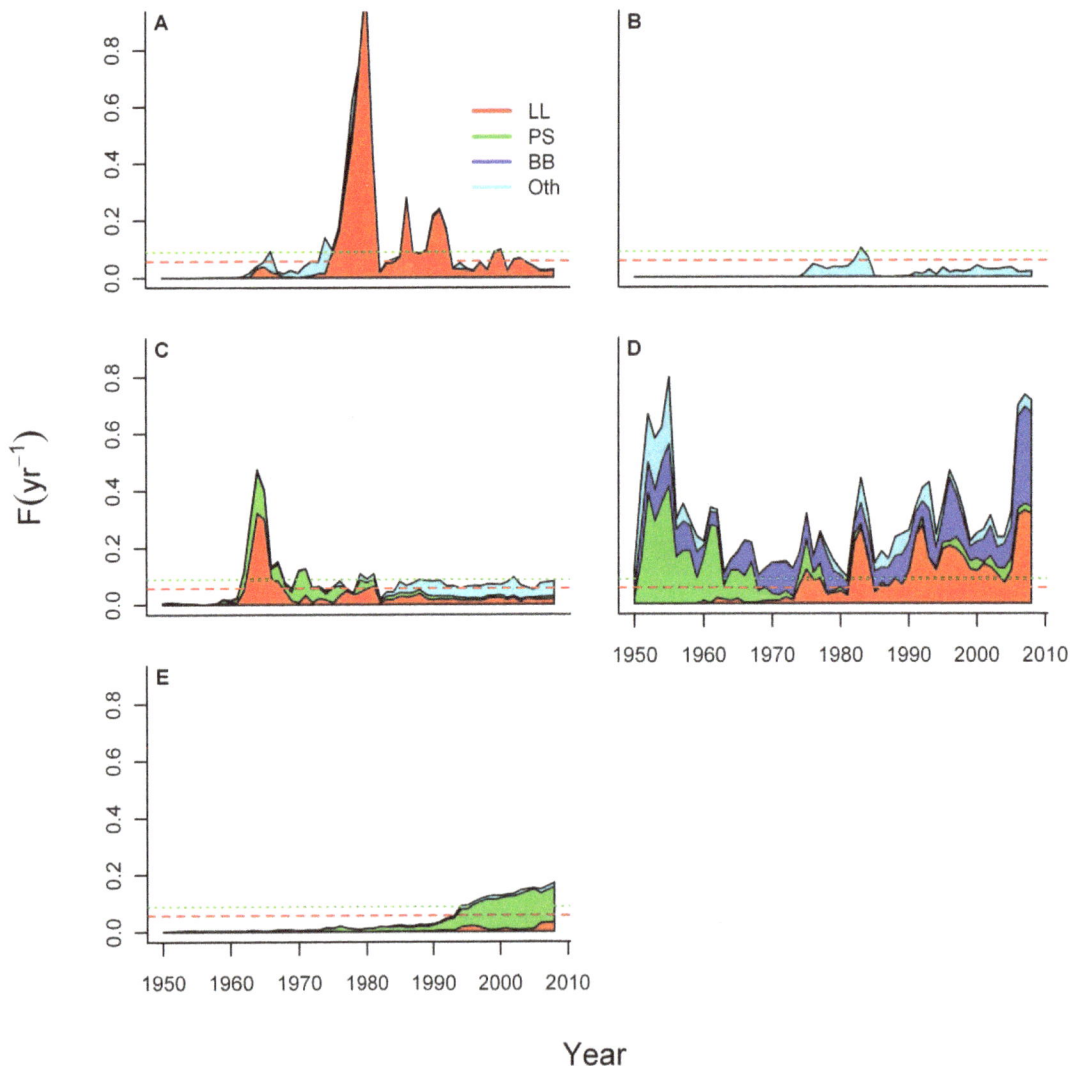

Figure 2. Mean annual fishing mortality rates (yr⁻¹) for Atlantic bluefin tuna by longline (LL), purse-seine (PS), bait boat (BB), and other (Oth) gear types in (A) the Gulf of Mexico, (B) the Gulf of St. Lawrence, (C) the western Atlantic Ocean, (D) the eastern Atlantic Ocean, and (E) the Mediterranean Sea. Red and green dotted lines represent F_{msy} for western and eastern stocks, respectively.

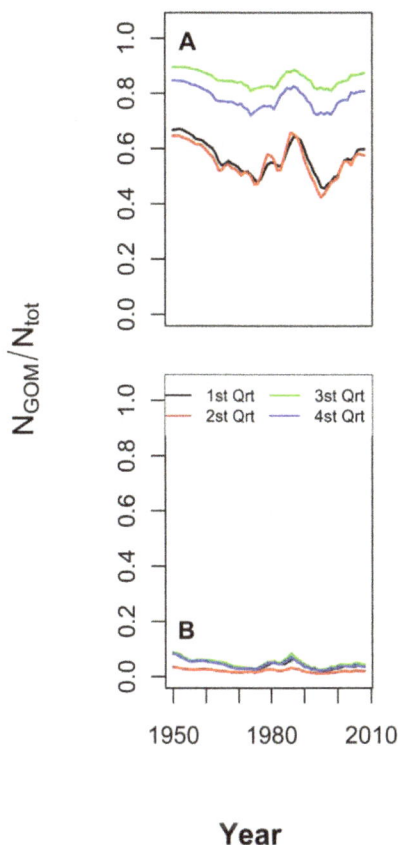

Figure 3. Predicted ratio of the numbers of western Atlantic bluefin tuna to total numbers of tuna from 1950 to 2008 in the (A) western Atlantic Ocean and (B) eastern Atlantic Ocean during the first quarter (black), second quarter (red), third quarter (green), and fourth quarter (blue).

and the model predicts that a small proportion of Nordic purse-seine catch (Fig. 2D) could have consisted of up to 10% western stock (Fig. 3B). However, the eastern-stock subsidy of western Atlantic bluefin tuna fisheries was substantial, with the western-stock ratio ranging between 50% and 90% western and 10–50% eastern during that period (Fig. 3A). Fishing mortality rates were well above the MSY rate (F_{msy}) for both stocks in all mixed-stock areas (Figs. 2C and D).

Relative to 1950–1970, eastern Atlantic bluefin tuna catches fell in the 1980s, but then increased again in the 1990s. Since 1990, the bulk of the tuna fishing effort has moved into the Mediterranean Sea in association with purse seining to populate tuna ranches. During this period, the largest Atlantic bluefin tuna catches in history occurred between 1998 and 2007, with maximum catches of approximately 60 kt occurring in 2007 in the eastern Atlantic and Mediterranean Sea [13]. With catches of this magnitude, the model suggests that a very large number of eastern-stock fish, up to 800–900 kt, must have existed in the Mediterranean Sea to have supported such removals. Fishing mortality rates were approximately double F_{msy} between 1995 and 2008 (Fig. 2E).

Future Projections of the Atlantic Bluefin Tuna Fishery

MAST predicts that western Atlantic bluefin tuna stock rebuilding depends on eastern Atlantic and Mediterranean catches. If oceanwide catches are zero (scenario i), the model

predicts that the western stock has a low probability of being at levels that produce maximum sustainable yield (median $B_{2020}/B_{msy} = 0.81$, Fig. 4A), but that eastern-stock rebuilding will be much faster (median $B_{2020}/B_{msy} = 1.4$, Fig. 4B). MAST also predicts that relative to no closure (scenario ii, Figs. 4C and D), western-stock rebuilding will not be faster with a Gulf of Mexico closure (scenario iii, Figs. 4E and F), and eastern-stock rebuilding will be unaffected. However, if eastern-stock rebuilding is slow, or if the stock declines, then western-stock growth must also be slow, or decline, without western quota adjustments to compensate for the loss in the eastern subsidy. In addition, MAST predicts western- and eastern-stock median B_{2020}/B_{msy} ratios to be 0.51 and 0.92, respectively, for the double eastern quota (scenario iv, Fig. 4G and H), and 0.36 and 0.35, respectively, for the historical overfishing case (scenario v, Fig. 4I and J). In all cases, the high variability of predicted spawning stock biomasses during the recovery may prevent the benefits of reduced fishing quotas from being statistically detectable for many years.

The redistribution of quota is likely to limit the effectiveness of large-scale closures. Under the Gulf of Mexico closure scenario (iii), the quota tonnage associated with the bycatch and dead or discarded bluefin in the Gulf is redistributed to the western Atlantic. It follows that the predicted landings there would include larger numbers of immature fish of both western and eastern stocks to compensate for their smaller size. The redistribution of quota would not require the Gulf of Mexico fleet to move to western Atlantic fishing areas, because unused quotas could simply be reallocated to other sectors (such as rod and reel) or even other countries through reallocations at ICCAT. There is additional uncertainty over the effects of a Gulf of Mexico closure, because bycatch and dead discard estimates for both inside and outside the Gulf of Mexico are unknown.

Discussion

We present a fisheries population assessment model that incorporates novel datasets on the spatial and seasonal dynamics of Atlantic bluefin tuna. This is the first assessment to incorporate fine-scale electronic tagging data for this species. Electronic tagging data can provide more precise and reliable seasonal movement and fishing mortality rate estimates than can be obtained from traditional mark-recapture data [23]. Furthermore, satellite tags reveal where tunas go independent of fisheries. By incorporating data that reveal how distinct stocks mix on foraging grounds and separate to breeding grounds in the eastern and western Atlantic, we improve our capacity to capture movement information in the population assessments and understand how movement and mixing may affect management decisions.

The MAST model may be used to conduct fisheries stock assessment and evaluate future management policies. For example, the results of our analysis indicate that eastern and western tuna stocks have experienced systematic declines in the twentieth and twenty-first centuries, with estimated spawning stock biomass depletions of 83% in the west and 67% in the east. The western stock has been severely depleted since the early 1970s, and in the past decade the eastern stock has been subjected to the largest Atlantic bluefin catches since the fishery began. However, rebuilding of the eastern stock is possible in the near future under certain quota scenarios, whereas western-stock rebuilding is predicted to take more than 15 years. MAST results indicate that the incorporation of mixing is critical for understanding historical breeding populations and the efficacy of future quota policies as applied to mixed-stock areas.

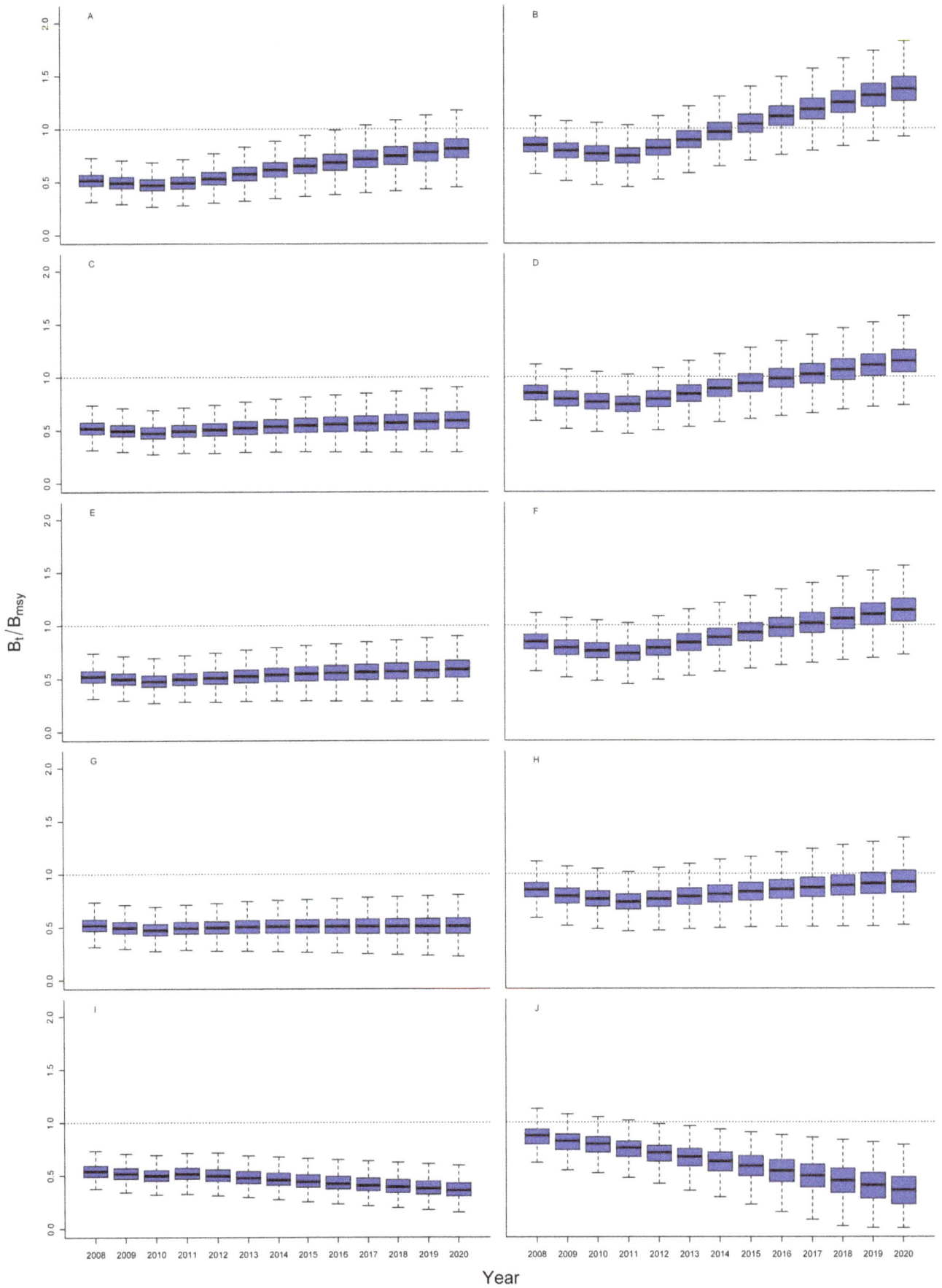

Figure 4. Box plots of the predicted ratio of biomass to the biomass that would produce maximum sustainable yield (B_t/B_{MSY}) under alternative management quotas for western (left column) and eastern (right column) Atlantic bluefin tuna stocks under various scenarios: total fisheries closures (A and B); catches at 1750 mt West and 12,900 mt East, with a Gulf of Mexico closure (C and D); catches at 1750 mt West and 12,900 mt East, with no Gulf of Mexico closure (E and F); catches at 1750 mt West and 25,800 mt East (G and H); and catches at 1750 mt West and 60,000 mt East (I and J). The horizontal lines within the blue bars represent the posterior median values, the blue bars represent the interquartile values, and whiskers are 1.5 times the interquartile range.

These results extend bluefin tuna stock assessments further into the past (i.e., to 1950) than recent ICCAT analyses (i.e., to 1970). Ignoring fishing that occurred before 1950 may, however, bias estimates of depletion levels and biomass reference points. While western Atlantic tuna fisheries began in the middle of the twentieth century, fishing had occurred for several hundred years in the Mediterranean Sea before official catches were recorded beginning in 1950 [24]. This suggests that eastern-stock depletion levels from the unfished state are underestimated. Assuming that the stock was already exploited by 1950 (i.e., not at B_0), unfished biomass, target reference points such as B_{msy}, and depletion levels relative to B_0 could be higher than those estimated by this study, which assumed that the stock was at B_0 in 1950. For example, predicted initial stock sizes calculated using deterministic estimates [15] of initial biomass, based on a range of assumptions about recruitment steepness, ranged between 1 and 11.7 million mt. A potential way of accounting for fisheries known to have occurred before reliable catch data were collected would be to consider alternative hypotheses about the initial fishing mortality rate experienced by the population.

The explicit consideration of mixing is likely to improve our understanding of how future Atlantic bluefin tuna populations will respond to alternative management scenarios. Because of the mixed-stock composition of western Atlantic fisheries, the successful rebuilding of the western population is tied to controlling the much larger fishing mortality rates that occur on the eastern stock. For example, continued high fishing mortality rates in the Mediterranean Sea and eastern Atlantic may compromise rebuilding efforts for the western Atlantic population. The converse, however, is not true. The eastern stock is both much larger and much more concentrated in the Mediterranean Sea. ICCAT could potentially increase the chances of successful western-stock rebuilding if it began to model and consider recovery plans [25] for eastern and western populations jointly rather than independently.

Modeling mixed-stock fisheries with complex population dynamics is challenging, and MAST exposes several key sensitivities of the results to model form and parameter inputs. In practice, movements are influenced by ontogeny, stock of origin, and environment, and may have interannual variability [18] or be dependent on oceanographic conditions [26]. In addition, there are alternative forms and time scales of modeling movement dynamics. For example, it is not known how the configuration of model areas and time-steps affects model results. The reliability of both growth and maturity parameters for bluefin are questionable, because samples have come from mixed-stock areas in either the Mediterranean [12] or the western Atlantic [11]. The corresponding estimates of growth and maturity parameters could therefore depend on the relative stock compositions encountered by fisheries and sampling programs at any given time and place.

One major issue for bluefin tuna stock assessment is that, in addition to the mixed-stock structure of the Atlantic Ocean, there is further population structure within the Mediterranean [12]. Recent genetic research indicates that there is a discrete eastern

Mediterranean population that is residential to the region [12]. These residential bluefin may be more productive than nomadic fish that move in and out of the Mediterranean Sea, potentially bolstering their capacity to withstand overfishing. Thus, it may be reasonable to consider a three-stock model to capture the additional mixed-stock dynamics. Considerable analytical work will be needed to capture how the violation of several assumptions could affect the reliability of MAST and other models in describing current and future population status.

New data and analytical techniques are revolutionizing our capacity to study the population structure and mixing of Atlantic bluefin tuna. The integration of multiple data types into a finer-scale spatial and temporal assessment of fish movement is a substantial advance in the development of tools for understanding bluefin tuna population dynamics. Incorporation of oceanographic data may enable tuning of models to discern seasonal aggregations in association with preferred ocean conditions. The data provide much needed biological information on how bluefin tunas utilize their entire range, and MAST allows us to synthesize these data. Many researchers have recognized the need to capture this new biological information in stock assessment models, and some have argued for the development of a management strategy evaluation (MSE) of Atlantic bluefin tuna fisheries [27]. The MAST model offers new directions for these cooperative efforts, and could be used as a reference model to simulate the performance of single-stock models, area-specific quotas, spatial management measures, and the interdependence of rebuilding the western and eastern stocks.

Supporting Information

Figure S1   Map of MAST spatial areas and electronic tag geolocations.

Figure S2   Annual catches by longline (LL), purse-seine (PS), bait boat (BB), and other (Oth) gears in (A) the Gulf of Mexico, (B) the Gulf of St. Lawrence, (C) the western Atlantic, (D) the eastern Atlantic, and (E) the Mediterranean Sea.

Table S1   Description of symbols and indices used in MAST

Table S2   Initialization of age-structured model assuming selectivity at age, natural mortality, age-specific fecundity, and Beverton-Holt recruitment

Table S3   Partial derivatives for the derivation of B_0 from MSY and F_{msy}

Table S4   Data, estimated parameters, and initial states

Table S5   State dynamics and observation model

Table S6   Objective function calculation

Table S7   Life-history parameters defining ϕ

Table S8   Electronic tag data state-transition equations in the MAST model

Table S9   Archival tag observation probabilities for state-space likelihoods in the MAST model for Atlantic bluefin tuna

Table S10   Summary of conventional tag cohorts of Atlantic bluefin tuna in the MAST model

Table S11   Summary of commercial CPUE data on relative abundance in the MAST model for Atlantic bluefin tuna

Table S12   Stock-composition data summary

Table S13   Summary of key MAST model output at the posterior mode for Atlantic bluefin tuna (A) base-case with time-invariant gear selectivity and normal reporting-rate priors; (B) estimated time-invariant gear selectivity and β(3,3) reporting-rate priors; (C) estimated time-varying gear selectivity and N(0.1,0.065) reporting-rate priors; (D) base-case with eastern age at 50% maturity at age 6; (E) base-case with bulk movement parameterization; and (F) single-stock model fit to estimated time-invariant gear selectivity. All projections

to 2025 were run using constant catches of 1750 and 12,900 tonnes West and East, respectively.

Table S14   Estimated tag reporting rates of the MAST model for Atlantic bluefin tuna by scenario (A) base-case with time-invariant gear selectivity and normal reporting-rate priors; (B) estimated time-invariant gear selectivity and β(3,3) reporting-rate priors; (C) estimated time-varying gear selectivity and N(0.1,0.065) reporting-rate priors; (D) base-case with eastern age at 50% maturity at age 6; and (E) base-case with bulk movement parameterization. Case F is omitted in this table because mark-recapture data were not used for single-stock-model fitting.

Acknowledgments

We thank Drs. Robert Ahrens, Hiroyuki Kurota, Andre Boustany, Shelley Clarke, and Daniel Pauly for reviewing early drafts of the manuscript. We thank Shana Miller for considerable help on all portions of the study. Mr. James Ganong and Mike Castleton helped extract and parse out electronic tag data. Funding for this project was provided by the Tag-A-Giant Foundation, the Lenfest Ocean Program, the Canadian Fisheries and Oceans International Governance Strategies Fund, the Monterey Bay Aquarium Foundation and the U.S. National Oceanic and Atmospheric Administration. Early drafts were greatly improved by the feedback of two anonymous reviewers.

Author Contributions

Conceived and designed the experiments: NT MM BB GL. Performed the experiments: NT MM BB GL TC. Analyzed the data: NT MM BB GL TC. Contributed reagents/materials/analysis tools: NT MM BB GL TC. Wrote the paper: NT MM BB GL TC.

References

1. Neilson JD, Campana SE (2008) A validated description of age and growth of western Atlantic bluefin tuna (*Thunnus thynnus*). Can J Fish Aquat Sci 65: 1523–1527.
2. Maigret J, Ly B (1986) Les Poissons de Mer de Mauritanie. Campiegne Sciences Nat. 213 p.
3. Gibbs R, Collette B (1967) Comparative anatomy and systematics of the tunas, genus *Thunnus*. US Fish Wild Seri Fish Bull 66: 65–130.
4. Mather F, Mason JM, Jones A (1995) Historical document: Life History and Fisheries of Atlantic Bluefin tuna. MiamiFlorida: United States Department of Commerce. 165 p.
5. Fromentin JM (2002) Descriptive analysis of the ICCAT bluefin tuna tagging database. Col Vol Sci Pap ICCAT 54: 353–362.
6. Block BA, Dewar H, Boustany A, Blackwell SB, Seitz A, et al. (2001) Migratory movements, depth preferences, and thermal biology of Atlantic bluefin tuna. Science 293: 1310–1314.
7. Block BA, Teo S, Walli A, Boustany A, Stokesbury MJW, et al. (2005) Electronic tagging and population structure of Atlantic bluefin tuna. Nature 434: 1121–1123.
8. Walli A, Teo SLH, Boustany A, Farwell CJ, Williams T, et al. (2009) Seasonal movements, aggregations and diving behavior of Atlantic bluefin tuna (*Thunnus thynnus*) revealed with archival tags. PloS ONE 4: e6151.
9. Boustany AM, Reeb CA, Block BA (2008) Mitochondrial DNA and electronic tracking reveal population structure of Atlantic bluefin tuna (*Thunnus thynnus*). Mar Biol 156: 13–24.
10. Dickhut RM, Deshpande AD, Cincinelli A, Cochran MA, Corsolini S, et al. (2009) Atlantic Bluefin Tuna (*Thunnus thynnus*) Population Dynamics Delineated by Organochlorine Tracers. Environ Sci Technol 43: 8522–8527.
11. Rooker JR, Secor DH, De Metrio G, Schloesser R, Block BA, et al. (2008) Natal homing and connectivity in Atlantic bluefin tuna populations. Science 322: 742–744.
12. Riccioni G, Landi M, Ferrara G, Milano I, Cariani A, et al. (2010) Spatio-temporal population structuring and genetic diversity retention in depleted Atlantic bluefin tuna of the Mediterranean Sea. Proc Natl Acad Sci USA 107: 2102–2107.
13. ICCAT (2008) Report of the 2008 Atlantic bluefin tuna stock assessment session. Madrid. 247 p.
14. Corriero A, Karakulak S, Santamaria N, Deflorio M, Spedicato D, et al. (2005) Size and age at sexual maturity of female bluefin tuna (*Thunnus thynnus* L. 1758) from the Mediterranean Sea. J Appl Ichthyol 21: 483–486.
15. ICCAT (2009) Extension of the 2009 SCRS Meeting to Consider the Status of Atlantic Bluefin Tuna Populations with Respect to CITES Biological Listing Criteria. Madrid.
16. ICCAT (2010) Report of the 2010 Atlantic bluefin tuna stock assessment session. Madrid. 132 p.
17. Teo SLH, Block BA (2010) Comparative influence of ocean conditions on yellowfin and Atlantic bluefin tuna catch from longlines in the Gulf of Mexico. PloS ONE 5: e10756.
18. Sibert JR, Lutcavage ME, Nielsen A, Brill RW, Wilson SG (2006) Interannual variation in large-scale movement of Atlantic bluefin tuna (*Thunnus thynnus*). Can J Fish Aquat Sci 63: 2154–2166.
19. Teo SLH, Boustany A, Dewar H, Stokesbury MJW, Beemer S, et al. (2007) Annual migrations, diving behavior, and thermal biology of Atlantic bluefin tuna, *Thunnus thynnus*, on their Gulf of Mexico breeding grounds. Mar Biol 151: 1–18.
20. Porch CE, Turner SC, Powers JE (2001) Virtual population analyses of Atlantic bluefin tuna with alternative models of transatlantic migration: 1970-1997. Col Vol Sci Pap ICCAT 52: 1022–1045.
21. Punt AE, Butterworth DS (1995) Use of tagging data within a VPA formalism to estimate migration ratesof bluefin tuna across the north Atlantic. Col Vol Sci Pap ICCAT 44: 166–182.
22. Gavaris S (1988) An adaptive framework for the estimation of population size Can Atl Fish Sci Adv Doc 88/2930.
23. Kurota H, Mcallister MK, Lawson GL, Nogueira JI, Teo SLH, et al. (2009) A sequential Bayesian methodology to estimate movement and exploitation rates using electronic and conventional tag data : application to Atlantic bluefin tuna (*Thunnus thynnus*). Can J Fish Aquat Sci 66: 321–342.
24. Fromentin JM (2009) Lessons from the past: investigating historical data from bluefin tuna fisheries. Fish Fish 10: 197–216.

25. ICCAT (2006) Recommendation by ICCAT to establisha multi-annual recovery plan for the bluefin tuna in the eastern Atlantic and Mediterranean. Madrid.
26. Senina I, Sibert J, Lehodey P (2008) Parameter estimation for basin-scale ecosystem-linked population models of large pelagic predators: Application to skipjack tuna. Prog Oceanogr 78: 319–335.
27. Anonymous (2001) ICCAT workshop on bluefin mixing. Col Vol Sci Pap ICCAT 54: 261–352.
28. Martell SJD, Pine WE, Walters CJ (2008) Parameterizing age-structured models from a fisheries management perspective. Can J Fish Aquat Sci 65: 1586–1600.
29. Goodyear CP (1977) Assessing the Impact of Power Plant Mortality on the Compensatory Reserve of Fish Populations.Conference on Assessing the Effects of Power-Plant-Induced Mortality on Fish Populations Gatlinburg, Tennessee. pp 184–194.
30. Diaz GA, Turner SC (2006) Size frequency distribution analysis, age composition, and maturity of western bluefin tuna in the Gulf of Mexico from the U.S. (1981-2005) and Japanese (1975-1981) longline fleets. Col Vol Sci Pap ICCAT 60: 160–170.
31. Cort JL (1991) Age and growth of bluefin tuna in the Northeast Atlantic. Col Vol Sci Pap ICCAT 35: 213–230.
32. Walters C, Ludwig D (1994) Calculation of Bayes Posterior Probability Distributions for Key Population Parameters. Can J Fish Aquat Sci 51: 713–722.
33. DeValpine P, Hastings A (2002) Fitting Population Models Incorporating Process Noise and Observation Error. Ecol Monogr 72: 57–76.
34. Gelman A, Carlin JB, Stern HS, Rubin DB (2004) Bayesian Data Analysis: Chapman Hall. 668 p.

5

Benefits of Rebuilding Global Marine Fisheries Outweigh Costs

Ussif Rashid Sumaila[1]*, **William Cheung**[2], **Andrew Dyck**[1], **Kamal Gueye**[3], **Ling Huang**[4], **Vicky Lam**[1], **Daniel Pauly**[2], **Thara Srinivasan**[5], **Wilf Swartz**[1], **Reginald Watson**[2], **Dirk Zeller**[2]

1 Fisheries Economics Research Unit, Fisheries Centre, University of British Columbia, Vancouver, British Columbia, Canada, 2 Sea Around Us Project, Fisheries Centre, University of British Columbia, Vancouver, British Columbia, Canada, 3 The United Nations Environment Programme, Geneva, Switzerland, 4 Department of Economics, University of Connecticut, Storrs, Connecticut, United States of America, 5 Pacific Ecoinformatics and Computational Ecology Lab, Berkeley, California, United States of America

Abstract

Global marine fisheries are currently underperforming, largely due to overfishing. An analysis of global databases finds that resource rent net of subsidies from rebuilt world fisheries could increase from the current negative US$13 billion to positive US$54 billion per year, resulting in a net gain of US$600 to US$1,400 billion in present value over fifty years after rebuilding. To realize this gain, governments need to implement a rebuilding program at a cost of about US$203 (US$130–US$292) billion in present value. We estimate that it would take just 12 years after rebuilding begins for the benefits to surpass the cost. Even without accounting for the potential boost to recreational fisheries, and ignoring ancillary and non-market values that would likely increase, the potential benefits of rebuilding global fisheries far outweigh the costs.

Editor: Julian Clifton, University of Western Australia, Australia

Funding: The Sea Around Us Project and the Global Ocean Economics Project are funded by the Pew Charitable Trusts, Philadelphia (http://www.pewtrusts.org/). The funders had no role in study design, data collection and analysis, decision to publish, or preparation of the manuscript.

Competing Interests: The authors have declared that no competing interests exist.

* E-mail: r.sumaila@fisheries.ubc.ca

Introduction

Fish are among the planet's most important renewable natural resources. Beyond playing a crucial role in marine ecosystems, fish support human well-being through employment in fishing, processing, and retail services [1–3], as well as food security for the poor, particularly in developing countries [4]. Overexploitation [3,5,6] and rising ocean temperatures threaten global fisheries [7–9]. As demonstrated by the collapse of northern cod off Newfoundland, the depletion of fish stocks can have devastating effects on human well-being [10,11]. As human populations continue to grow, the future benefits that fishery resources can provide will depend largely on how well they are rebuilt and managed. However, policy makers often perceive that rebuilding fisheries is too expensive in the short-term and therefore avoid taking the necessary actions to sustainably manage fish stocks. Therefore, a crucial question for policy makers is what is the potential net economic benefit of rebuilding global fisheries? Here, we address this question on a global scale.

Fisheries economists use resource rent (i.e., what remains after fishing costs and subsidies are deducted from revenue) as an indicator of fisheries performance [12], although others argue that this is inadequate because it does not capture all the benefits derived from marine fisheries [13]. Here, we adhere to using resource rent as our primary indicator of economic performance, but we also report payments to labor (i.e., wages) and earnings to fishing companies as additional indicators of fisheries benefits. With these additional indicators, we recognize that fishing capacity is not often converted to other uses easily (i.e., it is non-malleable)

and that the opportunity cost of fishing labor (i.e., the alternative wages that fishers can earn if they did not fish) in many fishing communities is low due to a dearth of alternative employment. Even with these additional indicators, other important contributions of fish populations to the economy, such as the value created through the production chain [1] and non-market values [14] are not captured.

Over the past decade, we have gathered data on the economics of global fisheries from a range of sources, including scientific, economic, governmental, inter- and non-governmental publications, to create several global databases on catch [15]; ex-vessel fish prices [16]; subsidies [17]; and fishing costs [18] (Tables S1, S2, S3, S4, S5 and S6). From these databases, we compile landed value of catch, cost of fishing, payments to labor, earnings of fishing companies, and fisheries subsidies for 144 maritime countries of the world. We then compute both current and potential maximum resource rent, wages, and earnings to fishing enterprises.

Results

1. Gains from Rebuilding

Global marine fisheries landings are projected to average 89 million t per year (range 83–99 million t) (Table 1) when rebuilt [19], with a corresponding mean landed value of US$101 billion per year (range US$93–116 billion). The wide ranges help to address uncertainties about the magnitude of global overfishing currently debated in the literature (as discussed in Materials and Methods). The cost of fishing in this rebuilt scenario is estimated at

US$37 (US$29–44) billion compared to US$73 (US$50–96) billion per year currently. Returns to capital invested (i.e., normal profit) and payments to labor would amount to US$3 (US$2–4) billion and US$16 (US$12–19) billion per annum, respectively, while resource rent from rebuilt global fisheries would be US$54 (US$39–77) billion per year (summary of current resource rent is displayed by country in Figure 1, with details in Table 1). (The *Sunken Billions* report of the World Bank [20], which estimated economic rent without addressing the cost of reform, arrived at a potential resource rent of US$50 billion per year, using a different approach.) Gains in resource rent from the current situation to a rebuilt global fishery would be US$66 (US$51–89) billion a year, while wages and returns to capital will decrease to US$16 and US$3 billion, respectively (Table 1). Figure 2 summarizes the net gains in resource rent by maritime country.

2. Cost of Reform

The real cost to society of rebuilding fisheries, once the elimination of an estimated US$19 billion per year of harmful and ambiguous subsidies is taken into account [17], is negative, implying that society as a whole will make money by engaging in rebuilding (Figure 3). However, fishing enterprises and fishers will lose profits and wages during rebuilding. Hence, to implement a rebuilding reform, governments may need to temporarily invest extra resources to mitigate these impacts.

The world's current fishing capacity is estimated to be up to 2.5 times more than what is needed to land the Maximum Sustainable Yield (MSY) [21]. This suggests that to rebuild global fisheries, we need to trim excess capacity from the current 4.3 million fishing boats [3]. Assuming that current capacity is between 1.5 and 2.5 times the level needed to maximize sustainable catch, fishing effort needs to be reduced by between 40 and 60 per cent, or up to 2.6 million boats. Fisheries currently employ more than 35 million people globally [3]. If we simplify by assuming linearity between boats and people, this implies that between 15 and 22 million fishers would need to be moved to other livelihood activities in order to rebuild global fisheries. This is a challenge, but one that is

surmountable. For instance, even though in some fisheries most fishers may see fishing as a way of life and therefore may not want to exit fishing [22], it has been reported that up to 75% of fishers in Hong Kong would be willing to leave the industry if suitable alternatives or compensation were available [23]. Similar sentiments are likely to also occur in many other countries. In any case, it is better to undertake this transition as part of a rebuilding policy rather than having it forced upon us through loss of resources [10,11].

Using the unit cost of reducing fishing effort calculated in Materials and Methods, the total amount that governments need to invest to rebuild world fisheries ranges between US$130 and US$292 billion in present value, with a mean of US$203 billion. This total transition cost would be spread over the time required to rebuild fisheries within each country.

3. Net Gain from Rebuilding

Global fisheries are not living up to their revenue potential; the total cost of fishing is too high and governments provide harmful subsidies to the sector, which results in a negative resource rent (i.e., economic loss to society) of about US$13 billion per year (Table 1). Rebuilding would result in a gain in resource rent of US$66 billion per year, which when discounted over the next 50 years using a 3 per cent real discount rate, generates a present value of between US$660 and US$1,430 billion (Table 1), i.e., between 3 and 7 times the mean cost of fisheries rebuilding reform. Furthermore, it would likely take just 12 years after rebuilding efforts begin for the gains to exceed the costs of adjustment (Figure 3). A higher discount rate will reduce the present value of gain from rebuilding and increase the time needed to balance the gain with the costs of adjustment, and *vice versa* (see Materials and Methods for the justification of a 3% discount rate). Our results suggest that, even without accounting for the potential boost to recreational fisheries, processing, retail and non-market values that would likely increase, there is a substantial net economic benefit to be derived from rebuilding global fisheries, with net gains large enough to compensate for uncertainties in our assumptions and

Table 1. Key economic figures of global fisheries.

Key indicators, annual data (unit)	Current	Rebuilt fisheries		
		Lower bound	Mean	Upper bound
Catch (t)	80.2	82.7	88.7	99.4
Catch value (US$ billions)	87.7	92.6	100.5	116.3
Variable fishing cost (US$ billions)	73.0	43.9	36.6	29.3
Normal profit (US$ billions)	6.1	3.7	3.0	2.4
Wages (US$ billions)	31.0	18.6	15.5	12.4
Subsidies (US$ billions)	27.2	10.0	10.0	10.0
Rent net of subsidies* (US$ billions)	−12.5	39.0	54.0	77.0
Rent increase over current values (US$ billions)	–	51.2	66.4	89.4
NPV of resource rent increases (US$ billions)	–	665.2	972.0	1,428.1
Transition costs** (US$ billions)	–	129.9	202.9	292.2
NPV net of transition costs (US$ billions)	–	535.3	769.1	1,135.9

NPV: Net Present Value.

*The (resource) rent is the return to 'owners' of fish stocks, which is the surplus from gross revenue after total cost of fishing is deducted and subsidies taken into account.

**Transition costs include the costs to society of reducing current fishing effort to levels consistent with maximum sustainable yield and the payments governments may decide to employ to adjust capital and labour to uses outside the fisheries sector. Such payments may include vessel buyback programs and alternative employment training initiatives for fishers.

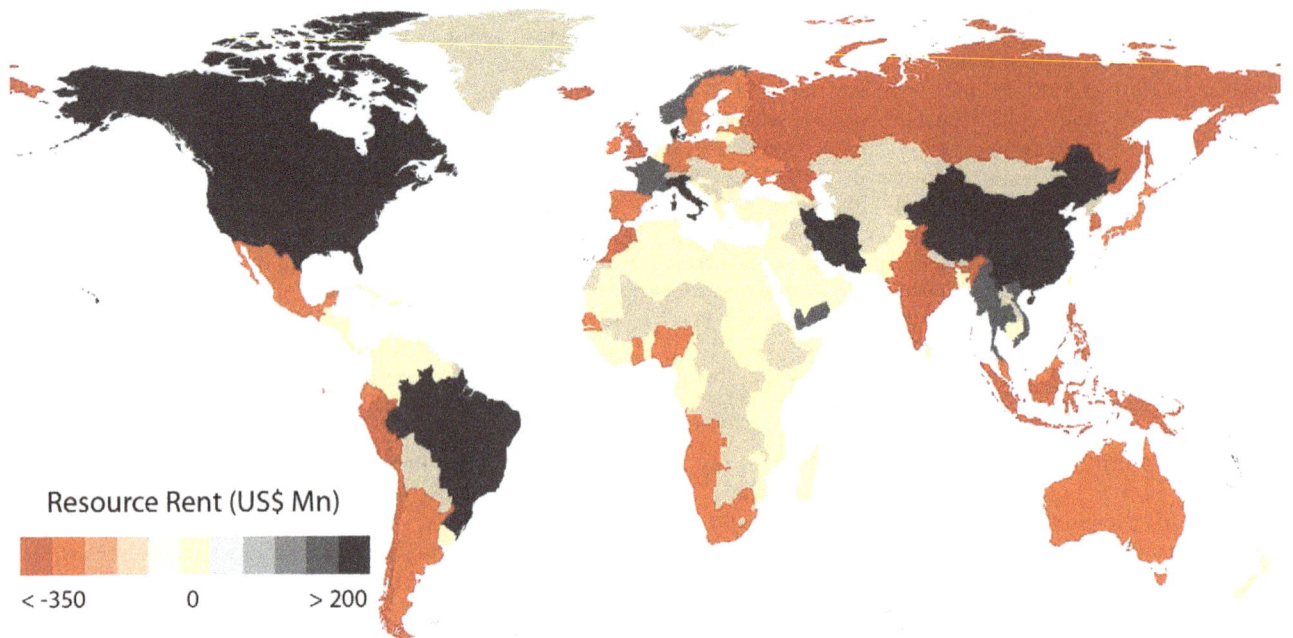

Figure 1. Summary of resource rent (adjusted for subsidies) from current fisheries. We see that several countries are in red once the full cost of fishing, including harmful subsidies are taken into account.

estimates. Rebuilding fisheries makes good business sense. The challenge is how to move global fisheries from their current dismal economic state to a more prosperous one.

Discussion

Even though the overall results we present are consistent with other estimates about the extent of subsidies, excess fishing pressure and the potential for increased biological yield, the country-by-country analysis (Tables S1, S2, S3, S4, S5 and S6) may reveal results that differ from expectations. This is not surprising, as our analysis produces estimates with ranges, and therefore computing midpoint estimates may over- or underestimate numbers for some countries. This is more likely to happen for small developing countries where observed data are limited, and we therefore had to rely on statistical methods to produce

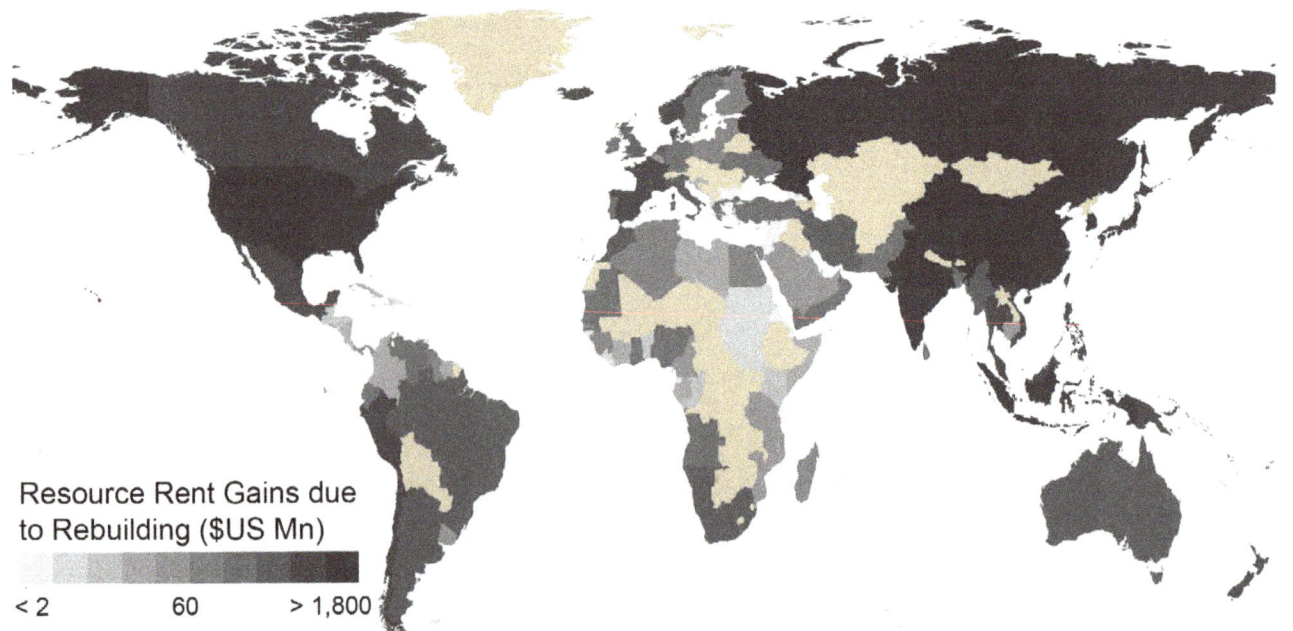

Figure 2. Summary of resource rent (adjusted for subsidies) from rebuilt fisheries (rent in all maritime countries increase after rebuilding).

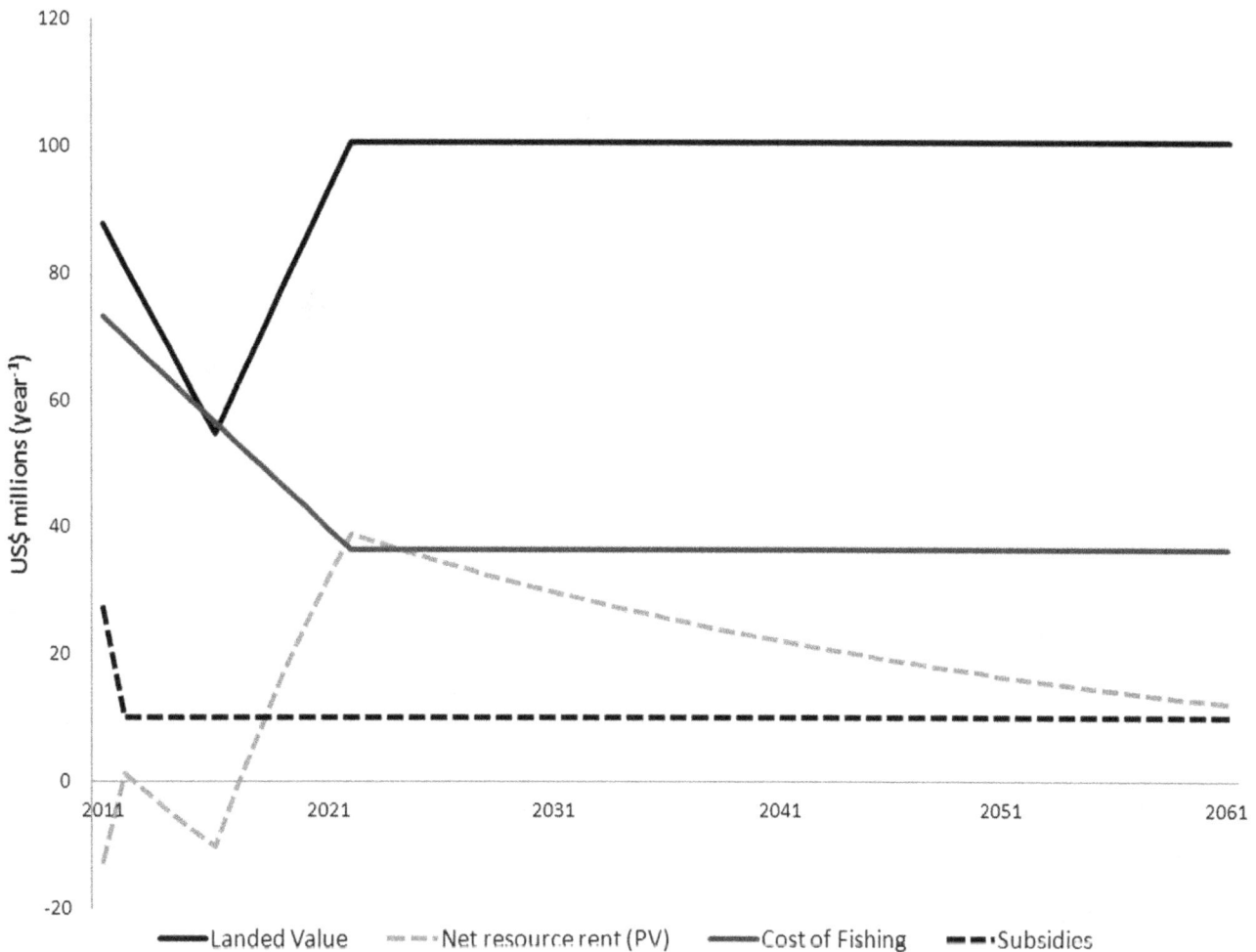

Figure 3. Transition time path of key rebuilding global fisheries variables.

estimates for these countries. The key to improving our estimates is for the collection of economic data for fisheries to be given priority by maritime countries.

There are also situations where large maritime countries (e.g., Peru, Chile, and Indonesia) may show counterintuitive estimates. Similarly unexpected results for countries such as Australia and Iceland, known to have good fisheries management regimes were found. In these cases too, the result of the statistical estimation required in the absence of observed or collected, publicly available country-specific data may be a reason. However, the reported results may indeed be correct yet unanticipated, as explained below.

A recent fishing industry report [24] provides financial data over 3 years (up to 2009), including pre-tax profit for the top 1000 commercial fishing companies worldwide. The numbers in this report support some of the counterintuitive outcomes of our study. These 1000 companies operate in 43 countries on all continents. The total annual sales value for all companies is about US$21 billion or 25% of estimated landed value worldwide. Of these 1000 companies, 339 reported negative annual pre-tax profits. Thirty-one of the 43 countries have at least one company reporting negative pre-tax profits, and of the 12 countries that report only positive pre-tax profits, nine countries had only one company in the dataset, suggesting that the optimistic results for these countries may be a result of limited data. Sixteen of the 43 countries for

which data were reported had negative average pre-tax profits at the aggregate national level, at which the average ratio of pre-tax profit to sales volume is only marginally greater than zero (Figure 4). These data present an interesting and more micro-level view of the industry that is complementary to our estimates, showing that within the same country, some firms may be quite profitable, while others are much less so, resulting in negative aggregate profit at the national level.

Materials and Methods

To estimate the potential gains from rebuilding global fisheries, we use estimates of catch loss [19], defined as the difference between current landings and Maximum Sustainable Yield (MSY) for those species that are considered to be over-exploited. It should be noted that MSY does not maximize economic yield (MEY) except when the stock size of fish does not affect the cost of fishing, and discount rate is zero. Still, we apply MSY in this analysis for practical and policy reasons, as it is a stipulated target or management reference point for many national legislations and international conventions. Other assumptions made in our analysis are: (i) the real ex-vessel fish price is constant through time (they have remained relatively stable since 1970) [16,25]; (ii) during rebuilding, the costs of fishing change in proportion to changes in effort; (iii) the costs of fisheries management increase by 25% to

A: Pre-tax profit share of sales for 1000 fishing companies

B: Pre-tax profit share of sales for 43 fishing countries

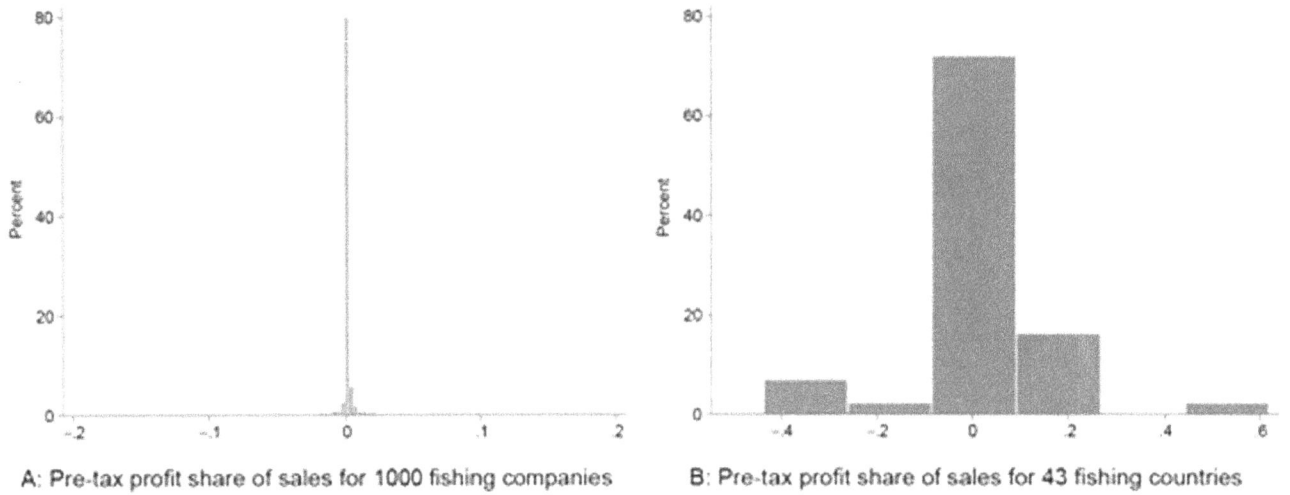

Figure 4. Histograms of pre-tax profit share of total sales for a sample of 1000 global fishing companies. Figure taken from [24].

US$10 billion a year, to support effective management under a rebuilt scenario; and (iv) the reported US$19 billion of annual harmful and ambiguous subsidies [17] are eliminated, since providing capacity-enhancing subsidies is fundamentally at odds with rebuilding fisheries. We also assume a rebuilding period of 10 years (e.g., Magnuson–Stevens Fishery Conservation and Management Act of the USA). Further support for this assumption is given by Costello et al. [26], who found that under an optimal rebuilding strategy, stock recovery requires between 4 and 26 years (with a mean of 11 years), depending on the fish species.

1. Estimating Global Fleet Size

The FAO estimates that there are currently 35 million people engaged in capture fisheries on either a part- or full-time basis [3]. The same report indicates that 90% of these fishers participate in the small-scale sector, while the remaining 10% can be classified as large-scale. The FAO also reports that the world's fishing fleet is comprised of 4.3 million vessels, 59% of which are motorized and 14% of motorized vessels (8% of all vessels) are greater than 12 meters in length [3]. In this study, we take a broad definition of large-scale vessels that includes all motorized vessels over 12 m in length, which is of sufficient size to represent considerable fishing pressure and potential impact on the environment. Under this criterion, we estimate the number of large-scale fishing vessels worldwide to be 355,000, with the remaining 3.94 million vessels are classified as small-scale.

2. Estimating Effort Reductions Required to Rebuild Global Fisheries

We model the global fishery using the Schaefer surplus-production model commonly applied to single-stock fisheries [27]. Since many fish stocks around the globe are either fully- or over-exploited, the global fishery is currently using more effort than needed to produce maximum sustainable yields (E_{MSY} in Figure 5), which we use as a global proxy for sustainable fisheries. We are cognizant of the diversity of fisheries management uses of biomass levels with MSY (either slightly above or below) as management reference or target points. In order to achieve maximum sustainable yields, effort will need to be reduced from current levels (e.g., E_0 in Figure 5) to a lower level that is consistent with maximum sustainable yield (E_{msy}). At E_{msy}, the total cost of fishing is reduced from TC_0 to TC_{msy}. For our calculations, we

make the simplifying assumption that there is no substitution between labor and capital, so the shares of components of fishing costs (i.e., fuel, wages, etc.) remain constant.

Recognizing that large- and small-scale fisheries have different fishing power, and in order to minimize the effect of effort reductions on fishers (labor), who are predominantly in the small-scale sector, we weigh effort reductions more heavily on large-scale operations. We express total fishing effort in the global fishery as:

$$LSF * P_l + SSF * P_s = \delta_0$$
$$P_l = \gamma P_s \tag{1}$$

where LSF and SSF are the number of large- and small-scale fishers, respectively. The parameters P_l and P_s represent the fishing power of large- and small-scale fishers, while γ represents the power of large-scale fishers relative to small-scale fishers. Total current fishing effort is δ_0. By re-expressing LSF, SSF and δ_0 as terms that are relative to the total current fishing effort (i.e., dividing both LHS and RHS of eq. 1 by δ_0), we have:

$$LSF' * P_l + SSF' * P_s = 1 \tag{2}$$

Pauly [28] reports an estimate of γ, which places the fish catching power of large-scale fishers at 18 times that of their small-scale counterparts. This leaves us with a system of two equations with two unknowns that can be solved for P_l and P_s, which are used to estimate the proportions of large- and small-scale fishers required to reduce overall fishing effort:

$$LSF' * P_l * x + SSF' * P_s * y = \delta$$
$$w_l * LSF' * P_l * x = w_s * SSF' * P_s * y \tag{3}$$

The parameters LSF', SSF', P_l and P_s are defined as in the system of equations (1 & 2) above, while δ represents the ratio of current effort required to rebuild fisheries, while w_l and w_s represent the weight of effort cuts levied on large- and small-scale fishers, respectively. The parameters x and y, which represent the proportion of large- and small-scale fishing activity to be cut, are

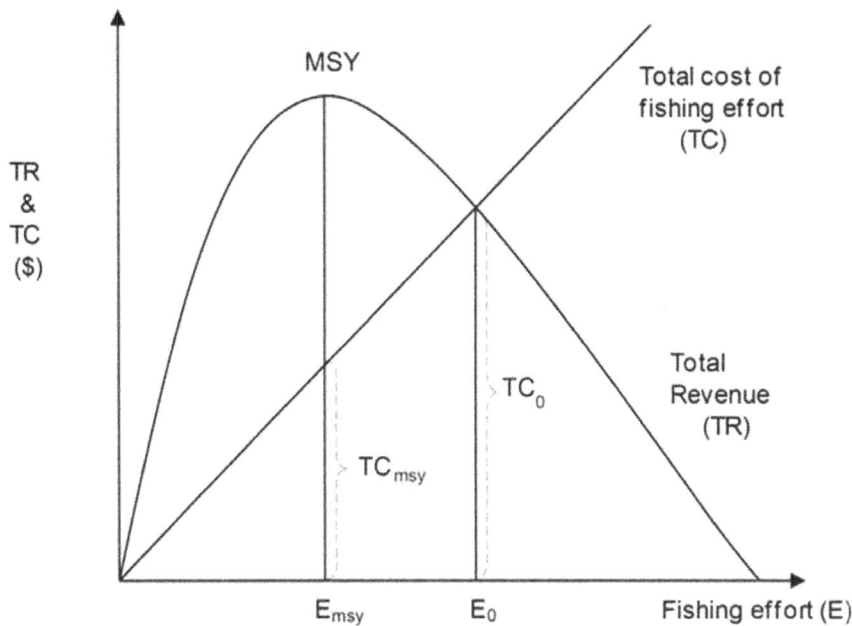

Figure 5. The Schaeffer surplus-production model, based on Gordon [27].

estimated from equations (3) and used to estimate the total reductions in large- and small-scale fishers as:

$$LSF_{cut} = LSF' * x$$
$$SSF_{cut} = SSF' * y$$

(4)

Lastly, we use our earlier estimates of the current number of large- and small-scale fishers and fishing vessels to estimate the number of large- and small-scale fishers and vessels that must be removed from the global fishery corresponding to our estimates of required reductions. We explore a range of weights (w_l and w_s) that represent equivalent total effort reductions. As can be seen in Figure 6, the trade-off between the cost of fishing effort reduction per fisher is non-linear, while the number of total fishers reduced is linear in the weighting placed on large-scale fishing effort. We suggest that by placing 80% of the weight of fishing effort reduction on large-scale fishing operations, it is consistent with cutting 60% of large-scale and 30% of small-scale fishing activity.

3. Estimating the Potential Value of Rebuilt Fisheries

For our present purposes, we assume that the estimated catch losses to overfishing reported by Srinivasan et al. [19] (Figure 7) may be fully regained after a period of rebuilding fisheries worldwide. To calculate potential catch losses, Srinivasan et al. [19] used catch time series from the *Sea Around Us* project for 1,066 taxa of fish and invertebrates in 301 EEZs, along with an empirical relationship they derived from catch data and stock assessments for 26 Northeast U.S. species from the U.S. National Oceanic and Atmospheric Administration (NOAA). The log-linear relationship that they found between a species' mean maximum catch C_{max} from catch data and its maximum sustainable yield (MSY) from stock assessment was robust ($R^2 = 0.84$, p<0.001), and has since been tested for 50 fully assessed stocks in the Northeast Atlantic, where variation in MSY accounted for 98% of the variability in C_{max} [29]. Therefore, given the dearth of detailed

stock assessments for the majority of species in the world's fisheries, Srinivasan et al. [19] applied the relationship they derived (with a 50% prediction interval) to estimate MSY levels for all stocks they identified as overfished. By comparing with reported catch levels, they arrived at estimates of lost catch by mass, reporting that without overfishing, potential landings worldwide in the year 2000 may have been 9.1 million t higher than current landings (50% prediction interval: 3.6 to 19 million t higher)

To calculate the value of these potential landings under rebuilt global fisheries, Srinivasan et al. [19] used a database of ex-vessel fish prices by Sumaila et al. [16]. For each taxon-EEZ pair designated as overfished, a price-per-tonne p for the maximum sustainable yield (MSY) was set by taking a weighted average of the actual prices corresponding to catches of the taxon within ±30, ±50, or ±100% of the estimated MSY level, in order of preference depending on data availability. This approach was used to account for the impacts of overfishing, and thus scarcity, on price levels.

There is debate among fisheries scientists as to the reliability of overfishing estimates based on catch trends rather than stock assessments, with some arguing that catch-based approaches are prone to overestimate depletion [30]. Srinivasan et al. [19] were careful to avoid the biases described by Branch et al. [30], with the result that the former's estimate of the percentage of overfished stocks worldwide (16–31%) was similar to, but more conservative than, that reported by Branch et al. (28–33%), and similar also to a recent assessment by the FAO [31]. Indeed, Froese et al. [29] demonstrated that both stock- and catch-based assessments of overfishing in the Northeast Atlantic show the same trends, although the catch-based methods were generally late to recognize declines in biomass. Thus, a catch-based method would underestimate lost catch, i.e., the direct opposite direction of the bias over which Branch et al. [30] have expressed concern. Moreover, Worm et al. [6] compared areas where there were both detailed stock assessment information and more general data including catch time series, and found that catches follow biomass trends, if belatedly.

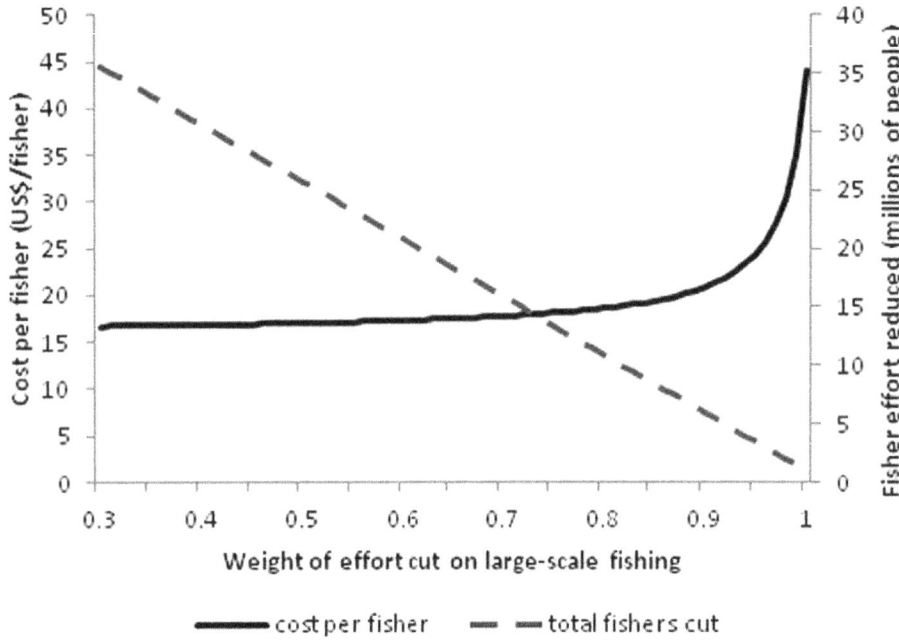

Figure 6. Trade-offs between reductions in cost of fishing effort and total fishing effort (in terms of number of fishers) reduced as the weight of effort cuts on large-scale fishing varies.

Based on Costello et al. [26], who estimated the recovery time for 18 simulated fish species to be 11 years on average, with a range of 4–26 years depending on the species, we assume a rebuilding period of 10 years $(t = 0–9)$ in this study. During this period, we assume that the only gains to occur are those from a reduction in the current net resource rent loss from negative US$13 billion per year to zero. Following modelling work reported in UNEP's Green Economy Report [32], we also assume that global fisheries landings decline linearly from ~80 to 50 million tonnes per year from $t = 0–5$ as fishing effort declines, but then rise linearly to the rebuilt level (~90 million tonnes) by $t = 9$. Once global fisheries have been rebuilt, this potential gain in resource rent would recur annually into perpetuity; here we consider only the flow for the subsequent 40 years after rebuilding $(t = 10–49)$.

We estimate R, resource rent adjusted for subsidies, as follows:

$$R = LV - (C + S) \qquad (5)$$

where LV represents the landed value of officially reported marine landings. The total variable cost of fishing is represented by C and subsidies are represented by S.

The computed resource rents for the six major Food and Agriculture Organization of the United Nations (FAO) regions (Africa; Asia; Europe; North America; Oceania; South and Central America plus the Caribbean) are summarized in Table 2.

We compute the gains from rebuilding (P_{gains}) as the value of the rebuilt resource rent $(R_{rebuilt})$ minus the value of current resource rent $(R_{current})$:

$$P_{gains_t} = R_{rebuilt_t} - R_{current_t} \qquad (6)$$

where t represents time. We assume that globally, rebuilt fisheries will be successful in avoiding subsequent unsustainable increases in effort.

We calculate the present value of net gains from rebuilding global fisheries as follows:

$$PV = \sum_{t=0}^{49} \frac{P_{gains}}{(1+r)^t} \qquad (7)$$

where PV is the present value of the net gain in resource rent, r is the prevailing rate of discount and t represents time from present $(t \in [0,49])$. In our analysis, we assume a fixed discount rate of $r = 0.03$ (i.e., 3%) and compute the present value of net gains in resource rent for 50 years after rebuilding. We use this discount rate because many environmental economists have argued for and applied lower-than-market rates due to the central role of environmental resources in ensuring sustainable economies through time [33–35] Changes in the value of fisheries landings, costs, subsidies and resource rent through the transition time period are summarized in Figure 3.

4. Estimating the Cost of Rebuilding Global Fisheries

In addition to differences between current resource rent and that which is captured during the period of rebuilding, we estimate the costs necessary to reduce fishing capacity to levels required to allow fish stocks to rebuild. These costs are estimated based on the cost of effort reductions described earlier in the methods. We estimate wages, profits, resource rent and increase in resource rent from rebuilding for the six major FAO regions (Africa; Asia; Europe; North America; Oceania; South and Central America plus the Caribbean) in Table 3.

Since the real cost of rebuilding fisheries is foregone resource rent that may occur as fishing effort is reduced initially, we estimate the cost of rebuilding global fisheries through the transition to rebuilt fisheries as the difference between current fisheries resource rent and that which is realized through the period of transition. We hold the assumption that all harmful capacity-enhancing and ambiguous subsidies (Table 4) must be cut

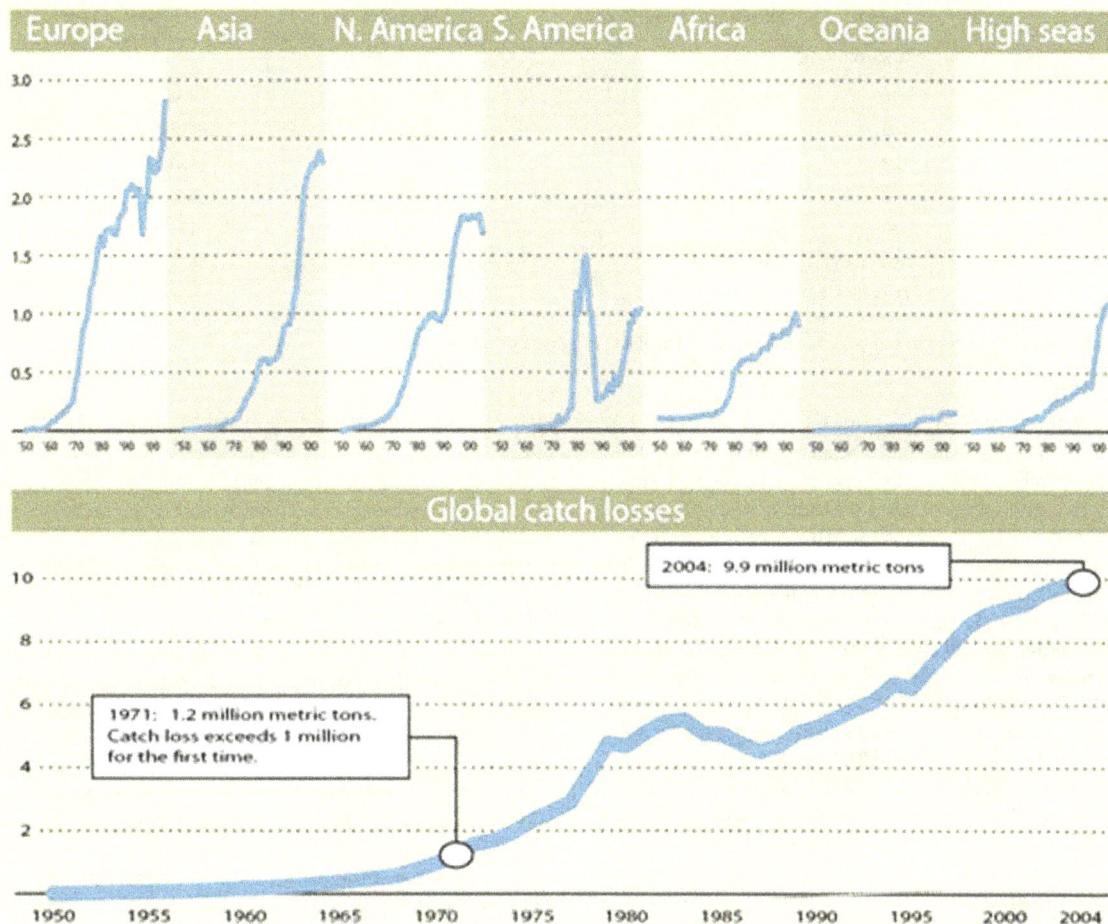

Figure 7. Lost catch potential due to overfishing for the six FAO regions of the world (top) and worldwide (bottom). Figure drawn using results reported in Srinivasan et al. [19].

Table 2. Wages, normal profit and resource rent for current fisheries by FAO region.

Region	Wages	Earnings/Normal Profit*	Resource rent
	(US$ billions)		
Africa	1.51	0.60	−2.63
Asia	14.93	3.13	−4.73
Europe	6.77	0.77	−4.32
North America	4.72	1.12	1.13
Oceania	1.38	0.16	−0.58
South America	1.65	0.30	−1.80
World Total	30.96	6.08	−12.93

*Profit is defined here as the return to capital or normal profit, i.e., payments to owners of capital.

immediately or re-directed to make them beneficial subsidies, e.g., by investing in managing the rebuilding process.

5. Calculating the Unit Cost of Reducing Fishing Effort

Policy makers generally prefer to minimize the employment impact of rebuilding fisheries. It would therefore be attractive to target effort reductions on large-scale vessels only, as they employ less people per unit of fish landed [36]. While the goal of matching global fishing capacity with the productive capacity of the resource by cutting only large-scale vessels seems theoretically possible, it would be ineffective in areas that are overfished but dominated by small-scale vessels. Available gear-related data [36,37] reveal that the split between large- and small-scale vessels in the developed world is about 50:50, while it is 25:75 in developing countries. Our analysis of fishing effort cuts show that permanently removing around 213,000 large-scale and 1.2 million small-scale vessels (60% and 30% reductions, respectively), would halve the world's fishing capacity. This weighting between large- and small-scale fishing capacity is supported by evidence that large-scale vessels currently land roughly two-thirds of the world's annual reported marine landings [28].

Table 3. Wages, normal profit, resource rent and increase in rent from rebuilt fisheries.

Region	Wages	Earnings/Normal Profit*	Resource rent	Increase in rent
	(US$ billions)			
Africa	0.76	0.30	0.85	3.48
Asia	7.46	1.56	30.82	35.54
Europe	3.38	0.39	8.80	13.12
North America	2.36	0.56	7.98	6.86
Oceania	0.69	0.08	2.73	3.31
South America	0.83	0.15	2.51	4.31
World Total	15.48	3.04	53.69	66.61

*Profit is defined here as the return to capital or normal profit, i.e., payments to owners of capital.

Cost data [18] reveal that crew from large- and small-scale fisheries earn, on average, wages of US$20,000 and US$10,000 - per year, respectively. Furthermore, vessels in large- and small-scale fisheries pay, on average, US$11,000 and US$2,500 per year for capital. Based on vessel and crew data from the European Union [38], we estimate that the average cost of a vessel buyback is roughly equal to the average interest payments on a vessel for five years, and the average cost of crew retraining is estimated at 1.5 times the average annual crew wages. Therefore, the average cost of decommissioning large- and small-scale fishing vessels would be US$55,000 and US$12,500, respectively. Likewise, payout/retraining costs for large- and small-scale fishers to leave fishing permanently would be US$30,000 and US$15,000 per person, respectively. Clearly, decommissioning costs for the extremely large industrial vessels with global roaming abilities would be higher than the above vessel averages.

6. Data and Databases

We utilize four interrelated global databases of fisheries statistics, namely, databases of fisheries landings, ex-vessel fish prices, subsidies, and fishing costs. Each database represents the work of an international team of fisheries scientists and economists, and collectively represents the world's most comprehensive collection of truly global fisheries (economic) data.

Our catch data, the main source of which is the FAO global capture production database supplemented by several more detailed regional catch data sources, allocates the reported fish landings to a

Table 4. Annual global fisheries subsidies by category [17].

Category	Subsidies (US$ billions)
Beneficial[a]	8
Harmful[b]	16
Ambiguous[c]	3
Total	27

[a]Lead to 'investment' in the natural capital of fishery resources. They enhance the growth of fish stocks through conservation programs, and control and surveillance measures.
[b]Lead to 'disinvestments' in the natural capital of the fishery resources, including all forms of capital inputs and infrastructure investments from public sources that reduce cost or enhance revenue.
[c]Have the potential to lead to either 'investment' or 'disinvestment' in the fishery resources, and lead to resource enhancement or to resource overexploitation.

global system of 30-minute latitude by 30-minute longitude cells (just under 180,000 marine cells globally) using the intersection of statistical reporting areas, biological taxon distributions of reported taxa, general habitat preferences, global fishing access agreements and fishing patterns of reporting countries. Details of the methods and procedures of this spatial allocation are described in Watson et al. [15], and country-specific data by FAO region are presented in Tables S1, S2, S3, S4, S5 and S6.

The global ex-vessel fish price database used here, described by Sumaila et al. [16], covers annual average ex-vessel prices for all marine fish taxa by country reported as caught from 1950 to 2006. Through their extensive search of publicly available, but widely scattered and incompatible, national and regional statistical reports and grey literature, Sumaila et al. [16] accumulated over 31,000 records of observed ex-vessel prices in 35 countries, representing about 20 percent of the global landings over the 60 year period. In order to 'fill the gaps' in the database, a series of rules were developed whereby all catches with no reported prices were inferred to have an estimated price computed from the reported prices from related taxa, similar markets or years. Since the database was first presented, new reported prices have been included from various additional sources, and rules as to how prices relate across taxa, markets or years have been modified to improve the quality of the estimated prices. The time series of landed values of the world's marine fisheries, computed through the combination of the spatially allocated catch data with the ex-vessel price database (country-specific landed values by FAO region) are presented in Tables S1, S2, S3, S4, S5 and S6, and have been used in various analyses, such as the estimation of global subsidies [17] and costs of marine protected areas [39].

The fisheries subsidies database defines subsidies as financial transfers, directly or indirectly, from government to the fishing industry [17]. This database is the most comprehensive collection of publicly available data on fisheries subsidies at the global level, spanning the years 1990 to 2006. Each record in the database represents expenditure in one of twenty-six identified subsidy categories for a given country and year combination. Where qualitative information indicates the presence of a subsidy program, yet quantitative data are not available, the database records the expenditure data as 'missing' for later estimation.

Estimation of 'missing' subsidy data follows the method of Sumaila et al. [17] who utilize the strong relationship between fisheries subsidies and landed value (Figure 8) to estimate subsidy expenditure in cases where programs are documented without quantitative information. We use this procedure to estimate existing but unquantified fisheries subsidies for any of the twenty-

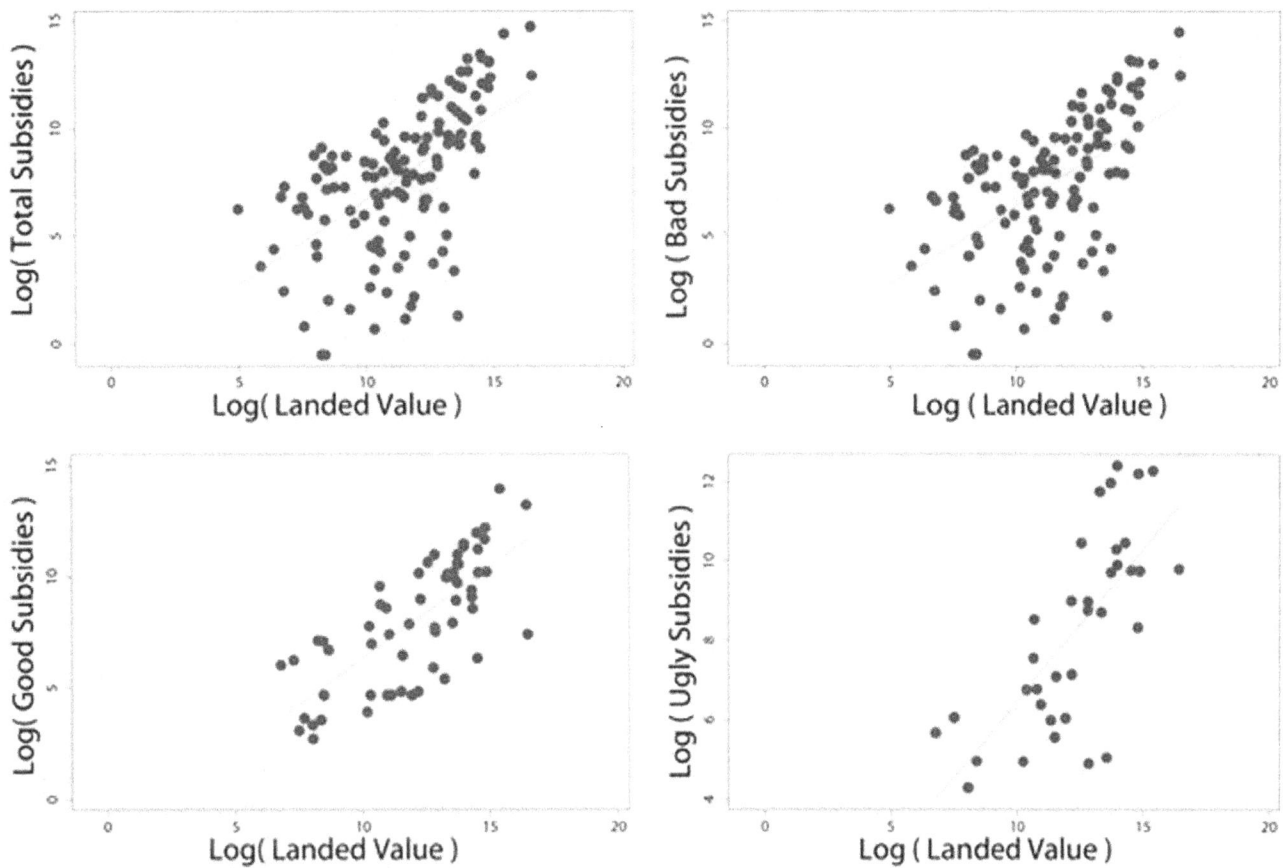

Figure 8. Correlation between reported subsidies [17] and landed-value [16].

six subsidy categories in any of the 144 maritime countries of the world, and summarize these globally by three general categories (Table 4): 'beneficial' (lead to 'investment' in the natural capital of fishery resources, thus enhancing growth of fish stocks through conservation programs, and control and surveillance measures), 'harmful' (lead to 'disinvestments' in the natural capital of the fishery resources, including all forms of capital inputs and infrastructure investments from public sources that reduce cost or enhance revenue) and 'ambiguous' (have the potential to lead to either 'investment' or 'disinvestment' in the fishery resources, and lead to resource enhancement or to resource overexploitation). Fuel- and non-fuel subsidies by country within FAO regions are summarized in Tables S1, S2, S3, S4, S5 and S6.

Lam et al. [18] developed a global database of fishing costs by country and gear types, capturing two types of fishing cost, variable (operating) and fixed costs in 144 maritime countries, representing approximately 98% of global landings in 2005. Results from this database are summarized in Table 5. Each record in the database represents a country and gear type combination. The gear types included in the database are based on the gear categorization system of the *Sea Around Us* project [40].

Fishing cost data were collected from secondary sources in major fishing countries in each of the six FAO regions. In order to include as many data of observed cost as possible, Lam et al. [18] accessed all available sources, irrespective of publication year, thus extending their efforts in collecting cost data from 1950 to the most

Table 5. Global cost (mean +/−95% CI) of fishing (Year 2005 US$ per t of catch), separated into variable and fixed cost component [18].

Cost category	Lower 95% CI	Mean	Upper 95% CI
	(US$ per t)		
Variable	639	928	1,413
Fixed	123	192	164
Total	**762**	**1,120**	**1,477**

Table 6. Sensitivity of present value of rebuilding costs to parameter assumptions.

Effort reduction to achieve MSY (%)	Adjustment costs (US$ billions)*		
	−20%	Mean	+20%
40	130	162	195
50	162	203	243
60	195	243	292

*Adjustment costs include the cost of vessel buybacks and payouts for fishers to ease the transition to alternate employment.

Table 7. Sensitivity of present value potential gain in resource rent from rebuilding to parameter assumptions.

Effort reductions to achieve MSY (%)	Potential gains (US$ billions)*		
	Lower 95% CI	Mean	Upper 95% CI
40	976	1,128	1,428
50	821	972	1,272
60	665	816	1,117

*Gains are denoted as the present value of increased resource rent from rebuilt fisheries.

recent year for which data were available. The data were then converted to 2005 real values using the consumer price index (CPI) for each country obtained from the World Bank. To make the comparison of fishing cost among different regions and countries possible, they converted all fishing costs from local currencies to US dollars by using currency exchange rates provided by the World Bank, and standardized the original cost to annual cost in US$ per tonne of catch.

A process of progressive refinement [16,18] was then used to estimate the cost of all gear types in each fishing country from the observed, collected cost. Therefore, Lam et al. [18] ensured that all gear types in each maritime country of the world were assigned a cost, either the observed value where available, or an appropriate average estimated cost. Variable fishing costs by country within FAO region are summarized in Tables S1, S2, S3, S4, S5 and S6.

7. Sensitivity Analysis

We test the sensitivity of our results both in terms of the benefits and costs of rebuilding. An estimate of the potential contribution of rebuilt fisheries from Srinivasan et al. [19] is 89 million t per year (50% prediction interval: 83 million to 99 million t per year) and US$100 billion per year in landed value (50% prediction interval: US$93 billion to US$116 billion per year). This represents an increase over the current value of fisheries landings of US$13 billion per year (range: US$5 billion to US$29 billion per year). We also test the sensitivity of our estimates to changes in the cost of fishing by allowing the needed effort reductions to attain maximum sustainable yield to be within a range of 40–60% of current fishing effort, with a mean value of 50% of current effort levels.

Table 6 presents the results of the sensitivity analysis with respect to the costs of adjusting factors of fisheries production to

maximum sustainable yield. The top-left and bottom-right cells of this table display the best- and worst-case scenarios for our estimates of adjustments costs. Table 7 presents the results of our sensitivity analysis on the estimates for rebuilding benefits in a three-by-three matrix, thus showing potential best- and worst-case scenarios of costs of rebuilding fisheries in the bottom-right and top-left cells of the table, respectively. It can readily be seen from these two tables that, even under drastic changes in our estimates, the combination of the worst-case scenarios result in large net present value of resource rent from rebuilt global fisheries.

Supporting Information

Table S1 Key fisheries data (annual averages for 2000s) for Africa.

Table S2 Key fisheries data (annual averages for 2000s) for Asia.

Table S3 Key fisheries data (annual averages for 2000s) for Europe.

Table S4 Key fisheries data (annual averages for 2000s) for North America.

Table S5 Key fisheries data (annual averages for 2000s) for Oceania.

Table S6 Key fisheries data (annual averages for 2000s) for South, Central America and the Caribbean.

Acknowledgments

This is a contribution from the Global Ocean Economics Project and the *Sea Around Us* Project, both of which are collaborations between the University of British Columbia and the Pew Environment Group. This contribution benefited from work carried out by contributors to the United Nations Environmental Programme Green Economy Chapter on fisheries. We also thank Jennifer Jacquet for reviewing an earlier version of the paper.

Author Contributions

Conceived and designed the experiments: URS DP. Performed the experiments: URS AD LH VL WS. Analyzed the data: URS AD ES. Contributed reagents/materials/analysis tools: WC RW KG LH TS. Wrote the paper: URS AD DP DZ.

References

1. Pontecorvo G, Wilkinson M, Anderson R, Holdowsky M (1980) Contribution of the ocean sector to the United States economy. Science 208: 1000–1006.
2. Dyck AJ, Sumaila UR (2010) Economic impact of ocean fish populations in the global fishery. Journal of Bioeconomics 12: 227–243.
3. FAO (2010) The State of World Fisheries and Aquaculture (SOFIA) 2010. Rome, Italy: Food and Agriculture Organization. 197 p.
4. Zeller D, Booth S, Pauly D (2006) Fisheries contributions to GDP: Underestimating small-scale fisheries in the Pacific. Marine Resource Economics 21: 355–374.
5. Pauly D, Christensen V, Guénette S, Pitcher TJ, Sumaila UR, et al. (2002) Towards sustainability in world fisheries. Nature 418: 689–695.
6. Worm B, Hilborn R, Baum JK, Branch T, Collie JS, et al. (2009) Rebuilding global fisheries. Science 325: 578–585.
7. Cheung WWL, Close C, Lam V, Watson R, Pauly D (2008) Application of macroecological theory to predict effects of climate change on global fisheries potential. Marine Ecology Progress Series 365: 187–197.

8. Sumaila UR, Cheung WWL, Lam V, Pauly D, Herrick S (2011) Climate change impacts on the biophysics and economics of world fisheries. Nature Climate Change 1: 449–456, doi:410.1038/nclimate1301.
9. Merino G, Barange M, Blanchard JL, Harle J, Holmes R, et al. (2012) Can marine fisheries and aquaculture meet fish demand from a growing human population in a changing climate? Global Environmental Change in press. Available: http://dx.doi.org/10.1016/j.gloenvcha.2012.1003.1003.
10. Mason F (2002) The Newfoundland cod stock collapse: a review and analysis of social factors. Electronic Green Journal 1.
11. SCFO (2005) Northern cod: a failure of Canadian fisheries management. Ottawa, Canada: Report of the Standing Committee on Fisheries and Oceans.
12. Clark CW (1990) Mathematical Bioeconomics: The Optimal Management of Renewable Resources. New York: Wiley.
13. Béné C, Hersoug B, Allison EH (2010) Not by rent alone: analysing the pro-poor functions of small-scale fisheries in developing countries. Development Policy Review 28: 325–358.
14. Heal G (2000) Valuing ecosystem services. Ecosystems 3: 24–30.

15. Watson R, Kitchingman A, Gelchu A, Pauly D (2004) Mapping global fisheries: sharpening our focus. Fish and Fisheries 5: 168–177.
16. Sumaila UR, Marsden AD, Watson R, Pauly D (2007) A global ex-vessel fish price database: construction and applications. Journal of Bioeconomics 9: 39–51.
17. Sumaila UR, Khan A, Dyck A, Watson R, Munro R, et al. (2010) A bottom-up re-estimation of global fisheries subsidies. Journal of Bioeconomics 12: 201–225.
18. Lam V, Sumaila UR, Dyck A, Pauly D, Watson R (2011) Construction and potential applications of a global cost of fishing database. ICES Journal of Marine Science 68: 1996–2004.
19. Srinivasan UT, Cheung WWL, Watson R, Sumaila UR (2010) Food security implications of global marine catch losses due to overfishing. Journal of Bioeconomics 12: 183–200.
20. World Bank, FAO (2009) Sunken Billions: The Economic Justification for Fisheries Reform: Case Study Summaries. Available: http://go.worldbank.org/MGUTHSY7U0. Accessed 2009 Apr 15.
21. Porter G (1998) Estimating overcapacity in the global fishing fleets. Washington, D.C.: WWF.
22. Daw TM, Cinner JE, McClanahan TR, Brown K, Stead SM, et al. (2012) To fish or not to fish: Factors at multiple scales affecting artisanal fishers' readiness to exit a declining fishery. PLoS ONE 7: e31460. doi:31410.31371/journal.pone.0031460.
23. Teh L, Cheung WWL, Cornish A, Chu C, Sumaila UR (2008) A survey of alternative livelihood options for Hong Kong's fishers. International Journal of Social Economics 35: 380–395.
24. Plimsoll (2011) The Commercial Fishing (Global) Industry: A comprehensive financial analysis of the top 1000 companies. England: Plimsoll Publishing Limited. 1256 p.
25. Delgado CL, Courbois C (1997) Changing fish trade and demand patterns in developing countries and their significance for policy research. MTID Discussion Paper 18, International Food Policy Research Institute (IFPRI).
26. Costello C, Kinlan BP, Lester SE, Gaines SD (2012) The economic value of rebuilding fisheries, OECD Food, Agriculture and Fisheries Working Papers, No. 55. Paris: OECD Publishing. 68 p.

27. Gordon HS (1954) The economic theory of a common property resource: the fishery. Journal of Political Economics 62: 124–142.
28. Pauly D (2006) Major trends in small-scale marine fisheries, with emphasis on developing countries, and some implications for the social sciences. Maritime Studies (MAST) 4: 7–22.
29. Froese R, Zeller D, Kleisner K, Pauly D (2012) What catch data can tell us about the status of global fisheries. Marine Biology (Berlin) 159(6): 1283–1292.
30. Branch T, Jensen O, Ricard D, Ye Y, Hilborn R (2011) Contrasting global trends in marine fishery status obtained from catches and from stock assessments. Conservation Biology 25: 777–786.
31. FAO (2009) The State of World Fisheries and Aquaculture (SOFIA) 2008. Rome, Italy: Food and Agriculture Organization. 176 p.
32. UNEP (2011) Towards a Green Economy: Pathways to Sustainable Development and Poverty Eradication. Nairobi: UNEP. 631 p. Available: www.unep.org/greeneconomy.
33. Weitzman ML (2001) Gamma discounting American Economic Review 91: 260–271.
34. Sumaila UR (2004) Intergenerational cost benefit analysis and marine ecosystem restoration. Fish and Fisheries 5: 329–343.
35. Stern R (2006) Stern Review on the Economics of Climate Change. London: Her Majesty's Treasury. ix +579 p.
36. Gabriel O, Lange K, Dahm E, Wendt T, editors (2005) Von Brandt's Fish Catching Methods of the World, 4th ed. Surrey, England: Fishing News Books. 521 p.
37. Anticamara JA, Watson R, Gelchu A, Pauly D (2011) Global fishing effort (1950–2010): trends, gaps and implications. Fisheries Research 107: 131–136.
38. European Commission (EC) (2006) Economic performance of selected European fishing fleets. 2005 Annual Report.
39. Cullis-Suzuki S, Pauly D (2008) Marine Protected Area costs as 'beneficial' fisheries subsidies: a global evaluation. Coastal Management 38: 113–121.
40. Watson R, Revenga C, Kura Y (2006) Fishing gear associated with global marine catches I. Database development. Fisheries Research 79: 97–102.

6

Structure, Composition and Metagenomic Profile of Soil Microbiomes Associated to Agricultural Land Use and Tillage Systems in Argentine Pampas

Belén Carbonetto[1]*, Nicolás Rascovan[1], Roberto Álvarez[2], Alejandro Mentaberry[3], Martin P. Vázquez[1]*

1 Instituto de Agrobiotecnología de Rosario (INDEAR), Predio CCT Rosario, Santa Fe, Argentina, 2 Facultad de Agronomía, Universidad de Buenos Aires, Buenos Aires, Argentina, 3 Departamento de Fisiología y Biología Molecular y Celular, Facultad de Ciencias Exactas y Naturales, Universidad de Buenos Aires, Buenos Aires, Argentina

Abstract

Agriculture is facing a major challenge nowadays: to increase crop production for food and energy while preserving ecosystem functioning and soil quality. Argentine Pampas is one of the main world producers of crops and one of the main adopters of conservation agriculture. Changes in soil chemical and physical properties of Pampas soils due to different tillage systems have been deeply studied. Still, not much evidence has been reported on the effects of agricultural practices on Pampas soil microbiomes. The aim of our study was to investigate the effects of agricultural land use on community structure, composition and metabolic profiles on soil microbiomes of Argentine Pampas. We also compared the effects associated to conventional practices with the effects of no-tillage systems. Our results confirmed the impact on microbiome structure and composition due to agricultural practices. The phyla *Verrucomicrobia, Plactomycetes, Actinobacteria*, and *Chloroflexi* were more abundant in non cultivated soils while *Gemmatimonadetes, Nitrospirae* and WS3 were more abundant in cultivated soils. Effects on metabolic metagenomic profiles were also observed. The relative abundance of genes assigned to transcription, protein modification, nucleotide transport and metabolism, wall and membrane biogenesis and intracellular trafficking and secretion were higher in cultivated fertilized soils than in non cultivated soils. We also observed significant differences in microbiome structure and taxonomic composition between soils under conventional and no-tillage systems. Overall, our results suggest that agronomical land use and the type of tillage system have induced microbiomes to shift their life-history strategies. Microbiomes of cultivated fertilized soils (i.e. higher nutrient amendment) presented tendencies to copiotrophy while microbiomes of non cultivated homogenous soils appeared to have a more oligotrophic life-style. Additionally, we propose that conventional tillage systems may promote copiotrophy more than no-tillage systems by decreasing soil organic matter stability and therefore increasing nutrient availability.

Editor: Kathleen Treseder, UC Irvine, United States of America

Funding: Funding for this work was provided by Agencia Nacional de Promoción Científica y Tecnológica, Argentina-PAE 37164. The funders had no role in study design, data collection and analysis, decision to publish, or preparation of the manuscript.

Competing Interests: The authors have declared that no competing interests exist.

* E-mail: martin.vazquez@indear.com (MPV); belen.carbonetto@indear.com (BC)

Introduction

Agriculture is facing major challenges nowadays. Production will have to double in the next 50 years in order to face growing food demand and bioenergy needs [1,2]. This must be done without increasing environmental threats such as climate change, biodiversity loss and degradation of land and freshwater. Achieving such a goal represents one of the greatest scientific challenges ever. This is in part because of the trade-offs among economic and environmental goals and because of the insufficient knowledge about the biological, biogeochemical and ecological processes that are relevant for sustainable ecosystem functioning [3,4]. Much has been done during the last decade to gain sufficient information on agricultural ecosystem biology, still, more work needs to be done to gain deeper comprehension and to be able to reduce the negative environmental impacts of agriculture [2,5,6]. The main focus should be oriented to soil degradation. Soil fertility, as the capacity to sustain abundant crop production, needs to be preserved. Nowadays soil fertility is maintained by dependence on external inputs; with increasing water contamina-

tion [7]. In this context, the key to understand the behavior of life-supporting elements in soil, such as carbon, nitrogen, and phosphorus lies in the fluxes between their various forms in the environment, which are modulated by biology [8]. Comprehension of soil microorganism dynamics is then essential to understand soil processes that affect fertility. Ecological approaches are being taken into account in soil microbial studies trying to address these questions. These approaches involve diversity and functional analyses of soil communities [9,10]. Scholes & Scholes point out that this complex view is necessary for the comprehension of soil systems and that soil restoration of biological processes is the key to achieving lasting food and environmental security [8].

Argentine Pampas is an important player in this scenario. With a plain area of 50 million ha., nearly 50% of the whole Pampas area is devoted to crop production [11]. Cultivation began in the 19th Century in the central humid portion of the region, in soils of high fertility, and spread in last decades to the south and the semiarid west [12]. Soil degradation (i.e. intense erosion and net loss of nutrients and organic carbon) caused by the use of conventional tillage systems were reported in the Pampas [13–17].

Nowadys between 60 and 80% of production is conducted under conservational no-till practices [18]. Extensive research was done to evaluate the effects of reduced tillage and no-tillage systems on soil physical properties, water content, fertility and crop yields[19]. The main outcome of these analyses points that the adoption of limited tillage systems led to soil improvement, by augmenting organic matter content and soil structure. Still, external fertilization is needed in order to restore nutrient levels and fertility regardless the tillage system employed.

Even though the effects of different tillage practices on soil physical and chemical characteristics have been deeply studied, changes in microbial biodiversity and functioning have been poorly reported in Argentinean Pampas. Most works have studied tillage effects on microbial biomass or specific microbial activities (i.e. utilization of specific substrates, extracellular enzyme production, mineralization, etc.) rather than on full microbiome [13,20,21]. Other studies have focused on the behavior of specific bacterial taxa [22,23]. Reports with an ecological approach (i.e. microbial community analysis) have usually focused on individual effects of land use such as the application of herbicides [24,25]. In these cases, biodiversity variability has been assessed using classical fingerprinting techniques (such as RFLP and DGGE) that lack information about microbial taxonomic identity and only capture the most dominant species in the environment [26,27]. In the last few years, 16S amplicon pyrosequencing has been largely implemented to determine microbial diversity and structure of many different ecosystems worldwide [28,29]. This strategy allows a more exhaustive characterization of community patterns and composition. Moreover, some works have incorporated the use of shotgun metagenomics to study the metabolic potential of soil microbiomes [10,30]. The shotgun approach generates a massive amount of data using random high-trhoughput sequencing of soil isolated DNA. This allows the identification of functional capabilities by gene annotation and the comparison of metabolic profiles between samples. To our knowledge, Figuerola et al. [31] were the only authors studying microbial communities in agronomical soils of Argentine Pampas using high throughput sequencing approaches. They observed differences in microbial community composition of soils under no-tillage systems using 16S pyrosequencing. As a novelty, our efforts focused on assessing the impact of long-term agriculture on Pampas soil microbiomes using both shotgun metagenomics and deep 16S amplicon sequencing approaches. We evaluated the effect of more than a hundred years of agronomical land use on both community features and metabolic profiles of soil microbiomes in comparison with nearby control soils with no agricultural records. We also addressed the differences between the effects of two tillage systems: conventional tillage vs. no-tillage on microbial communities.

Several previous studies of soil microbiomes from different parts of the world showed the effects of agronomical land use on soil microbial communities [10,32–35]. Some of these studies showed differences in trophic strategies between microbial communities related to tillage; and most of them were done in experimental plots. As a novelty, we tested the impact of long term agriculture in soils sampled in production fields in the Argentine Pampas, allowing a deeper insight to the effects of intense land use on soil ecosystems functioning. We confirmed the hypothesis that agronomical practices affected Pampas soil microbiomes by promoting a shift of life-history and trophic strategies. We also showed differences in the effects of contrasting tillage systems (i.e. conventional vs. no- tillage) on community taxonomic and metabolic composition on a long term experiment.

Materials and Methods

Sites description and sampling

Soil samples were taken in production and experimental fields between June and August 2010. To address the effects of agricultural land use on soil microbial communities, three different production farms were sampled in the Rolling Pampas area: "La Estrella", "La Negrita" and "Criadero Klein" (See Rascovan et al and Table S1 for details). Rolling Pampas soils are classified as Typic Argiudolls [36] and mean annual rainfall and temperature were1002 mm and 16.8°C respectively. Two treatments were defined: *cultivated* for production plots, and *no cultivated* for farmhouses parks. Production plots were under cultivation for at least one century under conventional tillage systems, with a mixed rotation of pastures and annual grain crops. During the last 15 years before sampling plots were subjected to continuous crop cultivation under no-tillage systems (i.e. minimal soil disturbance, permanent soil cover, rotations and fertilization). The last crop rotation before sampling was wheat-soybean. Nitrogen and phosphorus fertilizers were applied. Samples were collected one month after soybean harvest. Soil samples were also collected nearby the farmers' houses where no agricultural land use (no tillage nor cultivation) was recorded for the last 30 years except from grass mowing. Parks around farmers' houses are usually considered as undisturbed environments in Argentine Pampas [37]. Soils under no land use were covered with grass and other herbaceous (non-woody) plants common in the region such as *Cirsium sp, Trifolium sp, Micropsis sp, Festuca sp, Dichondra sp, Cyperus sp* and *Taraxacum officinale*. For numerical analyses purposes the three farms are treated as experimental replicates. Four soil samples were taken with an auger from the upper 20 cm soil layer in each farm and treatment. A total of 24 samples were collected in Rolling Pampa soils.

In order to compare effects of contrasting tillage systems, samples were also collected in a 34-year-old experiment located in Balcarce in the Southern Pampas (See Rascovan et al and Table S1 for more details). Samples were taken in experimental plots because no production fields are using conventional tillage for crop production nowadays in the Pampas. Soils in Balcarce are a complex of Typic Argiudolls and Petrocalcic Paleudols and mean annual rainfall and temperatures were 875 mm and 13.8°C respectively. The experiment was carried out in three (175 m²) experimental plots (n = 3). Treatments were defined as: *no tillage* (NT) and *conventional tillage* (CT). NT plots had minimal soil disturbance and permanent soil cover combined with rotations; which have included pastures and grain crops (soybean, corn, wheat) during the last 16 years. CT plots were managed with moldboard plough. Nitrogen fertilization was performed in NT and CT plots (60 kg N ha-1). Last rotation before sampling was corn-soybean. Two sub-samples were collected from all treatment and replicate plots a month after soybean harvest. A total of 12 samples were collected.

Samples were immediately sent to the lab after collection. Samples used for DNA purification were air dried and sieved through 1 mm mesh to thoroughly homogenize, break aggregates and remove roots and plant detritus, then stored at −80°C. DNA purification and library preparation was previously described in Rascovan. et al.[38].

None of the sampling sites is located in protected areas. Permissions were obtained directly from each farm owner or manager: Alejandro Cattaneo at La Negrita and La Estrella, Roberto Klein at Criadero Klein and Guillermo Studdert at Balcarce experiment.

Soil chemical and physical measurements

Soil organic carbon was determined by wet digestion and organic matter was estimated [39]. Nitrate-nitrogen was analyzed by 2 M KCL extraction and the phenoldisulfonic acid method [40]. Extractable phosphorus was determined by the Bray method [41]. The pH was measured in a soil:water ratio 1:2.5. Salinity was estimated by the determination of electrical conductivity [42]. Texture analysis was performed by the hydrometer method [43] and nitrogen was determined by Kjeldahl method [40].

16S amplicon sequencing and shotgun metagenomic datasets

To analyze the effect of agronomical practices on soil microbiomes, sequence data from the previously reported Pampa dataset [38] was used. In order to evaluate microbial community structure and taxonomic composition, a total of 112,800 high-quality filtered 16S rRNA gene amplicon sequences, obtained from the 42 soil DNA samples (replicates and subsamples were included). DNA shotgun metagenomic data was used to analyze metabolic profiles. Shotgun metagenomic data completed a total of 10,445,170 sequences. In this case sequences were obtained from one subsample per sampling replicated plot.

In brief, libraries were prepared as follows: DNA was isolated from 10 g of soil of each of the 42 soil samples using the Power MaxSoil DNA Isolation Kit following the manufacturer's instructions (MO BIO Laboratories, Inc.). For amplicon libraries the V4 hyper variable region of the 16s rRNA gene was amplified. Duplicated reactions were performed using barcoded bacterial universal primers containing Roche- 454 sequencing A and B adaptors and a nucleotide multiple identifier (MID) to sort samples: 563F: 5'-CGTATCGCCTCCCTCGCGCCATCA-GACGAGTGCGTAYTGGGYDTAAAGNG -3' (where AC-GAGTGCGT is an example, different MIDs for each sample were used) and 802R (5'-CTATGCGCCTTGCCAGCCCGCT-CAGTACCRGGGTHTCTAATCC, 5'-CTATGCGCCTTGC-CAGCCCGCTCAGTACCAGAGTATCTAATTC, 5'-CTAT-GCGCCTTGCCAGCCCGCTCAGCTACDSRGGTMTCTA-ATC, 5'-CTATGCGCCTTGCCAGCCCGCTCAGTACNVG-GGTATCTAATCC) [44]. All amplicons were cleaned using Ampure DNA capture beads (Agencourt- Beckman Coulter, Inc.) and pooled in equimolar concentrations before sequencing on a Genome Sequencer FLX (454-Roche Applied Sciences) using Titanium Chemistry according to the manufacturer's instructions.

Shotgun metagenomic libraries were prepared by nebulization, followed by tagging with GS-FLX-Titanium Rapid Library MID Adapters Kit (454-Roche Applied Sciences) and sequenced with a Genome Sequencer FLX (454-Roche Applied Sciences) using Titanium Chemistry according to the manufacturer's instructions. Sequencing runs were performed in INDEAR sequencing facility.

All the sequences used in the present study are available in The Sequence Read Archive (SRA) under accession number SRA058523 and SRA056866. See Rascovan et al. [38] for more information.

Amplicon sequence processing, OTU classification and taxonomic assignment

Sequence data were quality controlled and denoised with the ampliconnoise.py script of QIIME [45].This script also eliminated chimeras. Sequences obtained from Rolling Pampa soil libraries and Balcarce soil libraries were processed separately. Sequences were clustered into Operational Taxonomic Units (OTUs) using the pick_otus.py script with the Uclust method [46] at 97%

sequence similarity. Rolling Pampas samples yielded 2,591 sequences on average (ranging from 1,455 to 3,991 sequences). Balcarce samples yielded 2,329 reads on average (ranging from 1,211 to 4,755 reads).OTU representative sequences were aligned using PyNast algorithm [47] with QIIME default parameters. Phylogenetic trees containing the aligned sequences were then produced using FastTree [48]. All downstream analyses were determined after each sample was randomly rarefied to 70% the number of reads of the smallest sample (i.e. 1,080 reads for Rolling Pampa libraries and 850 reads for Balcarce libraries). Phylogenetic distances between OTUs were calculated using unweighted and weighted Unifrac [49]. Taxonomic classification of sequences was done with Ribosomal Database Project (RDP) Classifier using Greengenes database using a 50% confidence threshold [50,51].

Microbial community analyses

Unifrac phylogenetic pairwise distances among samples were visualized with principal coordinates analysis (PCoA). Analysis of similarity statistics (ANOSIM) was calculated to test a-priori sampling groups. BIOENV analysis was performed to elucidate which soil properties correlated with community patterns. All calculations were carried out with R packages 'BiodiversityR' and 'Vegan' [52,53]. T- tests were performed with QIIME script otu_category_significance.py, and R scripts in order to elucidate differences in read abundances.

Shotgun metagenomic sequence processing and analysis

SSF files obtained from shotgun sequencing runs were uploaded to the MG-RAST webserver [54] for sequence filter and analyses. Reads more than two standard deviations away from the mean read length were discarded. For dereplication removal MG-RAST used a simple k-mer approach to rapidly identify all 20 character prefix identical sequences. This step is required in order to remove artificial duplicate reads. We obtained an average of 1.28×10^6 filtered reads per sample for Rolling Pampa shotgun libraries, and an average of 304,258 filtered reads per sample for Balcarce libraries.

Filtered high quality sequences were assigned to Cluster of orthologous groups (COG) by the MG-RAST sever pipeline using a similarity-based approach. COGs were assigned with a maximum E value of 10^{-20}, an average alignment of 80 amino acids length and 70% average identity.

Relative abundances were calculated by dividing the number of hits for each COG or COG-category by the total number of filtered reads in each sample. Euclidean distances based on relative abundances were calculated between sample pairs. PCoA visualizations and ANOSIM calculations were performed. All calculations were carried out with R packages 'BiodiversityR' and 'Vegan'. T- tests were performed with QIIME script otu_category-y_significance.py, and R scripts in order to elucidate differences in COG relative abundances between samples.

Results

Microbiome community changes related to agricultural land use

The PCoA visualization revealed clear differences between cultivated and noncultivated soils (ANOSIM R = 0.8406, p≤ 0.001; Figure 1A).Similar results were obtained when using Bray Curtis distance matrices (Figure S1).

The soil properties (Table S1) that best explained the phylogenetic variation observed in microbial communities were determined using Clarke and Ainsworth's BIOENV analysis. Our results showed that variables that best correlated with community

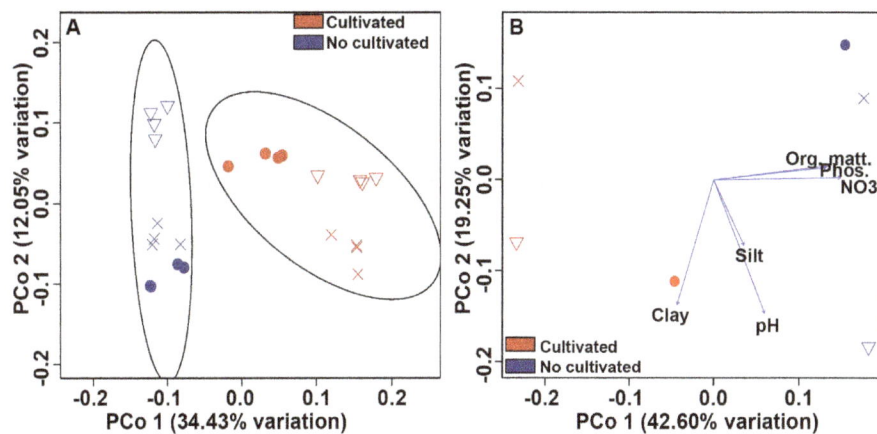

Figure 1. PCoA plots of Pampa production field soil microbiomes based on 97% similarity Weighted Unifrac distance matrices. A) PCoA of cultivated and non-cultivated soil microbiomes. All sub-samples are plotted. Standard error ellipses show 95% confidence areas. B) PCoA biplot of soil properties that best explained variation in community structure. Correlations were calculated using BIOENV on average data of each sampled site (Mantel r = 0.6107, p≤0.05). Circles represent samples from "La Estrella", crosses represent "Criadero Klein" samples and triangles represent "La Negrita" samples.

differences were organic matter, clay and silt content, nitrates, phosphorus and pH (Mantel r = 0.6107, p≤0.05).The PCoA biplot (Figure 1B) showed that organic matter, phosphorus and nitrate levels correlated with the first ordination axis that discriminates between cultivated and non cultivated soils. These three properties were higher in soils under no land use.

Regarding the taxonomic analyses, we observed that members of phyla *Verrucomicrobia*, *Planctomycetes*, *Actinobacteria* and *Chloroflexi* were more abundant in non cultivated soils (p≤0.05) (Figure 2). On the other hand, we found that sequences related to *Gemmatimonadetes*, candidate division WS3 and *Nitrospirae* were enriched in cultivated soils (p≤0.05) (Figure 2). No significant differences were found for *Proteobacteria* (for non of the Clases), *Acidobacteria* and *Bacteroidetes* phyla. The ten mentioned taxa represent on average 95% of total sequences of each sample.

Microbiome metagenomic profile changes related to agricultural land use

We found that cultivated and non cultivated soils also clustered apart when metagenomic functional categories were used for the analysis. The first two components of the PCoA explained over 60% of the variability between samples (Figure 3). Standard deviation ellipses overlapped in the ordination plot, indicating that some features are shared between metagenomes. Still, a positive correlation was observed between metabolic and weighted-Unifrac distance matrices (Mantel r: 0.5036 p≤0.05). The analyses of individual COG categories revealed that the relative abundances of COG categories associated with transcription, protein modification, nucleotide transport and metabolism, wall and membrane biogenesis and intracellular trafficking and secretion were higher in cultivated soils (Figure 4, p≤0.05). A deeper analysis inside COG categories revealed that COGs related to Coenzyme A and acetyl-Coa metabolism, energy storage and starvation or quiescence such as, pantothenate kinase, phospho-transacetylase, and trehalose utilization protein were more abundant in non cultivated than in cultivated soils (Figure S2, p≤0.05). On the other hand, COGs related to rapid regulation systems, tricarboxylic acid cycle and nitrogen assimilation such as urease, citrate synthase, glutamate synthase, fumarate hydratase, S-adenosyl- homocysteine hydrolase, S-adenosyl-methionine synthetase, cobalamin bio-

synthesis protein and ABC transporters, were more abundant in cultivated soils (Figure S2, p≤0.05).

Microbiome community structure and composition related to conventional tillage and no-tillage systems.

To compare the structure of microbiomes under different tillage systems, we collected samples from an experimental field located in Balcarce in the Southern Pampas. The 34-year-old experiment compared two tillage systems: no-tillage (NT) and conventional tillage (CT). Weighted Unifrac analysis showed differences in community structure associated to the tillage system employed

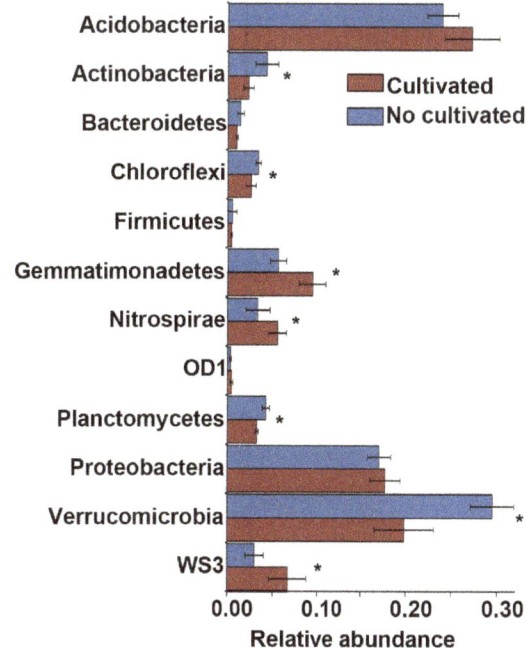

Figure 2. Relative abundances of taxonomic groups in Pampa production field soil microbiomes. Bars represent ± 1 standard error. (*) indicate significant differences (p≤0.05).

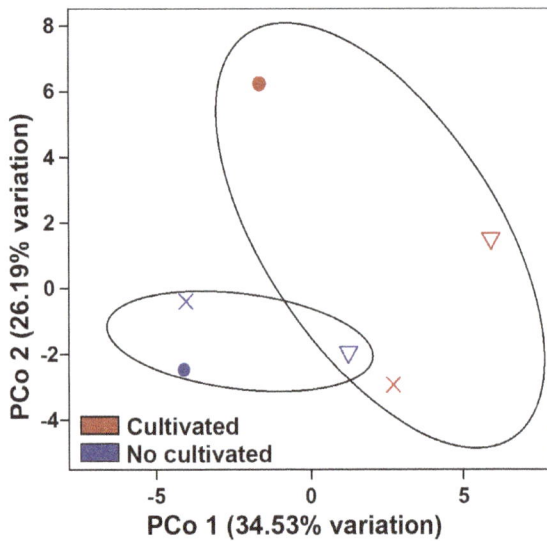

Figure 3. PCoA plot of metagenomic data based on Euclidean distance matrices of COG categories of Pampa production field soil microbiomes. Standard error ellipses show 95% confidence areas. Circles represent samples from "La Estrella", crosses represent "Criadero Klein" samples and triangles represent "La Negrita" samples.

(ANOSIM $R = 0.9009$, $p < 0.05$, Figure 5A). The first axis explained 31.56% of total variation and separated NT from CT. Additionally, we showed that nitrates were the only soil variable (Table S1) that significantly correlated with community structure (BIOENV analysis, Mantel $r = 0.7721$, $p \leq 0.01$, Figure 5B).

Moreover, microbiomes of CT and NT soils also differed in taxonomic composition. Members of *Acidobacteria*, *Gemmatimonadetes*, candidate division TM7 and class *Gammaproteobacteria* were more abundant in CT soils, while *Nitrospirae*, candidate divisionWS3 and *Deltaproteobacteria* were more represented in NT soils (Figure 6, $p \leq 0.05$).

Microbiome metabolic profiles related to conventional tillage and no- tillage systems

Variation in metagenomic profiles between CT and NT microbiomes was analyzed with PCoA based on Euclidean distance matrices of COG abundances. We could not find significant differences in overall profile metabolic structure between tillage systems (Figure S3). Additionally, we did not find significant correlation between Euclidean metabolic matrices and phylogenetic matrices. However, categories related to intracellular trafficking and secretion, amino acid transport and metabolism, and energy production and conversion were shown to be more abundant in soil under CT (Figure 7).

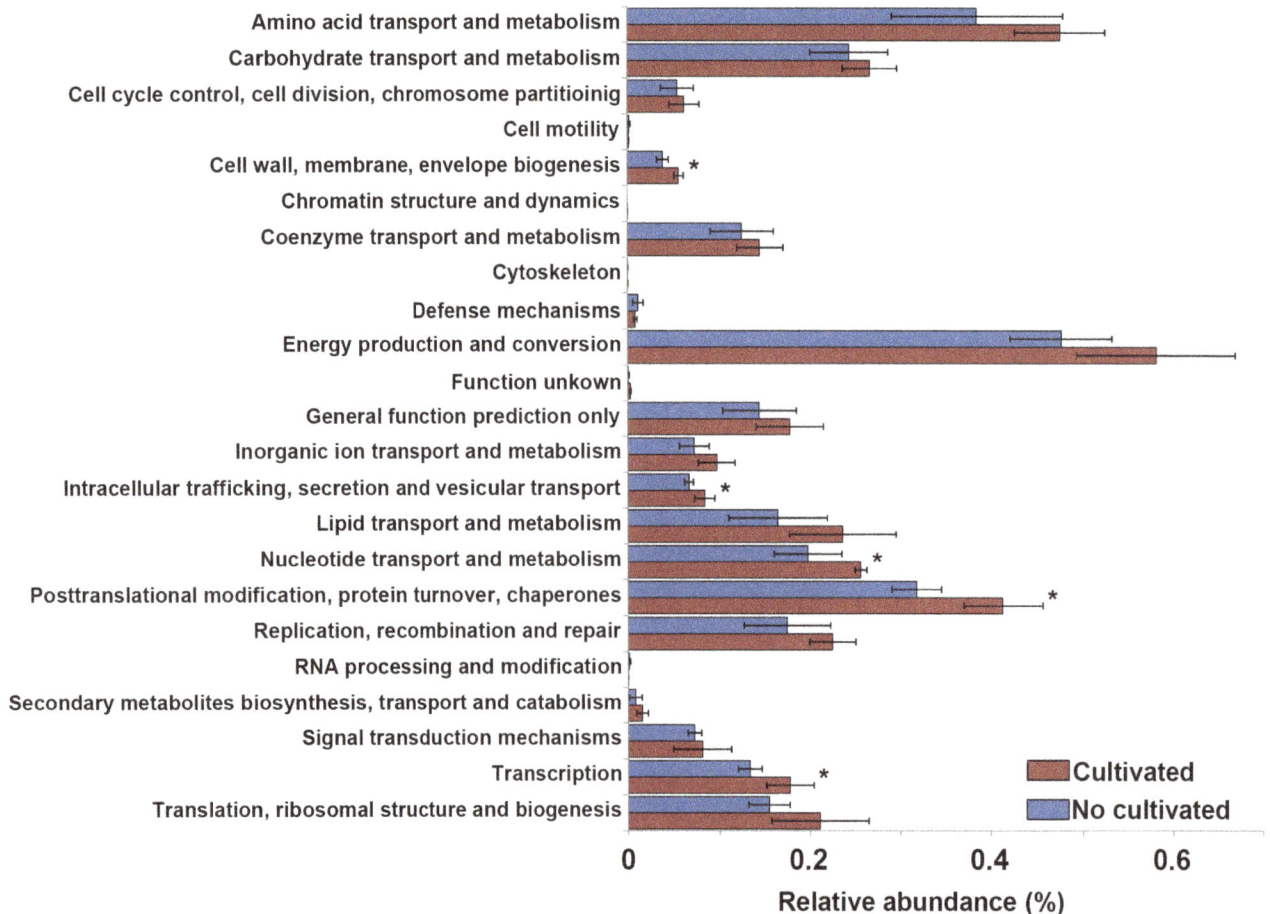

Figure 4. Relative abundances of COG categories in Pampas production field soil microbiomes. Bars represent ±1 standard error. (*) indicate significant differences ($p \leq 0.05$).

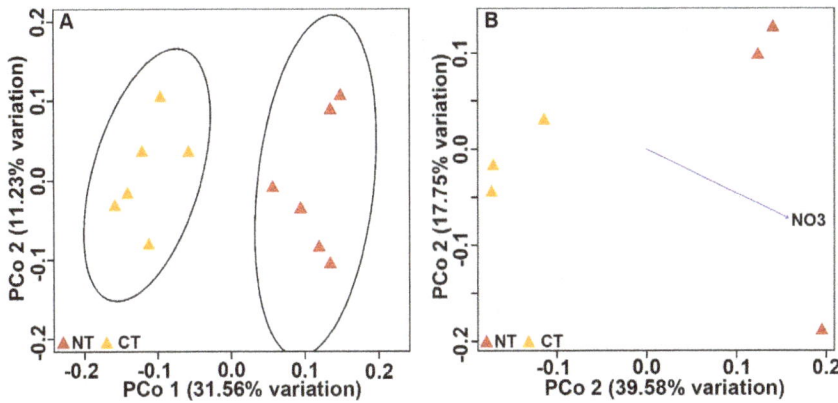

Figure 5. PCoA plots of Balcarce experimental field soil microbiomes based on 97% similarity Weighted Unifrac distance matrices. A) CA: conventional tillage; NT: no-tillage. All sub-samples are plotted. Standard error ellipses show 95% confidence areas. B) PCoA biplot, nitrate was the variable that best explained variation in community structure. Correlations were calculated using BIONEV on average data of each experimental plot (Mantel r = 0.7721, p≤0.01).

Discussion

Much work still needs to be done to get a comprehensive view of the soil microbiomes. Our work is one of the firsts done in the Argentine Pampas at this resolution level, with a combination of metagenomic and phylogentic approaches; and it is aimed to contribute to the comprehension of soil microbiomes function and dynamics. In that context, our results are in agreement with previous works that showed differences in soil microbial community structure and taxonomic composition due to the presence of agricultural land use [9,10,32,33].

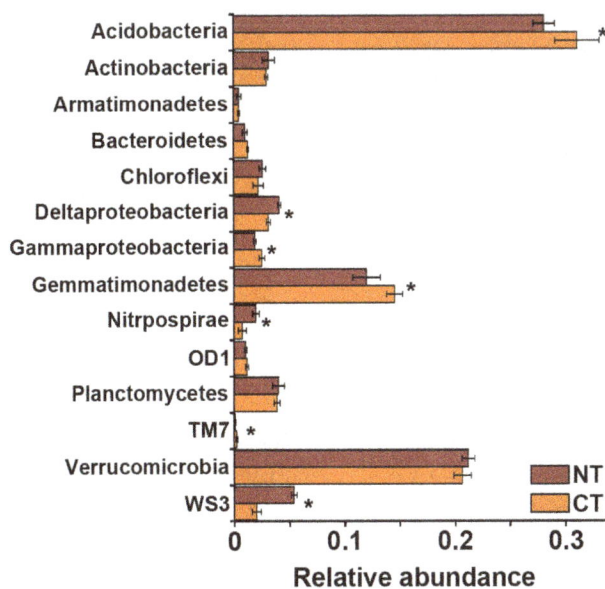

Figure 6. Comparison of relative abundances of taxonomic groups in NT and CT soils. CA: conventional tillage; NT: no-tillage. Bars represent ±1 standard error. (*) indicate significant differences (p≤ 0.05).

Effects of agricultural land use on community composition of soil microbiomes in Argentine Pampas

We observed effects on taxonomic composition at the phylum level. *Verrucomicrobia*, *Plactomycetes*, *Actinobacteria* and *Chloroflexi* were more abundant in soils that were never cultivated; while *Gemmatimonadetes*, *Nitrospirae* and WS3 were more abundant in crop cultivated soils. These results are in agreement with the copiotroph/oligotroph hypothesis [55], that propose that high number of oligotrophic prokaryotes may be found in bogs and soils with high amounts of recalcitrant organic matter [56]. On the other hand, copiotrophic organisms are able to use labile nutrient fractions and to grow at higher rates as a consequnece. In our study, non-cultivated soils are considered to be oligotrophic since they present high levels of organic matter highly rich in humic acids [12] and cultivated soils as copiotrophic environments due to fertilization and the seasonal presence of crop residues, which increases organic matter and nitrogen accessibility [57]. Under this assumption, bacteria in non-cultivated soils are expected to be K selected and to present low growth rates and very efficient nutrient uptake systems with higher substrate affinities. In contrast, bacteria in cultivated soils are expected to be r-selected and to have higher rates of activity per biomass unit, higher turnover rates and faster growth rates. Our results showed a trend toward these statements since a reduction in the abundance of taxa with oligotrophic characteristics, such as *Verrucomicrobia* [34], and *Planctomycetes* [58] were detected in cultivated fertilized soils. Moreover, this is in agreement with recent findings that confirm a correlation between *Verrucomicrobia* abundance patterns and conditions of limited nutrient availability in Prairie Soils in the United States [9]. On the other hand, the relative abundance of phylum *Gemmatimonadetes* was increased in fertilized cultivated soils as previously described for nitrogen-fertilized forest soils [59]. Consistently with our results, these authors observed that nitrogen fertilization was related to a higher abundance of *Gemmatimonadetes* and detected no presence of *Verrucomicrobia*. Little is known about *Gemmatimonadetes* ecology and metabolism since only one representative from this phylum has been isolated and characterized [60]. Even though, their presence in environments with a wide range of nutrient concentrations and redox states suggests versatile metabolisms [61].

We can also say that cultivated soils are more heterogeneous environments than non cultivated soils. Crop rotation, periodic fertilization and pesticide application generate temporal and

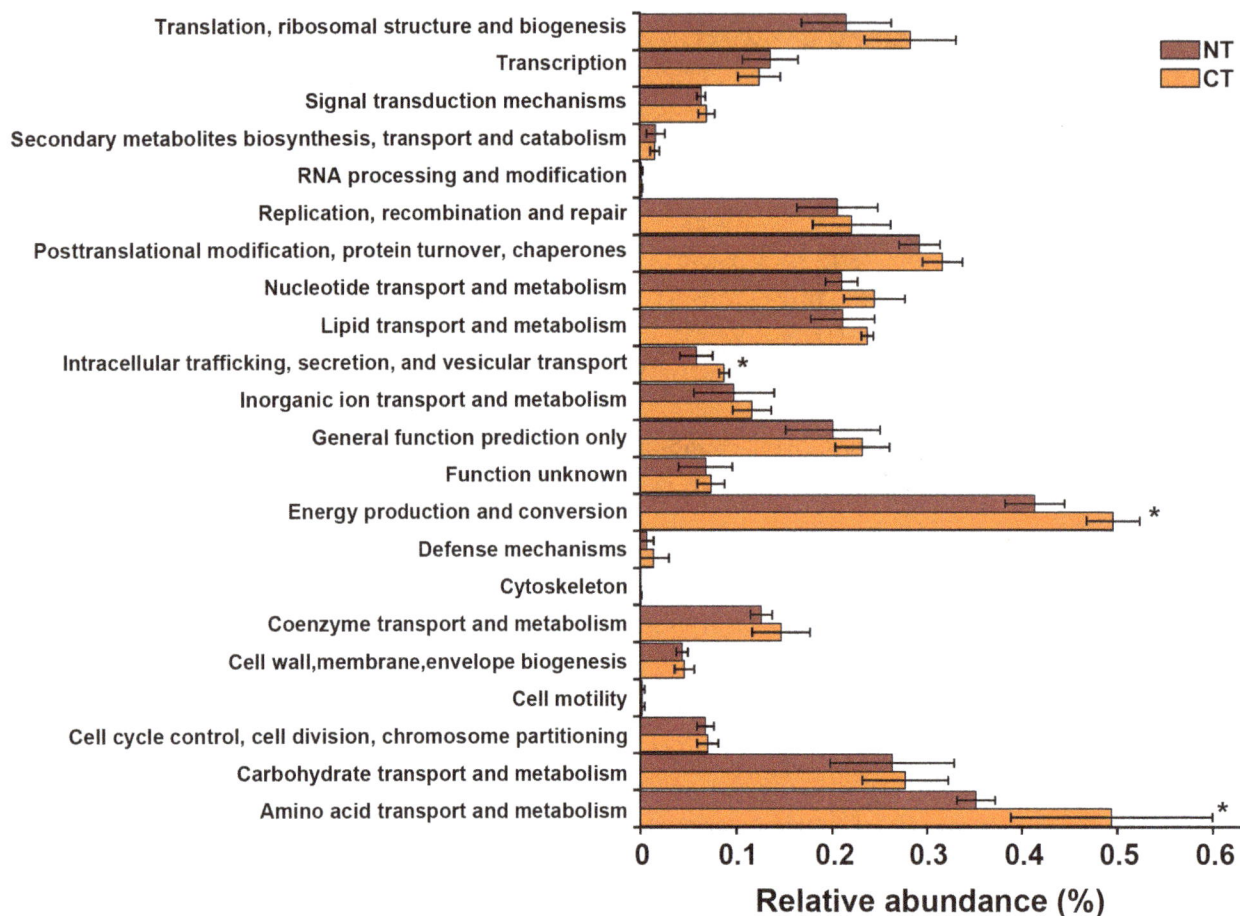

Figure 7. Relative abundances of COG categories in soil microbiomes of tillage systems comparison experiment in Balcarce. CA: conventional tillage; NT: no-tillage. Bars represent ±1 standard erro. (*)indicate significant differences (p≤0.05).

spatial changes in soil chemical properties, and therefore in nutrient availability and approachability for microorganisms. Bacteria dominating in this type of soils should be adapted to heterogeneity. These microorganisms should be able to fine-tune carbon and nitrogen intakes according to their metabolic needs under frequent external changes. This seems to be the case for *Gemmatimonadetes* suggesting a generalist ecological strategy. The high abundance of *Nitrospirae* may also respond to heterogeneous conditions. It has been proposed that *Nitrospirae* lineages occupy different positions on an imaginary scale reaching from K- to r-strategies [62]. We hypothesize that in cultivated soils, *Nitrospirae* K-strategists would be exploiting nitrite in high N microenvironments, while r-strategist would be mining low concentration areas in a nitrogen gradient enhanced by fertilization [63]. This competition would be less fierce in non cultivated soils due to the highest homogeneity and a less marked nitrogen gradient due to the lack of fertilization and more recalcitrant organic matter forms.

Effects of agricultural land use on metabolic profiles of soil microbiomes in Argentine Pampas

Our results from shotgun metagenomic data also indicated a tendency for the microbiomes of cultivated soils towards adaptation to nutrient heterogeneity. The highest relative abundances of sequences assigned to COGs related to transcription, protein modifications, nucleotide transport and metabolism, wall and

membrane biogenesis and intracellular trafficking and secretion in cultivated fertilized soils are consistent with a copiotrophic strategy (i.e. rapid tight metabolism regulation and fast grow rates).Moreover, some of these COG categories were previously shown to be up-represented in copiotrophic marine microorganism genomes [58]. In a deeper look we detected that the relative abundance of sequences assigned to riboswitch regulated genes was higher in cultivated soils (i.e. cobalamin biosynthesis protein, S-adenosyl-methionine synthetase, S-adenosyl- homocysteine hydrolase) [64–66]. These ancient regulators may be playing a central role in a life-style strategy adapted to nutrient heterogeneity since they were described to be the most 'economical' and fast-reacting regulatory systems (no intermediate factors involved) [67]. Moreover, the abundant riboswitch- COGs in cultivated soils were related to synthesis of B vitamins. The levels of these vitamins have already been linked to differences in community composition in marine ecosystems [68]. More studies are needed in order to address this relationship in soil environments; still, our results suggest an important role of B-vitamins in fertilized heterogeneous soils.

Another interesting observation was the highest abundance of glutamate synthase (GOGAT) related COGs in cultivated soils metagenomic profiles. The combined role of GOGAT with glutamine synthetase and glutamate dehydrogenase allows the cells to sense ammonia external levels [69]. The high abundance of this regulation system detected in cultivated soil metagenomes suggests its importance in the detection of N fluxes related with fertilization. Moreover COGs related to the tricarboxylic acid cycle (TCA), such as citrate synthase and succinate dehydrogenase, were more abundant in cultivated soil microbiomes. GOGAT nitrogen assimilation pathway and TCA are related. It has been shown that the concentration of compounds of both pathways changes considerably and rapidly upon nitrogen up shift; in contrast, the concentrations of glycolytic intermediates remains homeostatic [70]. In addition, abundance of urease related COGs were also higher in cultivated soils. The soils sampled in this study have a long history of agricultural land use and have been long fertilized with both type of nitrogen (i.e. ammonia-based and urea-based fertilizers). This kind of environmental pressure finally selected a microbiome adapted to changing N sources and availability.

It is important to mention that these results do not infer that expression or activity of these metabolisms will be necessarily increased in cultivated soil microbiomes, still, the highest abundance of these COGs could be reflected in a highest diversification and specialization. The presence of a highest copy number of strategic genes have already been linked to copiotrophic or oligotrophic life-styles [58].

Microbiomes of non cultivated soils showed a higher abundance of sequences related to Coenzyme A and acetyl-Coa metabolism than microbiomes of cultivated soils. It is known that acetyl-CoA is a fundamental building block and energy source [71]. High acetyl-CoA levels would indicate a "proliferative" or "fed" state, while low acetyl-CoA levels (and high CoA levels) would be indicative of a "quiescent" or "starved" state. Pantothenate kinase (PanK), the key enzyme in CoA syntehsis, was also highest in non cultivated soil microbiomes [72]. In addition, it was stated that some oligotrophs preferentially use lipids as immediate and stored sources of carbon and energy in marine environments [58]. The observed abundance of Pank genes may ensure a correct CoA intracellular level in fasted moments, allowing proper lipid utilization. In addition, non cultivated metagenomic profiles showed higher abundance of sequences related to trehalose utilization. Trehalose is known to serve as energy source in many microorganisms [73]. Members of Actinobacteria genera Mycobacterium and family Frankiaceae are known to produce and/or utilize trehalose [74–76]. As mentioned above, our results showed higher abundance of Actinobacteria-related sequences in non cultivated soils. Moreover, sequences classified within family Frankiaceae were only present in these soils and the abundance of sequences assigned as Mycobacterium was higher than in cultivated soils (not shown). These observations are congruent with an oligotrophic strategy based on the use of storage components for carbon and energy sources.

The responses of the phylogenetic structure and metagenomic profile to agronomic land use were significantly correlated, suggesting some degree of correspondence between these different microbiome features. These results agree with previous observations done in soils from different biomes [9,77]. Moreover, Fierer et al. found similar a similar correlation between metagenomic and phylogenetic data for microbial communities of agricultural soils under different nitrogen gradients in experimental plots [10].

Correlation with soil properties

In addition, soil properties such as organic matter, phosphorus and nitrate levels explained most of the variability observed between cultivated and non cultivated soils. As mentioned, the highest amount of organic matter is found mostly in a recalcitrant form in these soils. Moreover, it was already established that nitrate accumulation exhibits a negative correlation with organic carbon availability [78]. In addition it has been proved that

organic matter source and quality played an important role in regulating the magnitude of carbon metabolism and could be as important as nutrient abundance in water environments [79]. Our results are congruent with these observations since non cultivated soils, with highest levels of organic carbon, presented metagenomic profiles with tendencies to oligotrophy and the best explanation for this scenario is the low lability of the recalcitrant forms of carbon and nitrogen.

Assessing the effects of different tillage systems on Pampas soil microbiomes

Our results also showed differences in the structure and composition of soil microbiomes between no-till and conventional tillage soils. Sequences related to phyla Gemmatimonadetes, candidate division TM7 and Acidobacteria, were highest in CT soils, while the abundances of Nitrospirae and candidate division WS3 were highest in NT soils. Tillage is the principal agent producing soil disturbance and subsequent soil structure modification [80,81]. The negative effects of CT in soil stabilization and macroaggregate losses were previously registered in Pampas soils [82]. It has also been shown that NT increases macroaggregate abundance and organic matter content [83].These increments are related with a more recalcitrant organic matter, with increased humic acid contents and nutrient retention [84–86]. The higher abundance of Nitrospirae in NT soils reinforces the idea of a community adapted to a better N mining in these environments. Moreover, we observed very low abundance of Nitrobacter related sequences in Balcarce soils and no significant difference was observed between CT and NT soils (not shown). Changes in ammonia oxidation due to a decrease in ammonia availability by humic substances have already been proposed in microcosm experiments [87]. If this is the case of Balcarce soils, higher humic acids in NT soils would be decreasing ammonia availability for oxidation into nitrite and therefore decreasing nitrate availability, compared to CT soils. The predicted highest stability of organic matter and abundance of macroaggregates in NT soils are probably generating more marked nitrite gradients than in CT soils. The highest abundance of Nitrospirae in NT may be reflecting a highest lineage diversity that would be better adapted to these gradients.

In addition, communities under NT showed higher abundance of sequences related to the order Syntrophobacterales (Deltaproteobacteria, Figure S4). These are known anaerobic and syntrophic organisms [88,89]. These characteristics may be an advantage in stable soils with higher number of highly humic macro aggregates since syntrophy is known to be important for community functioning in micro-environments with low nutrient levels [90]. On the other hand, CT soils presented higher relative abundance of Gammaproteobacteria related sequences. The order Xanthomonadales was the main responsible for these differences (Figure S4). This taxon has already been associated to CT practices [32].

Even though we could not find significant differences associated to tillage systems at the community structure level for metabolic profiles, COG categories related to intracellular trafficking and secretion, amino acid transport and metabolism and energy production and conversion were more abundant in CT mirobiomes. These results suggest a tendency of CT microbiomes to a more copiotrophic life-style strategy than NT microbiomes.

Conclusion

Our results are consistent with the hypothesis that microbiomes exhibit different life- history and trophic strategies in Pampean soils under different land uses and tillage systems. Our data suggest that microbiomes of fertilized cultivated soils have more flexible metabolisms adapted to nutrient fluxes with tendencies to copiothropy while microorganisms in non cultivated soils are better adapted to lowest external nutrient availability and homogenous environment. The lowest nutrient accessibility in non cultivated soils may be explained by the higher amount of humic substances, recalcitrant organic matter and the lack of fertilizer amendments. Moreover, NT soils, with most stable structure and highest macroaggregate abundance, presented microbiomes better adapted to recalcitrant environments; while CT microbiomes presented a higher tendency to copiotrophy.

This work is of major contribution to understand how historical changes in soil properties due to agronomical land use have altered the diversity and function of below-ground communities. The importance of high-throughput characterization for the reconstruction of pre- agricultural microbiomes is being reinforced nowadays [9]. Following this direction, our findings will be very useful in future restoration and monitoring programs of Argentine Pampas ecosystems.

Supporting Information

Figure S1 PCoA plots of Pampa production field soil microbiomes based on average Bray Curtis distance matrices. A) PCoA of cultivated and non cultivated soil microbiomes. Standard error ellipses show 95% confidence areas. B) PCoA biplot of soil properties that best explained variation in community structure. Correlations were calculated using BIOENV on average data of each sampled site (Mantel r = 0.6214, p≤0.05). Circles represent samples from "La Estrella", crosses represent "Criadero Klein" samples and triangles represent "La Negrita" samples.

Figure S2 Relative abundances of Cluster of Orthologous groups (COGs) in Pampa production field soil

microbiomes. Bars represent ± 1 standard error. Only significant COGs are showed.

Figure S3 Comparison of tillage systems effects on the structure of metabolic profiles. PCoA plot based on Euclidean distance matrices. CA: conventional tillage; NT: no-tillage. Standard error ellipses show 95% confidence areas.

Figure S4 Relative abundances of reads assigned to orders within classes *Gammaproteobacteria* and *Deltaporteobacteria* in NT and CT soils.

Table S1 Soil chemical and physical properties.

Acknowledgments

We acknowledge Germán F. Domínguez and Guillermo A. Studdert from the "Manejo Sustentable del suelo" research group (from Facultad de Ciencias Agrarias, Universidad Nacional de Mar del Plata, Unidad Integrada Balcarce) for providing samples of the tillage systems experiment. We thank Gonzalo Berhongaray, Josefina de Paepe, and María Rosa Mendoza for his help during sampling in the Rolling Pampas. We also acknowledge Roxana Colombo and Marcelo Soria for helpful discussion and advice.

Author Contributions

Conceived and designed the experiments: BC NR RÁ AM MV. Performed the experiments: BC NR. Analyzed the data: BC. Contributed reagents/materials/analysis tools: BC NR RÁ AM MV. Wrote the paper: BC NR MV. Obtained permission for sampling: RÁ.

References

1. Foley JA, Ramankutty N, Brauman KA, Cassidy ES, Gerber JS, et al. (2011) Solutions for a cultivated planet. Nature 478: 337–342.
2. Tilman D, Balzer C, Hill J, Befort BL (2011) Global food demand and the sustainable intensification of agriculture. Proc Natl Acad Sci U S A 108: 20260–20264.
3. Tilman D, Cassman KG, Matson PA, Naylor R, Polasky S (2002) Agricultural sustainability and intensive production practices. Nature 418: 671–677.
4. Balmford A, Green RE, Scharlemann JPW (2005) Sparing land for nature: exploring the potential impact of changes in agricultural yield on the area needed for crop production. Global Change Biology 11: 1594–1605.
5. Power AG (2010) Ecosystem services and agriculture: tradeoffs and synergies. Philos Trans R Soc Lond B Biol Sci 365: 2959–2971.
6. Foley JA, Defries R, Asner GP, Barford C, Bonan G, et al. (2005) Global consequences of land use. Science 309: 570–574.
7. Bennett EM, Carpenter SR, Caraco NF (2001) Human Impact on Erodable Phosphorus and Eutrophication: A Global Perspective: Increasing accumulation of phosphorus in soil threatens rivers, lakes, and coastal oceans with eutrophication. BioScience 51: 227–234.
8. Scholes MC, Scholes RJ (2013) Dust Unto Dust. Science 342: 565–566.
9. Fierer N, Ladau J, Clemente JC, Leff JW, Owens SM, et al. (2013) Reconstructing the Microbial Diversity and Function of Pre-Agricultural Tallgrass Prairie Soils in the United States. Science 342: 621–624.
10. Fierer N, Lauber CL, Ramirez KS, Zaneveld J, Bradford MA, et al. (2012) Comparative metagenomic, phylogenetic and physiological analyses of soil microbial communities across nitrogen gradients. ISME J 6: 1007–1017.
11. Satorre EH, Slafer GA (1999) Wheat: Ecology and Physiology of Yield Determination: Taylor & Francis.
12. Hall AJ, Rebella CM, Ghersa CM, Culot JP (1992) Field crop systems of the Pampas. In: Pearson CJ, editor. Field Crop Ecosystems of the World. Amsterdam: Elsevier Science & Technology Books. pp. 413–450.
13. Alvarez R, Díaz RA, Barbero N, Santanatoglia OJ, Blotta L (1995) Soil organic carbon, microbial biomass and CO2-C production from three tillage systems. Soil and Tillage Research 33: 17–28.
14. Bernardos JN, Viglizzo EF, Jouvet V, Lértora FA, Pordomingo AJ, et al. (2001) The use of EPIC model to study the agroecological change during 93 years of

farming transformation in the Argentine pampas. Agricultural Systems 69: 215–234.
15. Alvarez R (2001) Estimation of carbon losses by cultivation from soils of the Argentine Pampa using the Century Model. Soil Use and Management 17: 62–66.
16. Hevia GG, Buschiazzo DE, Hepper EN, Urioste AM, Antón EL (2003) Organic matter in size fractions of soils of the semiarid Argentina. Effects of climate, soil texture and management. Geoderma 116: 265–277.
17. Quiroga AR, Buschiazzo DE, Peinemann N (1996) Soil Organic Matter Particle Size Fractions in Soils of the Semiarid Argentinian Pampas. Soil Science 161.
18. Kassam A, Friedrich T, Shaxson F, Pretty J (2009) The spread of Conservation Agriculture: justification, sustainability and uptake. International Journal of Agricultural Sustainability 7: 292–320.
19. Alvarez R, Steinbach HS (2009) A review of the effects of tillage systems on some soil physical properties, water content, nitrate availability and crops yield in the Argentine Pampas. Soil and Tillage Research 104: 1–15.
20. Gomez E, Bisaro V, Conti M (2000) Potential C-source utilization patterns of bacterial communities as influenced by clearing and land use in a vertic soil of Argentina. Applied Soil Ecology 15: 273–281.
21. Aon MA, Cabello MN, Sarena DE, Colaneri AC, Franco MG, et al. (2001) I. Spatio-temporal patterns of soil microbial and enzymatic activities in an agricultural soil. Applied Soil Ecology 18: 239–254.
22. Agaras B, Wall LG, Valverde C (2012) Specific enumeration and analysis of the community structure of culturable pseudomonads in agricultural soils under no-till management in Argentina. Applied Soil Ecology 61: 305–319.
23. Nievas F, Bogino P, Nocelli N, Giordano W (2012) Genotypic analysis of isolated peanut-nodulating rhizobial strains reveals differences among populations obtained from soils with different cropping histories. Applied Soil Ecology 53: 74–82.
24. Zabaloy MC, Garland JL, Gómez MA (2008) An integrated approach to evaluate the impacts of the herbicides glyphosate, 2,4-D and metsulfuron-methyl on soil microbial communities in the Pampas region, Argentina. Applied Soil Ecology 40: 1–12.
25. Zabaloy MC, Gómez E, Garland JL, Gómez MA (2012) Assessment of microbial community function and structure in soil microcosms exposed to glyphosate. Applied Soil Ecology 61: 333–339.

26. Deng W, Xi D, Mao H, Wanapat M (2008) The use of molecular techniques based on ribosomal RNA and DNA for rumen microbial ecosystem studies: a review. Molecular Biology Reports 35: 265–274.

27. Pontes D, Lima-Bittencourt C, Chartone-Souza E, Amaral Nascimento A (2007) Molecular approaches: advantages and artifacts in assessing bacterial diversity. Journal of Industrial Microbiology & Biotechnology 34: 463–473.

28. Sogin ML, Morrison HG, Huber JA, Welch DM, Huse SM, et al. (2006) Microbial diversity in the deep sea and the underexplored "rare biosphere". Proceedings of the National Academy of Sciences 103: 12115–12120.

29. Fortunato CS, Herfort L, Zuber P, Baptista AM, Crump BC (2012) Spatial variability overwhelms seasonal patterns in bacterioplankton communities across a river to ocean gradient. ISME J 6: 554–563.

30. Delmont TO, Prestat E, Keegan KP, Faubladier M, Robe P, et al. (2012) Structure, fluctuation and magnitude of a natural grassland soil metagenome. ISME J 6: 1677–1687.

31. Figuerola ELM, Guerrero LD, Rosa SM, Simonetti L, Duval ME, et al. (2012) Bacterial Indicator of Agricultural Management for Soil under No-Till Crop Production. PLoS ONE 7: e51075.

32. Souza RC, Cantão ME, Vasconcelos ATR, Nogueira MA, Hungria M (2013) Soil metagenomics reveals differences under conventional and no-tillage with crop rotation or succession. Applied Soil Ecology 72: 49–61.

33. Lauber CL, Ramirez KS, Aanderud Z, Lennon J, Fierer N (2013) Temporal variability in soil microbial communities across land-use types. ISME J 7: 1641–1650.

34. Ramirez KS, Craine JM, Fierer N (2012) Consistent effects of nitrogen amendments on soil microbial communities and processes across biomes. Global Change Biology 18: 1918–1927.

35. Ramirez K, Lauber CL, Knight R, Bradford MF (2010) Consistent effects of nitrogen fertilization on soil bacterial communities in contrasting systems. Ecology: 3463–3470

36. Lavado RS (2008) La Región Pampeana: historia, características y uso de sus suelos. In: Alvarez R, editor. Materia Orgánica Valor agronómico y dinámica en suelos pampeanos. Buenos Aires: Editorial Facultad de Ingeniería. pp. 1–11.

37. Berhongaray G, Alvarez R, De Paepe J, Caride C, Cantet R (2013) Land use effects on soil carbon in the Argentine Pampas. Geoderma 192: 97–110.

38. Rascovan N, Carbonetto B, Revale S, Reinert M, Alvarez R, et al. (2013) The PAMPA datasets: a metagenomic survey of microbial communities in Argentinean pampean soils. Microbiome 1: 21.

39. Nelson DW, Sommers LE, Sparks DLE, Page ALE, Helmke PAE, et al. (1996) Total Carbon, Organic Carbon, and Organic Matter. Methods of Soil Analysis Part 3-Chemical Methods: Soil Science Society of America, American Society of Agronomy. pp. 961–1010.

40. Bremner JME, Sparks DLE, Page ALE, Helmke PAE, Loeppert RH (1996) Nitrogen-Total. Methods of Soil Analysis Part 3-Chemical Methods: Soil Science Society of America, American Society of Agronomy. pp. 1085–1121.

41. Kuo SE, Sparks DLE, Page ALE, Helmke PAE, H LR (1996) Phosphorus. Methods of Soil Analysis Part 3-Chemical Methods: Soil Science Society of America, American Society of Agronomy. pp. 869–919.

42. Rhoades JDE, Sparks DLE, Page ALE, Helmke PAE, H. LR (1996) Salinity: Electrical Conductivity and Total Dissolved Solids. Methods of Soil Analysis Part 3-Chemical Methods: Soil Science Society of America, American Society of Agronomy. pp. 417–435.

43. Gee GW, Bauder JWEKA (1986) Particle-size Analysis. Methods of Soil Analysis: Part 1—Physical and Mineralogical Methods: Soil Science Society of America, American Society of Agronomy. pp. 383–411.

44. Cole J, Wang Q, Cardenas E, Fish J, Chai B, et al. (2009) The Ribosomal Database Project: improved alignments and new tools for rRNA analysis. Nucleic acids research: 141–145.

45. Caporaso JG, Kuczynski J, Stombaugh J, Bittinger K, Bushman FD, et al. (2010) QIIME allows analysis of high-throughput community sequencing data. Nat Meth 7: 335–336.

46. Edgar RC (2010) Search and clustering orders of magnitude faster than BLAST. Bioinformatics 26: 2460–2461.

47. Caporaso JG, Bittinger K, Bushman FD, DeSantis TZ, Andersen GL, et al. (2010) PyNAST: a flexible tool for aligning sequences to a template alignment. Bioinformatics 26: 266–267.

48. Price MN, Dehal PS, Arkin AP (2009) FastTree: Computing Large Minimum Evolution Trees with Profiles instead of a Distance Matrix. Molecular Biology and Evolution 26: 1641–1650.

49. Lozupone C, Knight R (2005) UniFrac: a New Phylogenetic Method for Comparing Microbial Communities. Applied and Environmental Microbiology 71: 8228–8235.

50. DeSantis TZ, Hugenholtz P, Larsen N, Rojas M, Brodie EL, et al. (2006) Greengenes, a Chimera-Checked 16S rRNA Gene Database and Workbench Compatible with ARB. Applied and Environmental Microbiology 72: 5069–5072.

51. Wang Q, Garrity GM, Tiedje JM, Cole JR (2007) Naïve Bayesian Classifier for Rapid Assignment of rRNA Sequences into the New Bacterial Taxonomy. Applied and Environmental Microbiology 73: 5261–5267.

52. Kindt R, Coe R (2005) Tree Diversity Analysis: A Manual and Software for Common Statistical Methods for Ecological and Biodiversity Studies: World Agroforestry Centre.

53. Dixon P (2003) VEGAN, a package of R functions for community ecology. Journal of Vegetation Science 14: 927–930.

54. Meyer F, Paarmann D, D'Souza M, Olson R, Glass EM, et al. (2008) The metagenomics RAST server - a public resource for the automatic phylogenetic and functional analysis of metagenomes. BMC Bioinformatics 9: 386.

55. Fierer N, Bradford MA, Jackson RB (2007) Toward an ecological classification of soil bacteria. Ecology 88: 1354–1364.

56. Dion P, Nautiyal CS (2008) Microbiology of Extreme Soils: Springer.

57. Galantini J, Rosell R (2006) Long-term fertilization effects on soil organic matter quality and dynamics under different production systems in semiarid Pampean soils. Soil and Tillage Research 87: 72–79.

58. Lauro FM, McDougald D, Thomas T, Williams TJ, Egan S, et al. (2009) The genomic basis of trophic strategy in marine bacteria. Proceedings of the National Academy of Sciences 106: 15527–15533.

59. Nemergut DR, Townsend AR, Sattin SR, Freeman KR, Fierer N, et al. (2008) The effects of chronic nitrogen fertilization on alpine tundra soil microbial communities: implications for carbon and nitrogen cycling. Environmental Microbiology 10: 3093–3105.

60. Zhang H, Sekiguchi Y, Hanada S, Hugenholtz P, Kim H, et al. (2003) Gemmatimonas aurantiaca gen. nov., sp. nov., a Gram-negative, aerobic, polyphosphate-accumulating micro-organism, the first cultured representative of the new bacterial phylum Gemmatimonadetes phyl. nov. International Journal of Systematic and Evolutionary Microbiology 53: 1155–1163.

61. DeBruyn JM, Nixon LT, Fawaz MN, Johnson AM, Radosevich M (2011) Global Biogeography and Quantitative Seasonal Dynamics of Gemmatimonadetes in Soil. Applied and Environmental Microbiology 77: 6295–6300.

62. Maixner F, Noguera DR, Anneser B, Stoecker K, Wegl G, et al. (2006) Nitrite concentration influences the population structure of Nitrospira-like bacteria. Environmental Microbiology 8: 1487–1495.

63. Attard E, Poly F, Commeaux C, Laurent F, Terada A, et al. (2010) Shifts between Nitrospira- and Nitrobacter-like nitrite oxidizers underlie the response of soil potential nitrite oxidation to changes in tillage practices. Environmental Microbiology 12: 315–326.

64. Edwards AL, Reyes FE, Héroux A, Batey RT (2010) Structural basis for recognition of S-adenosylhomocysteine by riboswitches. RNA 16: 2144–2155.

65. Loenen WA (2006) S-adenosylmethionine: jack of all trades and master of everything? Biochem Soc Trans 34: 330–333.

66. Winkler WC, Breaker RR (2005) Regulation of bacterial gene expression by riboswitches. Annual Review of Microbiology 59: 487–517.

67. Nudler E, Mironov AS (2004) The riboswitch control of bacterial metabolism. Trends in biochemical sciences 29: 11–17.

68. Sañudo-Wilhelmy SA, Cutter LS, Durazo R, Smail EA, Gómez-Consarnau L, et al. (2012) Multiple B-vitamin depletion in large areas of the coastal ocean. Proceedings of the National Academy of Sciences.

69. Yan D (2007) Protection of the glutamate pool concentration in enteric bacteria. Proceedings of the National Academy of Sciences 104: 9475–9480.

70. Doucette CD, Schwab DJ, Wingreen NS, Rabinowitz JD (2011) α-ketoglutarate coordinates carbon and nitrogen utilization via enzyme I inhibition. Nat Chem Biol 7: 894–901.

71. Cai L, Tu BP (2011) On Acetyl-CoA as a Gauge of Cellular Metabolic State. Cold Spring Harbor Symposia on Quantitative Biology 76: 195–202.

72. Leonardi R, Rehg JE, Rock CO, Jackowski S (2010) Pantothenate Kinase 1 Is Required to Support the Metabolic Transition from the Fed to the Fasted State. PLoS ONE 5: e11107.

73. Elbein AD, Pan YT, Pastuszak I, Carroll D (2003) New insights on trehalose: a multifunctional molecule. Glycobiology 13: 17R–27R.

74. Barabote RD, Xie G, Leu DH, Normand P, Necsulea A, et al. (2009) Complete genome of the cellulolytic thermophile Acidothermus cellulolyticus 11B provides insights into its ecophysiological and evolutionary adaptations. Genome Research 19: 1033–1043.

75. Lopez MF, Fontaine MS, Torrey JG (1984) Levels of trehalose and glycogen in Frankia sp. HFPArI3 (Actinomycetales). Canadian Journal of Microbiology 30: 746–752.

76. Tropis M, Meniche X, Wolf A, Gebhardt H, Strelkov S, et al. (2005) The Crucial Role of Trehalose and Structurally Related Oligosaccharides in the Biosynthesis and Transfer of Mycolic Acids in Corynebacterineae. Journal of Biological Chemistry 280: 26573–26585.

77. Fierer N, Leff JW, Adams BJ, Nielsen UN, Bates ST, et al. (2012) Cross-biome metagenomic analyses of soil microbial communities and their functional attributes. Proceedings of the National Academy of Sciences 109: 21390–21395.

78. Taylor PG, Townsend AR (2010) Stoichiometric control of organic carbon-nitrate relationships from soils to the sea. Nature 464: 1178–1181.

79. Apple JK, del Giorgio PA (2007) Organic substrate quality as the link between bacterioplankton carbon demand and growth efficiency in a temperate salt-marsh estuary. ISME J 1: 729–742.

80. Bayer C, Mielniczuk J, Amado TJC, Martin-Neto L, Fernandes SV (2000) Organic matter storage in a sandy clay loam Acrisol affected by tillage and cropping systems in southern Brazil. Soil and Tillage Research 54: 101–109.

81. Langdale GW, West LT, Bruce RR, Miller WP, Thomas AW (1992) Restoration of eroded soil with conservation tillage. Soil Technology 5: 81–90.

82. Bongiovanni MD, Lobartini JC (2006) Particulate organic matter, carbohydrate, humic acid contents in soil macro- and microaggregates as affected by cultivation. Geoderma 136: 660–665.

83. Plaza-Bonilla D, Cantero-Martínez C, Viñas P, Álvaro-Fuentes J (2013) Soil aggregation and organic carbon protection in a no-tillage chronosequence under Mediterranean conditions. Geoderma 193–194: 76–82.

84. Jiao Y, Whalen JK, Hendershot WH (2006) No-tillage and manure applications increase aggregation and improve nutrient retention in a sandy-loam soil. Geoderma 134: 24–33.

85. Tivet F, de Moraes Sá JC, Lal R, Borszowskei PR, Briedis C, et al. (2013) Soil organic carbon fraction losses upon continuous plow-based tillage and its restoration by diverse biomass-C inputs under no-till in sub-tropical and tropical regions of Brazil. Geoderma 209–210: 214–225.

86. Slepetiene A, Slepetys J (2005) Status of humus in soil under various long-term tillage systems. Geoderma 127: 207–215.

87. Dong L, Córdova-Kreylos AL, Yang J, Yuan H, Scow KM (2009) Humic acids buffer the effects of urea on soil ammonia oxidizers and potential nitrification. Soil Biology and Biochemistry 41: 1612–1621.

88. Sieber JR, McInerney MJ, Gunsalus RP (2012) Genomic Insights into Syntrophy: The Paradigm for Anaerobic Metabolic Cooperation. Annual Review of Microbiology 66: 429–452.

89. McInerney MJ, Sieber JR, Gunsalus RP (2009) Syntrophy in anaerobic global carbon cycles. Current Opinion in Biotechnology 20: 623–632.

90. Kim HJ, Boedicker JQ, Choi JW, Ismagilov RF (2008) Defined spatial structure stabilizes a synthetic multispecies bacterial community. Proceedings of the National Academy of Sciences 105: 18188–18193.

7

Does Fire Influence the Landscape-Scale Distribution of an Invasive Mesopredator?

Catherine J. Payne[1], Euan G. Ritchie[1], Luke T. Kelly[2], Dale G. Nimmo[1]*

1 Centre for Integrative Ecology, School of Life and Environmental Sciences, Deakin University, Melbourne, Victoria, Australia, **2** Australian Research Council Centre of Excellence for Environmental Decisions, School of Botany, University of Melbourne, Melbourne, Victoria, Australia

Abstract

Predation and fire shape the structure and function of ecosystems globally. However, studies exploring interactions between these two processes are rare, especially at large spatial scales. This knowledge gap is significant not only for ecological theory, but also in an applied context, because it limits the ability of landscape managers to predict the outcomes of manipulating fire and predators. We examined the influence of fire on the occurrence of an introduced and widespread mesopredator, the red fox (*Vulpes vulpes*), in semi-arid Australia. We used two extensive and complimentary datasets collected at two spatial scales. At the landscape-scale, we surveyed red foxes using sand-plots within 28 study landscapes – which incorporated variation in the diversity and proportional extent of fire-age classes – located across a 104 000 km^2 study area. At the site-scale, we surveyed red foxes using camera traps at 108 sites stratified along a century-long post-fire chronosequence (0–105 years) within a 6630 km^2 study area. Red foxes were widespread both at the landscape and site-scale. Fire did not influence fox distribution at either spatial scale, nor did other environmental variables that we measured. Our results show that red foxes exploit a broad range of environmental conditions within semi-arid Australia. The presence of red foxes throughout much of the landscape is likely to have significant implications for native fauna, particularly in recently burnt habitats where reduced cover may increase prey species' predation risk.

Editor: R. Mark Brigham, University of Regina, Canada

Funding: Funding for the site-scale study was provided by the Victorian government's Department of Environment and Primary Industries, under the Mallee Hawkeye project (Contract Number 313764). Funding and logistical support for the landscape-scale study was provided by Land and Water Australia, the Mallee Catchment Management Authority, Parks Victoria, Department Sustainability and Environment Victoria, Department Environment and Heritage SA, Lower Murray-Darling Catchment Management Authority, Department Environment and Climate Change NSW, Australian Wildlife Conservancy and Birds Australia. The funders had no role in study design, data collection and analysis, decision to publish, or preparation of the manuscript.

Competing Interests: The authors have declared that no competing interests exist.

* Email: dale@deakin.edu.au

Introduction

Predators shape ecosystems worldwide [1]. They can exert top-down regulation of lower trophic levels [2] and induce trophic cascades which flow through entire ecosystems [3]. Predators introduced to areas outside of their native range can have a particularly strong effect on native species [4], and have caused population declines and extinctions in a range of ecosystems [5]. Many invasive predators are 'mesopredators': smaller predator species that increase in abundance or activity following the removal of apex predators [6]. For example, in Australia, persecution of the native apex predator, the dingo (*Canis dingo*), has led to increases in the density or activity of invasive mesopredators (e.g. the red fox [*Vulpes vulpes*]) throughout large portions of the continent [3].

Fire is another globally significant process that affects environments worldwide [7]. Fire influences ecosystems via bottom-up control by altering the availability of key resources for biota. Fire incinerates plant matter, altering vegetation structure [8,9], which in turn affects the distribution and abundance of animals [10].

Invasive mesopredators and fire share an important characteristic from a conservation perspective: both can be manipulated through management interventions. Invasive mesopredators are managed using lethal control and exclusion fencing, and fire using suppression or prescribed burning. However, management of mesopredators and fire usually occurs in isolation, without consideration of the potential effects of fire *on* mesopredators [11]. It is important to rapidly address this significant knowledge gap because some fire regimes may exacerbate the effects of invasive mesopredators by simplifying vegetation and amplifying predation risk [12,13]. For example, interactions between fire regimes and invasive mesopredators have been hypothesised as a cause of lower survival of reptile species in recently-burned areas [14], and a contributor to the collapse of small mammal communities in northern Australia [15].

The red fox is one of the world's most widely distributed mesopredators. It is common in both the northern and southern hemispheres. Foxes, and a second introduced mesopredator, the feral cat (*Felis catus*), are widely regarded as the primary cause of extinctions and declines of Australia's marsupial fauna [5]. Evidence for the negative impact of foxes has been demonstrated through predator-control experiments that have shown that prey species increase in both range and activity when foxes are removed [16,17]. Further evidence comes from dietary studies showing foxes eat a wide range of native mammal, reptile, bird, and invertebrate prey [18–20].

Despite indications that foxes may inhibit the recovery of native species following fire [12,21], whether foxes are themselves influenced by fire remains poorly known. This knowledge gap limits the ability of land managers to consider the effects of fire management on red foxes, which could have negative ramifications for native biodiversity. While foxes are widely considered as habitat generalists, they do display local variability in occurrence related to habitat or landscape structure [22]. For example, in some regions, foxes prefer heterogeneous landscapes [22], as they are able to use multiple landscape elements on a daily or seasonal basis [23,24]. Fire management in many regions seeks to maximise landscape heterogeneity by creating mosaics of fire ages (i.e. 'patch mosaic burning'; [25]). Does such management inadvertently favour invasive mesopredators?

The few studies that have explored the topic have focused on relatively short temporal scales (<30 years and often <10 years post fire) or small spatial scales (but see [26]). However, in some ecosystems, post-fire vegetation recovery continues for a century or more after fire [27]. Consequently, animal species respond to fire over similarly long time-frames [28]. The effects of fire can also occur across multiple spatial scales [29]; while time since fire may affect a species' occurrence at any *point* in the landscape, the area and composition of fire-ages within a 'whole' landscape can play a critical role in affecting species' landscape-level distributions [30]. This is likely to be especially true for large, mobile species, such as the red fox.

In addition to the effects that fire may have on species' occurrence, other environmental factors may be locally important. With regard to foxes, this includes climate [26], the distribution of vegetation types [31], and the distance to roads [24] and agricultural land [22]. Foxes rely on free standing water for drinking, particularly when temperatures are high (>30°C), as is common in many semi-arid environments. Hence, as annual rainfall decreases (aridity intensifies) permanent water may be reduced in its availability and limit fox occurrence. Foxes are often thought of as edge specialists [22]. They often prefer to hunt in open areas such as resource-rich agricultural fields or structurally simple vegetation types adjacent to more complex vegetation which provides cover during the day [22,32]. Their ability to hunt may be further enhanced where roads create easy access and increased visibility in otherwise structurally complex habitats [24,33].

Here, we examine what drives the occurrence (reporting rate) of red foxes in semi-arid Australia at multiple spatial scales, with a particular emphasis on the role of fire. We conducted two large-scale natural experiments. First, we explored landscape-scale patterns of fox occurrence in relation to the properties of fire mosaics; namely, the amount and diversity of fire age-classes within each of 28 study landscapes (each 12.6 km²). Second, we explored site-scale patterns of fox occurrence in relation to fire history at 108 sites stratified along a century-long post-fire chronosequence. In both cases, we also quantified the influence of other environmental variables such as vegetation type and distance to agricultural land. Our aims were: 1) to determine the drivers of fox distribution in semi-arid Australia; and 2) to understand the specific role of fire in influencing fox occurrence at large scales relevant to fire and mesopredator management.

Materials and Methods

Study region

This study was undertaken in the Murray Mallee region of south-eastern Australia (Fig. 1). The climate in the region is semi-arid, with mean annual rainfall of 200–350 mm and average daily maximum temperatures are 30–33°C in summer and 15–18°C in winter (Australian Bureau of Meteorology; http://www.bom.gov.au). The vegetation is predominantly 'tree mallee' characterised by an overstorey of *Eucalyptus* species (<5–8 m) with a multi-stemmed growth form [34]. Two vegetation types are common throughout region [35]. 'Triodia Mallee' has a canopy of *Eucalyptus dumosa* and *E. socialis* with an understorey of *Triodia scariosa* and mixed shrubs, and occurs mainly on sandier soils typical of dunes. 'Chenopod Mallee' has a canopy of *E. oleosa* and *E. gracilis* with an open understorey of chenopod species, and occurs on heavier soils typical of swales.

Mallee vegetation is fire-prone with large fires (i.e. > 100,000 ha) occurring somewhere in the region on a bidecadal basis [36], although individual sites can go long periods without fire (i.e. >100 years; [37]). Fire is actively managed in the region through prescribed burning and suppression for both asset protection and conservation objectives [36]. Most wildfires are ignited by lightning strikes and are stand-replacing, essentially resetting vegetation succession to 'year-zero' (Fig. 2; [8]).

Site selection

We refer to two datasets in this study derived from two different natural experiments that differed in both their spatial grain and extent. We refer to these as 'landscape-scale' and 'site-scale' datasets throughout, in reference to the spatial scale of the response and predictor variables (i.e. the spatial grain) of the respective datasets.

Landscape-scale data. The landscape-scale dataset consists of 28 study landscapes, each with a 4 km diameter circle (12.6 km²; Fig. 1), distributed throughout a 104, 000 km² study area. These landscapes were selected as part of a broad-scale natural experiment: the Mallee Fire and Biodiversity Project. Study landscapes were selected to allow a comparison of the effects of different approaches to patch mosaic burning on biodiversity, with a particular emphasis on the role of the area and diversity of fire-ages ('pyrodiversity', see [25]). Thus, landscapes were stratified according to number and spatial extent of fire-age classes within the landscape [29]. The fire history of the region was mapped using the ENVI package [38] and then converted to shape files for use in ArcMap version 9.2 [39]. Only fires that occurred post-1971 were mapped due to limited availability of Landsat imagery prior to this time (see [36]).

Site-scale data. We collected site-scale data within a subset of the 28 study landscapes located within the region's largest national park; Murray Sunset National Park (6630 km²; Fig. 1). Ten sites were established within each of 10 of the original study landscapes. Sites were distributed to incorporate a range of fire-age classes (range = 7–105 years), as well as capturing geographic and topographic variation. We established an additional landscape, containing 12 sites, following an experimental burn during the study (fire age = 0 years), resulting in 11 landscapes containing 112 sites. We omitted four sites to comply with ethics permits due to their close proximity to active nesting sites of the endangered malleefowl (*Leipoa ocellata*). This resulted in a total of 108 sites being surveyed. All sites were a minimum of 200 m apart and typically >100 m from the edge of a fire-age class.

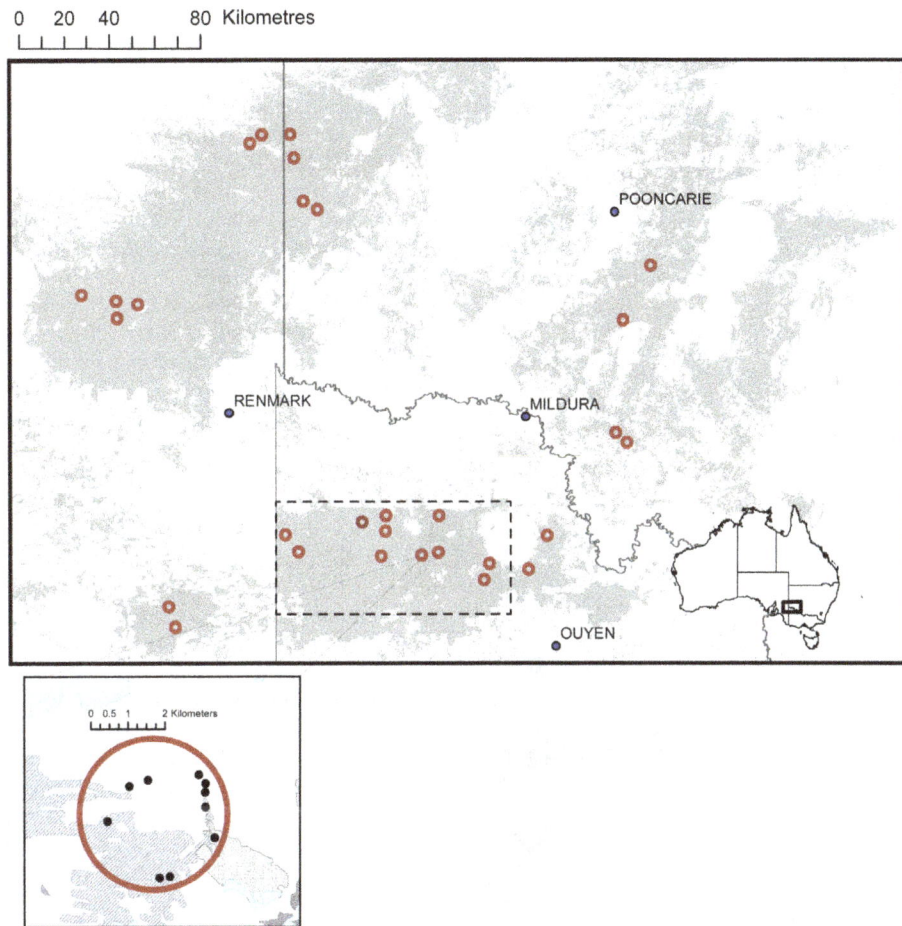

Figure 1. Map of study area showing all landscapes (circles) considered in this study (grey shading indicates mallee vegetation; majority of white areas indicates agricultural land used for grazing and cereal crops). The dashed box shows the spatial extent of the site-scale study. An inset shows an example of a study landscape including the position of 10 sites within where site-scale data were collected. Within the inset, different hatching represents different fire ages.

Predator surveys

Landscape-scale data. We surveyed large mammalian predators using track surveys from three sand-plots within each study landscape (n = 84 in total). Each sand-plot was a 100 m×2 m area smoothed out by dragging a weight along an unsealed vehicle track. The locations of sand-plots within landscapes were chosen to incorporate variation in the topography (dunes and swales) within each landscape. Sand-plots were typically >500 m apart. We checked each sand-plot for tracks once per day by walking along the transect and identifying tracks to species level for three consecutive days in spring (October–November 2007), and again in summer (January–March 2008), resulting in six survey nights for each sand-plot and thus 18 survey nights per landscape. Following checking, sand-plots were smoothed over in preparation for the following day. If the sand-plot was heavily disrupted on one day (due to weather or vehicle disturbance), it was surveyed for an additional day.

Site-scale data. We used camera traps (Passive ScoutGuard 550; ScoutGuard IR Cameras, Australia) to survey for mammalian predators at the site-scale during April–July 2012. We installed one camera per site and deployed each for a minimum of 15 nights. We attached cameras to a post at a height of 0.5 m and positioned them facing southward. Vegetation was removed within the immediate area of the camera to reduce false triggering. A

15 second video was taken each time the camera motion sensor was triggered. To attract predators to the front of the camera from the local vicinity, we placed a scent lure of tuna oil soaked into chemical wadding inside a bait holder made from PVC piping with steel mesh at one end. We positioned the lure 3 m from the base of the camera post, and secured it to the ground with a peg.

Predictor variables

Landscape-scale data. Six predictor variables were chosen to represent the properties of the study landscapes (Table 1). Three of these variables represent the fire history of the landscape: (1) the extent of recently burnt vegetation in the landscape (<10 years since fire; 'recently burned'); (2) the extent of long unburnt vegetation (unburned since 1972; 'long unburned'); and (3) the diversity of fire-ages within a landscape ('fire diversity'). Fire diversity was calculated as the Shannon-Wiener diversity index of the proportional cover of fire age-classes within each landscape.

Three predictor variables were chosen to describe properties of the study landscapes other than fire history. We used a measure of mean solar radiation ('solar radiation') as a surrogate for aridity across the region. The solar radiation variable represents the total amount of solar energy falling on a horizontal space per day (MJ/m^2). We derived these values from a gridded data set (5 km resolution) extending over 18 years (1990–2008; Australian

(a)

(b)

Figure 2. Examples of mallee vegetation with differing fire histories. (a) A recently burned site; (b) A long unburned site.

Bureau of Meteorology http://www.bom.gov.au, 2009). Solar radiation was the mean of the 18 yearly averages of the grids that overlaid each landscape. Solar radiation is negatively correlated with annual rainfall and positively correlated with temperature. We used the proportional extent of *Triodia* mallee vegetation ('Triodia Mallee') within the landscape to capture differences in vegetation types. The extent of mallee vegetation in the study area was mapped in previous work (see [35]). Finally, we used the distance from the centre point of each landscape to the closest area of contiguous non-mallee vegetation ('distance to agricultural land'), to capture the context of landscapes with respect to landscape modification. The area surrounding each reserve is comprised almost entirely of grazing land and grain crops. We calculated distance to agricultural land using ArcGIS [39].

Site-scale data. Eight predictor variables were chosen at the site level (Table 1). The fire history of sites was represented by the time since the last fire ('time since fire'; range: 0–105 years). This was determined using two methods. Recent fire history (since 1972) was calculated using the fire history maps (see [36]). Fire-ages for sites burnt prior to the availability of satellite imagery (i.e. before 1972) were estimated using regression models of the relationship between stem diameter and tree age, and then using stem diameter to estimate the age of trees in areas where fire history was unknown (see [37] for detailed methods). This extended the time since fire axis from 0–32 years to 0–105 years.

Vegetation type was considered as a categorical variable with two levels: Triodia Mallee or Chenopod Mallee ('vegetation type'). We again considered the effects of landscape modification by including the distance of sites to both the border of the National Park (1.72–21.28 km; 'distance to edge') and dirt roads (range: 28–1044 m; 'distance to road'). We used park boundary as a proxy for an edge habitat because the park forms abrupt boundaries with cleared agricultural land and other non-mallee vegetation. We calculated distance variables using ArcGIS [39]. Aridity (solar radiation) was not considered at this scale as the data were collected from a single reserve.

Four additional predictor variables were included to describe vegetation structure at the sites. We established vegetation transects in representative areas 15 m from each camera location.

Table 1. Predictor variables included in models using the landscape-scale and the site-scale datasets.

Dataset	Predictor variable	Description
Landscape-scale data	Recently burned	Extent of landscape burned within 10 years of surveys
	Long unburned	Extent of landscape not burned since 1972 (>35 years since fire)
	Fire diversity	Shannon-Wiener diversity index of the extent of three fire age classes (0–10 years, 11–35 years and >35 years)
	Solar radiation	Long-term average monthly gridded solar exposure (MJ/m^2) from 1990–2008 for each landscape
	Triodia Mallee	Extent of landscape comprised of vegetation type in which *Triodia scariosa* typically occurs
	Distance to agricultural land	Distance from the centre of each landscape to contiguous non-mallee vegetation (m)
Site-scale data	Time since fire	Amount of time since a site last experienced fire (years)
	Bare ground cover	Cover of bare ground present
	Triodia cover	Cover of *Triodia scariosa* <1 m
	Eucalypt cover	Cover of eucalypt shrubs <1 m
	Shrub cover	Cover of non-eucalypt shrubs <1 m
	Vegetation type	Broad vegetation classification (Triodia Mallee or Chenopod Mallee)
	Distance to edge	Distance from each site to the nearest park boundary (m)
	Distance to road	Distance from each site to the nearest road (m)

We recorded substrate type and vegetation structure at 1 m intervals along a 50 m transect using a 2 m structure pole (2 cm diameter) held vertically above the ground. The four variables considered in the analysis represent the cover of open, bare ground ('bare ground cover'), spinifex ('*Triodia* cover'), eucalypt shrubs (defined as Eucalypt trees <3 m in height 'eucalypt cover') and non-eucalypt shrubs ('shrub cover'). Bare ground was included because it gives an approximation of the 'openness' of the vegetation at the ground level. The cover of spinifex, eucalypt shrubs, and non-eucalypt shrubs were included as they form the majority of the ground and understorey structural complexity, and are known to drive fauna in the region [34,40].

Response variables

For both datasets, the response variable was the 'reporting rate' of foxes. At the landscape-scale, we defined reporting rate as the number of nights that a fox was recorded as 'present' and 'absent', respectively, at a sand pad over the 18 nights of sampling per landscape (i.e. three sand-plots surveyed for six nights in each landscape). Likewise, at the site-scale, reporting rate is the number of nights that foxes were and were not detected at the site, respectively, over the course of sampling (i.e. 15 nights).

Statistical analysis

We used generalised linear mixed models (GLMMs) with the Laplace approximation [41] to examine the relationship between response and predictor variables at both landscape and site-scales. In landscape-scale models, we included 'reserve' as a random effect to account for spatial clustering of landscapes in conservation reserves (Fig. 1). Similarly, in the site-scale models, we included 'landscape' as a random effect to account for potential spatial correlation due to the clustering of sites into landscapes. Because we were studying the reporting rate of red foxes, a proportion, we modelled the response variable (at both scales) using a binomial distribution of errors and a logit link function.

For the landscape-scale dataset, we developed a set of candidate models that included all combinations of the six landscape-scale predictor variables. At the site-level, we developed two separate sets of models. As fire affects the variables used to describe vegetation structure (e.g. *Triodia* cover, bare ground cover; [8]), including both fire and vegetation structure variables in the same model could result in unreliable parameter estimates due to colinearity between predictor variables [42]. Thus, one model set (model set 1) included time since fire, vegetation type, distance to edge and distance to road, and a second model set (model set 2) included the vegetation structure variables (bare ground cover, *Triodia* cover, eucalypt cover and shrub cover). All combinations of predictors within the two sets of models were considered, meaning all eight site-level variables were in the same number of models overall. All variables included within a model set had low levels of colinearity (i.e. r <0.5). We tested both datasets for overdisperson using Pearson's residuals [43], and found no evidence of overdispersion.

We compared each set of candidate models using Akaike's Information Criterion corrected for small sample sizes (AICc; [44]). To compare the level of support for each model relative to the most parsimonious model, we calculated the difference (Δ_i) between the AIC$_c$ value of the best model (lowest AIC$_c$ value) and the AIC$_c$ value of each candidate model [44]. We considered models with Δ_i<2 to have substantial support [44]. We also calculated the Akaike weight (w_i) for each model. By summing these weights to calculate predictor weights ($\sum w_i$) for each variable, we were able to explore the influence of individual predictor variables at both the landscape and site level.

When there was no clear 'best model' (i.e. the most parsimonious model was not strongly weighted [w_i<0.9]), we used model averaging to determine the direction and magnitude of the effect of each predictor variable [44]. We considered a variable as important when the associated 95% confidence interval of the averaged estimate did not overlap with zero. We performed all statistical analyses in R version 2.15.1 [45] using the lme4 package [41] and the MuMIn package [46].

Ethics statement

The landscape-scale data were collected with approval from animal ethics committees at La Trobe University (approval number AEC06/07[L]V2) and Deakin University (approval number A41/2006), and permits from the Department of Sustainability and Environment, Victoria (permit 10003791), the Department of Environment and Heritage, South Australia (permit 13/2006), and the National Parks and Wildlife Service, NSW (license number S12030). The site-scale data were collected in accordance with the regulations of the Deakin University Animal Ethics Committee (approval number B10-2012) and in accordance with Department of Sustainability and Environment, Victoria (approval number 10006279).

Results

At the landscape-scale, we recorded fox tracks in 24 of 28 (86%) study landscapes. We detected foxes on 3.32±0.49 (mean ± standard error) of 18 nights per landscape over the total sampling period. Other large-bodied, mammalian predators were uncommon: we detected cats at only 7 of 28 landscapes (25%). At the site-scale, we observed foxes at 62 of 102 (61%) sites (six cameras failed to reach the full 15 day survey period due to fault and were excluded from further analysis i.e. n = 102) and found the species to be widely distributed across the study area. We did not detect any cats at the site-scale over the 15 night sampling period.

At the landscape-scale, all models were a poor fit for the data and explained <6.5% of the variation in the data (% deviance explained). At the site-scale, all models explained <3.5% of the variation in the data. For both datasets, model selection indicated there was a similar level of support for several models (Δ_i<2; Table 2), including the intercept-only model (i.e. only an intercept terms, no predictor variables), which received substantial support at both scales. As no single model was supported as being clearly best (i.e. w_i>0.9; Table 2), we employed multi-model inference using model averaging to estimate the size, direction and uncertainty of parameter effects for fox explaining reporting rate in both datasets.

The model-averaged coefficients for each predictor variable, in both datasets, were small and uncertain. The 95% confidence intervals of all predictor variables overlapped with zero (Fig. 3). The $\sum w_i$ for all predictor variables was low: <0.5 and <0.6 for the landscape- and site-scale datasets respectively.

Graphical exploration of the data further highlights that fox activity was not strongly linked to key predictor variables (Fig. 4). In summary, the data shows that neither fire, nor any other predictor variable measured, affected the reporting rate of foxes at either the landscape- or site-scale.

Discussion

Introduced mesopredators and fire are two processes that shape ecosystems around the world [4,7]. Here, we have shown that a widespread and ecologically devastating mesopredator, the red fox [5], is largely unaffected by fire and is an extreme habitat

Table 2. Model selection results for red fox reporting rate for landscape-scale and sits-scale datasets.

Candidate model	df	LogLik	AIC$_c$	Δ_i	w_i	%Dev
Landscape-scale dataset						
Null model (intercept only)	2	−32.37	69.2	0.00	0.14	0.00
Distance to agricultural land	3	−31.21	69.4	0.21	0.12	3.57
Distance to agricultural land + Triodia Mallee	4	−30.53	70.8	1.58	0.06	5.68
Triodia Mallee	3	−31.99	71.0	1.76	0.06	1.17
Fire diversity	3	−32.10	71.2	1.99	0.05	0.82
Site-scale dataset						
Bare ground cover	3	−65.90	138.0	0.00	0.13	2.16
Bare ground cover + *Triodia* cover	4	−65.07	138.6	0.51	0.10	3.39
Triodia cover	3	−66.26	138.8	0.73	0.09	1.62
Null model (intercept only)	2	−67.36	138.8	0.79	0.09	0.00
Bare ground cover + eucalypt cover	4	−65.67	139.8	1.71	0.06	2.50

Models are shown for which $\Delta_i < 2.0$.

generalist in semi-arid Australia. This result was confirmed using two large, complementary datasets, collected at different times and characterised by differing spatial scales and sampling strategies.

Fire and the red fox

Our findings show that fire does not exert a strong influence on the distribution of the red fox in semi-arid mallee ecosystems. Despite conducting two intensive natural experiments across a broad geographic region, we did not detect a relationship between the reporting rate of foxes and fire history at either the landscape- or site-scale. At the landscape-scale, the red fox was recorded equally often in landscapes dominated by recently burned or long unburned vegetation, and in landscapes with a single fire age-class as those with a diversity of fire ages. At the site-scale, the red fox has a similar reporting rate in recently burned sites as in sites unburned for over a century. The post-fire preferences of the red fox are thus extremely broad, both spatially and temporally (also see [12,47]).

Fire causes significant changes to vegetation structure over century-long time frames in mallee ecosystems [8]. In doing so, fire affects the distribution of a large range of fauna species [34]. Indeed, work conducted within the same study landscapes has shown the large and long-term effects fire has on birds, reptiles, and small mammals [28–30]. The lack of a response to fire by foxes is therefore not typical of native fauna in the region. It also suggests that foxes are not restricted to areas with particular soil or vegetation attributes for denning. This is consistent with foxes not being affected by any of the vegetation attributes measured (e.g. *Triodia* cover, shrub cover etc.).

A related way that fire could influence foxes is by altering the distribution of prey resources. As mentioned above, the distribution of many prey species are significantly affected by fire in the study region (e.g. birds, mammals, reptiles). Thus, foxes occupy a range of post-fire ages despite the strong influence of fire on the type and abundance of prey available. Red foxes have a broad and generalist diet [48], being able to consume a wide range of prey including both vertebrates and invertebrates, and even vegetation [19,20]. Furthermore, foxes are capable of prey switching to capitalize on the most abundant prey source available [18,49], thereby reducing their reliance on any particular prey item. This flexibility in their diet is likely to be a key component of their life

history that allows them to occur within such a broad range of post-fire conditions.

One objection to our findings at the site-scale may be that the local site is not a relevant spatial scale to characterize the effects of fire, as foxes are a relatively large and mobile species. Given the large estimated home ranges of foxes in other parts of arid Australia (e.g. 8–33 km^2; [31]), foxes may select broader areas (i.e. kms^2) that capture their resource requirements across entire landscapes, and this might include a large area of a particular fire-age, or multiple fire ages. Such use of multiple habitat types by foxes has been demonstrated in other systems [23,24]. Our landscape-scale study characterized land mosaics at a large scale relevant to the home range of foxes (12.6 km^2), and still failed to detect any relationship between fox activity and fire history. Therefore, our results suggest that the lack of relationships between fox reporting rate and fire history does not stem from spatial scaling issues. Instead, foxes are resilient towards the effects of fire at multiple temporal and spatial scales.

Climate and distance to modified land

In addition to fire, we examined other variables that could influence the distribution of the red fox. Here, we again found red foxes to be flexible to a broad range of ecological conditions. Foxes displayed no response to an aridity gradient across the study region. This lack of response to aridity is unsurprising, as the geographic range of the red fox spans the northern hemisphere and much of Australia, suggesting the species is capable of coping with a range of climatic conditions.

Despite foxes occupying a broad climatic niche in space, fluctuations in populations do occur in response to extreme weather events. For example, fox populations in arid areas rise rapidly following high rainfall events, in response to increased prey availability [50]. Our site-scale study was carried out during a year of record high rainfall (Australian Bureau of Meteorology, Ouyen Station). Considered in isolation, this may suggest that the wide distribution of the fox was partly due to a productivity-related increase in food resources (predominantly populations of native and introduced rodents; [40]). However, the landscape-scale data were collected near the end of a severe, decade-long drought. Foxes were widely distributed across the region despite the drought. This indicates that, in semi-arid Australia, foxes can be

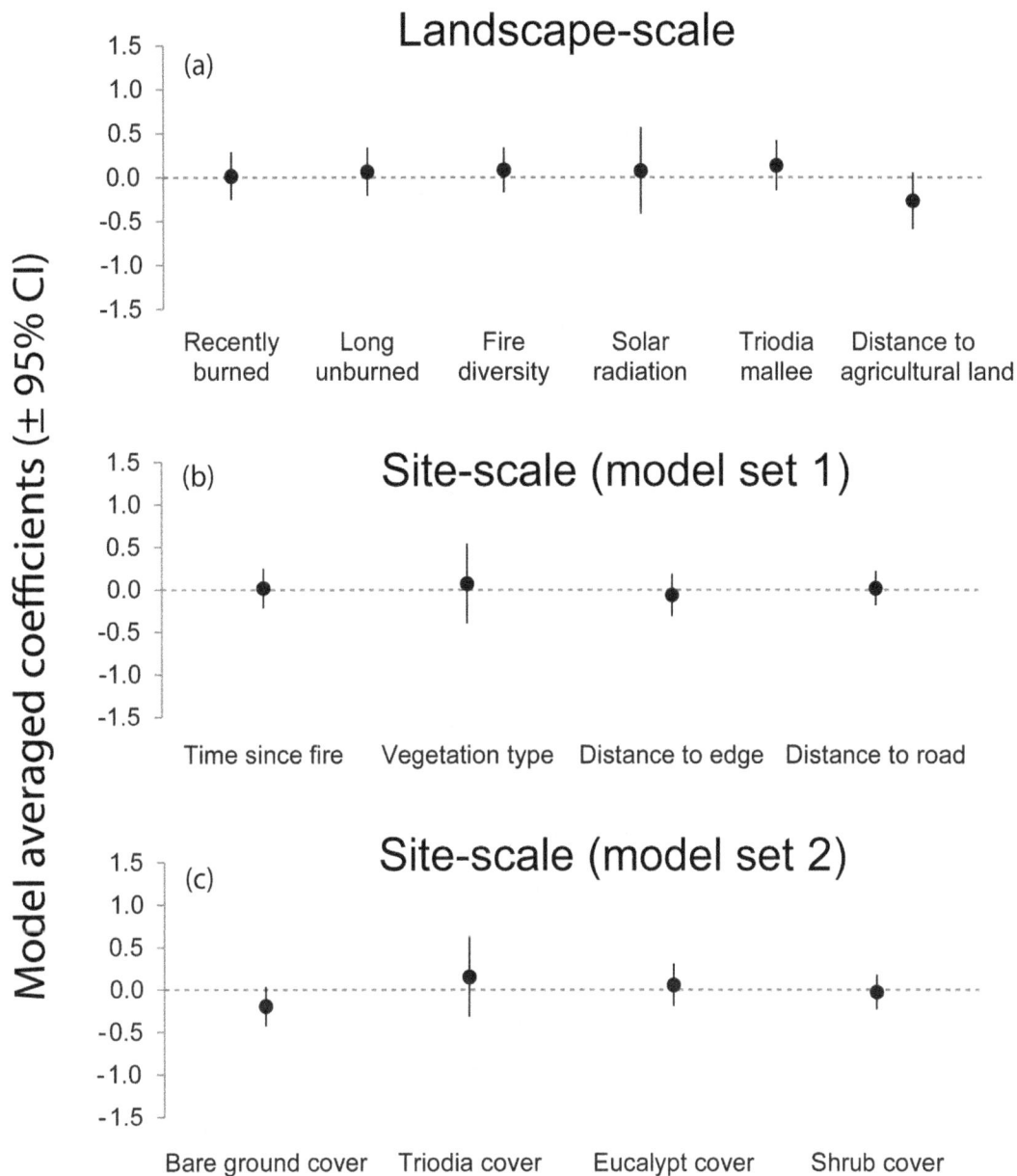

Figure 3. Model-averaged regression coefficients and 95% confidence intervals of models describing the reporting rate of foxes at both the landscape-scale (a) and site-scale (b and c).

widespread during a broad range of climatic conditions and despite fluctuations in their prey populations which accompany climactic extremes [40].

Some studies have found that foxes are positively associated with edges between fragments and modified land (e.g. agricultural land) [22,32]. Our results indicate that foxes do not show a preference for edge habitats in mallee ecosystems, despite our sites and landscapes capturing a broad gradient of distances to agricultural land, from <2 km to >30 km. Edge habitats may be more important for foxes in highly fragmented landscapes, where they occur with small remnant patches of wooded vegetation which provide the only available cover [22]. While the mallee region has been subject to large amounts of land clearing, there are still relatively large intact areas of native vegetation. Edges may be less important in this region because the interior mallee vegetation provides sufficient shelter and prey.

Nevertheless, it is also possible that edge effects occur closer to the agricultural boundary than we sampled (i.e. <2 km).

The use of roads and tracks by foxes is also well documented [51,52]. Foxes have been found to be more abundant along roadsides [33]. In the mallee system, however, we found similar reporting rates at varying distances (28–1044 m) from roads, indicating foxes use areas well away from roads equally as often as sites close to roads. One hypothesis for the use of roads by foxes is that they provide 'runways' which facilitate movement and allow access to foraging areas that would be otherwise difficult to reach [51,52]. In contrast to environments with a dense understory, mallee vegetation is relatively open, and is unlikely to limit the movement of foxes to roadsides. This may explain the lack of preference for sites near roads in the current study.

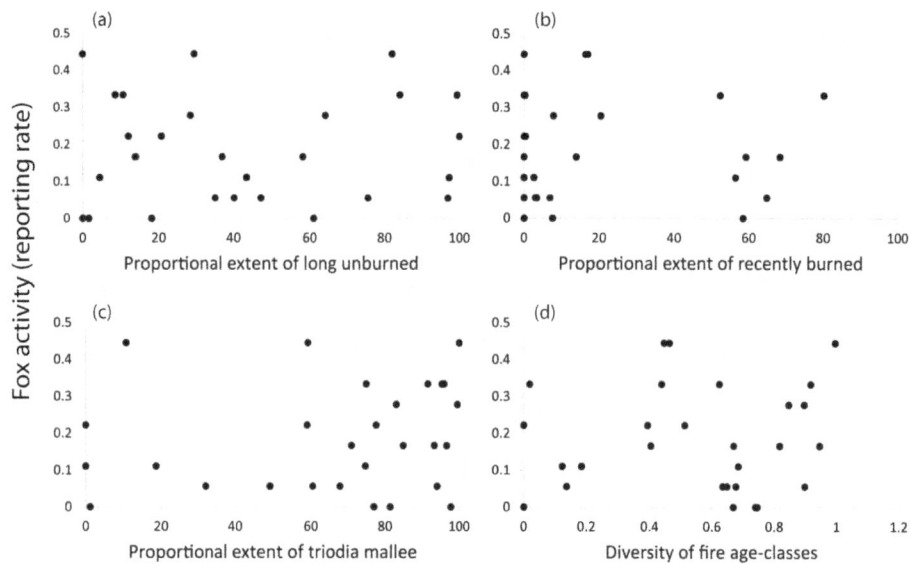

Figure 4. Relationships between the reporting rate of the red fox and the properties of fire mosaics. Circles are raw data points.

Implications

Fire is used as a conservation tool in Australia and around the world [25]. This study suggests it is unlikely that any particular approach to fire management will alter the reporting rate of the red fox in the semi-arid mallee systems of Australia. However, the presence of foxes in recently burned sites and landscapes is a concern. Predation by invasive mesopredators has been hypothesized as a cause of low post-fire survival in reptiles [14] and mammals [12,13], due to the reduced cover available in burnt habitats. Although we found no effect of fire history on red fox occurrence, it is possible that predation pressure differs across fire ages due to increased predation risk in recently burned areas. Thus, assessing predation pressure directly across a range of post-fire ages is an important area for further research.

The loss of apex predators can cause smaller predators to increase in abundance, expand their range, and change their temporal activity; this is known as 'mesopredator release' [2,6]. Red foxes have been shown to select particular habitats which may allow them to avoid dominant predators (e.g. coyotes; [53]). As such, one further explanation for the lack of obvious habitat selection by foxes in this system may be the lack of regulating predators. In other Australian systems, the presence of the dingo, Australia's largest terrestrial apex predator, has been shown to affect fox distributions [3]. Dingoes are largely extinct from the study area but were once common, and as such there is no direct

regulation of the abundance or distribution of foxes via biotic interactions. Thus, one potential way to control red foxes in mallee communities is by reinstating dingoes as the apex predator. As this is likely to be a controversial idea owing to the proximity of mallee vegetation to agricultural land and livestock, trialing reintroductions in a controlled and experimental way would be an important first step towards a proof of concept, and a potential solution to this complex conservation issue.

Acknowledgments

Thanks to all members of the Mallee Fire and Biodiversity Team, particularly Andrew Bennett, Mike Clarke, Lisa Farnsworth and Lauren Brown, and to the many volunteers, agency staff, and land owners who assisted with logistics and field work.

Author Contributions

Conceived and designed the experiments: DGN CJP EGR LTK. Performed the experiments: DGN CJP LTK. Analyzed the data: DGN CJP. Contributed to the writing of the manuscript: CJP DGN LTK EGR.

References

1. Estes JA, Terborgh J, Brashares JS, Power ME, Berger J, et al. (2011) Trophic downgrading of planet Earth. Science 333: 301–306.
2. Ritchie EG, Johnson CN (2009) Predator interactions, mesopredator release and biodiversity conservation. Ecology Letters 12: 982–998.
3. Letnic M, Ritchie EG, Dickman CR (2012) Top predators as biodiversity regulators: the dingo Canis lupus dingo as a case study. Biological Reviews 87: 390–413.
4. Salo P, Korpimäki E, Banks PB, Nordström M, Dickman CR (2007) Alien predators are more dangerous than native predators to prey populations. Proceedings of the Royal Society B: Biological Sciences 274: 1237–1243.
5. Johnson C (2006) Australia's Mammal Extinctions: a 50000 year history. New York: Cambridge University Press.
6. Crooks KR, Soule ME (1999) Mesopredator release and avifaunal extinctions in a fragmented system. Nature 400: 563–566.
7. Bowman DMJS, Balch JK, Artaxo P, Bond WJ, Carlson JM, et al. (2009) Fire in the Earth system. Science 324: 481–484.

8. Haslem A, Kelly LT, Nimmo DG, Watson SJ, Kenny SA, et al. (2011) Habitat or fuel? Implications of long-term, post-fire dynamics for the development of key resources for fauna and fire. Journal of Applied Ecology 48: 247–256.
9. Smit IPJ, Asner GP, Govender N, Kennedy-Bowdoin T, Knapp DE, et al. (2010) Effects of fire on woody vegetation structure in African savanna. Ecological Applications 20: 1865–1875.
10. Letnic M, Tamayo B, Dickman CR (2005) The responses of mammals to La Nina (El Nino Southern Oscillation) - associated rainfall, predation and wildfire in central Australia. Journal of Mammalogy 86: 689–703.
11. Driscoll DA, Lindenmayer DB, Bennett AF, Bode M, Bradstock RA, et al. (2010) Fire management for biodiversity conservation: Key research questions and our capacity to answer them. Biological Conservation 143: 1928–1939.
12. Arthur AD, Catling PC, Reid A (2012) Relative influence of habitat structure, species interactions and rainfall on the post-fire population dynamics of ground-dwelling vertebrates. Austral Ecology 37: 958–970.

13. Sutherland EF, Dickman CR (1999) Mechanisms of recovery after fire by rodents in the Australian environment: a review. Wildlife Research 26: 405–419.

14. Smith AL, Bull CM, Driscoll DA (2012) Post-fire succession affects abundance and survival but not detectability in a knob-tailed gecko. Biological Conservation 145: 139–147.

15. Woinarski JCZ, Legge S, Fitzsimons JA, Traill BJ, Burbidge AA, et al. (2011) The disappearing mammal fauna of northern Australia: context, cause, and response. Conservation Letters 4: 192–201.

16. Kinnear JE, Onus ML, Bromilow RN (1988) Fox control and rock-wallaby population dynamics. Wildlife Research 15: 435–450.

17. Risbey DA, Calver MC, Short J, Bradley JS, Wright IW (2000) The impact of cats and foxes on the small vertebrate fauna of Heirisson Prong, Western Australia. II. A field experiment. Wildlife Research 27: 223–235.

18. Catling PC (1988) Similarities and contrasts in the diets of foxes, *Vulpes vulpes*, and cats, *Felis catus*, relative to fluctuating prey populations and drought. Wildlife Research 15: 307.

19. Glen AS, Fay AR, Dickman CR (2006) Diets of sympatric red foxes *Vulpes vulpes* and wild dogs *Canis lupus* in the Northern Rivers Region, New South Wales. Australian Mammalogy 28: 101–104.

20. Risbey DA, Calver MC, Short J (1999) The impact of cats and foxes on the small vertebrate fauna of Heirisson Prong, Western Australia. I. Exploring potential impact using diet analysis. Wildlife Research 26: 621–630.

21. Letnic M and Dickman CR (2005) The responses of small mammals to patches regenerating after fire and rainfall in the Simpson Desert, central Australia. Austral Ecology 30: 24–39.

22. Graham CA, Maron M, McAlpine CA (2012) Influence of landscape structure on invasive predators: feral cats and red foxes in the brigalow landscapes, Queensland, Australia. Wildlife Research 39: 661–676.

23. Lucherini M, Lovari S, Crema G (1995) Habitat use and ranging behaviour of the red fox (Vulpes vulpes) in a Mediterranean rural area: is shelter availability a key factor? Journal of Zoology 237: 577–591.

24. Meek PD, Saunders G (2000) Home range and movement of foxes (*Vulpes vulpes*) in coastal New South Wales, Australia. Wildlife Research 27: 663–668.

25. Parr CL, Andersen AN (2006) Patch mosaic burning for biodiversity conservation: a critique of the pyrodiversity paradigm. Conservation Biology 20: 1610–1619.

26. Southgate R, Paltridge R, Masters P, Ostendorf B (2007) Modelling introduced predator and herbivore distribution in the Tanami Desert, Australia. Journal of Arid Environments 68: 438–464.

27. Gosper CR, Yates CJ, Prober SM (2013) Floristic diversity in fire-sensitive eucalypt woodlands shows a 'U'-shaped relationship with time since fire. Journal of Applied Ecology 50: 1187–1196.

28. Watson SJ, Taylor RS, Nimmo DG, Kelly LT, Haslem A, et al. (2012) Effects of time since fire on birds: How informative are generalized fire response curves for conservation management? Ecological Applications 22: 685–696.

29. Nimmo DG, Kelly LT, Spence-Bailey LM, Watson SJ, Taylor RS, et al. (2013) Fire mosaics and reptile conservation in a fire-prone region. Conservation Biology 27: 345–353.

30. Kelly LT, Nimmo DG, Spence-Bailey LM, Taylor RS, Watson SJ, et al. (2012) Managing fire mosaics for small mammal conservation: a landscape perspective. Journal of Applied Ecology 49: 412–421.

31. Moseby KE, Stott J, Crisp H (2009) Movement patterns of feral predators in an arid environment – implications for control through poison baiting. Wildlife Research 36: 422–435.

32. Catling P, Burt R (1995) Why are red foxes absent from some eucalypt forests in eastern New South Wales? Wildlife Research 22: 535–545.

33. Towerton AL, Penman TD, Kavanagh RP, Dickman CR (2011) Detecting pest and prey responses to fox control across the landscape using remote cameras. Wildlife Research 38: 208–220.

34. Bradstock RA, Cohn JS (2002) Fire regimes and biodiversity in semi-arid mallee ecosystems. In: R. A Bradstock, J. E Williams and M. A Gill, editors. Flammable Australia: The fire regimes and biodiversity of a continent. Cambridge: Cambridge University Press. 238–258.

35. Haslem A, Callister KE, Avitabile SC, Griffioen PA, Kelly LT, et al. (2010) A framework for mapping vegetation over broad spatial extents: A technique to aid land management across jurisdictional boundaries. Landscape and Urban Planning 97: 296–305.

36. Avitabile SC, Callister KE, Kelly LT, Haslem A, Fraser L, et al. (2013) Systematic fire mapping is critical for fire ecology, planning and management: A case study in the semi-arid Murray Mallee, south-eastern Australia. Landscape and Urban Planning 117: 81–91.

37. Clarke MF, Avitabile SC, Brown L, Callister KE, Haslem A, et al. (2010) Ageing mallee eucalypt vegetation after fire: insights for successional trajectories in semi-arid mallee ecosystems. Australian Journal of Botany 58: 363–372.

38. ITT (2005) ENVI. Version 4.2. Boulder, Colorado: ITT Industries.

39. Environmental Systems Research Institute (2007) Arc View. Version 9.2. Redlands, California: ESRI.

40. Kelly LT, Dayman R, Nimmo DG, Clarke MF, Bennett AF (2013) Spatial and temporal drivers of small mammal distributions in a semi-arid environment: The role of rainfall, vegetation and life-history. Austral Ecology 38: 786–797.

41. Bates D, Maechler M, Bolker B (2012) lme4: Linear mixed-effects models using S4 classes. R package (Version 0.999999-0). Available: http://CRAN.R-project.org/package=lme4.

42. Quinn GP, Keough MJ (2002) Experimental design and data analysis for biologists. New York: Cambridge University Press.

43. Zuur AF, Ieno EN, Walker N, Saveliev AA, Smith GM (2009) Mixed effects models and extensions in ecology with R. New York: Springer.

44. Burnham KP, Anderson DR (2002) Model selection and multi-model inference. New York: Springer.

45. R Development Core Team (2012) R: A language and environment for statistical computing. Vienna, Austria. Available: http://www.R-project.org/: R Foundation for Statistical Computing.

46. Bartoń K (2012) MuMIn: Multi-model inference. R package (Version 1.7.11). Available: http://CRAN.Rproject.org/package=MuMIn.

47. Catling PC, Coops N, Burt RJ (2001) The distribution and abundance of ground-dwelling mammals in relation to time since wildfire and vegetation structure in south-eastern Australia. Wildlife Research 28: 555–565.

48. White JG, Gubiani R, Smallman N, Snell K, Morton A (2006) Home range, habitat selection and diet of foxes (*Vulpes vulpes*) in a semi-urban riparian environment. Wildlife Research 33: 175–180.

49. Leckie FM, Thirgood SJ, May R, Redpath SM (1998) Variation in the diet of red foxes on Scottish moorland in relation to prey abundance. Ecography 21: 599–604.

50. Pavey CR, Eldridge SR, Heywood M (2008) Population dynamics and prey selection of native and introduced predators during a rodent outbreak in arid Australia. Journal of Mammalogy 89: 674–683.

51. Carter A, Luck GW, McDonald SP (2012) Ecology of the red fox (Vulpes vulpes) in an agricultural landscape. 2. Home range and movements. Australian Mammalogy 34: 175–187.

52. Frey SN, Conover MR (2006) Habitat use by meso-predators in a corridor environment. The Journal of Wildlife Management 70: 1111–1118.

53. Gosselink TE, Deelen TRV, Warner RE, Joselyn MG (2003) Temporal habitat partitioning and spatial use of coyotes and red foxes in east-central Illinois. The Journal of Wildlife Management 67: 90–103.

8

Demographic Diversity and Sustainable Fisheries

8

Masami Fujiwara*

Department of Wildlife and Fisheries Sciences, Texas A&M University, College Station, Texas, United States of America

Abstract

Fish species are diverse. For example, some exhibit early maturation while others delay maturation, some adopt semelparous reproductive strategies while others are iteroparous, and some are long-lived and others short-lived. The diversity is likely to have profound effects on fish population dynamics, which in turn has implications for fisheries management. In this study, a simple density-dependent stage-structured population model was used to investigate the effect of life history traits on sustainable yield, population resilience, and the coefficient of variation (CV) of the adult abundance. The study showed that semelparous fish can produce very high sustainable yields, near or above 50% of the carrying capacity, whereas long-lived iteroparous fish can produce very low sustainable yields, which are often much less than 10% of the carrying capacity. The difference is not because of different levels of sustainable fishing mortality rate, but because of difference in the sensitivity of the equilibrium abundance to fishing mortality. On the other hand, the resilience of fish stocks increases from delayed maturation to early maturation strategies but remains almost unchanged from semelparous to long-lived iteroparous. The CV of the adult abundance increases with increased fishing mortality, not because more individuals are recruited into the adult stage (as previous speculated), but because the mean abundance is more sensitive to fishing mortality than its standard deviation. The magnitudes of these effects vary depending on the life history strategies of the fish species involved. It is evident that any past high yield of long-lived iteroparous fish is a transient yield level, and future commercial fisheries should focus more on fish that are short-lived (including semelparous species) with high compensatory capacity.

Editor: Mark S. Boyce, University of Alberta, Canada

Funding: This work was funded in part by the United States National Oceanic and Atmospheric Administration (DOC Contract-NFFR7500-10-18114). The funders had no role in study design, data collection and analysis, decision to publish, or preparation of the manuscript. No additional external funding received for this study.

Competing Interests: The author has declared that no competing interests exist.

* E-mail: fujiwara@tamu.edu

Introduction

Organisms exhibit a wide range of life history strategies [1]. For example, some have early maturation while others delay maturation, some adopt semelparous reproductive strategies while others are iteroparous, and some are long-lived and others short-lived. Such demographic diversity is likely to have profound effects on population dynamics. As fisheries management worldwide faces the challenge of managing fish stocks that encompass broad demographic diversity, there is great interest in investigating the relationship between life history traits and population dynamics [2,3,4,5,6,7], and the need to adjust fish stock management based on fish life history strategies, e.g. [8,9,10,11,12]. In this study, a simple population model was used to assess the effects of demographic diversity on population dynamics under fishing mortality.

One of the most important concepts in fishery management is sustainable yield, see [13,14,15]. In a general sense, sustainable yield is a consistent catch over a long (often infinite) period of time, e.g. [16] and often equated to the catch level that results in a stable equilibrium abundance under a deterministic fishery model. When it is at the maximum level, the sustainable yield is called the maximum sustainable yield. When a model-based MSY is available, the desirable yield or fishing mortality can be determined after incorporating precautionary measures that reflect uncertainties, including fluctuations in the environment, errors in parameter estimates, and deficiencies in model formulations [17].

As sustainable yield plays an important role in fishery management, I investigate how sustainable yield and MSY varies with different life history traits of target fish and the fishing mortality rate.

Another objective of this study was to investigate how the sensitivity of transient population dynamics is affected by the life history traits of fish and the fishing mortality rate. Understanding transient dynamics is important in the management of natural resources [18,19,20] because a large part of what we actually observe are transient dynamics. In this study, transient population dynamics are measured in two ways. The first involves estimating the coefficient of variation (CV) of adult fish abundance [21] under stochastically fluctuating juvenile survival, which is commonly thought to be the fish population parameter most sensitive to fluctuating environmental conditions. A recent study demonstrated that increased fishing mortality also increases the CV of adult fish abundance [21]. There is great interest in understanding how this measure is affected by various factors, e.g. [22] because unpredictability associated with a large fluctuation in fish abundance will reduce the optimal fishing quota [23]. The second measure of transient dynamics involves calculation of the resilience of the population abundance near a stable equilibrium point. Resilience is a measure of the time that takes for a population to return to asymptotic dynamics after a perturbation, e.g. [24,25,26]. Resilience is a measure of intermediate-term transient dynamics, whereas the CV of adult fish abundance is a measure of short-term transient dynamics. In this study, the CV of adult fish abundance, resilience, and MSY were used to characterize short-, intermediate-, and long-term dynamics, respectively, of fishery models incorporating demographic diversity.

In addition to life history traits, a major factor affecting population dynamics is density- dependent regulation, which makes model equations non-linear. Density dependence is necessary for fishery sustainability, and the processes involved have been the subject of much research since Verhulst [27] developed the logistic model. Theoretical understanding of the potential dynamics that can arise from deterministic density-dependent population models is well established. The dynamics converge asymptotically to a stable equilibrium, cycle, aperiodic loop, or chaotic attractor, e.g. [28,29,30]. Two types of density-dependent processes are commonly used in fisheries population models. The first is an over-compensatory density-dependent process represented by the Ricker model [31], and the second is compensatory process represented by the Beverton-Holt model [32]. The focus of this study was on density-dependent regulation that results from resource limitation (e.g. competition for available food). This type of regulation is likely to be more common and tends to lead to the Beverton-Holt density-dependent process. Furthermore, the Beverton-Holt density dependence is discrete-time equivalent to the classic logistic model.

Methods

Model

The aim of the study was to investigate the effect of demographic diversity on transient and asymptotic population dynamics under various fishing mortality rates, and involved the use of a density-dependent two-stage matrix population model. Although the model is simple, it can incorporate a wide range of life history traits by varying parameter values. In particular, when the life history is semelparous, the model is discrete-time equivalent to a simple logistic (Shaffer) model, which is still widely used in fishery modeling. Therefore, the model presented herein is more general and applicable to real fish populations than the majority of existing fishery population models. Finally, because of the simplicity, asymptotic abundance, sustainable yield, MSY, and resilience can be calculated analytically. This allowed exploration of the model under a wide range of parameter values.

The model consists of two stages (Fig. 1). The first (the juvenile stage) is for reproductively immature individuals, and the second (the adult stage) is for mature individuals. The model includes four population parameters: juvenile survival rate (s), adult survival rate (p), maturation rate (m), and fertility (r). It was assumed that the time unit was one year, and therefore the rates were annual rates. Fertility (r) is the product of the survival of adults until the reproductive season and the annual per capita fecundity (the

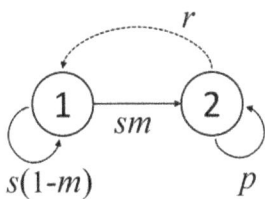

Figure 1. Lifecycle graph of a two-stage model with stage 1 for juveniles and stage 2 for adults. Arrows indicate possible contributions of one stage to the other. s: annual survival rate of juveniles, m: annual maturation rate, p: annual survival rate of adults, and r: annual fertility rate.

Table 1. Juvenile and adult stage duration after the first reproduction under the parameter values considered in the analysis.

		m=0.1			m=0.5			m=0.9		
		Figure Panel	Juvenile Duration (years)	Adult Duration (years)	Figure Panel	Juvenile Duration (years)	Adult Duration (years)	Figure Panel	Juvenile Duration (years)	Adult Duration (years)
s=0.1	p=0.0	4a	1.10	1	4b	1.05	1	4c	1.01	1
	p=0.5	4d	1.10	2	4e	1.05	2	4f	1.01	2
	p=0.9	4g	1.10	10	4h	1.05	10	4i	1.01	10
s=0.5	p=0.0	3a	1.82	1	3b	1.33	1	3c	1.05	1
	p=0.5	3d	1.82	2	3e	1.33	2	3f	1.05	2
	p=0.9	3g	1.82	10	3h	1.33	10	3i	1.05	10
s=0.9	p=0.0	2a	5.26	1	2b	1.82	1	2c	1.10	1
	p=0.5	2d	5.26	2	2e	1.82	2	2f	1.10	2
	p=0.9	2g	5.26	10	2h	1.82	10	2i	1.10	10

The panel indices correspond to those in Figures 2–4. Life histories on Figures 5, 8, and 11 are the same as Figure 2. Life histories on Figures 6, 9, and 12 are the same as Figure 3. Life histories on Figures 7, 10, and 13 are the same as Figure 4.

number of eggs). This type of matrix population model is called a post-breeding model [33], in which the population abundance is determined immediately after spawning event. Because the survival rate of eggs and juveniles are often substantially different for fish, it is assumed that the multiplicative difference between the egg and juvenile survival is also implicitly included in r. When the four parameters in the model are density independent, the densities of juveniles (n_1) and adults (n_2) are "projected" from year t to the next year, as follows:

$$\begin{bmatrix} n_1 \\ n_2 \end{bmatrix}_{t+1} = \begin{bmatrix} s(1-m) & r \\ sm & p \end{bmatrix} \begin{bmatrix} n_1 \\ n_2 \end{bmatrix}_t \quad (1)$$

where the subscripts of the vectors denote year. The matrix is in general termed a population matrix. The dominant eigenvalue of the matrix gives the annual asymptotic population growth rate, and an associated right eigenvector gives the relative asymptotic densities between the two stages, see [31].

By varying parameter values, various life history strategies can be incorporated into the model. For example, by increasing m from a low to a high value, the life history strategy is modified from maturing early (precocious) to maturing late (delayed maturation). Together m and s determine the mean age of maturation. By increasing p from 0 toward 1 the life history strategy changes from semelparous to iteroparous. It is noted that even if $p = 0$, a positive

value of r ensures that some individuals will reproduce once before their death. The generation time of the organisms is determined by m, s and p. Under this model, the average time individuals spend in the juvenile stage before maturing is given by $(1-s(1-m))^{-1}$, and the average time individuals spend in the adult stage is given by $(1-p)^{-1}$. These average stage durations can be used to approximate actual species of fisheries interest using modeled life history strategies.

Fishing mortality affects adult survival. It is incorporated into the model by multiplying the rate of surviving from fishing mortality (f) with parameters in the projection matrix. Thus, $1-f$ is the annual fishing mortality rate. With fishing mortality included, the equation becomes:

$$\begin{bmatrix} n_1 \\ n_2 \end{bmatrix}_{t+1} = \begin{bmatrix} s(1-m) & rf \\ sm & pf \end{bmatrix} \begin{bmatrix} n_1 \\ n_2 \end{bmatrix}_t. \quad (2)$$

The fertility term is also multiplied by f because r includes the adult survival rate as described previously. The sequence of event is that (1) fishing mortality and natural adult mortality and then (2) spawning. The use of a post-breeding matrix population model avoids spurious results in which a population is persistent with the adult fishing mortality rate of 1 because newly recruited adults always have at least one chance of reproduction. An alternative way of modeling is to have a pre-breeding model, and the <2,1>

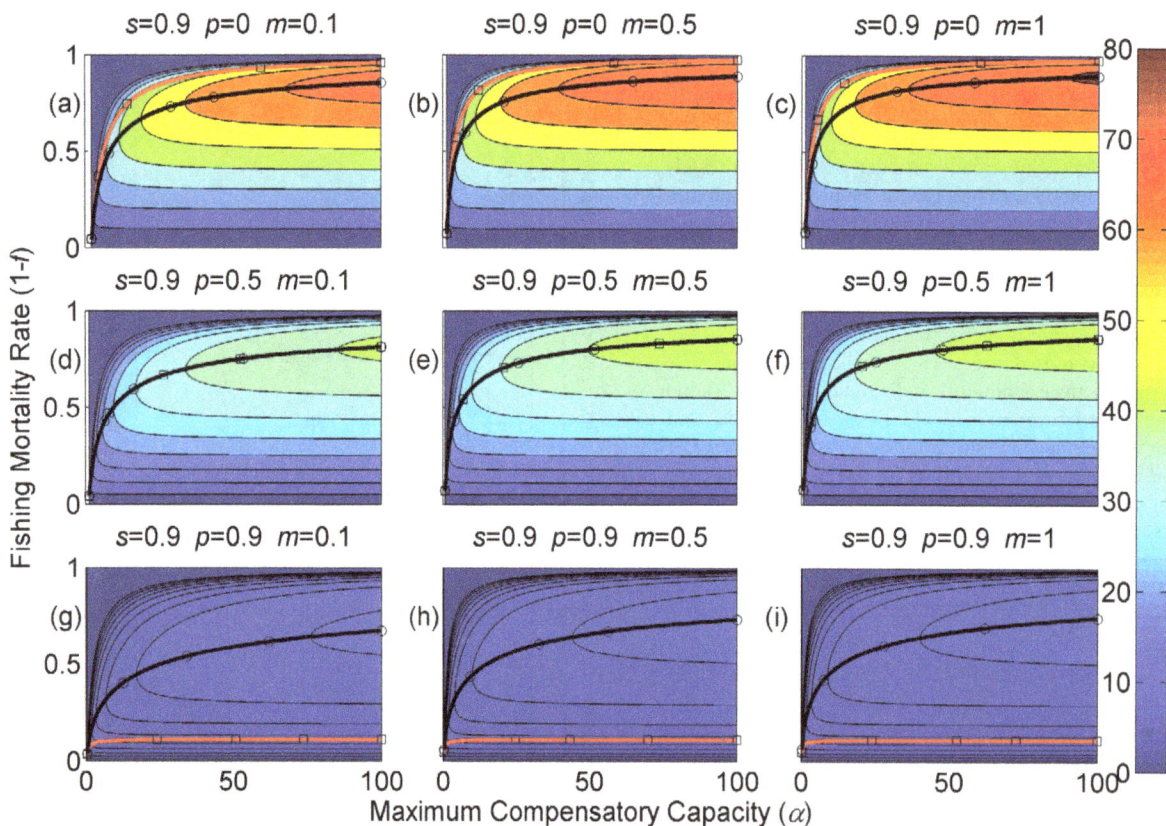

Figure 2. Sustainable yield (shown in contours) as a function of the annual fishing mortality rate and the maximum compensatory capacity when $s = 0.9$. The black curves with circle markers indicate the annual fishing mortality rate at the maximum sustainable yield as a function of the maximum compensatory capacity. The red curves with square markers indicate the annual fishing mortality rate when the equilibrium adult abundance is 50% of the carrying capacity (i.e. asymptotic adult abundance when there is no fishing mortality) as a function of the maximum compensatory capacity. Each panel represents a life history strategy defined by the parameters shown above.

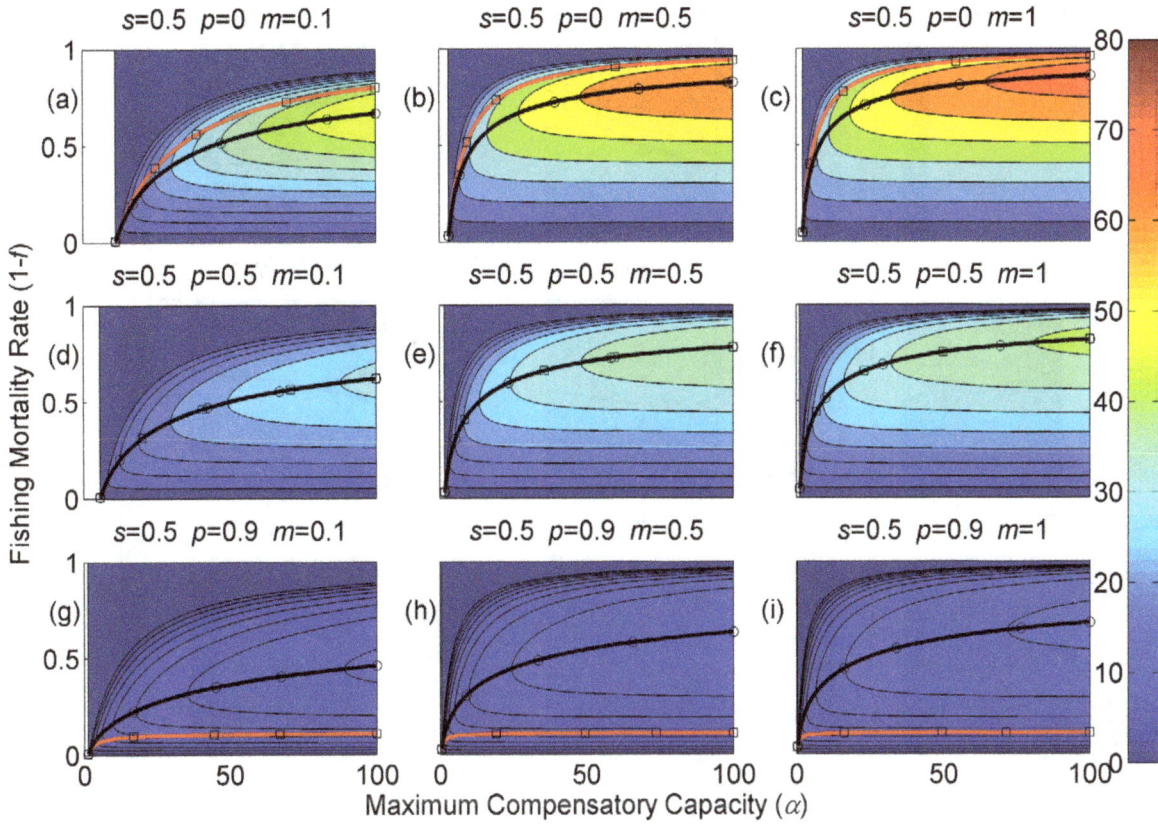

Figure 3. Sustainable yield (shown in contours) as a function of the annual fishing mortality rate and the maximum compensatory capacity when $s = 0.5$. The black curves with circle markers indicate the annual fishing mortality rate at the maximum sustainable yield as a function of the maximum compensatory capacity. The red curves with square markers indicate the annual fishing mortality rate when the equilibrium adult abundance is 50% of the carrying capacity (i.e. asymptotic adult abundance when there is no fishing mortality) as a function of the maximum compensatory capacity. Each panel represents a life history strategy defined by the parameters shown above.

entry of the population matrix is multiplied by f instead of the $<1,2>$ entry.

The Beverton-Holt density-dependent process is also incorporated into the model. Here, it was assumed that the effect of the density-dependent process is on fecundity and/or the survival during the first year of life. Therefore, it was incorporated into the fertility term, which includes both fecundity and the difference in survival rate between egg and juvenile stage, as:

$$r_t = \frac{\alpha}{1 + \beta f n_2^{(t)}}, \tag{3}$$

which replaces r in equation (2). Parameter α is the maximum per capita fertility rate at a low adult abundance, and it is referred to here as the maximum compensatory capacity (α). For a given value of α, parameter β determines the equilibrium adult abundance. It can be shown that under this model if there is an equilibrium point at which both juvenile and adult densities are positive, the equilibrium point is stable.

In equation (1), there are four population parameters. However, under the stable equilibrium point the dominant eigenvalue of the population matrix is 1. Therefore, the number of free population parameters is three under the equilibrium point without fishing mortality, and consequently the fertility rate at the equilibrium point was chosen to be varied as a function of m, s and p. Thus, by setting the eigenvalue to 1, the

fertility term at the equilibrium under no fishing mortality (r^*) can be solved as follows.

$$r^* = \frac{(1-p)(ms-s+1)}{ms} \tag{4}$$

There are two density-dependent parameters (α and β) in equation (3). However, for given values of α, r^* and an equilibrium adult abundance n_2^*, parameter β is expressed as:

$$\beta = \frac{1}{n_2^*}\left(\frac{\alpha}{r^*} - 1\right), \tag{5}$$

when fishing mortality rate is 0 ($f = 1$). In the subsequent analyses, the equilibrium adult abundance without any fishing mortality was set at 100 under all life history traits. Hereafter, this level is referred to as the carrying capacity. Setting the carrying capacity at 100 allows any reduction in the equilibrium abundance because of fishing mortality to be interpreted as a percent decline from the carrying capacity. Similarly, sustainable yield can also be related to the percent of the carrying capacity.

The use of a simple model means various simplifying assumptions were made. For example, instead of modeling both biomass and fish abundance (common practices in fishery

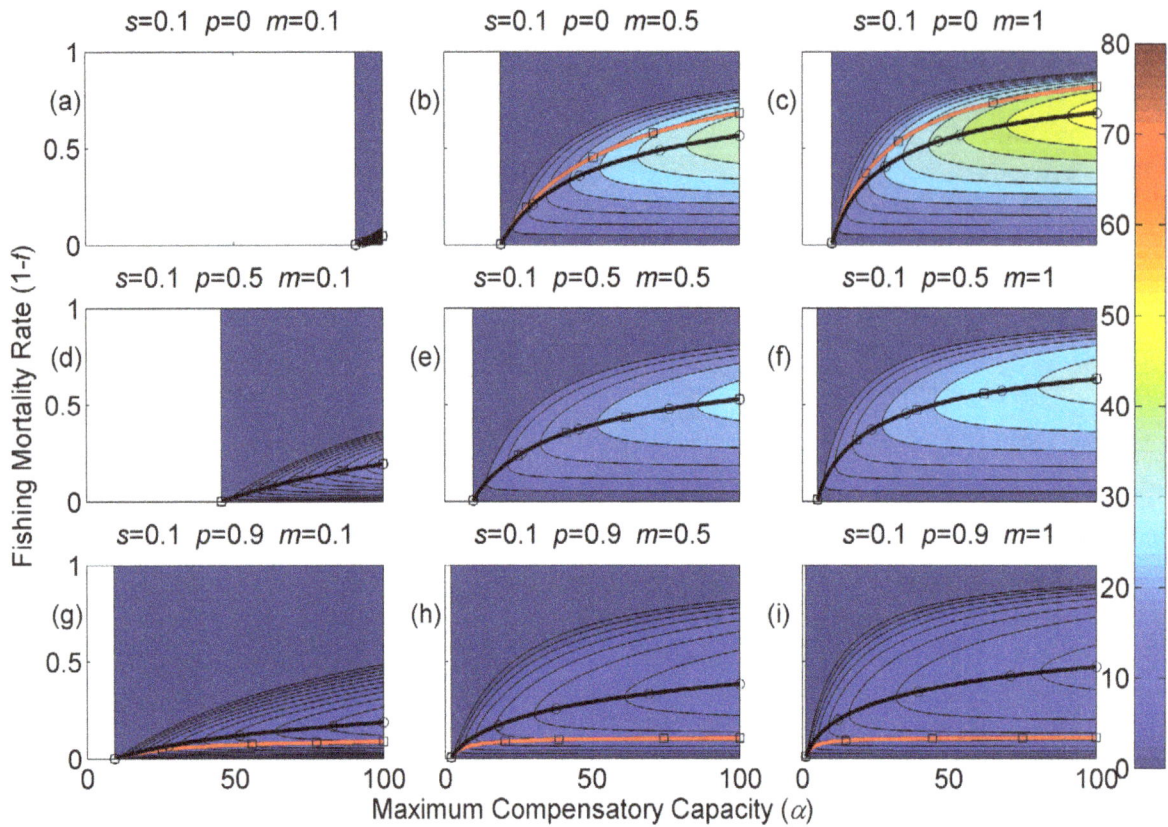

Figure 4. Sustainable yield (shown in contours) as a function of the annual fishing mortality rate and the maximum compensatory capacity when $s = 0.1$. The black curves with circle markers indicate the annual fishing mortality rate at the maximum sustainable yield as a function of the maximum compensatory capacity. The red curves with square markers indicate the annual fishing mortality rate when the equilibrium adult abundance is 50% of the carrying capacity (i.e. asymptotic adult abundance when there is no fishing mortality) as a function of the maximum compensatory capacity. Each panel represents a life history strategy defined by the parameters shown above.

modeling), the focus was placed solely on the latter. Similarly, instead of categorizing fish into different sizes as well as maturation status, the model only incorporates two stages; juveniles and adults. Consequently, although the model can incorporate a wide range of fish life history strategies, it does not incorporate all strategies. Instead, the model was designed to present a broad picture of the population dynamics of fish over a wide range of life history strategies, thus complementing existing efforts to build and analyze more complex fishery models, which are often stock specific.

The model in this study is closely related to stock recruitment/ replacement-line models, which are frequently used in fisheries management, e.g. [14,15]. In the current model, the stock recruitment relationship, e.g. [34] is given by the relationship between $n_{2,t}$ (stock) and $n_{1,t+1}sm$ (new recruitment). The slope (often the inverse of the slope) of the replacement line, which is stock abundance per new recruit, is given by the ratio between $n_{2,t+1}$ and ($n_{1,t}sm$). However, the model includes population parameters more explicitly, enabling the investigation of life history variations.

The model is also closely related to the surplus production model, which is often expressed in terms of ordinary differential equations. For example, the most basic surplus production model uses a logistic growth model with added instantaneous fishing mortality, see [14,35]. The current model is equivalent to the surplus production model except it includes life history strategy of fish and it is modeled with life history strategy events including

reproduction, compensatory density dependence, survival, maturation and fishing mortality occurring in a discrete sequence rather than simultaneously. The Beverton-Holt density dependence is also a discrete-time equivalent to the density dependence under the logistic equation.

Analyses

The ways in which equilibrium fishery yield, resilience, and the CV of adult abundance responded to changes in the parameters in the model (i.e. life history traits) were investigated. In particular, s, m and p were varied among low, intermediate and high values (s: 0.1, 0.5, 0.9; m: 0.1, 0.5, 1.0; p: 0, 0.5, 0.9). Average stage durations were calculated for each combination of parameters in the absence of fishing mortality. The equilibrium fishery yield, resilience, and CV of adult abundance were then calculated over ranges of α and fishing mortality ($1-f$), as described below.

Fishery yield at equilibrium (Y^*) is given by

$$Y^* = (1-f)n_2^*, \qquad (6)$$

where n_2^* is the equilibrium adult abundance under a fishing morality rate of $1-f$. The expression for the equilibrium adult abundance is obtained from equations (2) and (3) as

$$n_2^* = \frac{1}{\beta}\left(\frac{\alpha ms}{(1-pf)(ms-s+1)} - \frac{1}{f}\right), \qquad (7)$$

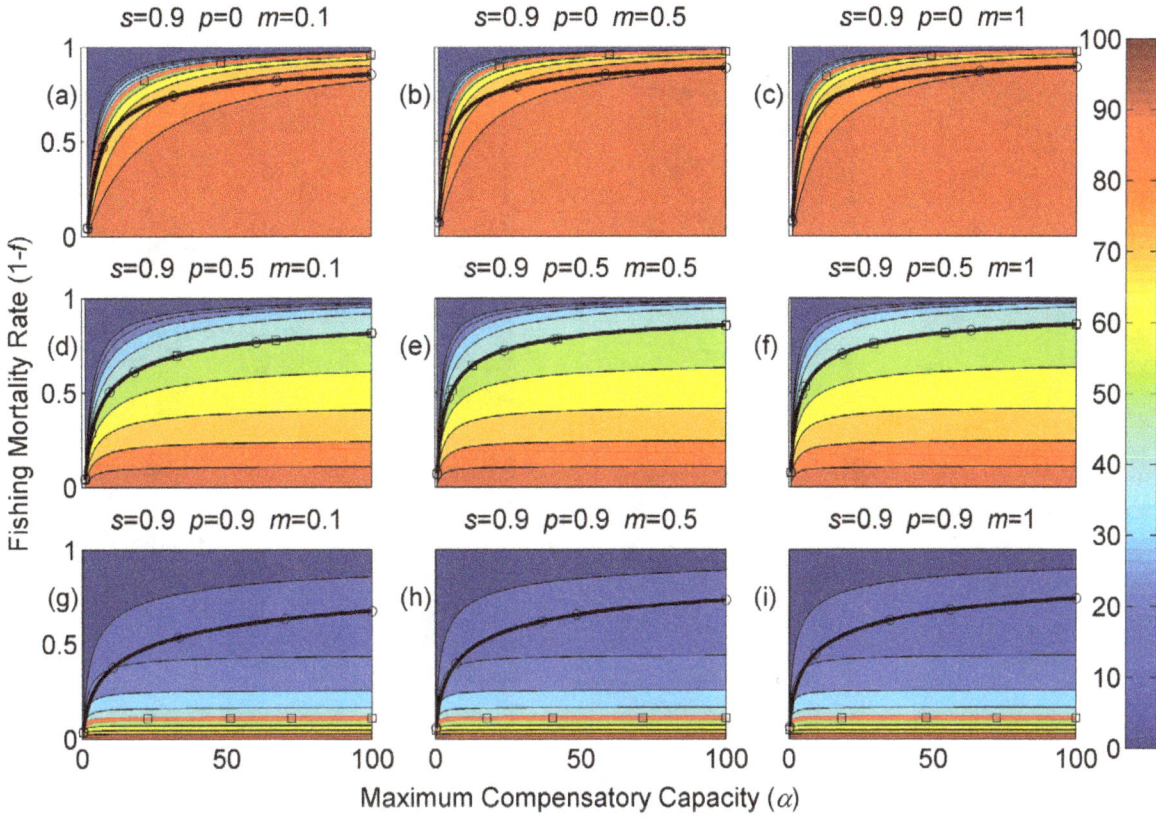

Figure 5. Equilibrium adult abundance (shown in contours) as a function of the annual fishing mortality rate and the maximum compensatory capacity when $s = 0.9$. The black and red curves are the same as shown in Figure 2. Each panel represents a life history strategy defined by the parameters shown above.

by setting the dominant eigenvalue of the population matrix (2) to be 1 and solving the equation for the adult abundance.

The fishing mortality rate at MSY satisfies the equation that is derived by taking the derivative of Y^* with respect to f, and setting the derivative to 0. In equation (6), n_2^* is also a function of f, and consequently the expression for the fishing mortality rate at MSY is not simple. However, the procedure for analytically taking the derivative is straightforward. The fishing mortality rate at MSY can be substituted into equations (6) and (7) to obtain the MSY.

The resilience (R) of a density-dependent matrix population model around the equilibrium point measures how quickly a population returns to an equilibrium point after a perturbation, and it is given by the negative slope of a log transformed difference between population abundance and the equilibrium point approaching 1. It is given by:

$$R = -\log(\lambda_J), \qquad (8)$$

where λ_J is the absolute value of the dominant eigenvalue of a linearized population projection matrix around the equilibrium point [33]. The linearized population projection matrix in this study was:

$$\mathbf{A}^* + \left[\frac{\partial \mathbf{A}}{\partial n_1} \mathbf{n}^* \quad \frac{\partial \mathbf{A}}{\partial n_2} \mathbf{n}^* \right] = \begin{bmatrix} s(1-m) & \frac{\alpha f}{\left(1 + \beta f n_2^*\right)^2} \\ sm & pf \end{bmatrix},$$

where * indicates that the population is at the equilibrium point and all of the derivatives are also evaluated at the equilibrium point. The linearized matrix is similar to the original population matrix because only one of the four parameters experience density dependence in the model. As it is a 2×2 matrix, obtaining the dominant eigenvalue of this matrix was also straightforward.

The CV of adult abundance [21] was estimated by incorporating stochastic perturbations into s, and simulating the population over 1000 time steps starting from the equilibrium abundance of a deterministic model. In this simulation, s was allowed to fluctuate according to the Beta distribution with means of 0.1, 0.5, and 0.9, and a constant variance of 0.01, without any serial autocorrelation. After each 1000 time step simulation, the CV of adult abundance was estimated.

Results

The average stage durations of various life history traits considered in this study are shown in Table 1 for reference. Hereafter, species with $p = 0.0$, 0.5, and 0.9 are referred to as semelparous, short-lived iteroparous, and long-lived iteroparous, respectively. Similarly, species with a juvenile stage longer than 1.5 years are referred to as having delayed maturation, while others are considered to be precocious.

Asymptotic fishery yields as a function of fishing mortality and the maximum compensatory capacity are shown in the contours in Figures 2, 3, 4 (for $s = 0.9$, 0.5, and 0.1, respectively). For example, Figure 2a shows asymptotic fishery yield for one of the semelparous precocious life history strategies. For a given value

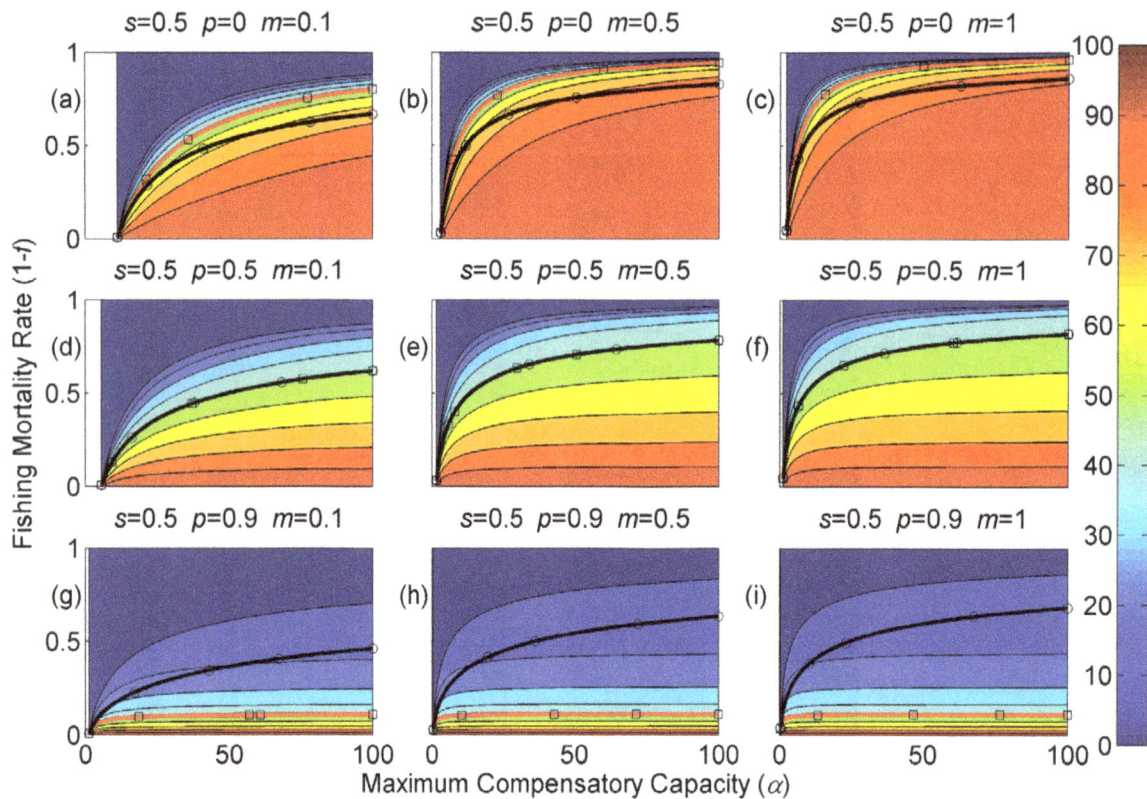

Figure 6. Equilibrium adult abundance (shown in contours) as a function of the annual fishing mortality rate and the maximum compensatory capacity when $s = 0.5$. The black and red curves are the same as shown in Figure 3. Each panel represents a life history strategy defined by the parameters shown above.

of maximum compensatory capacity, the sustainable yield initially increases as fishing mortality is increased from 0. This is because fishers catch more with increased effort. However, it peaks at some intermediate value of fishing mortality and declines thereafter because the equilibrium abundance of adults declines rapidly with increased mortality when the mortality rate is high. The grey area in the left side of some panels indicates that the population is not sustainable even without fishing mortality because of a low maximum compensatory capacity. As we go down the panels in each figure, life history is semelparous ($p = 0$) short-lived iteroparous ($p = 0.5$), and long-lived iteroparous ($p = 0.9$). As we move from the left to right panels, maturation rate increases.

The black curves with circles show the level of fishing mortality at the maximum sustainable yield for a given value of maximum compensatory capacity. As maximum compensatory capacity is increased, the population can be fished at a higher rate because the losses from the fishery are better compensated for, and the MSY also increases.

The red curves with square markers show the annual fishing mortality rate that reduces the adult abundance to 50% of the carrying capacity (i.e. the equilibrium abundance under no fishing mortality). This mortality level is referred to as the 0.5 K mortality. It should be noted, in the model, the equilibrium abundance is measured prior to fishing mortality but after recruitment. Under a traditional surplus production model with a logistic equation and instantaneous yield, which may be constant or proportional to fish abundance, the MSY is achieved when the population abundance is reduced to 50% of the carrying capacity. Figures 2, 3, 4 shows that the MSY fishing mortality is higher than the 0.5 K mortality

when the survival rate of adults is high ($p = 0.9$). Conversely, the MSY fishing mortality is lower than the 0.5 K mortality when the survival rate of adults is low ($p = 0$). They are equal when the survival rate of adults is at the intermediate value ($p = 0.5$) under all juvenile survival and maturation rates.

It is evident that the sustainable yield at a given value of fishing mortality and MSY declines from the semelparous ($p = 0$) to the iteroparous ($p = 0.5$) strategies, and also from short-lived iteroparous ($p = 0.5$) to long-lived iteroparous ($p = 0.9$) strategies. Therefore, the sustainable yield is the property associated with the adult parameter. This is despite the fact that the fishing mortality rate at MSY only slightly decreases from the semelparous to the long-lived iteroparous. For long-lived iteroparous species the maximum sustainable yield is always less than 10% of the carrying capacity, whereas for semelparous species the MSY can be greater than 50% of the carrying capacity under some parameter values. This results from differences in equilibrium abundance under fishing mortality among the various life history strategies.

The dark blue region in the upper left corner in each panel of Figures 2, 3, 4 indicates the region of unsustainable fisheries. As s is reduced (i.e. from Figure 2 to 3 and Figure 3 to 4), the region of unsustainable fisheries increases, and this is more pronounced when m is low. This occurs because, at a reduced s, few individuals are recruited into the adult stage. Consequently, only a slight increase in the adult mortality as a consequence of fishing will cause fishing to become unsustainable.

Equilibrium adult abundance as a function of fishing mortality and maximum compensatory capacity are shown in the contours in Figures 5, 6, 7 (for $s = 0.9$, 0.5, and 0.1, respectively). Figure 5a

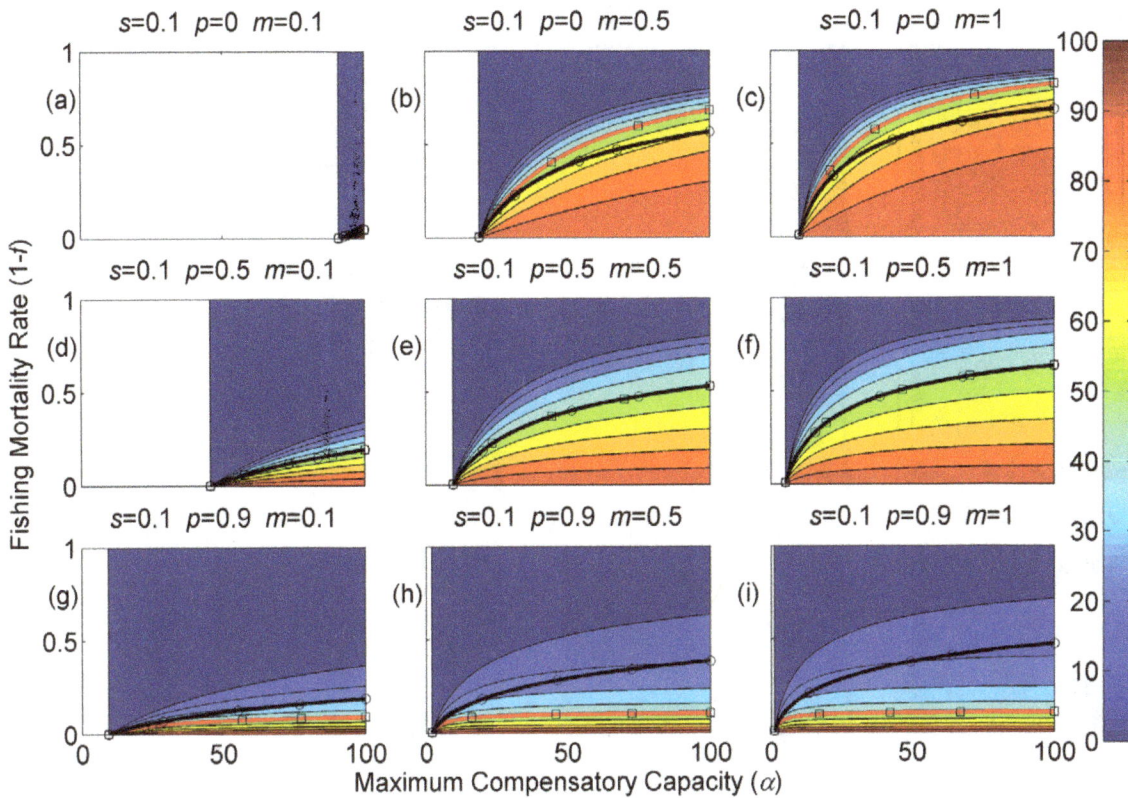

Figure 7. Equilibrium adult abundance (shown in contours) as a function of the annual fishing mortality rate and the maximum compensatory capacity when $s = 0.1$. The black and red curves are the same as shown in Figure 4. Each panel represents a life history strategy defined by the parameters shown above.

shows the equilibrium adult abundance for one of the semelparous precocious life history strategies. For a given value of maximum compensatory capacity, as fishing mortality rate is increased, equilibrium adult abundance declines. When maximum compensatory capacity is low, the decline starts early and quickly reaches very low abundance whereas, when maximum compensatory capacity is high, the population can maintain high abundance with higher fishing mortality rate. The black curve with circles and the red curves with squares are the same as before: fishing mortality rate at MSY and 0.5 K abundance, respectively. As adult survival rate (i.e. as we go down the panels) is increased, the equilibrium adult abundance declines substantially. On the other hand, maturation rate has almost no effect on the equilibrium adult abundance. These trends remain the same with lower juvenile survival rate (Fig. 6 and 7). However, as juvenile survival rate declines, the equilibrium abundance declines faster with increasing fishing mortality rate.

The resilience of a population around the equilibrium point is shown in Figures 8, 9, 10 (for $s = 0.9$, 0.5, and 0.1, respectively). With low maximum compensatory capacity, the population resilience always declines with increasing fishing mortality. This means that if there is a perturbation to the population, such as a natural or anthropogenic disaster, it will take longer to return to the equilibrium point as fishing mortality increases. Conversely, when maximum compensatory capacity is high, the resilience initially increases with increased fishing mortality, peaks, and then declines. However, irrespective of whether maximum compensatory capacity is high or low, when the fishing mortality exceeds the level for MSY, in most cases, the resilience declines with increasing fishing mortality.

Resilience is also affected by life history traits (Fig. 8, 9, 10). It increases with increasing maturation rate m and juvenile survival rate s. Although resilience decreases with increasing adult survival rate p, the change is by small amount. Therefore, the resilience is the property associated with parameters of juvenile stage. Finally, increased maximum compensatory capacity also increases resilience because of high compensatory capability.

The CV of the adult abundance increased with increased fishing mortality in all cases (Fig. 11, 12, 13). This implies that the population fluctuates more with higher fishing mortality rate. This occurs regardless of whether the fish have iteroparous, semelparous, precocious, or delayed maturation life history strategies.

Discussion

A simple stage-structured model was used to investigate how fishing mortality affects the population dynamics of fish with different life history strategies. Although simple, the model can encompass a wide variety of life history strategies. For example, Winemiller and Rose [5] investigated 10 life history traits of 216 North American fish species. The study revealed that the species can be categorized by three attributes: juvenile survivorship, generation time, and fecundity. The model in the present study also included three life history parameters that were varied to represent demographic diversity. Juvenile survival was explicitly included in the model, and generation time and fertility (instead of fecundity) were also functions of the three parameters in the model. Therefore, similar life history variation was investigated in the current study, but was parameterized in different ways.

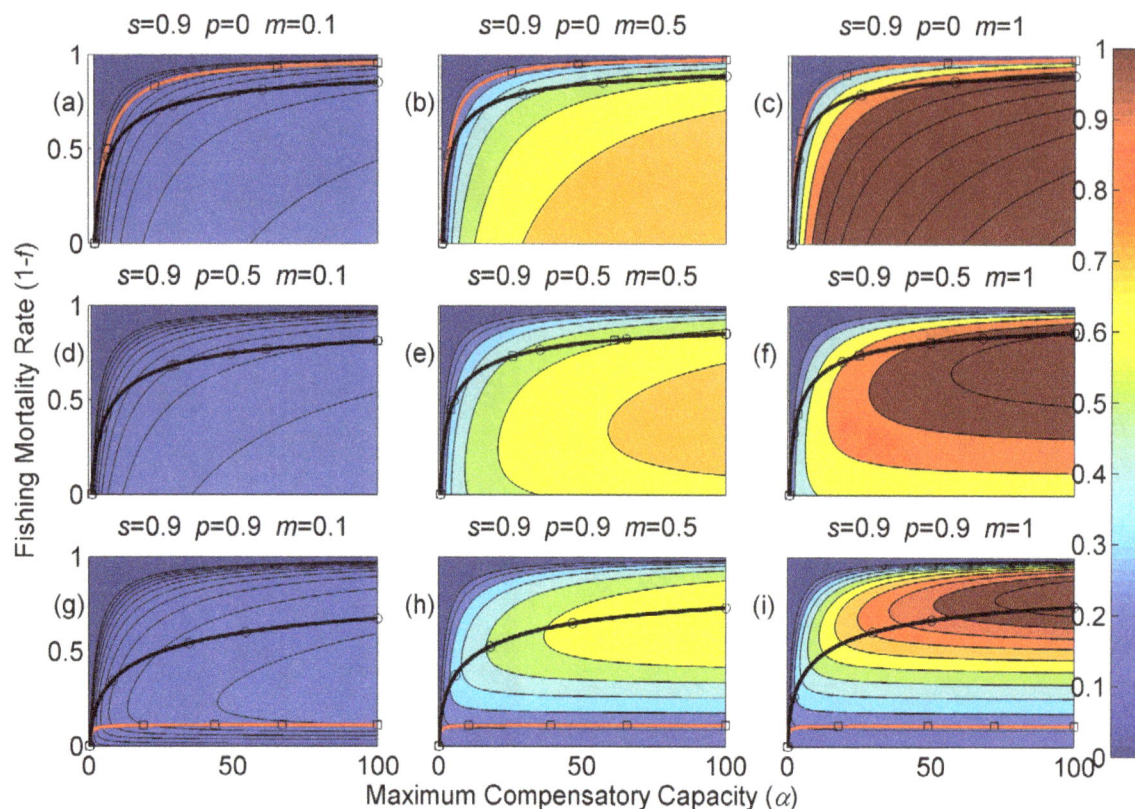

Figure 8. Resilience of the equilibrium densities (shown in contours) as a function of the annual fishing mortality rate and the maximum compensatory capacity when $s = 0.9$. The black and red curves are the same as shown in Figure 2. Each panel represents a life history strategy defined by the parameters shown above.

Fecundity (number of eggs produced) was not explicitly included in the model because the number of eggs that a fish produces is often a local/regional adaptation to the environment [36]. Of more importance to the overall population dynamics is the product of the survival rate of adults over the time-scale of a population model (1 year in the model in this study) and their fecundity (or fecundity and the survival of offspring over the time-scale of a population model in pre-breeding model).

The results suggest that the sustainable yield is reduced from long-lived iteroparous species (panels g, h, i in Fig. 2, 3, 4) to short-lived iteroparous species (panels d, e, f in Fig. 2, 3, 4) and also from short-lived iteroparous to semelparous (panels g, h, i in Fig. 2, 3, 4) under the same fishing mortality rate. This results from reduced adult equilibrium abundance, and suggests that the sensitivity of the equilibrium abundance increases as the duration of an adult stage increases. This means that, although we can rebuild over-exploited longer-lived fish by stopping its exploitation, after resuming exploitation, population abundance will decline again even with a low level of fishing mortality rate.

The MSY also exhibits large differences among different life history strategies. For example, there is a large difference in the MSY between long-lived iteroparous and semelparous fish (Fig. 2, 3, 4). For long-lived iteroparous species, the MSY was always less than 8% (often much less than that) of the carrying capacity. This result suggests that the large yields that fishers may have obtained in the past with some long-lived fish are transient yield and cannot be sustained at the same level. The only way to achieve those yields (although still transiently) is to rebuilt the stocks to historical levels. It is possible that the low MSY for long-lived iteroparous

species will not be economically viable for many fishery stocks. However, the MSY of semelparous fish can be 10 fold greater than that of long-lived iteroparous species under the same fishing mortality.

In contrast to MSY and sustainable yield, fishing mortality rate at MSY was similar among different life history strategies (Fig. 2, 3, 4). Semelparous fish have slightly higher fishing mortality at MSY than short-lived iteroparous species, and short-lived iteroparous fish have slightly higher fishing mortality at MSY than long-lived iteroparous species. However, the differences are small, and the factor that is differentially affected by fishing mortality among different life history strategies is the equilibrium abundance of adults. If we only want to know the model-based prediction of fishing mortality rate at MSY, it is not necessary to incorporate life history of organisms into the analysis. The information on the compensatory capacity of a fish population is sufficient. However, this does not mean that we should fish at the level of model-based MSY because, when population abundance is suppressed to a very low level, other factors such as environmental fluctuation and depensatory processes may affect the population.

In general, fishing mortality rate at MSY is the additional mortality rate that achieves the maximum production of the population. Under the logistic equation (Schaffer) model, it happens to be the level that suppresses the equilibrium abundance to a half of the carrying capacity (the 0.5 K level). However, under the two-stage model with the Beverton-Holt density dependence affecting a fertility term, mortality rates at 0.5 K and MSY levels are different. The exception was when the adult natural mortality rate was 0.5 (Fig. 2, 3, 4), but it should be noted that this exception

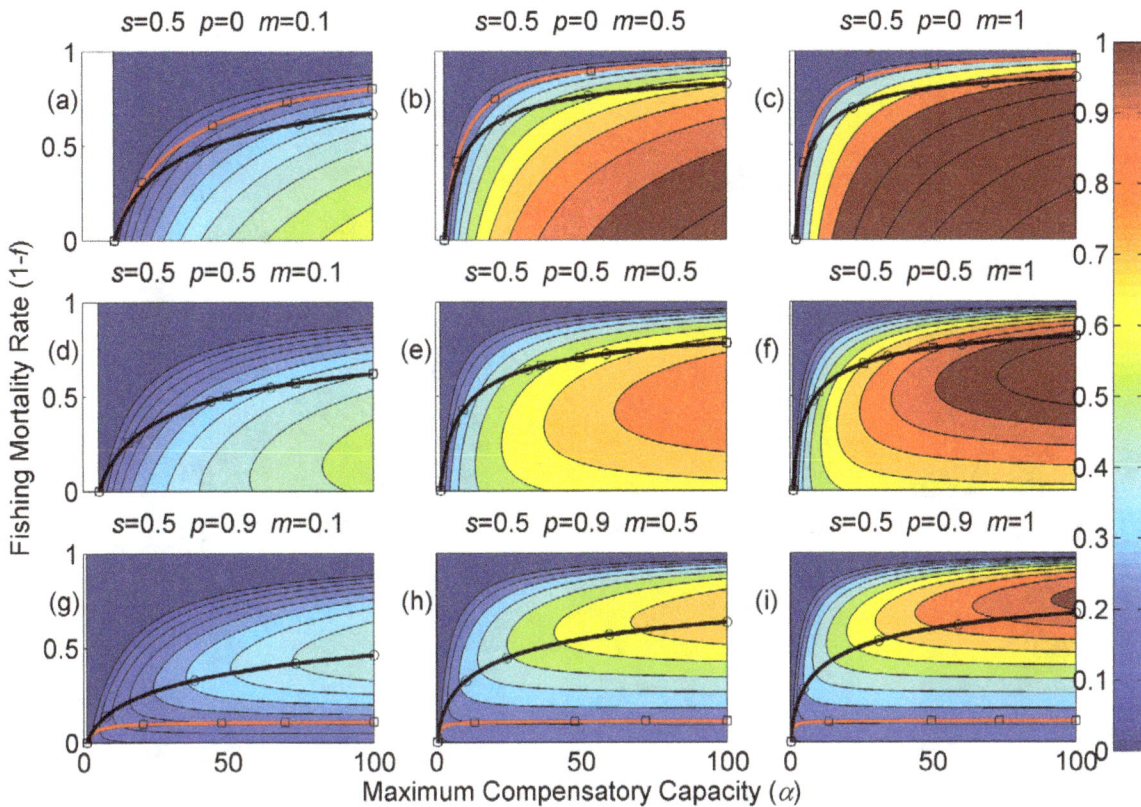

Figure 9. Resilience of the equilibrium densities (shown in contours) as a function of the annual fishing mortality rate and the maximum compensatory capacity when $s=0.5$. The black and red curves are the same as shown in Figure 3. Each panel represents a life history strategy defined by the parameters shown above.

is specific to the model used in the study. Therefore, in general, it cannot be assumed that MSY is achieved when the population abundance is at the 0.5 K level.

The maximum compensatory capacity (α) of a population affects the sustainable yield of fish; the greater α is, the greater the maximum sustainable yield is (Fig. 2, 3, 4). This makes intuitive sense, because, if α is high, a reduction in abundance because of fishing mortality can be better compensated for, and in turn allows higher fishing mortality rate. The mortality rate at MSY is also affected more by α than other parameters. Furthermore, a model for competition between stage-structured populations suggests the importance of this parameter [37]. The current result re-emphasizes the importance of accurately estimating density-dependent parameters, e.g. [38]. Unfortunately, α is probably the parameter in the model that is most difficult to estimate from field observations.

The effects of juvenile survival and maturation rate appear to have synergetic effects. For example, when the juvenile survival or maturation rate is high, reducing the other population parameter has only a small effect on the sustainable yield (Fig. 2, 3, 4). When the maturation rate is low, individuals tend to remain immature for longer. However, the low maturation rate is compensated for by increased reproduction. Consequently, individuals will accumulate in the juvenile stage as long as juvenile survival is high, and greater abundance in the juvenile stage will result in enough number of individuals maturing each year to maintain a population level. Similarly, when juvenile survival is low, a smaller proportion of individuals will survive to maturity, but reduced survival is compensated for by increased reproduction. As long as

the maturation rate is sufficiently high, the compensation is sufficient for enough number of individuals to mature each year to maintain equilibrium abundance. However, when juvenile survival or the maturation rate is low, reducing the other parameter has a pronounced effect on the sustainable yield. When both rates are low the population tends to be nonviable even though adults had high fertility rate, which is evidenced by the large grey areas in the figures.

The resilience of a population is an important population dynamics characteristic in fishery management (Fig. 8, 9, 10). The model suggests that fish with relatively low maximum compensatory capacity or those that are caught at levels above the MSY will have reduced resilience as fishing mortality is increased. This means that if there are additional mortalities caused by natural or anthropogenic events, these effects will last longer when the species are subject to high fishing mortality. This can have severe effects on economic sustainability of fisheries. Resilience is not a factor that is commonly incorporated into fishery management. I suggest it should be considered in future management decisions.

Resilience is also affected by life history traits, and fish stocks that exhibit the most resilience are those with an early maturing semelparous life history strategy and a high maximum compensatory capacity. Fish with such life history strategies will return to the equilibrium abundance more rapidly following a perturbation than stocks with a different life history strategy. Long-lived iteroparous species have the lowest resilience, but a change from short-lived iteroparous to long-lived iteroparous life history does not appear to change the resilience as much as the reduction that results from the change from delayed maturation to precocious

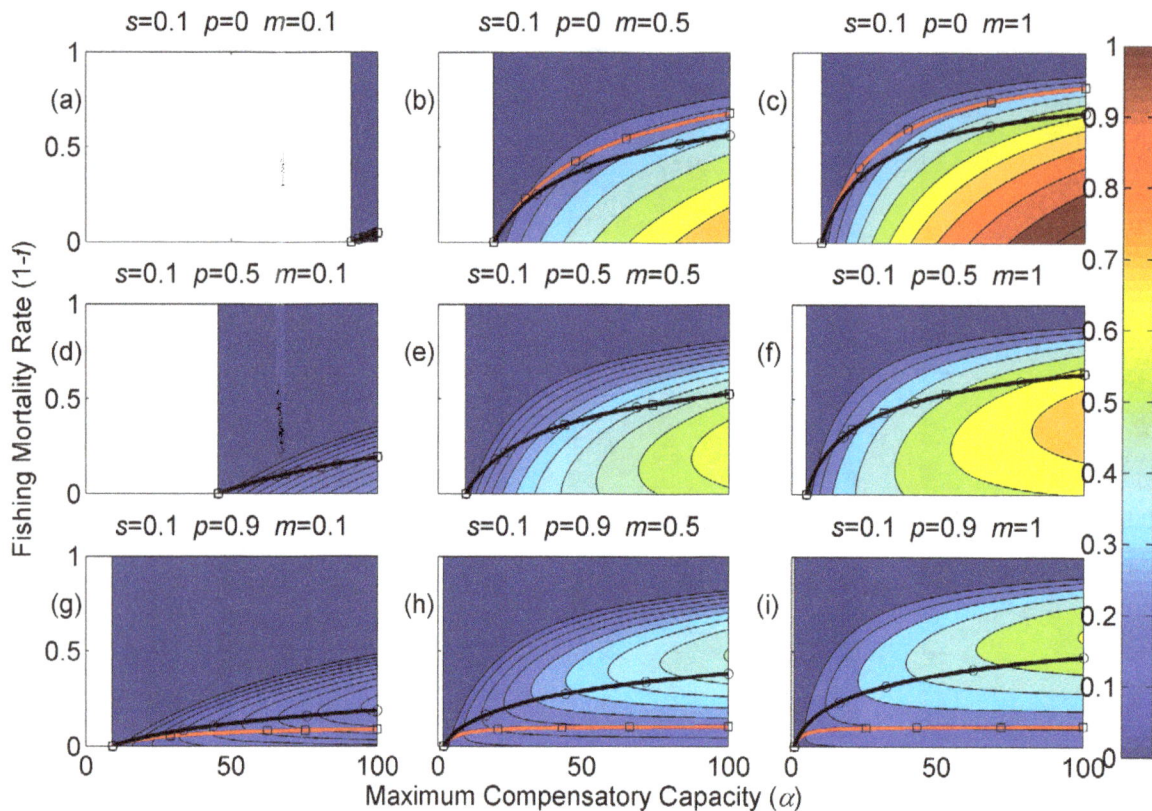

Figure 10. Resilience of the equilibrium densities (shown in contours) as a function of the annual fishing mortality rate and the maximum compensatory capacity when $s=0.1$. The black and red curves are the same as shown in Figure 4. Each panel represents a life history strategy defined by the parameters shown above.

reproductive strategies. This suggests that resilience is a quality of strategies associated with the juvenile stage. Thus, long-lived fish can have high resilience if they mature early and produce a large number of offspring. In order to determine the speed of recovery of over-exploited fish populations, we should focus on examining the early life-stage of the fish rather than how long adults can live.

The CV of adult abundance always increased with increasing fishing mortality when a stochastic perturbation was introduced into the juvenile survival rate (Fig. 11, 12, 13). The results were somewhat surprising because I hypothesized that a stock with low resilience would have a higher autocorrelation in adult abundance, which would in turn increase the variance of the abundance. Contrary to this hypothesis, the CV of adult abundance always increased with increasing fishing mortality. Examination of the mean and standard deviation of the adult abundance showed that fishing mortality reduces the mean abundance, which reduces its standard deviation. If both were reduced at the same rate, the CV of adult abundance would remain the same, but the mean was reduced faster than the standard deviation. Consequently, the CV of adult abundance increased with increasing fishing mortality. Resilience is associated with the dominant eigenvalue of the linearized projection matrix, which measures a longer-term transient dynamics. Consequently, resilience appears to be a measure of intermediate time-scale dynamics, whereas the CV of adult abundance may be viewed as a measure of short time-scale dynamics. However, the CV is not necessarily a measure of how much a population fluctuates because it appears to be affected more by the equilibrium abundance.

An increase in the CV of adult abundance with increasing fishing mortality has been observed, and its potential cause was attributed to an increase in newly recruited individuals in the adult stage [21], which causes increased variance in the adult abundance. However, the results presented here show that an increase in the CV of adult abundance also occurs with semelparous fish, which comprise only newly recruited individuals in the adult stage. The cause of the increased CV of adult abundance is that the mean adult abundance is more sensitive than its standard deviation, and the sensitivity increases with increasing adult duration.

The analyses in this study were based on a model containing various assumptions (see *Model* section). A simple model was intentionally used to provide general insights into how demographic diversity affects the response of fish stocks to fishing mortality. The model did not include factors that many fishery biologists may consider important in understanding fish population dynamics. Amongst these are the effect of age and/or size on population parameters, autocorrelation in environmental fluctuations, depensatory processes under low population density (the Allee effect), other types of compensatory density-dependent processes, differential effects of fisheries on different size classes, changes in parameters caused by interactions among populations (e.g. predation and competition), and changes in population parameters resulting from rapid evolution as a response to fishing pressures. The results presented here should form the basis for further investigations on how these other factors might affect the optimal management of fish stocks.

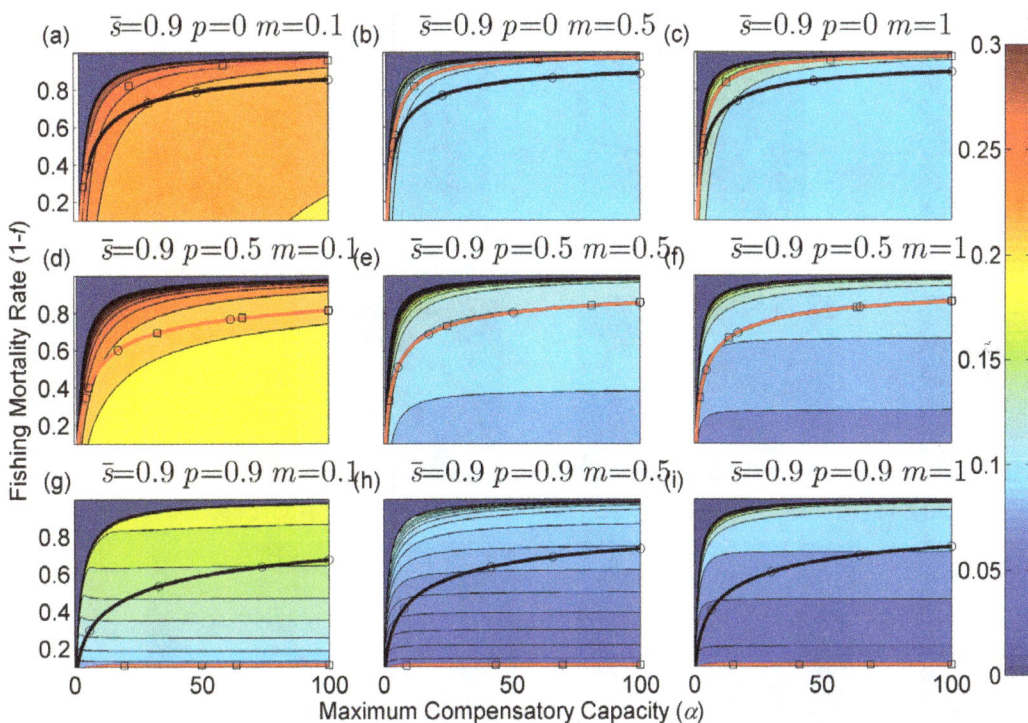

Figure 11. Coefficient of variation (CV) of adult abundance (shown in contours) as a function of the annual fishing mortality rate and the maximum compensatory capacity when juvenile survival fluctuates stochastically when $\bar{s}=0.9$. The black and red curves are the same as shown in Figure 2. Each panel represents a life history strategy defined by the parameters shown above.

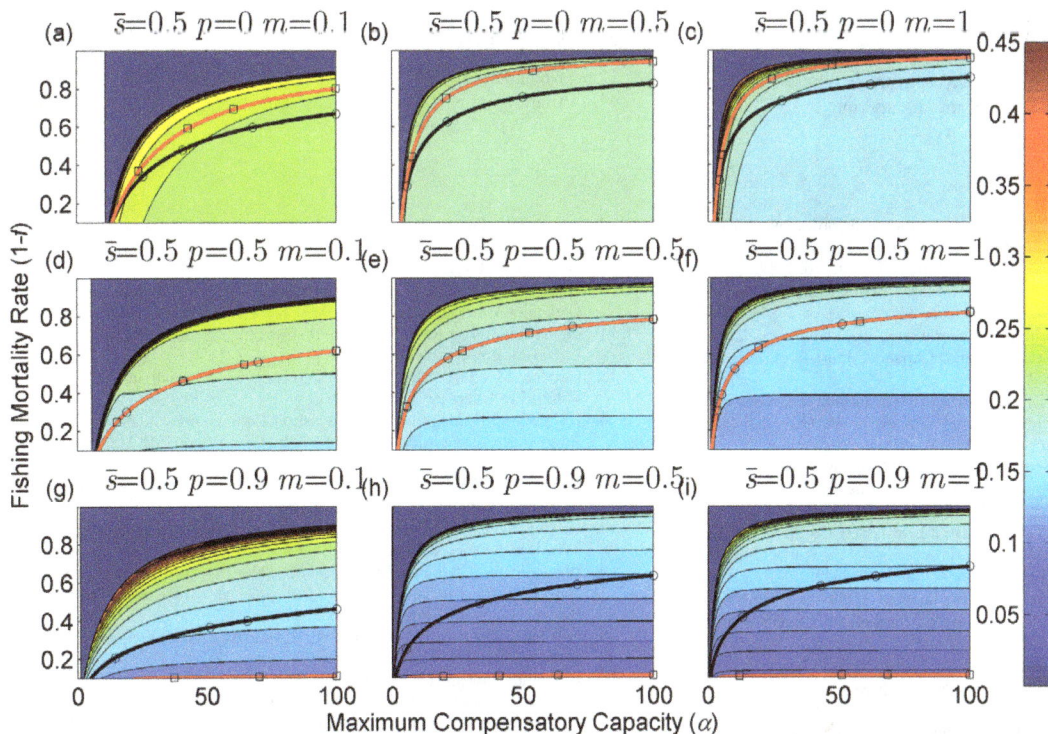

Figure 12. Coefficient of variation (CV) of adult abundance (shown in contours) as a function of the annual fishing mortality rate and the maximum compensatory capacity when juvenile survival fluctuates stochastically when $\bar{s}=0.5$. The black and red curves are the same as shown in Figure 3. Each panel represents a life history strategy defined by the parameters shown above.

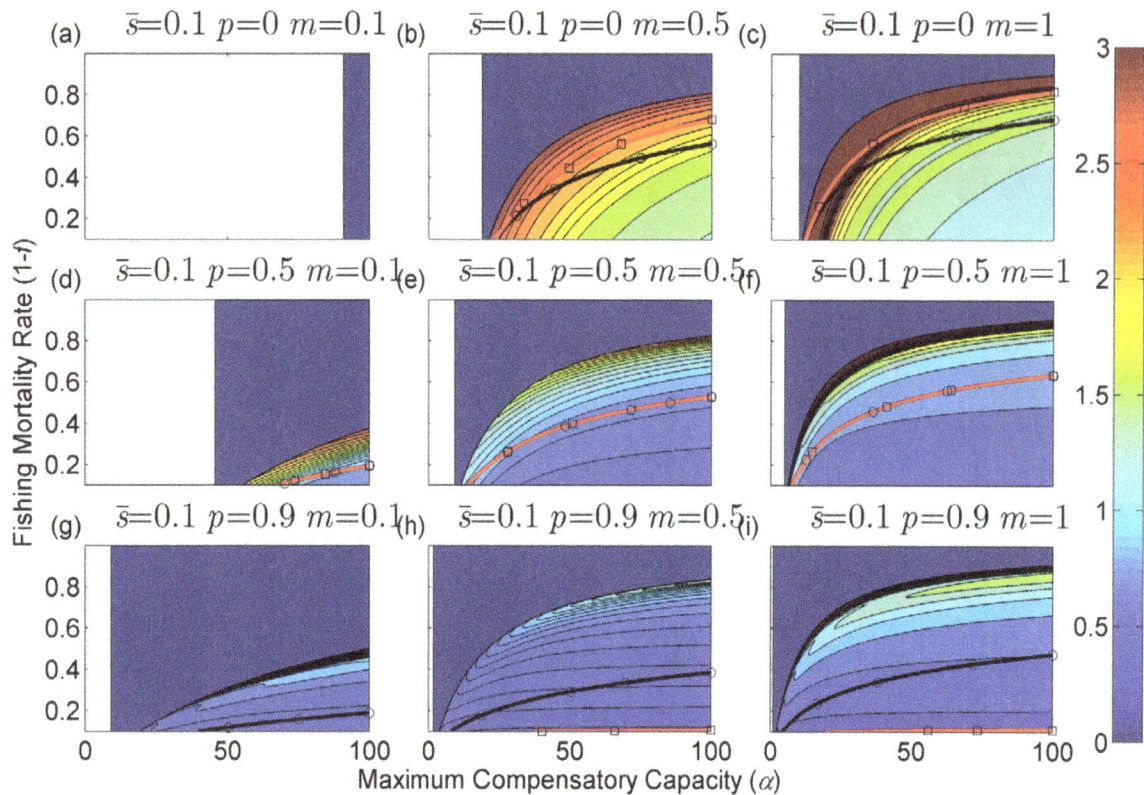

Figure 13. Coefficient of variation (CV) of adult abundance (shown in contours) as a function of the annual fishing mortality rate and the maximum compensatory capacity when juvenile survival fluctuates stochastically when $\bar{s}=0.1$. The black and red curves are the same as shown in Figure 4. Each panel represents a life history strategy defined by the parameters shown above.

Acknowledgments

I thank Mark Boyce and two anonymous reviewers for constructive comments on the previous version of the manuscript.

References

1. Stearns SC (1992) The evolution of Life Histories: Oxford University Press.
2. Adams PB (1980) Life-history patterns in marine fisheries and their consequences for fisheries management. Fishery Bulletin 78: 1–12.
3. Balon EK (1975) Reproductive guilds of fishes - proposal and definition. Journal of the Fisheries Research Board of Canada 32: 821–864.
4. Winemiller KO (2005) Life history strategies, population regulation, and implications for fisheries management. Canadian Journal of Fisheries and Aquatic Sciences 62: 872–885.
5. Winemiller KO, Rose KA (1992) patterns of life-history diversification in North-American fishes - implications for population regulation. Canadian Journal of Fisheries and Aquatic Sciences 49: 2196–2218.
6. Goodwin NB, Grant A, Perry AL, Dulvy NK, Reynolds JD (2006) Life history correlates of density-dependent recruitment in marine fishes. Canadian Journal of Fisheries and Aquatic Sciences 63: 494–509.
7. Pinsky ML, Jensen OP, Ricard D, Palumbi SR (2011) Unexpected patterns of fisheries collapse in the world's oceans. Proceedings of the National Academy of Sciences of the United States of America 208: 8317–8322.
8. Williams EH, Shertzer KW (2003) Implications of life-history invariants for biological reference points used in fishery management. Canadian Journal of Fisheries and Aquatic Sciences 60: 1037–1037.
9. Fromentin JM, Fonteneau A (2001) Fishing effects and life history traits: a case study comparing tropical versus temperate tunas. Fisheries Research 53: 133–150.
10. Shuter BJ, Abrams PA (2005) Introducing the symposium "Building on Beverton's legacy: life history variation and fisheries management". Canadian Journal of Fisheries and Aquatic Sciences 62: 725–729.
11. Schindler DE, Hilborn R, Chasco B, Boatright CP, Quinn TP, et al. (2010) Population diversity and the portfolio effect in an exploited species. Nature 465: 609-U102.

12. Brooks EN, Powers JE, Cortes E (2010) Analytical reference points for age-structured models: application to data-poor fisheries. Ices Journal of Marine Science 67: 165–175.
13. Hilborn R, Walters CJ (1992) Quantitative Fisheries Stock Assessment: Choice, Dynamics & Uncertainty. Boston, MA: Kluwer Academic Publishers.
14. Quinn TJ, Deriso RB (1999) Quantitative Fish Dynamics. New York: Oxford University Press, Inc..
15. Haddon M (2001) Modelling and Quantitative Methods in Fisheries. Boca Raton, FL: Chapman & Hall/CRC.
16. USDOC (2007) Magnuson-Stevens Fishery Conservation and Management Act: As amended through January 12, 2007. In: Commerce USDO. PL 94–265 and 109–479.
17. Cadrin SX, Pastoors MA (2008) Precautionary harvest policies and the uncertainty paradox. Fisheries Research 94: 367–372.
18. Hastings A (2004) Transients: the key to long-term ecological understanding? Trends in Ecology & Evolution 19: 39–45.
19. Wiedenmann J, Fujiwara M, Mangel M (2009) Transient population dynamics and viable stage or age distributions for effective conservation and recovery. Biological Conservation 142: 2990–2996.
20. Worden L, Botsford LW, Hastings A, Holland MD (2010) Frequency responses of age-structured populations Pacific salmon as an example. Theoretical Population Biology 78: 239–249.
21. Hsieh CH, Reiss CS, Hunter JR, Beddington JR, May RM, et al. (2006) Fishing elevates variability in the abundance of exploited species. Nature 443: 859–862.
22. Anderson CNK, Hsieh CH, Sandin SA, Hewitt R, Hollowed A, et al. (2008) Why fishing magnifies fluctuations in fish abundance. Nature 452: 835–839.
23. Hannesson R (1994) Bioeconomic analysis of fisheries. FAO Fisheries Report DN 0429–9337, no. 499. FAO, Rome, Italy.
24. Beddington JR, Free CA, Lawton JH (1976) Concepts of stability and resilience in predator-prey models. Journal of Animal Ecology 45: 791–816.

Author Contributions

Conceived and designed the experiments: MF. Performed the experiments: MF. Analyzed the data: MF. Contributed reagents/materials/analysis tools: MF. Wrote the paper: MF.

25. Harrison GW (1979) Stability under environmental-stress - resistance, resilience, persistence, and variability. American Naturalist 113: 659–669.

26. DeAngelis DL (1980) Energy-flow, nutrient cycling, and ecosystem resilience. Ecology 61: 764–771.

27. Verhulst PF (1838) Notice sur la loi que la population poursuit dans son accroissement. Correspondance Mathématique et Physique 10: 113–121.

28. May RM (1976) Simple mathematical-models with very complicated dynamics. Nature 261: 459–467.

29. Costantino RF, Cushing JM, Dennis B, Desharnais RA (1995) Experimentally-induced transitions in the dynamic behavior of insect populations. Nature 375: 227–230.

30. Costantino RF, Desharnais RA, Cushing JM, Dennis B (1997) Chaotic dynamics in an insect population. Science 275: 389–391.

31. Ricker WE (1954) Stock and rescruitment. Journal of the Fisheries Research Board of Canada 11: 559–623.

32. Beverton RJH, Holt SJ (1957) On the Dynamics of Exploited Fish Populations. : Ministry of Agriculture, Fisheries and Food, London (republished by Chapman & Hall in 1993).

33. Caswell H (2001) Matrix Population Models: Construction, Analysis, and Interpretation. Sunderland: Sinauer Associates, Inc..

34. Myers RA, Barrowman NJ (1996) Is fish recruitment related to spawner abundance? Fishery Bulletin 94: 707–724.

35. Jennings S, Kaiser MJ, Reynolds JD (2001) Marine Fisheries Ecology. Malden, MA: Blackwell Publishing.

36. Winemiller KO, Rose KA (1993) Why do most fish produce so many tiny offspring. American Naturalist 142: 585–603.

37. Fujiwara M, Pfeiffer G, Boggess M, Day S, Walton J (2011) Coexistence of competing stage-structured populations. Scientific Reports 1: DOI: 10.1038/srep00107.

38. Rose KA, Cowan JH (2003) Data, models, and decisions in US Marine Fisheries management: Lessons for ecologists. Annual Review of Ecology Evolution and Systematics 34: 127–151.

9

Spatial Variation in Carbon and Nitrogen in Cultivated Soils in Henan Province, China: Potential Effect on Crop Yield

Xuelin Zhang[1]*, Qun Wang[1], Frank S. Gilliam[2], Yilun Wang[1], Feina Cha[3], Chaohai Li[1]

1 The Incubation Base of the National Key Laboratory for Physiological Ecology and Genetic Improvement of Food Crops in Henan Province, Zhengzhou, China; Agronomy College of Henan Agricultural University, Zhengzhou, China, 2 Department of Biological Sciences, Marshall University, Huntington, West Virginia, United States of America, 3 Meteorological Bureau of Zhengzhou, Zhengzhou, China

Abstract

Improved management of soil carbon (C) and nitrogen (N) storage in agro-ecosystems represents an important strategy for ensuring food security and sustainable agricultural development in China. Accurate estimates of the distribution of soil C and N stores and their relationship to crop yield are crucial to developing appropriate cropland management policies. The current study examined the spatial variation of soil organic C (SOC), total soil N (TSN), and associated variables in the surface layer (0–40 cm) of soils from intensive agricultural systems in 19 counties within Henan Province, China, and compared these patterns with crop yield. Mean soil C and N concentrations were 14.9 g kg^{-1} and 1.37 g kg^{-1}, respectively, whereas soil C and N stores were 4.1 kg m^{-2} and 0.4 kg m^{-2}, respectively. Total crop production of each county was significantly, positively related to SOC, TSN, soil C and N store, and soil C and N stock. Soil C and N were positively correlated with soil bulk density but negatively correlated with soil porosity. These results indicate that variations in soil C could regulate crop yield in intensive agricultural systems, and that spatial patterns of C and N levels in soils may be regulated by both climatic factors and agro-ecosystem management. When developing suitable management programs, the importance of soil C and N stores and their effects on crop yield should be considered.

Editor: Dafeng Hui, Tennessee State University, United States of America

Funding: This study was supported by grants from Henan Science and Technology Department of China under the Key Research Project (30200051). The funder had no role in study design, data collection and analysis, decision to publish, or preparation of the manuscript.

Competing Interests: The authors have declared that no competing interests exist.

* Email: xuelinzhang1998@163.com

Introduction

Safeguarding food security and ensuring sustainable development are two fundamental goals of intensive agriculture in China [1,2]. Increasing soil C and N sequestration while reducing C and N emissions from agricultural fields are important aspects of sustainable farming and these goals can be achieved through improvement in soil quality [1,3]. This requires a better understanding of the functional relationship between crop yield and soil organic C and N stores.

Indeed, variations in soil C and N stores may closely regulate crop yield, although published data on the relationship between these parameters are inconsistent. Some studies have reported a positive correlation between soil C and N and crop yield [4,5], whereas other studies have found no significant relationship between these parameters [6,7]. Lal (2006) reported that the relationship between soil organic C and crop yield may vary between patterns that are sigmoidal, linear, or exponential [8]. Clearly, the existence of such variability warrants further investigation.

Soil C and N stores in crop lands, especially in the topsoil layer, are potentially greatly affected by human activity; thus, understanding the spatial pattern of soil C and N stores on a regional scale is crucial to developing a management strategy for improving soil fertility [1,2]. Spatial variation in soil C and N stores in agro-ecosystems has been widely reported [9,10,11], including from the northern [12,13], eastern [14], and southern [15,16] regions of China. Since these reports from China were based on two national surveys from 1960 and 1983, such data may have limited use in helping to develop management strategies based on current practices [17]. Therefore, in order to better understand the spatial patterns and their relationship to crop yield, it is necessary to update regional soil organic C and N information with contemporary measurements, especially for intensively-used crop land.

Henan Province is the second largest area of crop production in China (China National Bureau of Statistics). To produce an adequate supply of food for the domestic population, unsustainable production methods have often been used in this province. Historically, intensive production based on an annual wheat-maize system has been used to achieve high crop yield. This practice, however, has resulted in badly degraded agricultural soils, causing erosion and a loss of good soil structure. More than 600 kg N ha^{-1} annually has been applied in this production area, resulting in an increase in soil acidity [18]. Based on the determination that crop yields in China will need to increase from 50 billion in 2010 to 65 billion kg in 2020, the provincial crop lands in Henan Province will continue to play an important role in food production. Such goals create the challenge of improving soil quality, enhancing soil fertility, and mitigating C and N loss, while achieving food security and practicing sustainable agriculture. A better understanding of the spatial variability of soil organic C and N, and their relationship to crop yield, should help to develop management practices that are designed to meet this challenge [1,19].

The objective of the present study was to characterize the spatial distribution of C and N stores in intensively cultivated counties within the Henan Province of China and to determine the relationship between crop yield and soil organic C and N.

Materials and Methods

Statement: We have field permits for sampling soil in each of the field sites within each county of Henan Province, China. All of the sampling sites are privately owned, and there was no potential impact on any endangered or protected species among these sampling sites.

Study site

The study was carried out in 19 counties within Henan Province, located in central China (Figure 1). Map data were obtained from the National Geomatics Center of China (http://ngcc.sbsm.gov.cn/) using ArcGIS software. As of 2009, the human population of Henan was about 9.9×10^7 persons. The Province is

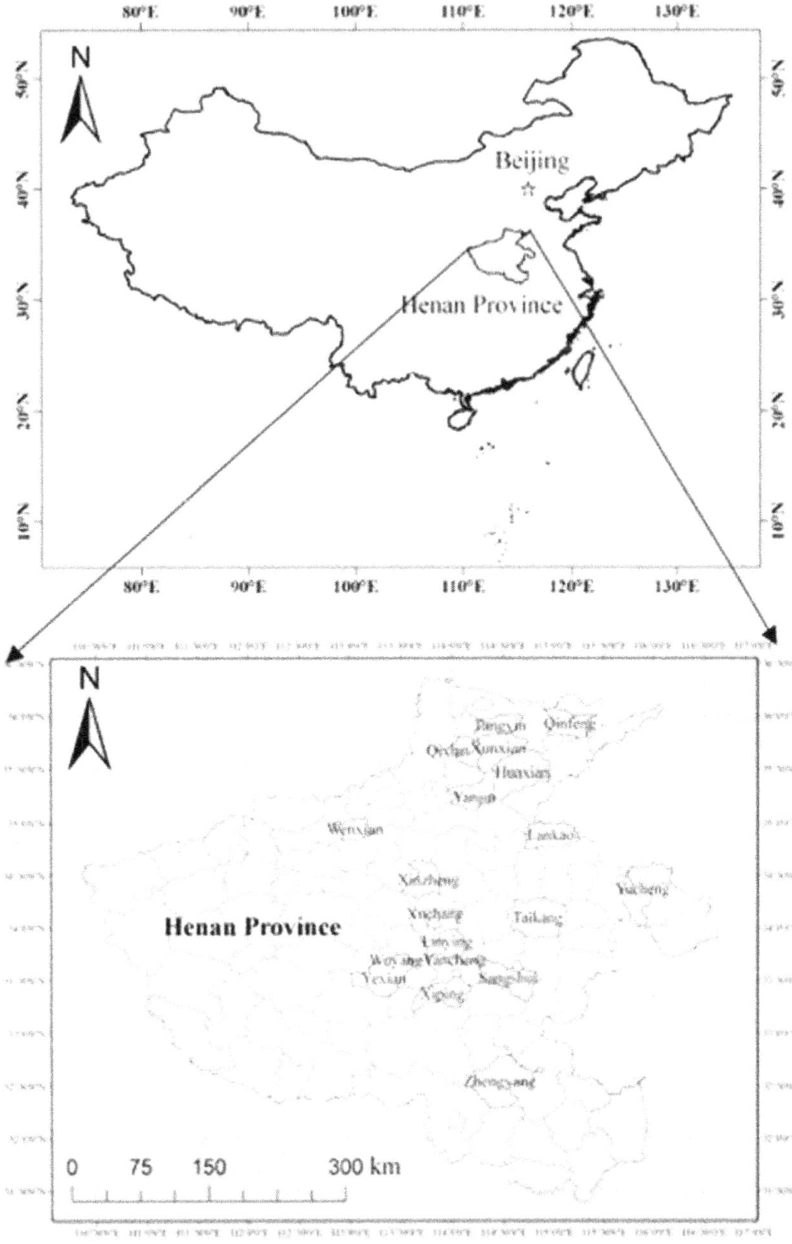

Figure 1. Map of China (top) showing location of Henan Province and counties (bottom) within Henan Province used in this study.

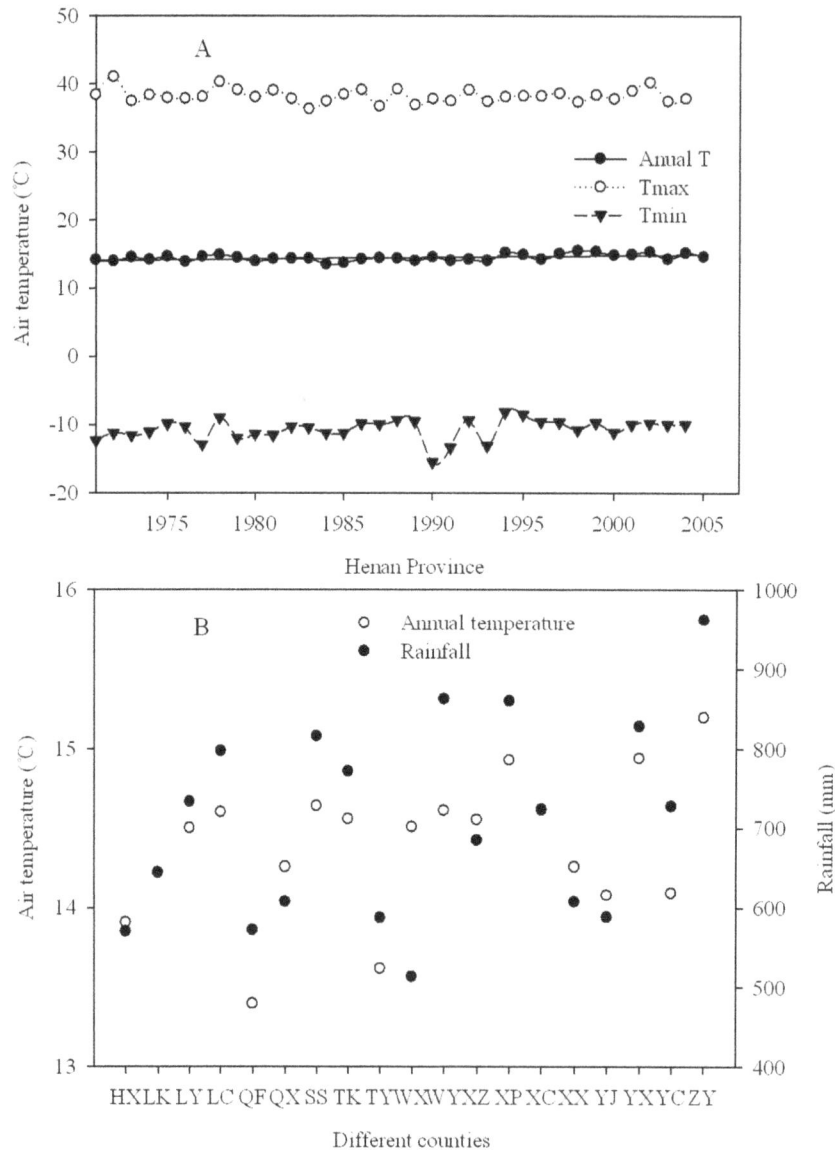

Figure 2. Average annual, maximum, and minimum temperature from 1971 to 2004 in Henan Province, China (A), and (B) average temperature and rainfall from 1975 to 2006 in different counties within Henan Province, China (B). See Table 1 for key to county name abbreviations. All these counties were arranged in English alphabetical order.

approximately 167,000 km^2 in land area, lying within the monsoonal temperate zone. It has a cultivated land area of 79, 260 km^2 for the production of wheat and maize. There are three dominant soil types in Henan Province: Yellow-cinnamon soil (Eutric Cambisols in FAO taxonomy), Sajiang black soil (Eutric Vertisols/Gleyic Cambisol), and Fluvo-aquic soil (Fluvisols in FAO taxonomy) [20]. Mean annual precipitation ranges from 400 to 1000 mm among the counties of the study, with ~70% of it occurring from July to September; mean annual temperature ranges from 13.6 to 15°C (Figure 2). Cultivated agricultural fields are the predominant land use, representing 60% of the total land area in Henan Province. A double cropping system of winter wheat (early October-early June) and maize (mid-June–later September) is the most common planting system used in this region.

Collection of crop yield and soil sampling and analysis

Data on total crop production (including wheat, maize and millet) and wheat yield from 1978–2009 (Figure. 3A) were obtained from the Henan Statistical Yearbook 2010 (13–17) (http://www.ha.stats.gov.cn/hntj/index.htm). Annual yield data for winter wheat and total crop production in 2009 were also obtained from Henan Statistical Yearbook 2010 (29-7) and the Agricultural Bureau of each of the 19 counties in which soil sampling took place (Figure. 3B). These counties, along with basic climatic information, are listed in Table 1. Climatic data of each county were obtained from Meteorological Bureau of Zhengzhou. All counties will be referred to by the two-letter codes presented in Table 1.

The 19 counties were selected as representative of the main agro-ecosystems of Henan Province. Soil samples were collected during June 1–15, 2009 following the wheat harvest but prior to the sowing of maize. Six representative, replicate field plots,

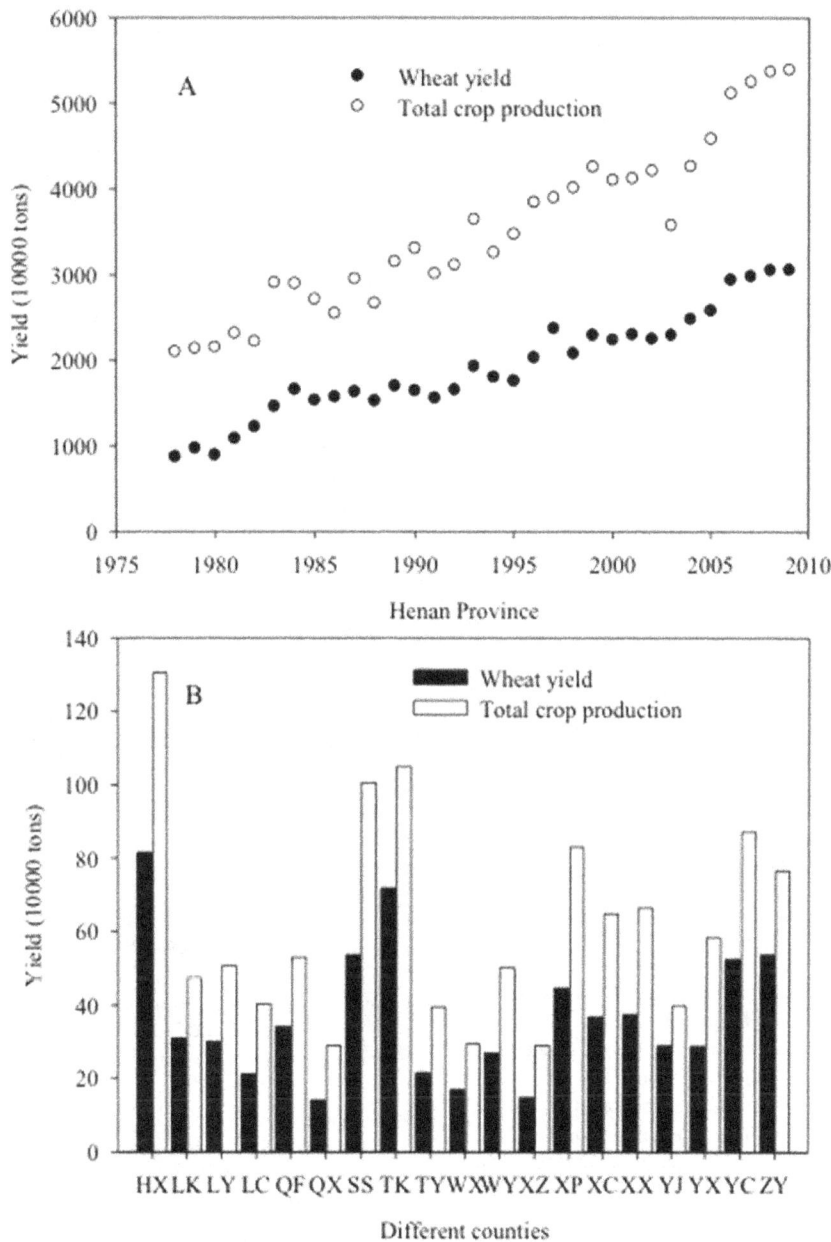

Figure 3. Wheat yield and total crop production (including wheat, maize, millet,) in Henan Province from 1978–2009 (A), and wheat yield and total crop production in different counties within Henan Province in 2009 (B). See Table 1 for key to county name abbreviations.

located at least 6 km apart, were selected within each county based on four criteria: (1) the field plots had been continuously cultivated for at least 30 yr with a native variety, (2) the cropland area was located within 5 km of native vegetation with a similar landscape, soil type and texture, and a relatively flat terrain, and (3) all of the sampling sites are privately owned, and (4) there was no potential impact on any endangered or protected species in the sampling site. Geographic coordinates of each sampling site was recorded by handed GPS of Magellan eXplorist 210(USA), and all of these data were attached in the supporting information.

Sample areas of ∼1300 m^2 were established in each plot, with sixteen sampling points taken at random in each of two layers (0–

20 cm and 20–40 cm) using a 70 mm - diameter auger. All of the soil samples taken at each layer within a sample plot were mixed together and treated as one sample to represent the value of the plot, yielding 114 soil samples at each layer.

Residual plant material was removed from the soil samples after the samples were air-dried at room temperature. The soil samples were then ground to pass a 2 mm sieve, and a portion of the ground sample was subsequently ground again in a porcelain mortar in order to pass through a 0.15–mm sieve. Organic C and total N measurements were obtained from the twice-ground soil samples. Soil organic C (SOC) was measured using a modified Mebius method. Briefly, 0.1 g soil samples were digested for 5 min

Table 1. Basic geographic coordinates for each county, along with climate data for 19 counties within Henan Province, China.

County	Latitude	Longitude	Sea level (m)	Average Temp (°C)	Rainfall (mm)	Sunshine (h)
Huaxian (HX)	35°44′	114°28′	68	13.9	570.0	2060.9
Lankao (LK)	34°55′	114°46′	70	14.2	644.5	2183.2
Linying (LY)	33°55′	113°55′	63	14.5	732.9	2141.3
Luoheyancheng (LC)	33°35′	114°02′	65	14.6	797.2	2273.0
Qingfeng (QF)	35°53′	115°06′	51	13.4	571.9	2209.1
Qixian (QX)	35°35′	114°12′	72	14.3	607.5	2133.8
Shangshui (SS)	33°39′	114°34′	52	14.6	815.8	1902.0
Taikang (TK)	34°05′	114°50′	53	14.6	770.9	1998.4
Tangyin (TY)	36°03′	114°19′	103	13.6	587.1	2159.3
Wenxian (WX)	35°01′	113°03′	109	14.5	513.2	2302.2
Wuyang (WY)	33°36′	113°32′	77	14.6	862.3	2060.4
Xinzheng (XZ)	34°30′	113°39′	159	14.6	684.6	2058.7
Xiping (XP)	33°29′	113°59′	65	14.9	859.8	2084.7
Xuchang (XC)	34°04′	113°52′	72	14.6	722.7	1959.8
Xunxian (XX)	35°40′	114°32′	59	14.3	607.5	2133.8
Yanjin (YJ)	35°13′	114°11′	69	14.1	588.0	2287.8
Yexiang (YX)	33°38′	113°21′	88	14.9	827.8	1972.4
Yucheng (YC)	34°25′	115°52′	46	14.1	727.3	2244.6
Zhengyang (ZY)	32°37′	114°24′	70	15.2	961.8	2004.4

Note: Counties are arranged in English alphabetical order.

with 5 mL of 1N $K_2Cr_2O_7$ and 10 mL of concentrated H_2SO_4 at 150°C, followed by titration of the digests with standardized $FeSO_4$. Total soil N (TSN) was measured using a modified Kjeldahl wet digestion procedure and a Tector Kjeltec System 1026 distilling unit. Soil available N was analyzed using a micro-diffusion technique after alkaline hydrolysis (1.8 mol L^{-1} NaOH). The Olsen method was used to determine available soil phosphorus (P), and available soil potassium (K) was measured in 1 mol L^{-1} NH_4OAc extracts by flame photometry (Table 2).

Three sampling points were used to determine soil bulk density in each plot. Samples were collected separately from four layers within a depth of 0–40 cm in each sampling point. Soil bulk density was measured using 100-cm^3 soil cores obtained from the four layers. Soil porosity was calculated from soil bulk density and specific gravity, with any stone material removed and not considered in bulk density calculations.

Calculation of soil organic C and N stores and SOC and TSN

Total soil organic C store (TSOCS) and total soil N stores (TSNS) at 0–40 cm depth were calculated as follows:

$TSOC(g.m^{-2}) =$

Soil organic $C(g.kg^{-1})$ × soil bulk density $(g.cm^{-3})$ × sampling depth(cm)

$TSN(g.m^{-2}) =$

Soil total $N(g.kg^{-1})$ × soil bulk density $(g.cm^{-3})$ × sampling depth(cm)

Given the cultivated area, the total cultivated topsoil (0–40 cm) C and N stocks of each county were estimated by the equation:

$$CS = \sum area_i \times TSOC$$
$$NS = \sum area_i \times TSN$$

where $area$ is the given total cultivated area of each county, and CS and NS are C and N stocks, respectively. SOC and TSN were means of six sampling sites of each county.

Statistics

Analysis of variance was used to assess the significance of location (county) on soil C and N concentration and storage; means were compared using Duncan's multi-range test at $\alpha = 0.05$. Linear regression was used to determine the relationships between C and N stock versus wheat and total crop production. Principle components analysis was used to assess patterns of similarity/dissimilarity among counties with respect to several environmental variables [21]. All statistical analyses were performed using SPSS 10.0 (Chicago IL, USA).

Results

Wheat yields increased more than 250% from 1978 to 2009 while total annual crop production in Henan Province increased from 21 to 54 million tons over the same time period (Figure 3A). Wheat yield varied from 143 to 729 thousand tons among the different counties in 2009 (Figure 3B).

The absolute value of SOC concentration in the top 40 cm of soil varied from 8.13 to 27.89 g kg^{-1} among the 19 counties in 2009 (Table 2) while TSN concentration varied from 0.84 to 2.2 g kg^{-1}. Soil C/N varied from 6.4 to 20 (Table 2). Soil organic C stores (TSOCS) in the 0–40 cm soil layer varied from 2,322 g m^{-2} to 8,038 g m^{-2}, whereas total N stores (TSNS) varied from 221 to

Table 2. Spatial variation in soil (0–40 cm depth) properties, soil organic C (SOC), total soil N (TSN) concentration (g kg⁻¹), and C/N in the 0–40 cm soil layer in 19 counties within Henan province, China.

	Alkaline-extractable N (mg kg⁻¹)	Olsen-extractable P (mg kg⁻¹)	NH₄OAc-extractable K (mg kg⁻¹)	Bulk density (g cm⁻³)	Soil porosity (%)	SOC (g kg⁻¹)	TSN (g kg⁻¹)	C/N
HX	48.9±3.2abc	1.8±0.7a	80.1±9.2abc	1.44±0.03de	38.3±1.4abcd	12.4±0.9abc	1.4±0.05abcd	8.8±0.7abc
LK	56.5±2.4abcd	7.6±1.9ab	71.9±11.9abc	1.42±0.02bcde	40.7±1.0bcdef	11.2±0.7ab	1.4±0.09abcd	7.9±0.5a
LY	49.9±3.1abc	4.2±0.7a	145.4±24.1ef	1.36±0.02abc	41.7±1.4cdefg	15.5±1.0bcd	1.1±0.07a	14.2±0.2efg
LC	49.0±1.6abc	11.5±2.1abc	103.6±8.9abcde	1.39±0.02bcd	38.7±1.6abcd	14.6±1.4abcd	1.4±0.14bcd	10.5±1.1abcd
QF	47.5±2.8abc	10.9±4.9abc	71.7±6.3abc	1.39±0.01bcd	41.8±0.4cdefg	11.8±0.6ab	1.4±0.08abcd	8.7±0.8abc
QX	51.9±3.7abc	6.3±2.2ab	82.1±10.8abc	1.44±0.01de	38.7±0.3abcd	21.1±1.8f	1.5±0.21cd	16.2±3.1g
SS	45.1±1.9abc	11.7±3.7abc	169.3±33.9f	1.35±0.02ab	37.9±1.3abcd	14.5±0.9abcd	1.3±0.08abc	11.4±0.8bcde
TK	59.3±4.5cd	17.7±8.2bcd	140.6±23.9def	1.35±0.02ab	41.5±1.5cdef	13.4±1.0abcd	1.1±0.05ab	11.9±0.7cde
TY	59.2±2.4cd	6.4±2.6ab	110.9±12.8bcde	1.45±0.03de	38.6±1.3abcd	15.0±0.4abcd	1.7±0.09de	8.9±0.5abc
WX	56.8±3.5abcd	11.0±1.9abc	82.1±7.7abc	1.3±0.03a	43.6±1.5fg	17.1±1.9de	1.5±0.11bcd	11.5±0.6bcde
WY	47.9±1.1abc	10.3±2.9ab	84.2±10.5abc	1.38±0.02bcd	36.8±1.2ab	14.9±1.6abcd	1.6±0.09cd	9.5±0.6abc
XZ	72.9±7.4e	7.8±2.1ab	95.8±18.9abcd	1.47±0.02e	38.3±0.7abcd	16.1±1.6cd	1.1±0.12ab	14.5±1.1efg
XP	72.2±4.2e	27.3±5.8d	117.5±17.7cde	1.43±0.01cde	37.8±1.2abc	19.9±1.9ef	1.3±0.07abc	15.4±1.3fg
XC	43.2±3.4a	17.2±5.8bcd	66.2±9.6ab	1.42±0.03bcde	42.9±0.9efg	16.6±2.3cde	1.3±0.18abc	12.7±0.7def
XX	53.1±4.2abcd	5.1±0.5ab	89.9±3.0abc	1.41±0.02bcde	35.3±1.0a	14.8±0.6abcd	1.9±0.11e	7.9±0.3a
YJ	49.5±2.9abc	12.8±4abc	77.0±9.6abc	1.39±0.02bcd	38.2±1.0abcd	10.8±0.6a	1.3±0.07abc	8.2±0.4ab
YX	44.5±2.1ab	22.9±5.9cd	89.5±11.4abc	1.3±0.04a	45.5±1.9g	15.4±0.9bcd	1.2±0.07ab	13.6±1.1defg
YC	66.4±11.3de	3.5±1.1a	59.9±8.3a	1.44±0.02de	39±1.3abcde	14.8±0.9abcd	1.1±0.08ab	13.1±0.2defg
ZY	58.7±2.4bcd	6.3±1.4ab	73.8±5.7abc	1.35±0.02ab	41.9±0.9defg	12.7±0.3abc	1.5±0.06bcd	8.8±0.5abc

Different letters indicate significant differences ($p = 0.05$) among the 19 counties. Counties are arranged in English alphabetical order.

Table 3. Total C (TSOCS) and N (TSNS) stores in the surface soil layer (0–40 cm) of soils in 19 counties in Henan Province, China.

	C store (g m^{-2})	N store (g m^{-2})
HX	3541±261.8 abcd	410.1±18.9 cde
LK	3118.9±189.3 ab	399.7±23.6 bcde
LY	4106.7±294.1 abcd	290.4±22.2 a
LC	4023.8±372 abcd	398.6±35.6 bcde
QF	3229.7±140 abc	381.8±22.3 abcd
QX	5977.9±524.3 e	429.1±63.1 de
SS	3881.8±219 abcd	348.3±24.0 abcd
TK	3605.1±328.1 abcd	303.4±19.2 ab
TY	4300.4±98.5 bcd	494.3±24.9 ef
WX	4396.2±451.9 cd	379.4±23.4 abcd
WY	4081.3±455.2 abcd	429.3±25.3 de
XZ	4528.2±516.5 d	320.3±40.1 abc
XP	5709.2±582.9 e	369.8±21.9 abcd
XC	4614.5±609.9 d	366.3±47.7 abcd
XX	4323.8±224.7 bcd	558.2±35.1 f
YJ	3072.5±178.3 a	378.7±19.2 abcd
YX	4190.7±241.3 abcd	315.4±17.4 abc
YC	3926.2±304.8 abcd	299.3±26.3 ab
ZY	3413.4±104.3 abcd	398.1±21.7 bcde

Counties are arranged in English alphabetical order.

659 g m^{-2}. The highest value was in XX County and the lowest in LY County in N reserves (Table 3).

Linear regression analysis indicated that total crop production was significantly and positively correlated with SOC and TSN (Figure 4A), soil C and N store (Figure 4B), and soil C and N stocks (Figure 4C). Soil bulk density was significantly and positively correlated with soil N concentration ($r = 0.25$, $p = 0.008$, n = 114), soil C ($r = 0.21$, $p = 0.03$, n = 114) and N store ($r = 0.43$, $p = 0.001$, n = 114). While soil porosity was significantly and negatively correlated with soil N concentration ($r = -0.19$, $p = 0.05$, n = 114), soil C ($r = -0.25$, $p = 0.007$, n = 114) and N store ($r = -0.32$, $p = 0.001$, n = 114).

Principle components analysis revealed that Axis 1, which explained 98% of the variation in all data (eigenvalue = 0.98), was highly correlated with soil C, whereas Axis 2, explaining 1% of the variation (eigenvalue = 0.09), was highly correlated with soil N. Thus, counties such as QX and XP located highly positive on Axis 1 with high levels of soil C, but other counties, such as LK, YJ, and QF, occupied positions toward the negative end of Axis 1 with low soil C (Figure 5).

Discussion

Potential influences on crop yield

It is notable that 14 environmental (e.g., mean annual temperature and precipitation –Table 1) and soil variables (including extractable nutrients-Table 2) examined in our analysis of the data from the 19 counties in Henan Province were correlated with either wheat or total crop yield (data not shown), and total crop production were significantly, positively related to SOC and TSN, soil C and N store, and soil C and N stock

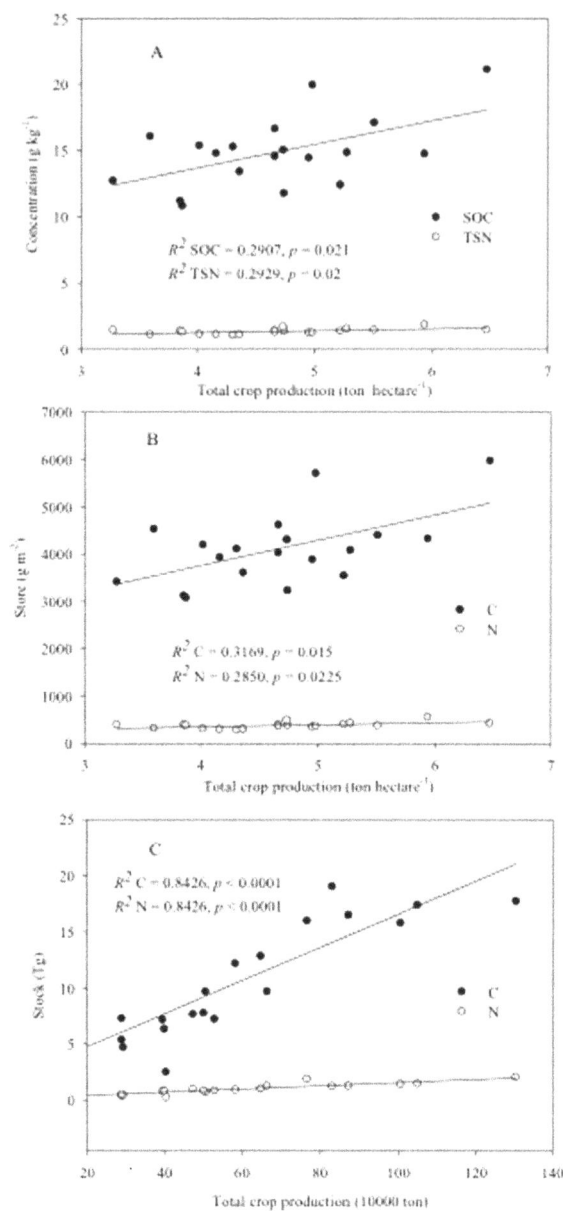

Figure 4. Linear regression analysis of total crop production in each county (ton ha^{-1}) with SOC and TSN (A) and with soil C and N store (0–40 cm) (B), and the total crop production of each county (10000 ton) with their soil C and N stock (C) (n = 19).

(Figure 4). Part of this is likely related to the highly integrated nature of the measures of C and N stocks, i.e., their calculations combine soil concentrations of C and N, soil bulk density, sampling depth, and area of cultivation. However, all of these have been shown to directly influence crop performance. For example, increases in soil C have been shown to increase crop yield in other studies. Lal (2004, 2006) reported increases in yield from 20 to 70 kg ha^{-1} and 10 to 300 kg ha^{-1} for wheat and maize, respectively, following increases of 1 MT of C in agricultural soils in Africa [1,8]. Similarly, loss of soil C has been shown to decrease yield in agricultural soils of Canada and the U.S. [4,5].

Soil C-mediated increases of crop yields also may arise from improvements in soil structure and available water-holding

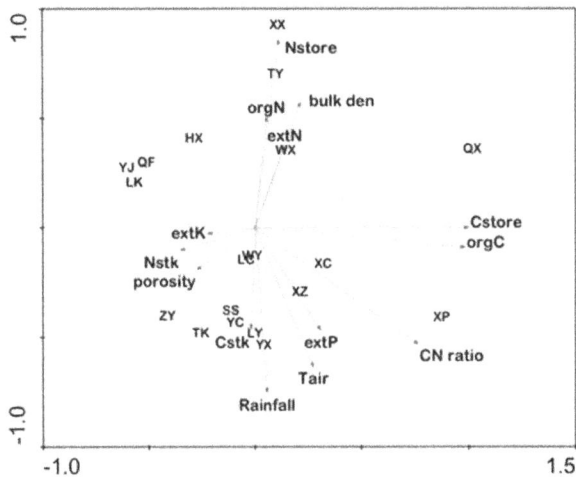

Figure 5. Principle components analysis of environmental and soil variables for agricultural soils in 19 counties within Henan Province. Length of arrows is directly proportional to their importance in explaining spatial patterns in the counties. Direction of the arrows indicates increasing values. Thus, the x-axis is primarily a gradient in soil C, whereas the y-axis is primarily a gradient in soil N and rainfall and secondarily a gradient in soil N. See Table 1 for key to county name abbreviations.

capacity. Enhanced soil structure, via increased soil C, generally arises from several processes, including increasing stability of soil aggregates [22,23,24]. As a result of the increased stability of the aggregates, soils become less prone to crusting, compaction, and erosion [25,26,28]. Emerson (1995) demonstrated that an increase of 1 g of soil organic matter (~50% of which is C) can increase available soil moisture by up to 10 g [27], which is enough to maintain crop growth between periods of rainfall of 5 to 10 days [8].

Spatial variation in cultivated soils

In this study, soil organic C concentration averaged 14.9 g kg^{-1} and total N averaged 1.4 g kg^{-1} in the 0–40 cm layer across all sites, while soil C and N stores averaged 4.1 kg C m^{-2} and 0.38 kg N m^{-2}, respectively. These values are comparable to published values from other regions of China, including 9–15 g C kg^{-1} and 1.2–1.8 g N kg^{-1} in northern China [12,29], and 16.1 g C kg^{-1} and 1.04 g N kg^{-1} in eastern and southern China [14,16,30]. Liu et al. (2011) reported soil C stores of 4.57 kg C m^{-2} in the Loess Plateau region in northwestern China [13].

Principal components analysis separated the 19 counties primarily along a gradient in soil C, with counties LK, YJ, QF, ZY, HX, and TK (mean soil C = 12.1 g C kg^{-1}) toward the lower end and XP and QX (mean soil C = 20.5 g C kg^{-1}) toward the

upper end of Axis 1, which accounted for nearly 80% of the variation in soil and environmental data (Figure 5). Spatial variation in soil organic C in agricultural systems can be influenced by several factors, including microclimate, soil type, topography, and especially human activity [31].

Spatial variation in soil N was essentially orthogonal to that of soil C. This was surprising since typically, the two are highly correlated in terrestrial ecosystems [32]. As a result, the secondary gradient (i.e., Axis 2) was one of soil N, with counties TK, YC, SS, LY, YX, and XP (mean soil N = 1.15 g N kg^{-1}) located toward the lower end of Axis 2 (accounting for <10% of variation) and XX and TY (mean soil N = 1.81 g N kg^{-1}) located toward the upper end of Axis 2 (Figure 5). Although C and N are often correlated through their organic forms in plant detritus, spatial variation of N in soils of agro-ecosystems can also be greatly influenced by the extensive use of N fertilizers.

Management methods used in crop production systems, including tillage practices and fertilizer use, can affect soil C and N on broad spatial scales, including that of an entire Province [33]. Over the course of repeated seasons of crop growth in Henan Province, agricultural fields are repeatedly subjected to soil tillage, planting, fertilization, irrigation, and harvest, all of which potentially influence soil C and N stores [30,34]. In contrast, Zhang et al. (2012) reported that raised-bed planting, a viable alternative to conventional tillage, can significantly enhance the yield of summer maize while simultaneously improving soil structure, as well as the structure and function of microbial communities essential to the quality of agricultural soils [22].

Results presented in the current study underscore the complexity of factors that can impact agricultural soils and their ability to produce crops to meet the ever-increasing demand in China resulting from population growth. Some of the spatial pattern exhibited in ordination space (Figure 5) is clearly related to regional factors, such as microclimate. For example, WY and LC are adjacent to each other in Henan Province (Figure 1) and are also closely clustered in ordination space, indicating that they are very similar with respect to environmental and soil characteristics. XP and SS, however, are also adjacent counties; yet occur distant from each other in ordination space, indicating great dissimilarity in environmental and soil factors. Agronomists should take into account the large spatial variability in important components of the soils in Henan Province, especially in the variation of soil C and N, when considering appropriate agronomic management practices.

Author Contributions

Conceived and designed the experiments: XLZ QW CHL. Performed the experiments: XLZ QW YLW. Analyzed the data: XLZ QW FSG. Contributed reagents/materials/analysis tools: XLZ QW YLW FNC. Contributed to the writing of the manuscript: XLZ FSG CHL.

References

1. Lal R (2004) Soil carbon sequestration impacts on global climate change and food security. Science 304: 1623–1627.
2. Liu DW, Wang ZM, Zhang B, Song KS, Li XY, et al. (2006) Spatial distribution of soil organic carbon and analysis of related factors in croplands of the black soil region, Northeast China. Agriculture, Ecosystems and Environment 113: 73–81.
3. Smith WN, Desjardins RL, Pattey E (2000) The net flux of carbon from agricultural soils in Canada 1970–2010. Global Change Biology 6: 557–568.
4. Bauer A, Black AL (1994) Quantification of the effect of soil organic matter content on soil productivity. Soil Science Society of America Journal 58: 185–193.
5. Larney FJ, Janzen HH, Olson BM, Lindwall CW (2000) Soil quality and productivity response to simulated erosion and restorative amendments. Canadian Journal of Soil Science 80: 515–522.

6. Hairiah K, Van Noordwijk M, Cadisch G (2000) Crop yield, C and N balance of the three types of cropping systems on an Ultisol in northern Lampung. Netherland Journal of Agricultural Science 48: 3–17.
7. Duxbury JM (2001) Long-term yield trends in the rice-wheat cropping system: results from experiments in Northwest India. Journal of Crop Production 3: 27–52.
8. Lal R (2006) Enhancing crop yields in the developing countries through restoration of the soil organic carbon pool in agricultural lands. Land Degradation and Development 17: 197–209.
9. Batjes NH (2002) Carbon and nitrogen stocks in the soils of Central and Eastern Europe. Soil Use and Management 18: 324–329.

10. Maia SMF, Ogle SM, Cerri CC, Cerri CEP (2010) Changes in soil organic carbon storage under different agricultural management systems in the Southwest Amazon Region of Brazil. Soil and Tillage Research 106: 177–184.

11. Piao SL, Fang JY, Ciais P, Peylin P, Huang Y, et al. (2009) The carbon balance of terrestrial ecosystems in China. Nature 458, doi:10.1038/nature 07944.

12. Wang ZM, Zhang B, Song KS, Liu DW, Ren CY (2010) Spatial variability of soil organic carbon under maize monoculture in the Song-Nen plain, Northeast China. Pedosphere 20: 80–89.

13. Liu ZP, Shao MA, Wang YQ (2011) Effect of environmental factors on regional soil organic carbon stocks across the Loess Plateau region, China. Agriculture, Ecosystems and Environment 142: 184–194.

14. Liao QL, Zhang XH, Li ZP, Pan GX, Smith P, et al. (2009) Increase in soil organic carbon stock over the last two decades in China's Jiangsu Province. Global Change Biology 15: 861–875.

15. Zhang HB, Luo YM, Wong MH, Zhao QG, Zhang GL (2007) Soil organic carbon storage and changes with reduction in agricultural activities in Hong Kong. Geoderma 139: 412–419.

16. Feng S, Tan S, Zhang A, Zhang Q, Pan G, et al. (2011) Effect of household land management on cropland topsoil organic carbon storage at plot scale in a red earth soil area of South China. Journal of Agricultural Science 149: 557–566.

17. Harper RJ, Gilkes RJ (1995) Some factors affecting the distribution of carbon in soils of a dry land agricultural system in southwestern Australia. In: Lal R, Kimble JM, Follett RF, Stewart BA (editors). Assessment Methods for Soil Carbon. CRC Press. Boca Raton, FL, USA. PP.577–591.

18. Guo JH, Liu XJ, Zhang Y, Shen JL, Han WX, et al. (2010) Significant Acidification in Major Chinese Croplands. Science 327: 1008–1010.

19. Pan GX, Li LQ, Wu LS, Zhang XH (2003) Storage and sequestration potential of topsoil organic carbon in China's paddy soils. Global Change Biology 10: 79–92.

20. Wu HB, Guo ZT, Gao Q, Peng CH (2009) Distribution of soil inorganic carbon storage and its changes due to agricultural land use activity in China. Agriculture, Ecosystems and Environment 129: 413–421.

21. Gilliam FS, Saunders NE (2003) Making more sense of the order: A review of Canoco for Windows 4.5, PC-ORD version 4 and SYN-TAX 2000. Journal of Vegetation Science 14: 297–304.

22. Zhang XL, Ma L, Gilliam FS, Wang Q, Liu T, et al. (2012) Effects of raised-bed planting for enhanced summer maize yield on soil microbial functional groups and enzyme activity in Henan Province, China. Field Crops Research 130: 28–37.

23. Feller C, Beare MH (1997) Physical control of soil organic matter dynamics in tropics. Geoderma 79: 69–116.

24. Haynes RJ, Naidu R (1998) Influence of lime, fertilizer and manure applications on soil organic matter content and soil physical conditions: a review. Nutrient Cycling in Agroecosystems 51: 123–137.

25. Diaz-Zorita M, Grosso GA (2000) Effect of soil texture, organic carbon and water retention on the compatibility of soils from the Argentinean Pampas. Soil and Tillage Research 54: 121–126.

26. Schertz DL, Moldenhauer WC, Livingston SJ, Weeisies GA, Hintz AE (1989) Effect of past soil erosion on crop productivity in Indiana. Journal of Soil and Water Conservation 44: 604–608.

27. Emerson WW (1995) Water-retention, organic-carbon and soil texture. Australian Journal of Soil Research 33: 241–251.

28. Powlson DS, Hirsch PR, Brookes PC (2001) The role of soil micro-organisms in soil organic matter conservation in the tropics. Nutritional Cycling in Agroecosystems 61: 41–51.

29. Du ZL, Ren TS, Hu CS (2010) Tillage and residue removal effects on soil carbon and nitrogen storage in the North China Plain. Soil Science Society of American Journal 74: 196–202.

30. Pan GX, Li LQ, Zhang Q, Wang XK, Sun XB, et al. (2005) Organic carbon stock in topsoil of Jiangsu Province, China, and the recent trend of carbon sequestration. Journal of Environmental Sciences 17: 1–7.

31. Post WM, Pastor J, Zinke PJ, Stangenberger AG (1985) Global patterns of soil nitrogen storage. Nature 317: 613–616.

32. Gilliam FS, Dick DA, Kerr ML, Adams MB (2004) Effects of silvicultural practices on soil carbon and nitrogen in a nitrogen saturated Central Appalachian (USA) hardwood forest ecosystem. Environmental Management 33: S108–S119.

33. Pan GX, Zhao QG (2005) Study on evolution of organic carbon stock in agricultural soils of China: facing the challenge of global change and food security. Advances in Earth Science 20: 384–393 (in Chinese).

34. Dersch G, Böhm K (2001) Effects of agronomic practices on the soil carbon storage potential in arable farming in Austria. Nutrient Cycling in Agroecosystems 60: 49–55.

10

Roles of Extension Officers to Promote Social Capital in Japanese Agricultural Communities

Kosuke Takemura[1]*, Yukiko Uchida[2], Sakiko Yoshikawa[2]

1 Graduate School of Management, Kyoto University, Kyoto, Japan, **2** Kokoro Research Center, Kyoto University, Kyoto, Japan

Abstract

Social capital has been found to be correlated with community welfare, but it is not easy to build and maintain it. The purpose of the current study is to investigate the role of professional coordinators of social relationships to create and maintain social capital in a community. We focused on extension officers in Japanese agricultural communities, who help farmers in both technical and social matters. A large nation-wide survey of extension officers as well as two supplementary surveys were conducted. We found that (1) social capital-related activities (e.g., assistance for building organizations among farmers) were particularly effective for solving problems; (2) social capital (trust relationships) among community residents increased their life quality; (3) social capital in local communities was correlated with extension officers' own communication skills and harmonious relationships among their colleagues. In sum, social capital in local communities is maintained by coordinators with professional social skills.

Editor: Kimmo Eriksson, Mälardalen University, Sweden

Funding: This work was supported by grants from the Kokoro Research Center at Kyoto University, Japan (Project Number: 08-1-01, 10-1-07); and the Japan Agricultural Development and Extension Personnel Association. The funders had no role in study design, data collection and analysis, decision to publish, or preparation of the manuscript.

Competing Interests: The authors have declared that no competing interests exist.

* E-mail: boz.takemura@gmail.com

Introduction

As humans live in societies, they are more or less social and interwoven in the society [1]. Given that humans significantly rely on social relationships to survive [2], mutual cooperation within community is essential. However, a state of mutual cooperation is not necessarily easy to establish. Tragedy of the commons, which was initially proposed by Hardin [3], is a typical example that highlights the conflicting nature of individual self-interests versus public goods. It has been suggested that one of the key factors to deal with this difficulty of social life and establish mutual cooperation efficiently is social capital [4]. The purpose of the current research is to investigate consequences (e.g., social capital brings improvement of the living condition in a community) as well as antecedents (e.g., personal social skills or positive social interactions with others) of social capital, by focusing on roles of professional social coordinators in agricultural communities.

Social capital

The term social capital captures the idea that social bonds and norms are important for people and communities [4–5]. Social capital can be broadly defined as the benefits of investing in social relationships [6], similar to financial capital and human capital (investing in individual capacities such as education). An extensive literature shows that human welfare depends heavily on social capital and also that social capital varies widely among human social environments. A recent study by Gutiérrez, Hilborn, and Defeo [7] have found, for instance, that social capital was a contributing factor in preventing the tragedy of the commons and leaded to the success of community-based co-management fisheries. Also, Sampson, Raudenbush, and Earls [8] showed that

neighborhood-level collective efficacy, comprised of social capital and informal social control, reduced violent acts such as homicide. Social capital has also been suggested to increase regional incomes, life satisfaction, and life expectancy (see [4,9,10], for reviews).

Why does social capital or social cohesion improve the quality of life in the community? For example, a trust relationship, one feature of social capital [4,9], enhances cooperation, and so reduces transaction costs between people. Instead of having to invest in monitoring others, individuals are able to trust them to act as expected, thus saving money and time [4]. Accordingly, it becomes possible for them to invest money and time in other things such as health (see also [11–12], for reviews on how social relationships promote physical health). Also, mutual cooperation that social capital promotes is essential to build and/or maintain public goods such as irrigation systems in agricultural communities and sustainable resources [7,13].

Although it is becoming well-known and accepted that social capital is important for people and communities, it is not easy to build and maintain it. Imagine someone has altruistic intention and is willing to have reciprocal relationships with others. Even in this instance, it is not always easy to promote reciprocal relationships because his or her intention is not transparent to others. He or she needs to prove his or her good will so that others feel safe to step into the relationship. Norms, which are another important component of social capital that help establish mutual cooperation [9], generally take time to be established. Sampson et al. [8] have found that residential instability is negatively associated with neighborhood-level collective efficacy, which, as mentioned above, comprised mutual control against norm violation. Thus, in neighborhoods where residents change frequently, collective efficacy tends to be low and consequently violence tends to be more prevalent. Moreover, even if mutual cooperation is established, maintaining it is not easy. It is generally observed in public goods game experiments that cooperation decreases as the game proceeds [14–15]. Trust relationships, which promote cooperation, are also very sensitive and easy to be broken. Once installing a monitoring and sanctioning system against free-riders, group members' trust ironically becomes even lower than initial levels after removing the sanctioning system [16].

Role of the intermediary

The current paper examines roles of individuals playing as an *intermediary* or *coordinator* of social relationships, which has been suggested to create and maintain social capital. For example, it has been suggested that the existence of an intermediary promotes the building of trust relationships [17]. Suppose that one person needs to decide whether he or she trusts another person whom he or she has just met. In this situation, trust tends to be promoted if there is someone whom both the trustor and trustee have already established relationships with. The first reason is that uncertainty for the trustor about the trustee's personality is reduced through confidence in the intermediary as a good judge of people. Second, the existence of the third person changes incentive structures for the trustee. If the trustee cheats the trustor, the relationship between the trustor and the intermediary may be damaged. Thus, cheating the trustor will bring negative consequences not only to the trustor, but also to the intermediary. As a result, the relationship between the trustee and the intermediary will also be potentially damaged by the trustee's ill-intended behavior against the trustor. Knowing such potential consequences, the trustee has no (or at least smaller) incentive to cheat the trustor. Given this incentive structure, the trustor now feels secure to trust this person (there is *assurance* in Yamagishi's [18] terminology). After all, the existence of an intermediary promotes the building of trust between two parties who have no established relationship. These mechanisms pointed out by Coleman [17] are examples of how intermediaries serve in constructing social capital.

In line with Coleman's argument, Harada et al. [19] found that public health nurses, professionals that support community residents' health, played the role of intermediary in social relationships. They suggested that public health nurses helped to strengthen relationships within local communities (i.e., bonding social capital; [9]), and contributed to establishing relationships between community residents and the government as well as other types of professionals such as nursery school teachers (i.e., bridging social capital; [9]). Though an untrained third person, such as the mutual acquaintance in Coleman's [17] argument, may help social ties to be formed, the findings of Harada et al. [19] suggests that the existence of professionals who have skills for construction and maintenance of social capital is an important key for community welfare. However, Harada et al.'s [19] study, which employed a semi-structured interview method, was limited because of sample size (only 20 public health nurses). Further investigations about roles of professional intermediaries are undoubtedly needed. To this end, the current study examined roles played by a different professional coordinator of social relationships: extension officers working in an agricultural community.

Extension officers in Japanese agricultural communities

According to Zakaria and Nagata [20], since the end of World War II, Japan started to embark on a concerted effort to revitalize its agriculture sector in order to boost production to meet the escalating demand for food. The central and prefectural governments worked closely to enhance the training of farmers to promote their technical and managerial skills and to ensure sustainability, and this was carried out through the activities and programs by the agricultural extension services.

The Japanese extension system for agriculture, which started in 1948, was meant to help farmers acquire useful, appropriate, and practical knowledge in the domain of agriculture. This system was adapted from the American extension system but modified when applying it in Japan, so that the extension works could well function in Japanese culture for local needs and requirements [21].

As of 2010, there were approximately 7,000 officers in Japan, each of them belonging to a prefectural government. To become an extension officer, one has to pass a national exam. The job of extension officers is to help farmers in person to acquire technical and managerial knowledge and other skills in the domain of agriculture [22]. They do this job in collaboration with the Ministry of Agriculture, Forestry and Fisheries, research institutions, agricultural universities, and the prefectural government.

The Japanese Ministry of Agriculture, Forestry and Fisheries officially states that extension officers are expected to serve two functions: 1) "specialist" function, and 2) "coordinator" function [23–24]. Specialist function means "extension activities to provide farmers with advanced techniques and related knowledge (including managerial knowledge and skills), according as appropriate to local environments." On the other hand, coordinator function means to "help local farmers and related parties share future goals, clarify tasks they need to address, develop an approach to the tasks, and conduct it, under the cooperation with leading farmers as well as relevant organizations within and around local communities." Thus, extension officers are supposed to help farmers not only with skills and knowledge directly related to agriculture, but also help farmers build and maintain bonding-type social capital (social capitals within local communities) as well as bridging-type social capital (connections with related parties outside of communities).

In fact, Zakaria and Nagata [20] pointed out that extension officers in Japan "as intermediaries and catalysts, are the key links between farmers and the relevant agencies in terms of providing personalized and need-based information for decision making by all parties concerned" (in abstract [20]). Also, "the agricultural extension organizations naturally provide the place or the Japanese concept of 'ba' [25], which means 'a shared space that serves as a foundation for knowledge creation' for the promotion of active interactions, consultations and exchanges between extension and farmers" (p. 34 [20]). Also, Fukuda [26,27] points out that one of the roles extension officers play in Japanese agricultural communities is to facilitate communications and interactions among farmers as well as between farmers and related agencies in order to help innovate techniques and spread knowledge. These arguments suggest the significance of social ties within and around agricultural communities, and roles played by extension officers to construct and maintain them.

Importance of social capital in agricultural communities

Prior research has suggested that residents of agricultural communities are particularly interdependent and rely on social relationships compared to communities engaged in other types of economic activities such as herding [28]. This implies that social capital is highly important for agricultural communities. Indeed, there are several studies showing the association between social capital and collective actions as well as welfare of agricultural communities. Fukushima et al. [29], for example, found that trust toward other residents in the same local community was positively related with participation in collective management of local resources such as irrigation systems and commons (see also [30], for similar findings in Thailand). Also, trust relationships within agricultural communities have been found to be associated with residents' self-rated health [31–32] as well as settlement in the communities [33]. It was also shown that damages by animals, such as monkeys, which brought serious harms to agricultural communities, could be efficiently mitigated by routinely employing community level cooperation to scare away the animals [34]. More generally, Gyawali, Fraser, Bukenya, and Banerjee [35] found that increased human well-being (composite of income, education, and employment) was associated with social capital, through a survey conducted in the west-central Black Belt region of Alabama, which contains vast amounts of forest resources and fertile agricultural land. As a whole, in line with the social capital literature, social capital such as trust relationships have been found to relate to human welfare in agricultural communities.

There are at least two issues awaiting empirical study. First, though it has been suggested that social capital is associated with human welfare in agricultural communities, much of the prior research did not prove that the observed associations are causal. The current study addressed this by analyzing panel data collected through multiple surveys targeting extension officers in agricultural communities (Analyses 1 and 2). Second, we explored what kind of characteristics and situations the extension officers should possess for building social capital efficiently, by analyzing large survey data of extension officers all over Japan (Analysis 3).

Overview

We conducted three survey studies, and performed several analyses by combining the datasets when necessary. Data 1 constituted our major dataset, which was created by a nation-wide survey for more than 4,000 extension officers all over Japan, conducted in 2010. Data 2 and 3 were studies that included responses of the extension officers in two different areas (Data 2: Kinki area in 2009 and Data 3: Aichi prefecture in 2011). In Data 2 and Data 3, we could identify those who were also included in Data 1 in 2010 and thus we can examine that data as a panel data for a time-series analysis.

We conducted analyses mainly based on Data 1, but used combined datasets as well, depending on the purpose of the analysis. Analyses 1 and 2 addressed the effect of social capital in communities, such as trust relationships among community residents. In Analysis 1 with Data 1, we investigated the effect of extension service activities related to establishing social capital for solving farmers' problems. Though previous studies on extension officers in Japan [20] and the literature on the importance of social capital suggest the significant roles of the officers' coordinator function, this is not necessarily a widely shared notion even among the officers themselves [36], and thus needs to be examined empirically. Analysis 1 compared the performance of extension activities related to social capital and the other types of activities. In Analysis 2, we examined whether trust relationships among community residents (one of the important components of social capital) would increase life quality or not through a time-series analysis based on combined datasets of Data 1, 2, and 3.

In Analysis 3, we investigated what kinds of extension officers were more likely to contribute to trust relationships among community residents. We analyzed Data 1 to examine effects of personal traits of extension officers such as communication skills and the social relationships they had.

We first describe the three datasets we used. Then, we report the results of our Analyses 1 to 3.

Method

Ethics statement

The survey for Data 1 was approved by the Japan Agricultural Development and Extension Personnel Association. The survey for Data 2 was approved by the Kinki Regional Agricultural Administration Office. The survey for Data 3 was approved by the Aichi Agricultural Development and Extension Personnel Association. All participants gave consent by completing the survey. Data files are not available due to the consent agreement with the organization of extension officers. Upon request, however, we could provide the detailed information about the data.

Data 1

Respondents. With the cooperation of the Japan Agricultural Development and Extension Personnel Association, we called all the extension officers in Japan ($N = 7,241$) for the study, and 4,355 extension officers participated in the study (response ratio was 60.0%). Respondents from all but two prefectures completed the online survey (for Saitama and Wakayama prefectures, the surveys were sent and returned through the mail because computers at their workplaces could not access the survey website due to access limitations). Data collection was conducted between September and October 2010. Table 1 provides information on gender, age, and years of working experience as extension officers of respondents in Data 1.

Measures. Extension activities they had conducted. The respondents were asked to recall one of their recent experiences in which they were faced with a difficulty in the case they were charged with. Then they were instructed to check all the extension activities they had conducted in that agricultural situation from the list of 11 types of activities, which were derived from a previous study [37]: 1) Assistance to foster the sustainable workforce, 2) Assistance to establish the desirable area of productions, 3) Assistance to conduct eco-friendly agriculture, 4) Assistance regarding food safety, 5) Assistance for the development of agricultural communities, 6) Introduction of agricultural techniques, 7) Assistance for sales promotion, 8) Collaboration and coordination with relevant organizations, 9) Assistance for building organizations and collaboration among farmers, 10) Providing a vision for the future, and 11) Identifying specific problems the community has.

After indicating all the conducted activities, they completed the following four items, all of which were considered to reflect their assessment of their activities; the first two items asked how satisfied they were, and how satisfied they thought the community residents were with their activities as a whole in that situation. For both items, they indicated the level of satisfaction using the same scale ranging from 0 to 100. The other two questions were about positive feedbacks from the community residents. The first item asked how often the community residents showed their gratitude to the respondents (from $0 =$ Never to $3 =$ Very often). The second one asked how pleased the community residents were about the respondent's activities in total (from $-3 =$ Not pleased at all to $3 =$ Fairly pleased). To give equal weights to the four items, we rescaled the items so that each ranged from 0 to 1. The average of the rescaled items forms our measure of performance ($\alpha = .87$).

Perceived state of the community. Another part of the survey was about the state of the community where the respondents were working at that time. The respondents were instructed to indicate life quality of the community residents (2 items, "The community residents are satisfied with their circumstances of life," "The living conditions of the community residents were all right"; $\alpha = .78$). The second set of this part was about trust

Table 1. Summary of respondent characteristics.

		Data 1	Data 2	Data 3[a]
Study period		September-October, 2010	July-August, 2009	October, 2011
Population		Extension officer in Japan	Extension officer in Kinki area (6 prefectures)	Extension officer in Aichi prefecture
Sample size		4,355	319	101
Response rate		60%	52%	54%
Gender	Female	23%	18%	68%
	Male	60%	63%	26%
	No response	17%	19%	6%
Age	20s	6%	3%	6%
	30s	18%	18%	16%
	40s	37%	35%	61%
	50s	24%	24%	16%
	60s	1%	2%	0%
	No response	13%	18%	0%
Years of experience working for the current job	3 or less	11%	6%	16%
	Between 4–10	16%	15%	16%
	Between 10–15	15%	17%	6%
	15 or longer	38%	44%	58%
	No response	20%	18%	3%

[a]Data 3 was planned to be merged with Data 1. To reduce the burden of respondents, we did not ask respondents in Data 3 on their gender, age, or years of working experience. For Data 3, we report information on these variables based on 31 respondents who were identified in Data 1.

relationships within community (7 items, e.g., "I think the community residents trust each other," "The interpersonal relationships among the community residents are generally smooth"; $\alpha = .82$). Response options were provided on 7-point scales (1 = Strongly disagree to 7 = Strongly agree).

The respondent's skills and social relationships. We administered a modified version of Tsutsui's [38] collaboration activity scale, which was designed to measure a behavioral tendency for collaborations and communications with other related parties such as agricultural cooperatives and local government. Sample items included "I hear about services and actual situations of other related organizations (including resident organizations) from themselves," "I know what kind of professionals are in other related organizations (including resident organizations)," "I ask for cooperation of other related organizations (including resident organizations)," ($\alpha = .84$). Response options were provided on 4-point scales (1 = Not at all to 4 = Very much). We call this the collaboration index hereafter.

We also asked about their self-evaluation of communication skills ("What do you feel about your own current communication skills as an extension officer?") and self-evaluation of their knowledge and technical skills ("What do you feel about the current level of your knowledge and techniques which are directly related to extension activities?"), which related to the major functions of extension officers ("coordinator" function and "specialist" function, respectively). For both items, response options were provided on 7-point scales (-3 = Far from enough to 3 = Good enough). They also worked on a 10-item scale of extraversion developed by Goldberg [39] including sample items such as "I am the life of the party" and "I start conversations" (1 = Strongly disagree to 7 = Strongly agree; $\alpha = .91$). Extraversion was

measured in this study as one of the control variables of personality traits that might promote their coordinator function.

Two questions were administered to assess social relationships the respondents had. The first one was about their relationships with the community they were working for (hereafter referred to as "tie with community"), ranged from -3 (I am independent and separate from the community) to 3 (I am connected with the community). The second one was about relationship harmony at their workplace (i.e., extension center) (-3 = Very bad to 3 = Very good).

Data 2

Respondents. With the cooperation of the Research Group of Specialists in Extension Activities of Kinki, we called all the extension officers in six prefectures of the Kinki area ($N = 616$), and 319 extension officers participated in this study (response ratio was 51.8%). The URL of the survey website was announced and the Excel form of the questionnaire was also sent to them via email just in case they could not work on the survey online. Data collection was conducted between July and August, 2009, approximately one year before Data 1.

For both Data 1 and 2, respondents were asked to create a unique identification code for themselves. Out of the 319 respondents of Data 2, 61 people were identified in Data 1 as well, and among them, 29 people had worked for the same community between the two data collections, and completed all the relevant scales.

Measures. To measure the perceived trust relationships among the community residents as well as the perceived life quality of the community residents, the identical measures to those of Data 1 were used for Data 2 (αs = .83 and .76 for trust relationships and life quality of the community residents, respectively). The respondents in Data 2 also indicated how long they had been working for the same community. This information helped us to select the respondents who worked for the same community in Data 1 and 2.

Data 3

Respondents. With the cooperation of the Aichi Agricultural Development and Extension Personnel Association, we called all the extension officers in Aichi prefecture ($N = 188$) for the study, and 101 of them participated in the study (response ratio was 53.7%). The URL of the survey website was announced to all the extension officers in Aichi. They worked on the survey during October 2011, approximately one year after Data 1.

Respondents of Data 3 were asked to create a unique identification code based on the same rule from Data 1 and Data 2. Out of the 101 respondents of Data 3, 47 people were found to be identical in Data 1 as well, and 31 of them had worked for the same community between the two data collections, completed all the relevant scales, and did not learn about the results of Data 1.

Measures. Just like Data 2, the same measures for the perceived trust relationships among the community residents and the perceived life quality of the community residents were used for Data 3 (αs = .77 and .73, respectively). Also the respondents in Data 3 indicated how long they had been working for the same community so that we could select the respondents who worked for the same community in Data 1 and 3.

Results

Analysis 1: Types of extension activities and their effects

In Analysis 1, we analyzed Data 1 to see what kind of extension activities were efficient to solve problems that farmers faced. We predict that activities related to social capital (e.g., activities promoting coordination among farmers) are effective to solve the problems among farming communities.

As aforementioned, respondents of Data 1 were asked to recall one of their recent experiences in which they had faced a difficulty, and indicate all the extension activities they conducted in the situation. The most frequently conducted activity was collaboration and coordination with relevant organizations (63%), followed by introduction of agricultural techniques (61%; see Table 2 for the other extension activities).

What kinds of activities have a positive effect in solving problems? We first performed an exploratory factor analysis (principal factor solution) with varimax rotation on the 11 activity items to see convergence among the activities. The scree test suggested two factors, which together accounted for 38.4% of the variance. Factor 1 accounted for 24.7% of the variance, and Factor 2 accounted for 13.6% of the variance. As shown in Table 3, the six activity items that were related to social capital (e.g., "assistance for building organizations and collaboration among farmers") had high factor loadings on Factor 1 (hereafter called social capital-related activity). On the other hand, the other five activity items that were related to agricultural skills and business management (e.g., "assistance regarding food safety", "introduction of agricultural techniques") had high loadings on Factor 2 (hereafter called agricultural business management activity). We constructed the social capital-related activity indicator and the agricultural business management activity indicator by taking the mean of the respective items.

We then examined correlations of these two types of activity indicators with the performance score. The analysis revealed that social capital-related activity was positively related with the performance score, $r = .39$, $p < .001$. In addition, agricultural business management activity was also positively correlated with the performance score, $r = .28$, $p < .001$. As expected, however, the effect of social capital-related activity was greater than the other $t(3899) = 6.62$, $p < .001$. This suggests that extension activities which enhance social capital are especially effective and lead to good performance to solve problems in communities.

Analysis 2: Social capital (trust relationships) and life quality of community residents

To examine how social capital promotes the quality of life in agricultural communities, Analysis 2 focused on the effect of perceived trust relationships among the community residents (the indicator of social capital) on perceived life quality of the community residents.

First, we examined simple correlation between these two variables (Table 4). As predicted, they were found to be positively correlated with each other consistently across three datasets. This suggests that trust relationships among the community residents have a positive effect on their quality of life.

However, the analyses above do not indicate the causal association. To examine the causal relationships showing if

Table 2. Extension activities and implementation rate in difficult situations respondents experienced.

Extension activities	Implementation rate
Assistance to foster the sustainable workforce	50%
Assistance to establish the desirable area of productions	44%
Assistance to conduct eco-friendly agriculture	24%
Assistance regarding food safety	24%
Assistance for the development of agricultural communities	31%
Introduction of agricultural techniques	61%
Assistance for sales promotion	26%
Collaboration and coordination with relevant organizations	63%
Assistance for building organizations and collaboration among farmers	44%
Providing a vision for the future	36%
Identifying specific problems the community has	38%

Table 3. Rotated factor matrix of the extension activity items.

	Factor loading	
Item	**Social capital-related (Factor 1)**	**Agricultural business management (Factor 2)**
Providing a vision for the future	**.57**	.09
Assistance for building organizations and collaboration among farmers	**.54**	.06
Identifying specific problems the community has	**.51**	.17
Assistance for the development of agricultural communities	**.49**	.06
Collaboration and coordination with relevant organizations	**.41**	.17
Assistance to foster the sustainable workforce	**.37**	.09
Assistance regarding food safety	.05	**.65**
Assistance to conduct eco-friendly agriculture	.05	**.64**
Assistance to establish the desirable area of productions	.27	**.35**
Introduction of agricultural techniques	.08	**.34**
Assistance for sales promotion	.24	**.32**

Factor loadings .30 or greater are shown in bold.

perceived trust relationships really had an effect on perceived life quality, we conducted a time-series analysis by combining Data 1, 2, and 3. As mentioned above, 29 respondents in Kinki area completed the relevant scales both in Data 1 and 2. Similarly, 31 respondents in Data 3 (Aichi prefecture) were also found in Data 1, and completed the relevant scales. For both Kinki area and Aichi prefecture, the first data collection was conducted approximately one year before the second data collection. For Kinki area, responses at Time 1 were from Data 2, and responses at Time 2 came from Data 1. For Aichi prefecture, Data 1 and 3 were used for Time 1 and 2, respectively. Finally, we combined these two datasets about the two areas into single datasets to see if perceived trust relationships at Time 1 has significant effect on perceived life quality at Time 2 even after controlling for the effect of perceived life quality at Time 1.

We regressed perceived life quality at Time 2 on perceived trust relationships at Time 1, perceived life quality at Time 1, a dummy-coded variable of area (0 = Kinki, 1 = Aichi), and the interaction effect terms between perceived trust relationships at Time 1 and area, as well as perceived life quality at Time 1 and area (see Table 5; adjusted R^2 = .08, p = .085). As expected, perceived trust relationships at Time 1 had marginally significant positive effect on perceived life quality at Time 2. Standardized regression coefficient of perceived trust relationships at Time 1 suggests that an increase of one standard deviation in this variable led to an increase of 0.34 of one standard deviation in perceived life quality measured at approximately one year later. Area did not moderate the effect of perceived trust relationships. The other effects were also found to be non-significant. In sum, the results suggest the positive effect of trust relationships on life quality of community residents.

Analysis 3: Correlates of social capital (trust relationships) of communities

As shown in Analysis 2, trust relationships among community residents, one aspect of social capital, has a positive effect on the residents' life quality. The next question is, "what enhances trust relationships among community residents?"

By analyzing Data 1, we examined correlations of perceived trust relationships among community residents with respondent's collaboration index, extraversion, communication skills, knowledge and technical skills, tie with community, and interpersonal relationships at the workplace. As shown in Table 6, collaboration index, extraversion, and communication skills were positively correlated with perceived trust relationships. Though knowledge and technical skills also had a positive association with perceived trust relationships, the effect size was quite small.

Tie with community and interpersonal relationships at the workplace also had positive correlations with perceived trust relationships. Furthermore, these variables, which were about social relationships surrounding respondents, had positive correlations with perceived trust relationships among community residents even after controlling for the respondent's own internal traits on social relationships such as the collaboration index, extraversion, and communication skills (tie with community: r_ps = .31, .32, .30, ps <.001; interpersonal relationships at the workplace: r_ps = .22, .22, .21, ps <.001, by controlling for the collaboration index, extraversion, communication skills, respectively). These results indicate that the effects of social relationships surrounding extension officers are not spurious correlations caused by the extension officer's personal traits.

Discussion

The concept of social capital including trust relationships and social networks has served to bring researchers' attention to the significance of social bonds for human welfare. Prior research has actually demonstrated the associations of social capital with several domains of human life, such as financial incomes, life expectancy, life satisfaction, decreased violence, maintenance of public goods, and so on [4,9,10].

The current study was conducted to investigate consequences and antecedents of social capital in Japanese agricultural communities by focusing on roles of professional extension officers. Extension officers are involved in many kinds of activities to help farmers, such as introducing agricultural techniques, providing managerial knowledge, and building and maintaining trust

Table 4. Perceived trust relationships and perceived life quality of community residents (means, standard deviations, and Pearson's r coefficients).

	N	Perceived trust relationships among community residents		Perceived life quality of community residents		Correlation	
		M	(SD)	M	(SD)	r	p
Data 1	3268	5.27	(0.75)	3.83	(1.26)	.14	.000
Data 2	163	5.10	(0.65)	3.83	(1.16)	.19	.016
Data 3	97	5.36	(0.68)	4.40	(1.19)	.31	.002

Note. For this analysis, we included not only respondents who had data both in Data 1 and Data 2 (or Data 3) but also respondents who completed the relevant scales only in Data 2 or 3.

relationships and collaboration inside and around agricultural communities.

Our analyses, based on data collection including a nation-wide survey of extension officers, showed that extension activities related to social capital are particularly important. Analysis 1 revealed that to solve problems that farmers are faced with extension activities for enhancing social capital had greater effects compared to other activities such as the introduction of agricultural techniques. This finding suggests social capital plays essential roles for life in agricultural communities.

In line with the results of Analysis 1, trust relationships (one important aspect of social capital) and life quality of community residents were found to have positive association across three survey data (Analysis 2). Moreover, by analyzing panel data, we validated the causal relationship between them: Trust relationships among community residents promote their life quality.

Furthermore, to study antecedents of social capital, we explored which factors or skills of extension officers were associated with trust relationships among community residents. Analysis 3 revealed that extension officers' collaboration with related parties and communication skills were positively correlated with trust relationships within communities. In addition, interpersonal relationships at extension officers' workplace (i.e., extension centers) were positively connected with trust relationships in the local communities where the extension officers worked. This suggests a "chain effect" of social capital, meaning positive relationships in one place (extension officers' workplace) also facilitate positive relationships in another place (an agricultural community) presumably through extension officers' activities. Taken together, the current research demonstrates the importance of extension officers' work in promoting social capital in agricultural communities.

Extension activities related to agricultural techniques must not be viewed as unimportant. In fact, the demand for "specialist" function is still high. Fukuda's [27] research that collected farmers' opinions found that one of the highest-priority needs of farmers is extension of innovative techniques. Yet, the current study suggests that the other function of extension officers that have not received broad attention—"coordinator" function—has to do with a very important resource of agricultural communities, namely, social capital.

Limitations and future directions

It is important to emphasize that our data was collected through self-report and perceptions about states of communities by extension officers. This means that the current paper relies only on the service providers' point of view, rather than the service recipients (i.e., farmers). However, it is also important to note that relying only on the farmers' point of view is not sufficient either to investigate roles of extension activities. Some extension officers we interviewed emphasized that, to motivate farmers, it is sometimes important to hide the roles of their activities from farmers. Thus, the farmers may not be aware about the functions of extension activities. It is therefore of importance to investigate associations between extension activities and communities' welfare from *both* sides. In addition to this, it would be an important future work to include objective measures, such as objective indices of farmers' health, economic success, and the actual number of cooperative interactions among farmers within communities. It is suggested that relying only on the same type of measurement (e.g., self-report likert scale) from the same source may exaggerate observed correlations due to the common method variance bias [40]. Some of our findings, however, cannot be explained solely by this bias. We found the predicted associations even when we had covariates

Table 5. Effects of trust relationships among community residents (Time 1), life quality of community residents (Time 1), area, and their interactions on life quality of community residents (Time 2), $N = 60$.

	b	(SE)	β	t	p
Perceived trust relationships (Time 1)	0.65	(0.36)	.34	1.82	.075
Perceived life quality (Time 1)	0.29	(0.20)	.29	1.49	.142
Area (0 = Kinki, 1 = Aichi)	−0.19	(0.32)	−.08	−0.61	.543
Perceived trust relationships (Time 1) x Area	−0.61	(0.49)	−.23	−1.25	.217
Perceived life quality (Time 1) * Area	0.07	(0.26)	.05	0.28	.778

that should share the same bias (see Analyses 2 and 3). Yet, it is undoubtedly desirable to have objective measures as well and to examine the robustness of the findings.

Another limitation of the current study is the small sample size of the panel data for Analysis 2. We needed to include only respondents who had worked for the same community for both Time 1 and Time 2 in the analysis. We scarcely had a fair amount of respondents since extension officers' working terms for one community are generally short (in Data 1, we asked the respondents how long they had been working for the same community; the length of the mean time was 2.18 years, median was 1.50 years, and mode was 0.50 years). Future studies collecting panel data through those who stay in the same community for a longer period of time (e.g., farmers) are needed. It is also important to point out that our finding from the time-series analysis (Analysis 2) is not conclusive about the causality. By controlling for the effect of life quality at Time 1, we could show that the opposite causality (i.e., life quality promotes trust relationships among community residents) cannot fully explained the observed association. However, it is still possible that a third variable explains the association between trust relationships and life quality. For example, it may be possible that existence of strong leadership in a community promotes both trust relationships among residents and their life quality. Future studies that examine effects of such potential third variables are needed.

Additionally, it is also important to investigate potential negative effects of social capital on welfare of agricultural communities. For example, it has been suggested that excessive levels of bonding-type social capital (social capitals within a group) may promote

distrust toward outsiders and inhibit the group's economic growth [41]. Future studies need to investigate what kinds of extension activities promote (or inhibit) social networks crossing a boundary of local communities.

There is another important question that future research needs to address: How can we train good coordinators? Though it is suggested by the current study that good coordinators (e.g., extension officers who have high communication skills) can help communities enhance trust relationships and collaborations, knowledge on how to foster such good coordinators is requisite to keep communities benefiting from them in the future. Thus, we need to know, for example, how to obtain ability for collaboration, how to acquire good communication skills, and how to recruit those who are (or have potential to be) good coordinators. Also, though the current study targeted extension officers and social capital in Japanese agricultural communities, presumably other types of communities, organizations, and groups face similar problems and thus coordinators may play crucial roles. How to achieve efficient problem solving in groups is one of the questions social psychological research has extensively addressed. From the findings of the current study, skilled coordinators are expected to play significant roles in groups and organizations that need cooperation and collaboration among members, such as medical institutions and educational institutions. Future research is needed to find ways to build systems that can sustainably provide coordinators who support building connections between people.

Table 6. Collaboration index, extraversion, communication skills, knowledge and technical skills, tie with community, and interpersonal relationships at the workplace (means, standard deviations, and Pearson's r coefficients with perceived trust relationships among community residents).

	M	(SD)	Correlation with perceived trust relationships among community residents	
			r	p
Respondent's personal traits				
Collaboration index	2.63	(0.34)	.17	.000
Extraversion	3.87	(1.08)	.14	.000
Communication skills	0.28	(1.41)	.17	.000
Knowledge and technical skills	−0.50	(1.58)	.08	.000
Social relationships surrounding respondent				
Tie with community	0.63	(1.15)	.34	.000
Interpersonal relationships at the workplace	1.03	(1.29)	.24	.000

Note. Scales ranged from 1 to 4 for Collaboration index, from 1 to 7 for Extraversion, and from −3 to 3 for the other scales.

Acknowledgments

The authors would like to thank the Japan Agricultural Development and Extension Personnel Association, the Kinki Regional Agricultural Administration Office, the Research Group of Specialists in Extension Activities of Kinki, and the Aichi Agricultural Development and Extension Personnel Association, for their help with data collection. We also would like to thank researchers at the Kokoro Research Center for their helpful comments on the projects.

Author Contributions

Conceived and designed the experiments: KT YU SY. Performed the experiments: KT YU. Analyzed the data: KT. Wrote the paper: KT YU SY.

References

1. Aronson E (1972) The social animal. New York: Viking Adult. 338 p.
2. Baumeister RF, Leary ML (1995) The need to belong: Desire for interpersonal attachments as a fundamental human motivation. Psychol Bull 17: 497–529.
3. Hardin G (1968) The tragedy of the commons. Science 162: 1243–1248.
4. Pretty J (2003) Social capital and the collective management of resources. Science 302: 1912–1914.
5. Coleman JS (1988) Social capital in the creation of human capital. Am J Sociol 94: S95–S120.
6. Wilson DS, O'Brien DT, Sesma A (2009) Human prosociality from an evolutionary perspective: Variation and correlations at a city-wide scale. Evol Hum Behav 30: 190–200.
7. Gutiérrez NL, Hilborn R, Defeo O (2011) Leadership, social capital and incentives promote successful fisheries. Nature 470: 386–389.
8. Sampson RJ, Raudenbush SW, Earls F (1997) Neighborhoods and violent crime: A multilevel study of collective efficacy. Science 277: 918–924.
9. Putnam RD (2000) Bowling alone: The collapse and revival of American community. New York: Simon and Schuster. 541 p.
10. Wilkinson RG (1999) Health, hierarchy, and social anxiety. Ann N Y Acad Sci 896: 48–63.
11. House JS, Landis KR, Umberson D (1988) Social relationships and health. Science 241: 540–545.
12. Uchino BN, Cacioppo JT, Kiecolt-Glaser JK (1996) The relationship between social support and physiological processes: A review with emphasis on underlying mechanisms and implications for health. Psychol Bull 119: 488–531.
13. Pretty J, Ward H (2001) Social capital and the environment. World Dev 29: 209–227.
14. Fehr E, Gächter S (2002) Altruistic punishment in humans. Nature 415: 137–140.
15. Puurtinen M, Mappes T (2009) Between-group competition and human cooperation. Proc. R. Soc. B 276: 355–360.
16. Mulder LB, van Dijk E, De Cremer D, Wilke HAM (2006) Undermining trust and cooperation: The paradox of sanctioning systems in social dilemmas. J Exp Soc Psychol 42: 147–162.
17. Coleman JS (1990) Systems of trust and their dynamic properties. In: Coleman JS, editor. Foundations of social theory. Cambridge, MA: Harvard University Press. pp. 175–196.
18. Yamagishi T (2011) Trust: The evolutionary game of mind and society. New York: Springer. 153 p.
19. Harada H, Konishi M, Teraoka S, Ura M (2011) Professional skills in the process of establishing human relations within a support framework: Focusing on building community systems supported by public health nurses. Jpn J Exp Soc Psychol 50: 168–181 (in Japanese with English Summary).
20. Zakaria S, Nagata H (2010) Knowledge creation and flow in agriculture: The experience and role of the Japanese extension advisors. Libr Manag 31: 27–35.
21. Japan Agricultural Development and Extension Association (1992) Susumeyo jiko kenshu - shokuba kenshu [Let's do self-development training and on-job training]. The Extension Information Center, Japan Agricultural Development and Extension Association (in Japanese, the title translated by the current authors). 212 p.
22. Ministry of Agriculture, Forestry and Fisheries, Japan (n.d.) What is extension service? Available: http://www.maff.go.jp/j/seisan/gizyutu/hukyu/h_about/index.html (in Japanese). Accessed 2012 August 18.
23. Ministry of Agriculture, Forestry and Fisheries, Japan (2010) Guideline for management of agricultural extension services: Ministerial Notification No. 590 of MAFF. Available: http://www.maff.go.jp/kinki/seisan/keieishien/fukyu/pdf/h22_shishin.pdf (in Japanese). Accessed 2012 August 18.
24. Ministry of Agriculture, Forestry and Fisheries, Japan (2012) Guideline for management of agricultural extension services: Ministerial Notification No. 848 of MAFF. Available: http://www.maff.go.jp/j/seisan/gizyutu/hukyu/h_tuti/pdf/h24_shishin.pdf (in Japanese). Accessed 2012 August 18.
25. Nonaka I, Konno N (1998) The concept of "ba": Building a foundation for knowledge creation. Calif Manage Rev 40: 40–54.
26. Fukuda K (2003) A study on the functions of agricultural extension: Focusing on recent studies of agricultural extension theory overseas and in Japan. J Rural Community Stud 97: 82–90 (in Japanese with English Summary).
27. Fukuda K (2007) The direction of extension activities in formulating rural agriculture from the viewpoint of the needs of farmers: Based on the case of vegetable production areas in Yamagata prefecture and Chiba prefecture. J Rural Community Stud 105: 25–40 (in Japanese with English Summary).
28. Úskül AK, Kitayama S, Nisbett RE (2008) Ecocultural basis of cognition: Farmers and fishermen are more holistic than herders. Proc Natl Acad Sci U S A 105: 8552–8556.
29. Fukushima S, Yoshikawa G, Saizen I, Kobayashi S (2012) Analysis on rerated (sic) factors of participation in regional resource management in rural areas in northern Kyoto prefecture: A comparison of the two factors: Bonding and bridging social capital. J Rural Plan Assoc 31: 84–93 (in Japanese with English Summary).
30. Matsushita K, Asano K (2007) The effect of social capital on the efficiency of irrigation water management. Proc Agric Econ Soc Jpn 2007 482–489. (in Japanese).
31. Fukushima S, Yoshikawa G, Ichida Y, Saizen I, Kobayashi S (2009) Comparison of the influence between the generalized trust and the trust in members of one's community on self-rated health. Pap Environ Inf Sci 23: 269–274 (in Japanese).
32. Fukushima S, Yoshikawa G, Saizen I (2012) Determining the relationship between the frequency of contact with friends and self-rated health based on the proximity of residence: Aiming at the integration of individual and area level social capital studies. Environ Inf Sci 40: 31–39 (in Japanese with English Summary).
33. Yamaguchi S, Nakatsuka M, Hoshino S (2007) Study on region characteristic and settlement in rural area: A case of Sasayama City, Hyogo prefecture. J Rural Plan Assoc 26: 287–292 (in Japanese with English Summary).
34. Yamabata N (2010) Mitigate effect on damage to food crops achieved by collaboration of a whole village for chase-off of monkeys. J Rural Plan Assoc 28: 273–278 (in Japanese with English Summary).
35. Gyawali BR, Fraser R, Bukenya J, Banerjee SB (2010) Spatial relationship between human well-being and community capital in the Black Belt region of Alabama. J Agric Ext Rural Dev 2: 167–178.
36. Research Group of Specialists in Extension Activities of Kinki (2009) "Wakate fukyu shidoin no ikusei shuho" ni kansuru chosa kenkyu [Research on training methods for junior extension officers]. Research Group of Specialists in Extension Activities of Kinki (in Japanese, the title translated by the current authors). 40 p.
37. Uchida Y, Takemura K, Yoshikawa S (2011) The coordination roles of extension officers within Japanese agricultural communities. Sociotechnica 8: 194–203 (in Japanese with English Summary).
38. Tsutsui T (2005) Chiiki hoken sa-bisu no tantou shokuin ni okeru renkei hyouka shihyou kaihatsu ni kansuru toukei teki kenkyu [Statistical research to develop collaboration index of community health services officials]. Research Report of Health Labour Sciences Research Grant (the Integrated Research Project for Health Science) (in Japanese, the title translated by the current authors). 117 p.
39. Goldberg LR (1992) The development of markers for the Big-Five factor structure. J Pers Soc Psychol 4: 26–46.
40. Podsakoff PM, Organ DW (1986) Self-reports in organizational research: Problems and prospects. J Manage 12: 531–544.
41. Svendsen GLH, Svendsen GT (2004) The creation and destruction of social capital. Cheltenham, UK: Edward Elgar Publishing. 207 p.

11

Fish Product Mislabelling: Failings of Traceability in the Production Chain and Implications for Illegal, Unreported and Unregulated (IUU) Fishing

Sarah J. Helyar[1¤]*, Hywel ap D Lloyd[1¤], Mark de Bruyn[1], Jonathan Leake[2], Niall Bennett[3], Gary R. Carvalho[1]

1 Molecular Ecology and Fisheries Genetics Laboratory, Bangor University, Bangor, Wales, United Kingdom, **2** Sunday Times, London, United Kingdom, **3** Greenpeace UK, London, United Kingdom

Abstract

Increasing consumer demand for seafood, combined with concern over the health of our oceans, has led to many initiatives aimed at tackling destructive fishing practices and promoting the sustainability of fisheries. An important global threat to sustainable fisheries is Illegal, Unreported and Unregulated (IUU) fishing, and there is now an increased emphasis on the use of trade measures to prevent IUU-sourced fish and fish products from entering the international market. Initiatives encompass new legislation in the European Union requiring the inclusion of species names on catch labels throughout the distribution chain. Such certification measures do not, however, guarantee accuracy of species designation. Using two DNA-based methods to compare species descriptions with molecular ID, we examined 386 samples of white fish, or products labelled as primarily containing white fish, from major UK supermarket chains. Species specific real-time PCR probes were used for cod (*Gadus morhua*) and haddock (*Melanogrammus aeglefinus*) to provide a highly sensitive and species-specific test for the major species of white fish sold in the UK. Additionally, fish-specific primers were used to sequence the forensically validated barcoding gene, mitochondrial cytochrome oxidase I (COI). Overall levels of congruence between product label and genetic species identification were high, with 94.34% of samples correctly labelled, though a significant proportion in terms of potential volume, were mislabelled. Substitution was usually for a cheaper alternative and, in one case, extended to a tropical species. To our knowledge, this is the first published study encompassing a large-scale assessment of UK retailers, and if representative, indicates a potentially significant incidence of incorrect product designation.

Editor: Konstantinos I. Stergiou, Aristotle University of Thessaloniki, Greece

Funding: This study was jointly funded by Greenpeace and The Sunday Times. The funder Greenpeace provided support in the form of salary for author NB, but did not have any additional role in the study design, data collection and analysis, decision to publish, or preparation of the manuscript. The funder The Sunday Times provided support in the form of salary for author JL, but did not have any additional role in the study design, data collection and analysis, decision to publish, or preparation of the manuscript. The specific roles of these authors are articulated in the 'author contributions' section.

Competing Interests: The authors have the following interests: This study was jointly funded by Greenpeace and The Sunday Times. Co-author Jonathan Leake is employed by The Sunday Times. Co-author Niall Bennett is employed by Greenpeace. There are no patents, products in development or marketed products to declare.

* E-mail: sarah.helyar@matis.is

¤ Current address: Food Safety, Environment & Genetics, Matís, Reykjavík, Iceland

Introduction

In recent years, concerns about the health of the oceans and the effects of over-exploitation of fisheries have increased. Consumer demand for seafood is growing with the contribution of fish to the average annual diet reaching a record of 18.8 kg per person per year in 2011 [1], as compared to 17.1 Kg in 2008 [2]. This is partly due to an increase in the range of species consumed, and an increase in aquaculture. Fish products were worth a record $217.5 billion in 2010, up over 9% from 2009, and these trends are expected to continue. The increasing demand for fish highlights the need for the sustainable management of aquatic resources; 87.3% of world fish stocks are classed as overexploited, depleted or recovering: a number which continues to increase [1], with 29.9% of stocks classed as overexploited and unlikely to meet the targets

of the Johannesburg Plan of Implementation to restore them to a level that can produce maximum sustainable yield by 2015 [3].

A major threat for the sustainable management of these valuable resources is Illegal, Unreported and Unregulated (IUU) fishing. Current estimates suggest that globally up to 25% of fisheries catches fall within IUU practices [4–6], identifying it as the single largest threat to achieving sustainability. Both the FAO [7] and the European Union [8] have placed increasing emphasis on the use of trade measures to prevent IUU-sourced fish and fish products from entering international trade. One component of this increased regulation has required the inclusion of binomial species nomenclature on catch labels throughout the distribution chain [9].

In addition to top down pressure for improved labelling and traceability of fish products, many consumers are increasingly aware of nutritional and environmental issues regarding fisheries, leading to shifts in attitude regarding acceptable species, catch location and catch methods [10]. In parallel, due to globalization of the industry, consumers are encountering an increasing number of fish species and/or an escalation in common names applied to the same species. Such drivers have led to a greater demand for informative labelling, including the use of 'eco-labelling'. Although labelling to provide additional ecological information about a product is often voluntary, the FAO recognised that it could contribute to improved fisheries management and convened a Technical Consultation in 1998, which resulted in their Guidelines for the Eco-labelling of Fish and Fishery Products from Marine Capture Fisheries [11]. Informative labelling is particularly important for processed items because any recognizable external morphological features are typically removed, leaving consumers reliant on product labelling for content information. However, it has been argued that any such labelling scheme, whether voluntary or legislated, requires policing in order to prevent misuse and fraud [12].The mislabelling of a fish product may be unintentional if, for example, species that are morphologically similar are caught together, such as in many tropical or coral reef fisheries [13–17]. Alternatively, mislabelling may not be accidental, such as where product substitutions are from species that do not occur in the same ocean [18–20], or for lesser value species [21,22]. However, whether intentional or not, the outcome can be serious for management and sustainability targets. In addition to the direct impacts of depletion from IUU fishing, substitutions and misidentification that occur before fish are landed will inflate the inaccuracies in catch and forecast statistics.

Several recent studies of mislabelling have been undertaken in Europe [19,22,23], yielding rates of mislabelling of up to 32% [19]. Most mislabelled products have originated from small-scale retailers and convenience food outlets (e.g. fish and chip shops) but the major supermarkets have not hitherto been thoroughly investigated. Supermarket chains account for 72% of the total fish retail market in the UK (excluding canned products) [24]. If comparable rates of mislabelling occur in supermarket products it is thereby likely to have a substantial impact on efforts to manage the respective fisheries sustainably. It is therefore necessary to establish to what extent mislabelling of fish products occurs in the major retailers of the fish food supply chain, which is addressed in this study.

The current study uses two DNA-based methods to identify the species of origin for 386 samples collected from major supermarket chains around the UK. Species-specific real-time PCR probes [25] for cod (*Gadus morhua*), and haddock (*Melanogrammus aeglefinus*) were used to provide a highly sensitive test for the major species of white fish sold in British supermarkets. Additionally, DNA barcoding [26] using fish-specific COI primers [27] was employed. The COI mitochondrial gene has been validated for forensic species identification [28] to determine its reproducibility and limitations by testing its ability to provide accurate results under a variety of conditions. To our knowledge, the current findings represent the first large-scale assessment of fish product authentication across major UK supermarket retailers.

Materials and Methods

Sample collection

386 samples of processed white fish, ranging from fillets to fish fingers and fish cakes, were collected from six leading supermarket chains, at multiple locations across England, Scotland and Wales (Table S1in File S1). Approximately 20 mg of tissue was taken from the centre of each product to ensure minimal DNA damage from production, processing, or contamination. These were placed into numbered tubes filled with 96% ethanol. Sample details including the place and date of purchase, species designation, and eco-labelling were entered into a database linked to photographs of the packaging. Sample identities were not disclosed until completion of molecular genetic analyses, when molecular and sample IDs were cross-referenced.

Molecular methods

DNA was extracted with the E-Z 96 Tissue DNA kit (Omega-biotek), then quantified with a Nanodrop 1000 (Thermo Scientific), and standardised to either 5 ng/μL or 2 ng/μL depending on original concentration. Real-time PCR assays were carried out on all samples on an Applied Biosystems 7700 real-time sequence detection system. The 25 μL reactions contained 200 nM of each of the two species specific probes (see Table 1), 300 nM of the GAD-F and GAD-R primers (Taylor *et al.* 2002), 9.163 μl 2X Taqman Universal PCR Master Mix (UNG+ROX and passive reference) (Applied Biosystems), 15 ng of DNA, and (depending on DNA concentration) either 10.417 or 6.917 μL PCR grade H_2O (Sigma). Reactions were run in optical 96-well reaction plates using optical adhesive covers (Applied Biosystems). Plates were analysed under real-time conditions on two dye layers with eight 'no template controls' (NTCs) per 96-well plate, and 2 positive controls for each of the two target species. The assay was run using the default cycling conditions [25].

In addition to the real-time PCR, all samples were sequenced for approximately 655 bp from the 5' region of the COI gene from mitochondrial DNA using primers developed by Ward [27]. Tests were run with all combinations of the four available primers, but the combination of FishF1/FishR2 produced consistently good PCR products in the species tested, and was therefore used throughout (see Table 1). PCRs were carried out in 30 μL reactions containing 15 μL of 2 x PCR Mastermix (containing 0.75 U of *Taq* polymerase (buffered at pH 8.5), 400 μM each dNTP, 3 mM $MgCl_2$ (Promega), 9 μL PCR grade H_2O (Sigma), 15 pmol each primer, and 3.0 μL of DNA template. The PCRs consisted of a denaturation step of 2 min at 95°C followed by 35 cycles of 30 seconds at 94°C, 30 seconds at 54°C, and 1 min at 72°C, followed by a final extension of 10 min at 72°C and then held at 4°C. PCR products were visualized on 1.2% agarose gels. If a single clear band was produced, PCR products were sent to GATC (Germany, http://www.gatc-biotech.com) for sequencing. DNA from 48 samples was re-extracted as independent replicates of real-time PCR and sequencing, including all samples where molecular data contradicted species designations, and an additional randomly chosen 33 samples to test repeatability of DNA-based species ID.

Species identification

Real-time PCR. The results were analysed using the Sequence Detection Software version 1.71 (Applied Biosystems). The ΔRn values for each cycle and dye layer were then exported to MS Excel and additional manual processing was carried out. First, the mean and standard deviation of the endpoint (PCR cycle 40) ΔRn values of the NTCs were calculated for each dye layer. $z*M$-values ($z*M = M+(3.89 \times SD)+C$) were then calculated where M = mean of the NTC ΔRn, SD is the standard deviation of the NTC ΔRn and 3.89 is the one tailed Z-value for the 99.999% confidence interval, C is a constant (0.3) introduced to overcome the slight increase in fluorescence of samples above the NTC fluorescence due to spectral bleeding between dye layers. Samples which had ΔRn values larger than the value of $z*M$ were considered to have a fluorescence significantly greater than the NTCs, and therefore to be positive reactions.

Table 1. list of all primers used.

	Sequence 5'-3'	Reporter	Quencher
COD P	CTTTTTACCTCTAAATGTGGGAGG	-	-
HAD P	CTTTCTTCCTTTAAACGTTGGAGG	-	-
GAD-F	GCAATCGAGTYGTATCYCTWCAAGGAT	FAM	Non-fluorescent
GAD-R	CACAAATGRGCYCCTCTWCTTGC	TET	Non-fluorescent
FishF1	TCAACCAACCACAAAGACATTGGCAC	-	-
FishR2	ACTTCAGGGTGACCGAAGAATCAGAA	-	-

COD P, HAD P, GAD-F, and GAD-R were used in the real-time-PCR, and FishF1 and FISHR2 were used for the sequencing PCRs.

COI sequencing. Successfully sequenced COI amplicons were manually checked and edited to remove ambiguous base calling in BioEdit (Ibis Biosciences). Sequences were tested against the Barcode of Life database (BOLD) [29]. In addition, reference sequences for all species genetically identified and all species indicated on sample packaging, were downloaded from BOLD and aligned with the sample sequences in Clustal X [30], the Neighbour-joining tree was constructed in MEGA5 [31] with 1000 bootstrap replicates.

Results

For consistency, all samples are referred to by the labelled species unless otherwise stated. Of 386 samples, 371 (97.4%) produced DNA of sufficient quality for further analysis. Label designations indicated primarily cod (179), haddock (155) and pollock (32).

Real-time-PCR. All samples labelled as hake or Alaskan pollack showed negative results for probes designed to identify cod and haddock. The sample labelled as whiting was positive for cod. For the samples labelled as haddock (155), the haddock probe was positive in 134 samples (86.5%), while the cod probe gave a positive result for 6 samples, both probes were amplified in 7 samples (inconclusive result) and neither were amplified in 8 samples (negative). All cod labelled as originating from the Pacific were negative for both the cod and haddock probes. Out of the Atlantic cod samples (57), the cod specific probe amplified in 47 samples (82.5%), both probes were positive in 3 samples (inconclusive result) and neither in 7 samples (negative). For the cod samples which did not indicate a catch location (102), the cod specific probe was positive in 80 samples, the haddock specific probe was positive in 2 samples, both amplified in 8 samples (inconclusive result) and neither in 12 samples (negative). Real-time-PCR results are presented in Table 2.

COI sequencing. All sequence data has been submitted to NCBI, under accession numbers KJ614671 to KJ615069 (Table S2 in File S1). 48 samples have two sequences listed as these samples were re-extracted as independent replicates to ensure the repeatability of the methods.

The majority of sequences were identified with a sequence identity greater than 99.5% in the BOLD database, with sequences from two samples falling below this threshold. Additionally, two samples could not be matched unambiguously due to 100% sequence identity at COI at the taxon-pairs involved. The sequence data matched with *Gadus chalcogramma/G. finnmarchica* (Alaskan and Norwegian Pollock respectively; previously *Theragra* sp.), or *Gadus macrocephalus* and *Gadus ogac* (Pacific and Greenland cod respectively): these are both instances where the (sub-) species designation is debatable (see Discussion).

Of 179 samples labelled as cod, 57 were specified as Atlantic cod (*Gadus morhua*) and 20 as Pacific cod (*Gadus macrocephalus*), while for the remainder (102) there was no specification for either the species or catch area. In total, 9 (5.03%) of these cod samples were not verified as cod by DNA data, including 1 (0.56%) identified as *Melanogrammus aeglefinus* (haddock), and 2 (1.11%) highly processed samples that were found to have a mixed species composition (see Table 2: #1892; *G. morhua/G. chalcogramma* and #1886; *G. morhua/M. aeglefinus*). From the 57 samples labelled specifically as Atlantic cod, 51 had congruent label and DNA-based designations, while 6 (10.5%) were genetically identified as Pacific cod (*G. macrocephalus*).

155 samples were labelled as haddock (*M. aeglefinus*). Of these, 146 generated a molecular ID in agreement with labelling (5.81% mislabelled), with 6 (3.87%) identified as *G. morhua* (Atlantic cod), 1 (0.65%) as *G. macrocephalus* (Pacific cod) and 2 (1.29%) exhibited a mixed species composition (see Table 2: #1452 and #1847; *G. morhua/M. aeglefinus*).

In addition, one of the four hake (labelled as *Merluccius capensis*) samples was identified as *Merluccius paradoxus* (cape hake), one whiting (*Merlangius merlangus*) sample was identified as *Micromesistius poutassou* (blue whiting), and one Alaskan Pollack was also found to contain the Vietnamese catfish *Pangasius hypophthalmus*. Overall, our survey indicated a rate of mislabelling of 5.66%. All samples and results are presented in Tables S1 and S2 in File S1, with detailed results of the mislabelled samples in Table 2 and Table S3 in File S1. Sequence similarity with all reference samples is demonstrated in Figure 1, and the details of the reference sequences used are in Table S4 in File S1.

Discussion

Our study represents, to our knowledge, the largest published survey to date of mislabelling within the fish products sold by UK supermarkets. Samples were taken of products from leading brands and supermarket "own brands" from 6 major supermarket chains across the UK. Previous studies have examined the food retail sector and found high rates of mislabelling, particularly in restaurants and fast-food outlets [22,32]. Within our study of supermarket-sourced samples the overall inconsistency between product label and genetic species identification was 5.66%. This is considerably lower than observed in other sectors: 25% within mixed sectors [22]; 25% within markets and restaurants [32]; 32% within fishmongers [19]. Nevertheless, if our data are representative of overall trends, with over 4 billion fish products consumed (C. Roberts, unpublished data) the incidence of mislabelling could exceed 200 million products annually in the UK alone. This level of misinformation raises considerable concern in terms of consumer information and protection. It also presents substantial

Table 2. Summary of all mislabelled samples.

Identification Code	Species reported (type)	Area of Catch	real-time PCR	First sequence identity	Second sequence identity
1415	Cod (breaded fillet)	Atlantic	Negative	*Gadus macrocephalus*	*Gadus macrocephalus*
1426	Cod (breaded fillet)	Atlantic	Negative	*Gadus macrocephalus*	*Gadus macrocephalus*
1446	Cod (breaded fillet)	Atlantic	Negative	*Gadus macrocephalus*	*Gadus macrocephalus*
1747	Cod (precooked meal)	Atlantic	Negative	*Gadus macrocephalus*	*Gadus macrocephalus*
1889	Cod (precooked meal)	Atlantic	Negative	*Gadus macrocephalus*	*Gadus macrocephalus*
1975	Cod (breaded fillet)	Atlantic	Negative	*Gadus macrocephalus*	*Gadus macrocephalus*
1886	Cod (fish cakes)	NA	Inconclusive	*Gadus morhua*	*Melanogrammus aeglefinus*
1765	Cod (fish cakes)	NA	*Melanogrammus aeglefinus*	*Melanogrammus aeglefinus*	*Melanogrammus aeglefinus*
1892	Cod (fish fingers)	NA	*Gadus morhua*	*Gadus chalcogrammus*	*Gadus morhua*
1470	Haddock (precooked meal)	Atlantic	*Gadus morhua*	*Gadus morhua*	*Gadus morhua*
1812	Haddock (fish cakes)	Atlantic	*Gadus morhua*	*Gadus morhua*	*Gadus morhua*
1888	Haddock (precooked meal)	Atlantic	*Gadus morhua*	*Gadus morhua*	*Gadus morhua*
1977	Haddock (breaded fillet)	Atlantic	*Gadus morhua*	*Gadus morhua*	*Gadus morhua*
1989	Haddock (precooked meal)	Atlantic	*Gadus morhua*	*Gadus morhua*	*Gadus morhua*
1868	Haddock (precooked meal)	Atlantic	*Gadus morhua*	*Gadus morhua*	*Gadus morhua*
1851	Haddock (precooked meal)	Atlantic	Negative	*Gadus macrocephalus*	*Gadus macrocephalus*
1452	Haddock (fish cakes)	Atlantic	Inconclusive	*Gadus morhua*	*Melanogrammus aeglefinus*
1847	Haddock (fish cakes)	Atlantic	Inconclusive	*Gadus morhua*	*Melanogrammus aeglefinus*
1763	Alaskan Pollack (fish cakes)	Pacific	Negative	*Pangasius hypophthalamus*	*Gadus chalcogrammus*
1813	Hake (*M. capensis*) (breaded fillet)	NA	Negative	*Merluccius paradoxus*	*Merluccius paradoxus*
1848	Whiting (precooked meal)	NA	Inconclusive	*Micromesistius poutassou*	*Micromesistius poutassou*

NA: Not available from packaging. Negative: neither of the real-time PCR probes amplified. Inconclusive: both real-time PCR probes amplified. First and second sequence identities are the result of independent DNA extractions and sequencing (see methods for details).

challenges for the sustainable management of the respective fisheries.

Genetic identification of products was carried out with species specific real-time PCR, and by matching sample COI sequences with those of known species in the BOLD database with high (≥ 99.5%) sequence identity [33]. Such independent testing yields a high degree of certainty to the identifications, as more than 98% of species pairs have shown greater than 2% COI sequence divergence [34]. The BOLD database was used in preference to the nucleotide sequence database in GenBank (www.ncbi.nlm.nih. gov/), to ensure that the queried sequences were matched to taxonomically-validated specimens. Of all the sequences submitted, only two returned a match with less than 99.5% identity. Both of these were from highly processed samples (one labelled as cod, the other as haddock), and also returned inconclusive results for the real-time-PCR (both cod and haddock probes amplified). Both sequences were genetically identified as *M. aeglefinus* (haddock), although with relatively low sequence similarity (99.49% and 98.6%). For both of these sequences, the next closest match was *G. morhua*, rather than the next closest relative of haddock, *Merlangius merlangus* (see Figure 1), supporting the conclusion that the DNA amplified was a mix of more than one species, and therefore that these products had a mixed species composition.

Ambiguous results occurred when a sample matched with both Alaskan and Norwegian pollock (*Gadus* (= *Theragra*) *chalcogramma* and *G. finnmarchica*, respectively), or with Pacific and Greenland cod (*Gadus macrocephalus* and *G. ogac*, respectively), because congeners have 100% sequence identity at COI. However, in the case of Pacific and Greenland cod, catches of *G. ogac* are

thought to be extremely low, and currently only of local importance. The total reported catch for this stock from 2009–2011 was 586 metric tons [35], while for the same three years, the total reported catch for *G. macrocephalus* was 1,165,420 metric tons. Greenland cod is also no longer considered a separate species, but is now classed as a subspecies of Pacific cod, *G. macrocephalus* [36,37]. In the case of the Pollack species, *G. finnmarchica* was identified from a few samples from the northern tip of Norway [38] and recent molecular evidence has shown it to be indistinct from the Alaskan Pollock (*G. chalcogramma*) [39–41].

From all samples labelled as Atlantic cod, the majority of those found to be mislabelled were genetically identified as Pacific cod. This category of mislabelling could not originate at the pre-landing stage; as is evident from their common names; these species are harvested from different oceans. The implication, therefore, is that intentional mislabelling has occurred at a later stage in the supply chain. The incentive could be to supply products that mirror the preferences of the buying public, and so presumably fetch a higher price. This class of mislabelling may have little direct impact on the Atlantic cod stocks but it may influence efforts to sustainably manage stocks of Pacific cod. More importantly perhaps for this particular case of mislabelling is the issue of consumer misinformation and protection as it indicates that at some point in the supply chain there appears to be either negligence or a wilfully fraudulent attempt to provide inaccurate product information. Such instances erode consumer confidence and can undermine trust in product labelling, including any associated eco-labels.

Samples labelled as *M. aeglefinus* (haddock) show a different pattern of mislabelling. The majority of mislabelled products were

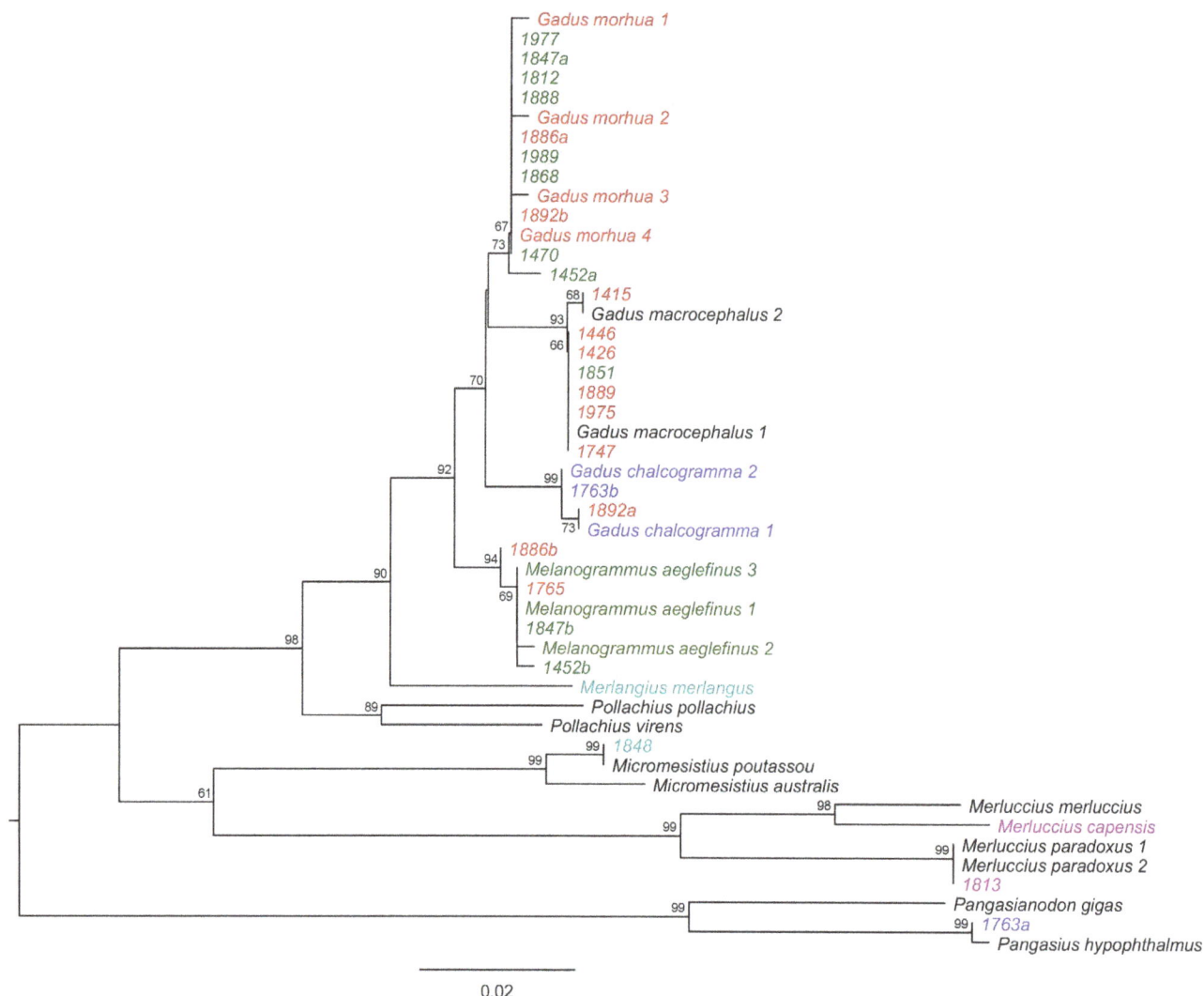

Figure 1. Neighbour-joining tree showing all mislabelled samples together with representative reference sequences taken from BOLD. Reference sequences are colour coded according to species and samples tested are colour coded according to the species stated on the packaging. Samples that have two sequences are labelled a and b.

identified as *G. morhua* (Atlantic cod). Haddock and Atlantic cod are frequently caught together in a mixed fishery and have similar market values, with cod slightly more valuable on average. As a result there is minimal direct benefit to intermediaries in the production chain to encourage such mislabelling. Alternatively, it has been suggested that such mislabelling may arise by an accidental consequence of the mixed fishery [23]. However, while we accept such possibility, mislabelling undeniably benefits the primary producer. Mislabelling *G. morhua* (Atlantic cod) as *M. aeglefinus* (haddock) enables fishermen to land undersized or over quota Atlantic cod and so profit from fish that should currently be discarded. Irrespective of the underlying cause, if the mislabelling occurs before the fish are landed (for example, if filleted and frozen at sea), such IUU activities will likely exceed catch quotas (TAQ) for a major North Atlantic fishery. The rate of mislabelling (3.87%) is comparatively low compared to other recent studies [22]. However, if we extrapolate such incidence to the TAQ for 2011, it represents an additional 2188 tonnes of Atlantic cod (or an excess 2.9% of the Atlantic cod TAQ for 2011) being landed and recorded as haddock.

In addition to the mislabelling of cod and haddock presented here, other mislabelling instances were found. One highly processed (fish cake) sample labelled as containing Alaskan Pollack (*G. chalcogramma*) was found to also contain *Pangasius hypophthalmus*. *P. hypophthalmus*, or Vietnamese catfish, is a freshwater species from Southeast Asia, legally described in the UK as Basa, Panga(s), *Pangasius*, River cobbler or any of these combined with 'catfish' [42]. Without performing a quantitative test for the presence of *P. hypophthalmus*, we were unable to estimate the relative quantities of the 2 species in this product (made of minced fish). It was therefore not possible to determine whether this reflected inadvertent contamination through inadequate cleaning of the production line between products, or deliberate substitution of a cheaper product. In either case it is unlikely to significantly affect catch data or to contribute to IUU. However, this accidental or fraudulent behaviour is a serious issue for consumer misinformation and trust, given the concerns over potentially increased contaminant levels in *Pangasius* species (such as mercury) [43], which may result in avoidance by some consumer groups.

Four hake samples were tested and one, labelled as *M. capensis*, was identified as *M. paradoxus* (25% mislabelled, although the low sample size requires caution). Historically, hake has been assessed as a single species, as separation of catches has not always been possible [44,45]. However species-specific assessments are now being conducted. The shallow water *M. capensis* stock is above sustainable levels, with catches below maximum sustainable levels and is certified by the Marine Stewardship Council (MSC). The deep-water *M. paradoxus* stock is below precautionary levels, and a rebuilding plan is in place [46]. The mislabelling of this species, whether intentional or not, at a rate even well below that observed here is a cause for serious concern, as such a practice would compromise restoration of *M. paradoxus* to sustainable levels.

One noteworthy pattern to emerge is the variation in amount of mislabelling found among the different levels of processing: within the fresh/frozen fillets (n = 84) no mislabelling was identified; in battered/breaded fillets (n = 84), fish fingers (n = 31), and precooked meals (n = 128), the respective mislabelling rates were 7.14%, 6.45% and 5.47% respectively. In fishcakes (n = 44), which are composed of minced fish, mislabelling rates of 13.6% were identified. However, these data are insufficient to identify where in this production chain, pre- or post-landing, yield higher rates of illegal activity. Targeted sampling at discrete stages across the supply chain is required: from on-board during sample catch to final retailer outlet. Alternatively it may be an inadvertent consequence of the particular processing activity, such as inadequate cleaning of processing machinery. Huxley-Jones et al. [23] found lower levels of mislabelling in processed products, such as fish fingers, than filleted products, and suggested that this may be due to greater economic gains associated with the mislabelling of fillets. In contrast, our study included more diverse forms of processing (from fresh fillets through fish fingers and precooked meals to fish cakes consisting of minced fish), and has demonstrated a clear pattern of mislabelling, from zero in unprocessed fish fillets to the highest levels of mislabelling in the most highly-processed category.

The main trends highlighted here have been the substitution of *G. morhua* (Atlantic cod) with *G. macrocephalus* (Pacific cod) in primarily filleted products, and the substitution of *M. aeglefinus* (haddock) with *G. morhua* (Atlantic cod) in precooked meals and fish cakes. Both aspects of mislabelling have a detrimental effect on *G. morhua*: substitution with *G. microcephalus* creates an erroneous impression of the abundance of the former, undermining work carried out by seafood awareness campaigns such as Seafood Watch and the Marine Stewardship Council, to educate consumers and provide tools for informed purchasing decisions. However, cod is one of the species for which there are now sufficient genomic resources to move beyond species identification and allow traceability to population level [47]. Testing by regulatory and certifying bodies would improve consumer confidence in products that are proven to fulfil claims of having been sourced from sustainably harvested stocks. In addition, as suggested here, if the substitution of *M. aeglefinus* with *G. morhua* is occurring at sea, the implications of such IUU activity would compromise the recovery of these heavily exploited species.

Previous studies have reported relatively high rates of mislabelling of seafood products globally [13,48], in Europe [19,22,49], and South Africa [21,50]. However, many studies have focused on smaller convenience food outlets and/or restaurants. Actions such

as increasing media attention, the importance of consumer confidence in the fisheries sector and revised EU legislation [51,52] will collectively highlight and tackle mislabelling practices. Nevertheless, only genetic testing across the supply chain can assess the scale and likely key stages of highest risk. It also appears increasingly likely that such practices are more frequent at the more highly processed end of the market, where opportunities for detection and/or levels of discrimination are reduced. As witnessed recently in the wake of the horsemeat scandal across Europe, the complexities of the modern food production chain demand close scrutiny at all stages to ensure authenticity and compliance. A forensic framework of genetic testing using validated reference databases [47,53–55] is expected to provide an increasingly effective approach for detection, prosecution and ultimate deterrence of illegal activity. Such actions are likely not only to protect policy compliant end-users and the wider fishing industry but importantly also enhance prospects for achieving sustainability of exploited marine resources.

Supporting Information

File S1 Supporting Information. Table S1. Summary of sampling effort for samples which produced a DNA of sufficient quality for testing. Samples are recorded by reported species and are split by supermarket (own brand/other brand items). **Table S2.** Genetic analyses for all samples. The sample identification number, product labelling (reported species and catch location), processing level* and results from the real time PCR and COI sequencing are given for each sample. Second sequence identity is the result of new DNA extraction and sequencing for mislabelled, ambiguous, and control samples. Genbank accession numbers are provided for each sequence. * classification of processing level (1: fresh or frozen fillets, 2: battered or breaded fillets, 3: fish fingers, 4: pre-cooked meals, 5: fishcakes). **Table S3.** Additional data for the mislabelled samples. Query sample details, including species labelled on packaging, COI sequence and the Genbank accession number are given. Reference species is the closest sequence match in the BOLD database, together with the catch location, BOLD ID number and Genbank accession number of the reference sample. Sequence similarity (% identity) between sample and reference is shown (* indicates sequence matches lower than 99.5%). Sequence similarity to the next closest match is also shown. **Table S4.** BOLD and Genbank identifiers for the reference sequences used in Figure 1.

Acknowledgments

Help with sampling was provided by Amy Sherborne, Willie McGee, and Charlotte Hunt-Grubbe. Martin Taylor, Delphine Lallias and Wendy Grail all provided technical advice. We would also like to thank the two reviewers for their insightful comments, which have greatly improved the clarity.

Author Contributions

Conceived and designed the experiments: SJH GRC JL. Performed the experiments: HapDL MdB. Analyzed the data: SJH HapDL. Wrote the paper: SJH HapDL GRC MdB JL NB.

References

1. FAO (2012) The State of the World's Fisheries and Aquaculture. Rome: Food and Agriculture Organization of the United Nations.

2. FAO (2011) The State of the World's Fisheries and Aquaculture. Rome: Food and Agriculture Organization of the United Nations.

3. WSSD (2002) United Nations Plan of Implementation of the World Summit on Sustainable Development A/CONF.199/20.

4. Pauly D, Christensen V, Guénette S, Pitcher TJ, Sumaila UR, et al. (2002) Towards sustainability in world fisheries. Nature 418: 689–695.

5. MRAG (2008) the Global Extent of Illegal Fishing. London: MRAG. Available: http://www.mrag.co.uk/Documents/ExtentGlobalIllegalFishing.pdf .Accessed: 20 September 2013.

6. Agnew DJ, Pearce J, Pramod G, Peatman T, Watson R, et al. (2009) Estimating the Worldwide Extent of Illegal Fishing. PLoS ONE 4(2): e4570.

7. FAO. (2001) International Plan of Action to prevent, deter and eliminate illegal, unreported and unregulated fishing. Rome, FAO. 24p. Available: http://www.fao.org/fishery/ipoa-iuu/en. Accessed: April 2012.

8. European Union (2008) Regulation (EC) No 1005/2008.

9. European Union (2009) Regulation (EC) No 1224/2009.

10. Potts T, Brennan R, Pita C, Lowrie G (2011) Sustainable Seafood and Eco-labelling: The Marine Stewardship Council, UK Consumers, and Fishing Industry Perspectives. SAMS Report: 270–211 Scottish Association for Marine Science, Oban. 78 pp. ISBN: 0-9529089-2-1.

11. FAO (2009) Guidelines for the ecolabelling of fish and fishery products from marine capture fisheries. Revision 1. Rome Available: http://www.fao.org/docrep/012/i1119t/i1119t00.htm. Accessed: April 2012.

12. Stokstad E (2010) To fight illegal fishing, forensic DNA gets local. Science 330: 1468–1469.

13. Marko PB, Lee SC, Rice AM, Gramling JM, Fitzhenry TM, et al. (2004). Mislabelling of a depleted reef fish. Nature 430: 309–310.

14. Ardura A, Pola IG, Ginuino I, Gomes V, Garcia-Vazquez E (2010) Application of barcoding to Amazonian commercial fish labelling. Food Research International 43: 1549–1552.

15. Crego-Prieto V, Campo D, Perez J, Garcia-Vazquez E (2010) Mislabelling in megrims: implications for conservation. In:Tools for Identifying Biodiversity Progress and Problems (Proceedings of the International Congress, Paris, eds. EUT Edizioni Universita di Trieste, Trieste, pp. 315–322.

16. Iglesias SP, Toulhoat L, Sellos DY (2010) Taxonomic confusion and market mislabelling of threatened skates: important consequences for their conservation status. Aquatic Conservation 20(3):319–333.

17. Gold JR, Voelker G, Renshaw MA (2011) Phylogenetic relationships of tropical western Atlantic snappers in subfamily Lutjaninae (Lutjanidae: Perciformes) inferred from mitochondrial DNA sequences. Biological Journal of the Linnaean Society 102(4):915–929.

18. Barbuto M, Galimberti A, Ferri E, Labra M, Malandra R, et al. (2010) DNA barcoding reveals fraudulent substitutions in shark seafood products: the Italian case of "palombo" (Mustelus spp.). Food Research International 43: 376–381.

19. Filonzi L, Chiesa S, Vaghi M, Marzano FN (2010) Molecular barcoding reveals mislabelling of commercial fish products in Italy. Food Research International 43: 1383–1388.

20. Armani A, Castigliego L, Tinacci L, Gianfaldoni D, Guidi A (2011) Molecular characterization of icefish, (Salangidae family), using direct sequencing of mitochondrial cytochrome b gene. Food Control 22: 888–895.

21. von der Heyden S, Barendse J, Seebregts AJ, Matthee CA (2010) Misleading the masses: detection of mislabelled and substituted frozen fish products in South Africa. ICES Journal of Marine Science 67: 176–185.

22. Miller DM, Mariani S (2010) Smoke, mirrors and mislabelled cod: poor transparency in the European seafood industry. Frontiers in Ecology and the Environment 8: 517–521.

23. Huxley-Jones E, Shaw JLA, Fletcher C, Parnell J, Watts PC (2012) Use of DNA Barcoding to Reveal Species Composition of Convenience Seafood. Conservation Biology 26: 367–371.

24. FAO (2004) Fishery and Aquaculture Country Profile UK. Available: ftp://ftp.fao.org/FI/DOCUMENT/fcp/en/FI_CP_UK.pdf. Accessed: April 2012.

25. Taylor MI, Fox C, Rico I, Rico C (2002) Species-specific TaqMan probes for simultaneous identification of (Gadus morhua L.), haddock (Melanogrammus aeglefinus L.) and whiting (Merlangius merlangus L.). Molecular Ecology Notes 2(4): 599–601.

26. Hebert PDN, Ratnasingham S, DeWaard JR (2003) Barcoding animal life: cytochrome c oxidase subunit 1 divergences among closely related species. Proceedings of the Royal Society B: Biological Sciences 270: S96–9.

27. Ward RD, Zemlak TS, Innes BH, Last PR, Hebert PDN (2005) DNA barcoding Australia's fish species. Philosophical Transactions of the Royal Society B 360: 1847–57.

28. Dawnay N, Ogden R, McEwing R, Carvalho GR, Thorpe RS (2007) Validation of the barcoding gene COI for use in forensic genetic species identification. Forensic Science International 173: 1–6.

29. Ratnasingham S, Hebert PDN (2007) BOLD: The Barcode of Life Data System (www.barcodinglife.org) Molecular Ecology Notes 7(3): 355–364.

30. Higgins DG, Sharp PM (1988). CLUSTAL: a package for performing multiple sequence alignment on a microcomputer. Gene 73: 237–244.

31. Tamura K, Peterson D, Peterson N, Stecher G, Nei M, Kumar S (2011) MEGA5: Molecular Evolutionary Genetics Analysis using Maximum Likelihood, Evolutionary Distance, and Maximum Parsimony Methods. Molecular Biology and Evolution 28: 2731–2739.

32. Wong EHK, Hanner RH (2008). DNA barcoding detects market substitution in North American seafood. Food Research International 41(8): 828–837.

33. Ward RD, Holmes BH (2007) An analysis of nucleotide and amino acid variability in the barcode region of cytochrome c oxidase I (cox1) in fishes. Molecular Ecology Notes 7(6): 899–907.

34. Hebert PDN, Ratnasingham S, de Waard JR (2003). Barcoding animal life: cytochrome c oxidase subunit 1 divergences among closely related species. Proceedings of the Royal Society London Series B 270: S96–99.

35. FAO (2012) Species Fact Sheet. Available: http://www.fao.org/fishery/species/2219/en. Accessed: 12 April 2012.

36. Carr SM, Kivlichan DS, Pepin P, Crutcher DC (1999) Molecular systematics of gadid fishes: Implications for the biogeographic origins of Pacific species. Canadian Journal of Zoology 77: 19–26.

37. Coulson MW, Marshall HD, Pepin P, Carr SM (2006) Mitochondrial genomics of gadine fishes: implications for taxonomy and biogeographic origins from whole-genome data sets. Genome 49 (9): 1115–1130.

38. FAO (1990) FAO species catalogue. Vol.10. Gadiform fishes of the world (Order Gadiformes). An annotated and illustrated catalogue of cods, hakes, grenadiers and other gadiform fishes known to date. FAO Fisheries Synopsis S125 Vol.10.

39. Ursvik A, Breines R, Christiansen JS, Fevolden SE, Coucheron DH, et al. (2007) A mitogenomic approach to the taxonomy of pollocks: Theragra chalcogramma and T. finnmarchica represent one single species. BMC Evolutionary Biology 7 1): 86.

40. Byrkjedal I, Rees DJ, Christiansen JS, Fevolden SE (2008) The taxonomic status of Theragra finnmarchica Koefoed, 1956 (Teleostei: Gadidae): perspectives from morphological and molecular data. Journal of Fish Biology 73 5): 1183–1200.

41. Carr SM, Marshall HD (2008) Phylogeographic analysis of complete mtDNA genomes from walleye pollock (Gadus chalcogrammus Pallas, 1811) shows an ancient origin of genetic biodiversity. Mitochondrial DNA 19 6): 490–496.

42. Fish Labelling (England) Regulations (2010) Statutory instrument No. 420 FOOD, ENGLAND.

43. Ferrantelli V, Giangrosso G, Cicero A, Naccari C, Macaluso A, et al. (2012) Evaluation of mercury levels in Pangasius and Cod fillets traded in Sicily (Italy) Food Additives & Contaminants 29 (7): 1046–1051.

44. Butterworth DS, Rademeyer RA (2005) Sustainable management initiatives for the Southern African hake fisheries over recent years. Bulletin of Marine Science 76(2): 287–319.

45. Johnsen E, Kathena J (2012) A robust method for generating separate catch time-series for each of the hake species caught in the Namibian trawl fishery. African Journal of Marine Science 34(1): 43–53.

46. Rademeyer RA, Butterworth DS, Plaganyi EE (2008) Assessment of the South African hake resource taking its two-species nature into account. African Journal of Marine Science 30(2): 263–290.

47. Nielsen E, Cariani A, Mac Aoidh E, Maes G, Milano I, et al. (2012) Gene-associated markers provide tools for tackling illegal fishing and false eco-certification. Nature Communications 3:851.

48. Logan CA, Alter SE, Haupt AJ, Tomalty K, Palumbi SR (2008) An impediment to consumer choice: overfished species are sold as Pacific red snapper. Biological Conservation 141: 1591–1599.

49. Pepe T, Trotta M, Di Marco I, Anastasio A, Bautista JM, Cortesi ML (2007) Fish species identification in surimi-based products. Journal of Agricultural and Food Chemistry 55(9): 3681–3685.

50. Cawthorn DM, Steinman HA, Witthuhn RC (2012) DNA barcoding reveals a high incidence of fish species misrepresentation and substitution on the South African market. Food Research International 46(1): 30–40.

51. EC (European Commission) (2000) European Council Regulation No 104/2000 of 17 December 1999 on the common organization of the markets in fishery and aqua-culture products. Official Journal of the European Communities, L17, 22–52.

52. EC (European Commission) (2001) Commission Regulation (EC) No 2065/2001 of 22 October 2001 laying down detailed rules for the application of Council Regulation (EC) No 104/2000 as regards informing consumers about fishery and aquaculture products. Official Journal of the European Communities, L278, 6–8.

53. Glover KA (2010) Forensic identification of fish farm escapees: the Norwegian experience. Aquaculture Environment Interactions 1: 1–10.

54. Glover KA, Haug T, Oien N, Walloe L, Lindblom L, et al. (2012) The Norwegian minke whale DNA register: a data base monitoring commercial harvest and trade of whale products. Fish and Fisheries 13: 313–332.

55. Ogden R (2008) Fisheries forensics: the use of DNA tools for improving compliance, traceability and enforcement in the fishing industry. Fish and Fisheries 9(SI): 462–472.

12

A Concept of Bayesian Regulation in Fisheries Management

Noél Michael André Holmgren[1]*, Niclas Norrström[1], Robert Aps[2], Sakari Kuikka[3]

1 Systems Biology Research Centre, School of Bioscience, University of Skövde, Skövde, Sweden, **2** University of Tartu, Estonian Marine Institute, Tallinn, Estonia, **3** Fisheries and Environmental Management Group, Department of Environmental Sciences, University of Helsinki, Helsinki, Finland

Abstract

Stochastic variability of biological processes and uncertainty of stock properties compel fisheries managers to look for tools to improve control over the stock. Inspired by animals exploiting hidden prey, we have taken a biomimetic approach combining catch and effort in a concept of Bayesian regulation (BR). The BR provides a real-time Bayesian stock estimate, and can operate without separate stock assessment. We compared the performance of BR with catch-only regulation (CR), alternatively operating with N-target (the stock size giving maximum sustainable yield, MSY) and F-target (the fishing mortality giving MSY) on a stock model of Baltic Sea herring. N-targeted BR gave 3% higher yields than F-targeted BR and CR, and 7% higher yields than N-targeted CR. The BRs reduced coefficient of variance (CV) in fishing mortality compared to CR by 99.6% (from 25.2 to 0.1) when operated with F-target, and by about 80% (from 158.4 to 68.4/70.1 depending on how the prior is set) in stock size when operated with N-target. Even though F-targeted fishery reduced CV in pre-harvest stock size by 19–22%, it increased the dominant period length of population fluctuations from 20 to 60–80 years. In contrast, N-targeted BR made the periodic variation more similar to white noise. We discuss the conditions when BRs can be suitable tools to achieve sustainable yields while minimizing undesirable fluctuations in stock size or fishing effort.

Editor: Jeffrey Buckel, North Carolina State University, United States of America

Funding: The study is part of the IBAM project "Integrated Bayesian risk analysis of ecosystem management – Gulf of Finland as a case study", which received funding from the European Community's Seventh Framework Programme under grant agreement 217246 made with the joint Baltic Sea research and development programme BONUS (http://www.bonusportal.org/), from FORMAS, Sweden, the Academy of Finland, and an Estonian Science Foundation grant 7609, Estonian target financed theme SF0180104s08. The funders had no role in study design, data collection and analysis, decision to publish, or preparation of the manuscript.

Competing Interests: The authors have declared that no competing interests exist.

* Email: noel.holmgren@his.se

Introduction

Fisheries managers are challenged with two widely permeated properties of their study system, uncertainty [1–3] and variability [4–6]. These can confound each other, e.g. imprecise spawning stock size estimates can generate apparent variability in the stock – recruitment relationship [7]. Reduced inter-annual variability in effort has socio-economic benefits with yields matching the capacity of the processing industry [8], a more stable job market, and fewer years with over-dimensioned fleets [9]. Often there is a trade-off between maximizing yield and stabilizing yield and fishing effort that makes management objectives ambiguous [10–12]. There are also concerns that fishing increases the temporal variability of harvested stocks, [4,13–15].

The problem with temporal variability in fisheries has led to an increased interest in the performance of alternative harvest control rules [16–18]. Stephenson et al. [19] argue that the total allowable catch (TAC) could be used to prevent unsustainable use, but is insufficient to control spatial and temporal variability, and cannot be used to achieve socioeconomic objectives. In a reflection over proposed alternative harvest controls, May et al. [20] conclude that further mathematical refinement is probably not as important

as developing "robustly self-correcting strategies that can operate with only fuzzy knowledge about stock levels and recruitment curves". If our belief of the stock size is a probability function (in contrast to a point estimate), Bayes' theorem postulates that harvesting information can be used to calculate a conditional probability function [21]. Here we propose Bayesian regulation (BR) that uses catch- and effort data from the ongoing fisheries to make real-time estimates of stock size and fishing mortality. The years can be linked by using the posterior distribution as a prior in the sequential year, and hence exploitation is combined with stock size assessment. Real-time assessment of BR would be advantageous to management routines relying on forecasts of stock abundances. Given that most commercial fisheries with CR-based control of biomass and fishing mortality (e.g. TAC) routinely monitor effort and catch, this additional information is surprisingly poorly utilized during the CR fishing season. We suggest a methodology where this information can be used for updating population size estimates to facilitate in-season management decisions.

The BR is derived from Bayesian foraging theory in behavioral ecology, which describes a giving-up rule for patch foraging animals [22–24]. The rule is a relaxation of the full information

Table 1. Characteristics and settings for the evaluation of eight different harvest controls: CR being a simple catch-only regulation; BR is the proposed regulation combining catch and effort.

Harvest control	Target	Supervision	Prior μ	Prior v
$CR_F{}^s$	F_{MSY}	Supervised	$N+\varepsilon$	0
$CR_F{}^u$	F_{MSY}	Unsupervised	N_{MSY}	0
$CR_N{}^s$	N'_{MSY}	Supervised	$N+\varepsilon$	0
$CR_N{}^u$	N'_{MSY}	Unsupervised	N_{MSY}	0
$BR_F{}^s$	F_{MSY}	Supervised	$N+\varepsilon$	$var(N_{MSY})$
$BR_F{}^u$	F_{MSY}	Unsupervised	N_{MSY}	$var(N_{MSY})$
$BR_N{}^s$	N'_{MSY}	Supervised	$N+\varepsilon$	$var(N_{MSY})$
$BR_N{}^u$	N'_{MSY}	Unsupervised	N_{MSY}	$var(N_{MSY})$

The subscript denotes type of target and the superscript the existence of supervision. The target column denotes the targets used: F_{MSY} is the fishing mortality giving highest yield in the MSY-analysis of the SOM, and N'_{MSY} is the corresponding post-harvest stock size. The management is either supervised with a separate assessment or unsupervised (see methods). For unsupervised management we use the mean, μ, and the variance, v, of the pre-harvest stock size from MSY-analyses: N_{MSY}. In supervised management μ is instead the actual pre-harvest stock size, N, with an added randomly generated error term ε drawn from the normal distribution, (mean = 0, var = $var(N_{MSY})$ where $var(N_{MSY})$ is taken from an MSY-analysis).

assumption of the marginal value theorem [25], with search time and number of prey caught as information variables. It applies to patches in which the prey are hidden, such as a woodpecker feeding on pupae under bark [26,27]. Given some prior information available to the forager, e.g. experience from foraging in patches of its territory, a Bayesian posterior distribution can represent the forager's continuously changing belief about the prey density in the current patch [28]. The animal maximizing its intake rate would thus leave the patch for a new one when the anticipated intake rate in the current patch drops below the one expected from patches on average in the territory. The expected number of remaining prey in a patch can be expressed with a fairly simple equation derived by Iwasa et al. [29], but the decision to leave should include a discounting of the value of further bits of information [30].

For fisheries applications, we have modified Iwasa's equation for Bayesian prey density estimation for recurrent exploitation of one population. Both are cases of Bayesian update of mean and variance of a population size estimate. In the original scenario, the exploiter is informed by the mean and variance of several exploited patches within its foraging area. In the fishery scenario, the exploiter (manager) is informed by previous exploitations and surveys in the same area, followed by quantitative assessment and stochastic forecast simulations. We compare CR with BR, and show how they perform in relation to management objectives and targets, by simulating fishing on a model of the main basin Baltic Sea herring (Appendix S2) [31]. We present levels and temporal variability in yield, fishing mortality, stock abundance, spawning stock biomass (SSB), and finally how frequently it surpasses the MSY related reference point $B_{trigger}$ [32].

Methods

We look at three levels of decision-making in fisheries management as they are described in the common fisheries policy of the Council of the European Union [33]: (i) The *management objective*, which is the ultimate goal of fisheries management. The objective can be simple or more complex, for example weighing incompatible goals [34,35]. We have chosen the maximum sustainable yield (MSY) because it is the current objective of fisheries management in the EU [36]. (ii) The *target* of

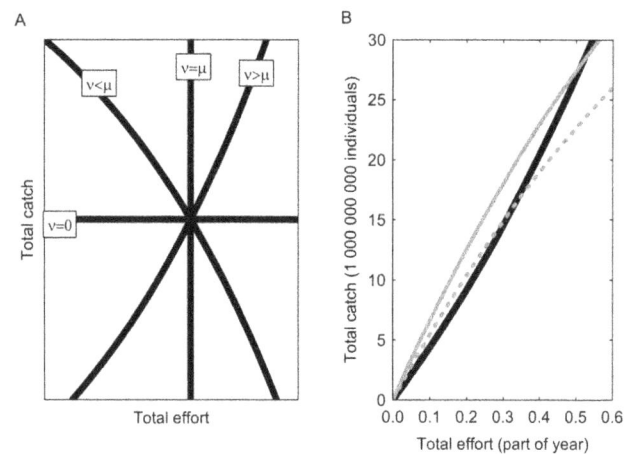

Figure 1. (A) Bayesian regulation curves indicating the combined total catch (y-axis) and total effort (x-axis) when fishing should be terminated for the season. The curves are given by equation 1 when r is set to the target number of fish, and solved for total catch. Hence, the curves show when the fishery target is achieved in catch-effort space, given by the ratio of variance (v) to mean stock size (μ). There are four qualitative cases of the v-μ ratio: when the ratio is zero, the fishery is regulated by catch only because there is no uncertainty around the mean. When the ratio equals 1, regulation is by effort only. Apart from these special cases, the fishery should be regulated by both catch and effort. If the variance is lower than the mean, the posterior mean will decrease with accumulating catches and effort. If the variance is higher than the mean, the posterior mean will increase with increasing catches but decrease with effort. These effects can be deduced from analyzing equation 1. (B) Here the solid curve denotes the Bayesian regulation when $v>\mu$, which is determined prior to the fishing season (see Methods). The grey curves are idealized trajectories of how cumulative catch and effort develops from the origin during the fishing season. When the trajectories cross the regulation curve, the Bayesian posterior is on target, and the fishery should be closed for the season. The hatched grey line indicates when the initial stock size is at the prior mean (μ), the solid grey line when the stock size is at the mean plus one standard deviation (see F_{MSY} -analysis). The BR thus allows larger catches when the stock size has been underestimated, and vice versa.

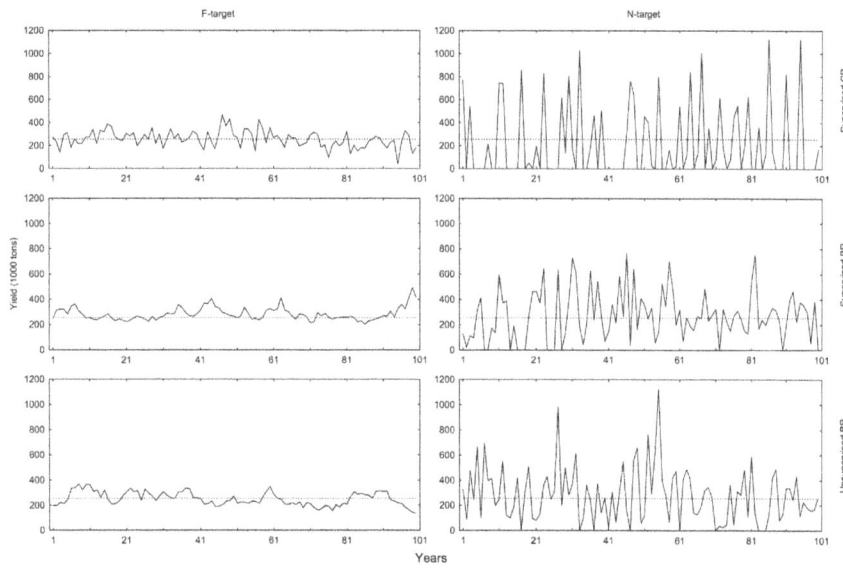

Figure 2. Yield over 100 years for different HCRs. The dashed reference line is the MSY = 254.4 thousand tons obtained from the F_{MSY}-analysis during controlled F.

exploitation, which should be given in a quantitative unit that relates to the management objective. We compare the efficiency of two targets: fishing mortality (F_{MSY}) and post-harvest stock size (N_{MSY}). (iii) The *harvest rules*, which "[lay] down the manner in which annual catch and/or fishing effort limits are to be calculated and provide for other specific management measures, taking account also of the effect on other species" [33]. On this level we choose to use other terms. We use the term *regulation type* to denote the use of catch and/or effort regulation, whereas the term *harvest control rule* (HCR) is used for the combined objective, target and regulation type.

The harvest control rules under development by the International Council for the Exploration of the Sea (ICES) involve a fishing mortality target (F^*) and the spawning stock biomass reference point $B_{trigger}$ [32,37]. If the SSB goes below $B_{trigger}$ the fishing mortality is reduced below target, a situation that should warrant a reinvestigation of the stock condition and the harvest control [32]. Based on a stock assessment and a short-term forecast, the fishing mortality target is recalculated as a TAC, which is how the recommendation is presented to the European Commission [38]. ICES harvest control rules use precautionary reference points below which fishing mortality is reduced [32],

something we do not use in our simulations. We have focused on the principle differences between CR and BR, e.g. the elaborate data collection and assessment procedure is simplified by using the actual size of the simulated stock with a random error.

Harvest control

Altogether we compared eight different harvest controls achieved by the combinations of two exploitation targets, two regulation types and two stock assessment modes (Table 1). The Bayesian regulation $BR(\gamma, f \mid \mu, v)$ is a function of catch (γ) and effort (f) given the prior information of the estimated stock size before harvest, μ, and the uncertainty of that estimate given as variance, v. The Bayesian estimate of the number of remaining individuals (r) in a harvested stock is:

$$r = \frac{\mu - \gamma\left(1 - \frac{v}{\mu}\right)}{(e^{qf} - 1)\frac{v}{\mu} + 1} \tag{1}$$

where q, the catchability, is defined as the fraction of the population captured by one unit of fishing effort and is the scalar

Table 2. Mean, standard deviation and CV (%) of annual yield.

	100 yrs			20,000 yrs		
	mean	**SD**	**CV**	**mean**	**SD**	**CV**
CR$_F^s$	256.6	72.0	28.1	254.3	83.9	33.0
CR$_N^s$	220.3	322.4	146.3	244.5	343.0	140.3
BR$_F^s$	282.8	50.6	17.9	258.6	58.9	22.8
BR$_N^s$	272.5	198.8	73.0	262.3	195.2	74.4
BR$_F^u$	251.8	55.0	21.9	253.5	57.6	22.7
BR$_N^u$	281.9	218.9	77.7	262.1	196.5	75.0

Two time series, one of 100 years and one of 20,000 years are presented. Mean values are given in thousands of tons. MSY from the F_{MSY}-analysis is 254.4 thousand tons.

Table 3. ANOVA-table showing effects of supervision, regulation type, target type, and target type interacting with supervision/regulation type on annual yield.

100 yrs	SS	DF	MS	F	p
Supervision	11,691	1	11,691	0.35	0.556
Regulation type	153,773	1	153,773	4.56	0.033
Target type	2,831	1	2,831	0.08	0.772
Superv.×Target	40,995	1	40,995	1.22	0.271
Regul.×Target	16,783	1	16,783	0.50	0.481
Error	20,018,836	594	33,702		

Data is from model simulations over 100 yrs.

Table 4. ANOVA-table showing effects of supervision, regulation type, target type, and target type interacting with supervision/regulation type on annual yield.

	SS	DF	MS	F	p
Supervision	141,961	1	141,961	4.16	0.041
Regulation type	1,582,157	1	1,582,157	46.40	<0.001
Target type	58,951	1	58,951	1.73	0.189
Superv×Target	120,612	1	120,612	3.54	0.060
Regul×Target	1,582,157	1	1,582,157	46.40	<0.001
Error	4,091,815,000	119,994	34,100		

Data is from model simulations over 20,000 yrs.

between catch per unit effort indices and average population abundance, hence $q = \frac{\gamma}{fN}$. When fishing is closed by fulfillment of the target condition, $f = f^*$, then the fishing mortality $F = q f^*$. This is a generalization of Iwasa's [29] Bayesian estimates of remaining prey population, see Appendix S1 for details on how Equation (1) is derived from some specific probability distributions. With a known prior probability distribution and a given catch and effort, a posterior distribution can be calculated with r being the mean. It is very unlikely that the prior mean is equal to the actual population size. The posterior mean will approach the actual value with accumulating catch and effort, but there will always be a bias towards the prior [39]. For large deviations of the prior mean in relation to the real population size, it may take more than one year to track the population size more closely. Hence, there can be temporal correlations in estimate biases. This depends on the random perturbations displacing the population size from the prior mean and the harvesting information making the posterior mean approach the real value.

When Equation (1) is solved for catch as a function of f, and r is given a target value, it can be visualized as harvest control curves (Fig. 1A). These curves can be calculated prior to the fishing

season and define the combination of total catch and total effort at which the Bayesian information indicates that the fishery has reached its target. The real-time accumulation of catch and effort during the fishing season is responsive to over- and under-estimation of stock size and will develop different trajectories. The intersection of the real-time accumulation of catch and effort with the BR-curve gives the total catches when an initial over- or under-estimation is compensated for (Fig. 1B). Note that if we have no uncertainty in our prior estimate, i.e. $v = 0$, Equation (1) simplifies to $r = \mu - \gamma$, in other words the remaining individuals in the population is our prior estimate minus the catch size. This means that catch-only regulation is a special case of BR (Fig. 1A). To be used in fisheries, Equation (1) needs to account for the simultaneous removal by predators (M), which is done by introducing two scaling parameters α and β:

$$r = \frac{\mu - \beta\gamma\left(1 - \frac{v}{\mu}\right)}{(e^{\alpha\hat{q}f} - 1)\frac{v}{\mu} + 1}, \tag{2}$$

where parameter $\alpha = 1 + M/\hat{q}$ simply scales up the effort with the total mortality in proportion to $\hat{q}$. The notation $\hat{q}$ denotes the assumed value of the catchability. As default $\hat{q}$ is constant and unbiased, but we make separate runs to explore the impact of biases and stochastic errors in $\hat{q}$. Catchability is difficult to determine accurately and can be affected by aggregation behavior of fish and technical enhancement of fishing gear [40]. Parameter β scales the catch γ to the total number of casualties from fishing and natural mortality:

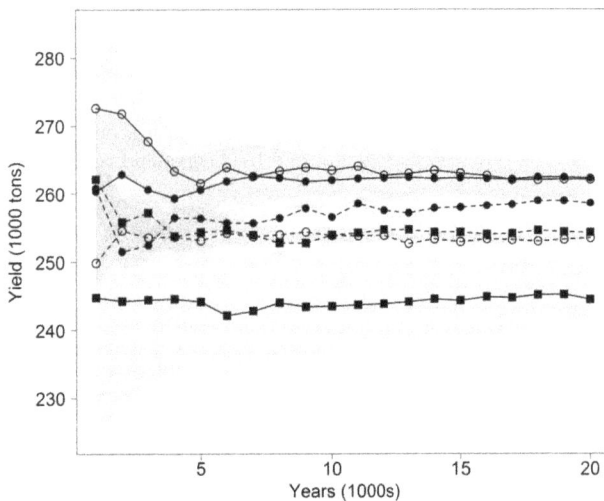

Figure 3. Average yields and confidence intervals as functions of increasing length of simulation in years. Filled symbols denote supervised and open symbols unsupervised harvest control. Squares denote CR and circles BR. Solid lines denote N-target management whereas hatched lines denote F-targeted management.

Table 5. Pair-wise test of CV in yield for the 100 year series, applying the F-statistics of testing differences in the variance of annual yield.

	CR_F^s	CR_N^s	BR_F^s	BR_N^s	BR_F^u	BR_N^u
CR_F^s		<0.001	<0.001	<0.001	0.007	<0.001
CR_N^s	27.205		<0.001	<0.001	<0.001	<0.001
BR_F^s	2.455	66.785		<0.001	0.024	<0.001
BR_N^s	6.764	4.022	16.605		<0.001	0.268
BR_F^u	1.647	44.809	1.490	11.141		<0.001
BR_N^u	7.661	3.551	18.807	1.133	12.619	

F-values are presented in the lower left triangle and p-values in the upper right.

Table 6. Mean, standard deviation and CV (%) of annual fishing mortality (F) of 100 years of simulated fishing.

	mean	SD	CV
CR_F^s	0.17	0.043	25.2
CR_N^s	0.20	0.325	158.4
BR_F^s	0.17	0.000	0.1
BR_N^s	0.17	0.118	70.1
BR_F^u	0.17	0.000	0.1
BR_N^u	0.17	0.119	68.4

F_{MSY} from the F_{MSY}-analysis is 0.17.

$$\beta = \frac{1 - e^{-(M+\hat{q})f}}{1 - e^{-\hat{q}f}} \tag{3}$$

Note that β depends on effort such that Equation (2) can work as a real-time estimator of the population size. The Bayesian estimator is used with the different targets (Table 1). When the target is formulated as a post-harvest population size (N'_{MSY}), harvesting should stop when the Bayesian estimate of remaining stock size (r) equals target N'_{MSY}:

$$re^{-(1-f)M} = N'_{MSY} \tag{4}$$

The target N'_{MSY} denotes the stock size at the end of the year, we therefore need to take into account the removal of individuals due to natural mortality for the remaining part of the year after fishing has ended, $e^{-(1-f)M}$. When our target is defined as fishing mortality, F_{MSY}, harvesting is aborted when.

$$\frac{\gamma}{\beta\gamma + r} = 1 - e^{-F_{MSY}} \tag{5}$$

Note that the prior μ cannot be used in the denominator, because the estimated size of the pre-harvest population changes as we receive information from catch and effort. When the conditions in Equation (4), or (5) are fulfilled, we extract the fishing effort (f^*) from Equation (2). Equations (2), (4) and (5) together define the control rules both for CR and BR.

Assessment and priors

Estimates of the pre-harvest stock size, the prior μ, are produced in two ways: by supervised assessment and unsupervised assessment (Table 1). Supervised assessment represents the current annually revised assessment practiced by ICES and other stock assessors. In this case, μ is calculated by adding an assessment error to the actual pre-harvest stock size, N. The error is a random value drawn from a normal distribution with mean = 0, and variance$_v$ which is equal to the variance of the pre-harvest stock size in the F_{MSY}-analysis (see below). The coefficient of variance in the F_{MSY}-analysis is 21%. This represents an ideal assessment where the error solely stems from the yearly variation of the population when fished at F_{MSY}. The unsupervised assessment uses μ equal to the average, and v the variance of the pre-harvest stock size from the F_{MSY}-analysis (i.e. when fishing at F_{MSY}; see below). The rationale of using a constant μ is that the Bayesian information of stock size when a fishery has been closed in the foregoing year indicates N'_{MSY} at the end of the year when using N-target (Equation 4). Similarly for an F-targeted fishery, fishing is closed every year when at estimated F_{MSY}, which is associated with the mean $N' = N'_{MSY}$. The variance is used for the BR, whereas $v = 0$ for the CR.

The stochastic operating model

We have used a stock model of the herring population in the main basin of the Baltic Sea (ICES catch area subdivisions 25–27, 28.2, 29 and 32) for harvest control evaluation. The model is a stochastic operating model (SOM) parameterized from statistical analyses of ICES catch data and outputs from ICES XSA runs (Appendix S2).

MSY analysis

We performed an MSY-analysis on the SOM by stepping the fishing mortality in steps of 0.01, and for each F value simulating 40,000 years and rejecting the first 500 to minimize the effects of initiation values. In contrast to simulated management, this algorithm executes perfectly-controlled constant fishing mortality. The relationship between yield and fishing mortality was used to identify the F_{MSY} and its associated average post-harvest stock size. These were used as targets for F-targeted and N-targeted management, respectively. The average pre-harvest stock size, given fishing mortality F_{MSY}, was used as the prior (μ) in the unsupervised simulations, and the variance in the pre-harvest stock size was used as a measure of the uncertainty (v) of μ in the BR. The lower 2.5% percentile of the associated SSB was used as the $B_{trigger}$. Targets, priors and assessment modes were used in

Table 7. ANOVA-table showing effects of supervision, regulation type, target type, and target type interacting with supervision/regulation type on annual fishing mortality.

	SS	DF	MS	F	p
Supervision	0.00047	1	0.00047	0.02	0.886
Regulation type	0.03166	1	0.03166	1.40	0.236
Target type	0.02945	1	0.02945	1.31	0.254
Superv×Target	0.00046	1	0.00046	0.02	0.886
Regul×Target	0.03265	1	0.03265	1.45	0.229
Error	13.39279	594	0.02255		

Data is from model simulations over 100 yrs.

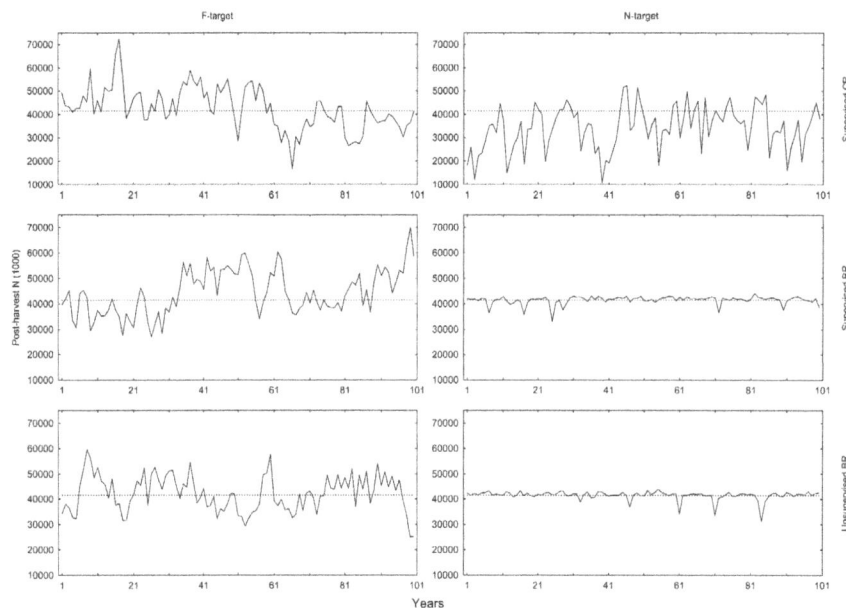

Figure 4. Fishing mortality (F) for 100 years for different HCRs. The dashed reference line is the $F_{MSY} = 0.17$ obtained from the F_{MSY}-analysis during controlled F.

Evaluation of harvest control rules

The eight HCRs were applied to Monte Carlo simulations of the SOM. A simulation started with a number of years of controlled fishing with F_{MSY} to get away from the initial population size and age-structure. After the initial period, data was collected for a number of years of applied HCR. We ran shorter time series with an initiation period of 1,000 years and 100 years of applied HCR. The time period of 100 years reflects a reasonably long time-horizon for management. We also ran longer series with an initiation period of 100 years and 20,000 years of applied HCR, in order to establish more accurate estimates of mean yields. We collected annual data on yield, fishing mortality, pre-harvest and post-harvest population size, and SSB. Cod SSB and year-specific growth were kept at a mean level ($C_y = 100,000$ tons, $k_y = 0$) during the simulations with the addition of random noise, the size of which was extracted from historical data after removing long-term changes [31]. Technically, fishing on the SOM is performed by applying the effort, f^*, from the solution of the quitting rules in Equation (4) and (5). f^* is calculated by halved-

distance iterations until the estimate (the left side of Equation 4 and 5) differed from the target with less than 0.1‰. The population is harvested by stepping the SOM one year with the catch being determined by Baranov's catch equation:

$$\gamma = \frac{qf^*}{M+qf^*} \mu \left(1 - e^{-(M+qf^*)}\right), \qquad (6)$$

in which we use the actual q of the SOM.

Sensitivity to error in catchability

We keep the estimated catchability $\hat{q}$ in Equation (2) and (3) constant and unbiased in our base runs, but fishery dependent catchability may in reality be biased and vary over time. Such uncorrelated errors and biases in catchability will contribute to error or bias in the assessment. Our base model already has assessment error (CV = 21%, as described in *Assessment and priors*) for supervised HCRs, but unsupervised BRs do not (Table 1). Unsupervised BRs do not use separate assessments, but on the other hand they will be affected by error or bias in $\hat{q}$. We chose to compare supervised F-targeted CR and unsupervised F-targeted BR because the former is only affected by the error in

Table 8. Pair-wise test of CV-values applying the F-statistics of testing differences in the variance of annual fishing mortality.

	CR_F^s	CR_N^s	BR_F^s	BR_N^s	BR_F^u	BR_N^u
CR_F^s		<0.001	<0.001	<0.001	<0.001	<0.001
CR_N^s	39.520		<0.001	<0.001	<0.001	<0.001
BR_F^s	67,742	2,677,131		<0.001	0.066	<0.001
BR_N^s	7.731	5.112	523,678		<0.001	0.404
BR_F^u	49,987	1,975,470	1.355	386,425		<0.001
BR_N^u	7.361	5.369	498,613	1.050	367,930	

F-values are presented in the lower left triangle and p-values in the upper right.

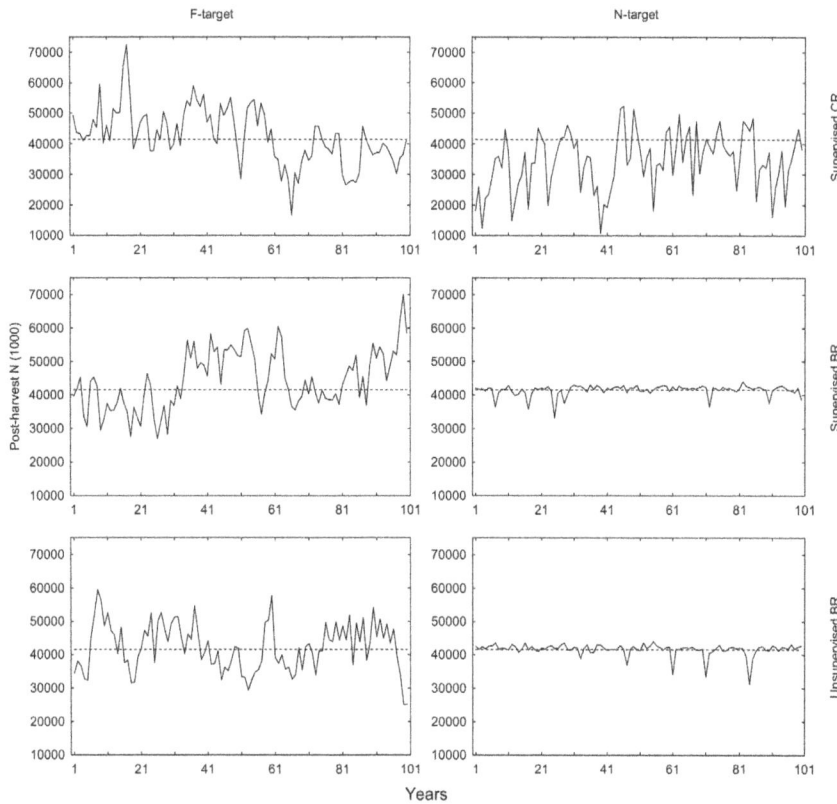

Figure 5. Post-harvest N-values for different HCRs in simulations over 100 years. The dashed reference line is the N-target = 41,491 million obtained from the F_{MSY} analysis during controlled F.

the assessed stock size, μ, whereas the latter is affected only by the error in $\hat{q}$. We also explored the sensitivity to correlated biases in $\hat{q}$ and μ by changing their values ±20%. Simulations ran for 41,000 generations and data from the last 40,000 were used for the analyses.

Statistical analyses

Unsupervised CR expectedly led to population crashes within a few decades or less. We therefore excluded them in the statistical analyses, which were performed using STATISTICA software (Statsoft, Tulsa, Oklahoma). General linear models (GLM) were used for three-way ANOVAs, testing the effects of supervision mode, regulation type and target type on yield, fishing mortality, spawning stock biomass, and post-harvest population size. Fourier analyses were performed on 1,000 year data series, tapered by 15% and padded to the length power of 2. The period length in years with the highest spectral density was identified after applying Hamming weights with a data window of size 7. Test of significant deviation from white noise was performed using Kolmogorov-Smirnov deviation (d) statistics, Table Y, $\rho = 0.5$, n>100 in Rohlf & Sokal [41].

Results

Yield

Mean yield in 100-year simulations is significantly higher with BR than with CR although the differences are small (Fig. 2, Table 2, 3). With an increasing length of the time series the confidence intervals narrow (Fig. 3). Analysis of 20,000 years reveals more clearly that BRs operated with N-target give the

highest average yields, even though the largest differences in means are less than 10% (Table 2, 4). Supervised and unsupervised BRs give higher yields (262 thousand tons) than the reference MSY from the F_{MSY}-analysis (254 thousand tons; Table 2, 3, 4, 5). Contrary to the small differences in mean yields, the CV in yields is clearly affected by the choice of HCR. In general, using F-targets gave less temporal variation than using N-targets (Fig. 2, Table 2, 3, 4, 5). There is also an effect of regulation type with BR, giving less variation than CR (Fig. 2, Table 3, 4). However, supervised BR and unsupervised BR have the same CV regardless of target (Table 5).

Table 9. Mean (Million), standard deviation and CV (%) of annual post-harvest population size, N', of 100 years of simulated fishing.

	mean	SD	CV
CR_F^s	42,620	9,201	21.6
CR_N^s	34,282	9,470	27.6
BR_F^s	44,391	8,798	19.8
BR_N^s	41,508	1,616	3.9
BR_F^u	42,270	7,248	17.1
BR_N^u	41,569	1,840	4.4

N-target from the F_{MSY}-analysis is 41,491 million.

Table 10. ANOVA-table showing effects of supervision, regulation type, target type, and target type interacting with supervision/regulation type on annual post-harvest population size.

	SS	DF	MS	F	p
Supervision	1.06E+08	1	1.06E+08	2.1	0.153
Regulation type	2.02E+09	1	2.02E+09	39.1	<0.001
Target type	2.44E+09	1	2.44E+09	47.2	<0.001
Superv×Target	1.19E+08	1	1.19E+08	2.3	0.130
Regul×Target	7.44E+08	1	7.44E+08	14.4	<0.001
Error	3.07E+10	594	5.17E+07		

Data is from model simulations over 100 yrs.

Table 12. ANOVA-table showing effects of supervision, regulation type, target type, and target type interacting with supervision/regulation type on annual SSB.

	SS	DF	MS	F	p
Supervision	1 054 971	1	1 054 971	19.8	<0.001
Regulation type	6 757 878	1	6 757 878	126.9	<0.001
Target type	2 008 542	1	2 008 542	37.7	<0.001
Superv×Target	164 153	1	164 153	3.1	0.080
Regul×Target	2 271 205	1	2 271 205	42.6	<0.001
Error	31 642 602	594	53 270		

Data is from model simulations over 100 yrs.

Fishing mortality

There is no significant difference in mean fishing mortality (F) due to any effect over 100 years (Table 6, Table 7). All HCRs reach the F-target of 0.17 as their mean, except supervised CR with N-target, which has a mean F of 0.20. Although the mean Fs are very similar across HCRs, the differences in CV are more pronounced. Supervised CR with N-target exhibits the highest CV of 158% (Table 6). F-targeted supervised CR has a CV of 25% (Table 6, Fig. 4). Since catchability is constant here, F is proportional to fishing effort. It is therefore not surprising that F-targeted HCRs result in less variation than N-targeted HCRs. F-targeted BRs, supervised and unsupervised, show very high precision in reaching the target (Fig. 4), the CV being as low as 0.1% (Table 6). As expected, N-targeted BRs exhibit higher CVs (68% unsupervised and 70% supervised; Table 6). In addition, it is only the CV between supervised and unsupervised BRs that does not differ significantly from each other (Table 8).

Post-harvest population size and SSB

Looking at post-harvest population size (N'), the harvest controls operating on N-targets are aiming for 41.5 billion individuals at the end of the year. The two N-targeted BRs reach this target with little variation between years (4% CV; Fig. 5, Table 9, 10, 11), whereas N-targeted CR leads to over-fishing in the sense that N' is on average 17% below target. In contrast to the N-targeted regulation types, there is a small difference in N' between F-targeted BR and CR. This explains the significant interaction between regulation type and target type in the ANOVA

Table 11. Pair-wise test of CV-values applying the F-statistics of testing differences in the variance of annual post-harvest population size.

	CR$_F$^{s}	CR$_N$^{s}	BR$_F$^{s}	BR$_N$^{s}	BR$_F$^{u}	BR$_N$^{u}
CR$_F$^{s}		0.007	0.198	<0.001	0.011	<0.001
CR$_N$^{s}	1.637		0.001	<0.001	<0.001	<0.001
BR$_F$^{s}	1.187	1.943		<0.001	0.076	<0.001
BR$_N$^{s}	30.736	50.318	25.903		<0.001	0.103
BR$_F$^{u}	1.585	2.595	1.336	19.387		<0.001
BR$_N$^{u}	23.800	38.964	20.058	1.291	15.012	

F-values are presented in the lower left triangle and p-values in the upper right.

(Table 10). Supervision has no significant effect on the mean N' (Table 10), nor the CV (Table 11).

The mean SSBs differ significantly by regulation type, harvest type and supervision (Table 12). BR SSBs are on average larger than CR SSBs (Fig. 6, Table 13, 13). Supervised CR with N-target has the lowest mean and is below $B_{trigger}$ in 50% of the years (Table 13, Fig. 6). Supervised BRs has on average larger SSBs than unsupervised, and the BRs operating with N-targets surpass the $B_{trigger}$ less frequently than when operating with F-targets (Table 13, Fig. 6). The lower CV for BRs operating with N-target (Table 13, 14) seems to be the main reason for this result.

Bias and error in harvest rates

So far we can summarize the difference between CR and BR as being BR's ability to reduce the variance in the target variable (F or N'; Tables 3, 4). Yield is partially dependent on F, and SSB on N', and thus the variance of these variables is also reduced, although to a lesser degree (Tables 2, 5). These results are based on the assumption of a constant and un-biased estimate of catchability ($\hat{q}$). When random error is added to the catchability (q) of the same magnitude as the error in μ, the CV in γ and F for the unsupervised F-targeted BR become similar to supervised F-targeted CR (Fig. 7). The two F-targeted harvest rules also perform very similar in response to the off-set bias in μ and $\hat{q}$. Hence, supervised CR and unsupervised BR behave similarly when error and bias in μ and q are of the same relative magnitude.

Temporal variation in population size

We found that all HCRs decrease the amplitude of temporal variation in pre- and post-harvest population numbers, except for post-harvest numbers under supervised N-targeted CR, at which the variation is indifferent from the unexploited population (Table 15). The BR is effective in reaching the N-target, and hence reduces the CV markedly in the post-harvest population size (Fig. 8, Table 15). The effect is still evident in the pre-harvest population size, in which the inclusion of recruits from the foregoing year increases the variation (Table 15).

The unexploited population exhibited a very pronounced periodicity of about 20 years (Fig. 8, Table 16). There is only uncorrelated variation in our operating model, which means that any periodicity is due to the demographic parameters and structure of the population. The BRs with N-target remove much of this periodicity and puts it closer to white noise (Lower K-S d-values, Table 16). HCRs with F-target increase the dominating period length to about 70–80 years (Table 16). Supervised CR with F-target, a simplified version of the HCR currently used,

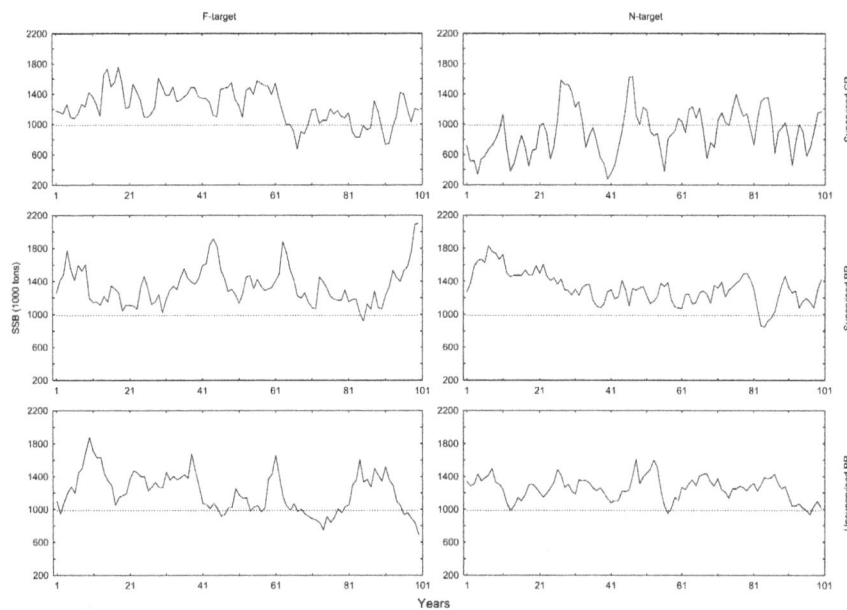

Figure 6. Spawning stock biomass (SSB) for different HCRs in simulations over 100 years. The dashed reference line is the $B_{trigger} = 988$ thousand tons. It is the 2.5% lower percentile of SSB obtained from the F_{MSY}-analysis during controlled F.

reduces the amplitude of variation in both pre- and post-harvested population size (Table 15). At the same time, it increases the dominating period length from 20 years in the unexploited population to about 80 years in the exploited (pre-harvest population size, Table 16).

Discussion

We have suggested a new approach to fisheries regulation, which we name *Bayesian regulation* (BR), and made a basic evaluation of its properties in relation to catch-only regulation. The idea of BR is adopted from Bayesian foraging theory, in which patch-foragers are assumed to rely on a giving-up rule based on prior information and sampling information [29,30,42]. The Bayesian foraging theory has been studied extensively, including the derivation of the probability function used and that cumulative catch and effort are sufficient information variables [29]. The BR uses the uncertainty of the stock size estimate as a parameter, which is rarely the case in current management [43–45], and which has been requested for some time [46]. The general applicability of the BR is emphasized by the fact that catch-only regulation, (CR), is a special case of BR _ i.e. when there is no uncertainty in the prior estimate of the stock size from an

assessment or survey. We have analyzed three aspects of harvest controls: (i) the relevance of supervision (with or without stock size assessment), (ii) operational target type (constant fishing mortality, F, or constant stock size, N), (iii) type of regulation (catch only = CR, or Bayesian evaluation of combined catch and effort = BR), and their effect on mean and variance in yield and stock size.

The effect of fishing on stock dynamics

There is an ongoing discussion about whether fishing leads to increased temporal variability in stock size compared to unexploited populations [4,14]. Observations from California, USA, corroborate the view that fishing magnifies fluctuations in fish abundance, driven by increased intrinsic growth rates as a response to higher mortality [4]. Models without age-structure show an increased variability in population abundance as a consequence of exploitation [14]. In a structured model, fishing can either increase or decrease the stock size variability [15]. We find that the CV of population size has been reduced in all investigated HCRs, and especially with N-targeted BR. The different effects of fishing between our and the previous studies

Table 13. Mean in thousand tons, standard deviation and CV (%) of annual SSB of 100 years of simulated fishing.

	mean	SD	CV	$<B_{trigger}$ (%)
CR_F^{s}	1 240	230.0	18.6	26.1
CR_N^{s}	906	308.6	34.1	49.9
BR_F^{s}	1 349	232.8	17.3	18.3
BR_N^{s}	1 317	192.4	14.6	5.6
BR_F^{u}	1 206	245.1	20.3	20.6
BR_N^{u}	1 255	142.3	11.3	5.9

The percentage of years the SSB surpassed the $B_{trigger}$ ($B_{trigger} = 998$ thousand tons), is given in the rightmost column.

Table 14. Pair-wise test of CV in yield for the 100 year series, applying the F-statistics of testing differences in the variance of annual SSB.

	CR_F^s	CR_N^s	BR_F^s	BR_N^s	BR_F^u	BR_N^u
CR_F^s		<0.001	0.236	0.009	0.183	<0.001
CR_N^s	3.368		<0.001	<0.001	<0.001	<0.001
BR_F^s	1.156	3.894		0.049	0.053	<0.001
BR_N^s	1.614	5.434	1.396		0.001	0.006
BR_F^u	1.200	2.807	1.387	1.936		<0.001
BR_N^u	2.678	9.018	2.316	1.660	3.213	

F-values are presented in the lower left triangle and p-values in the upper right.

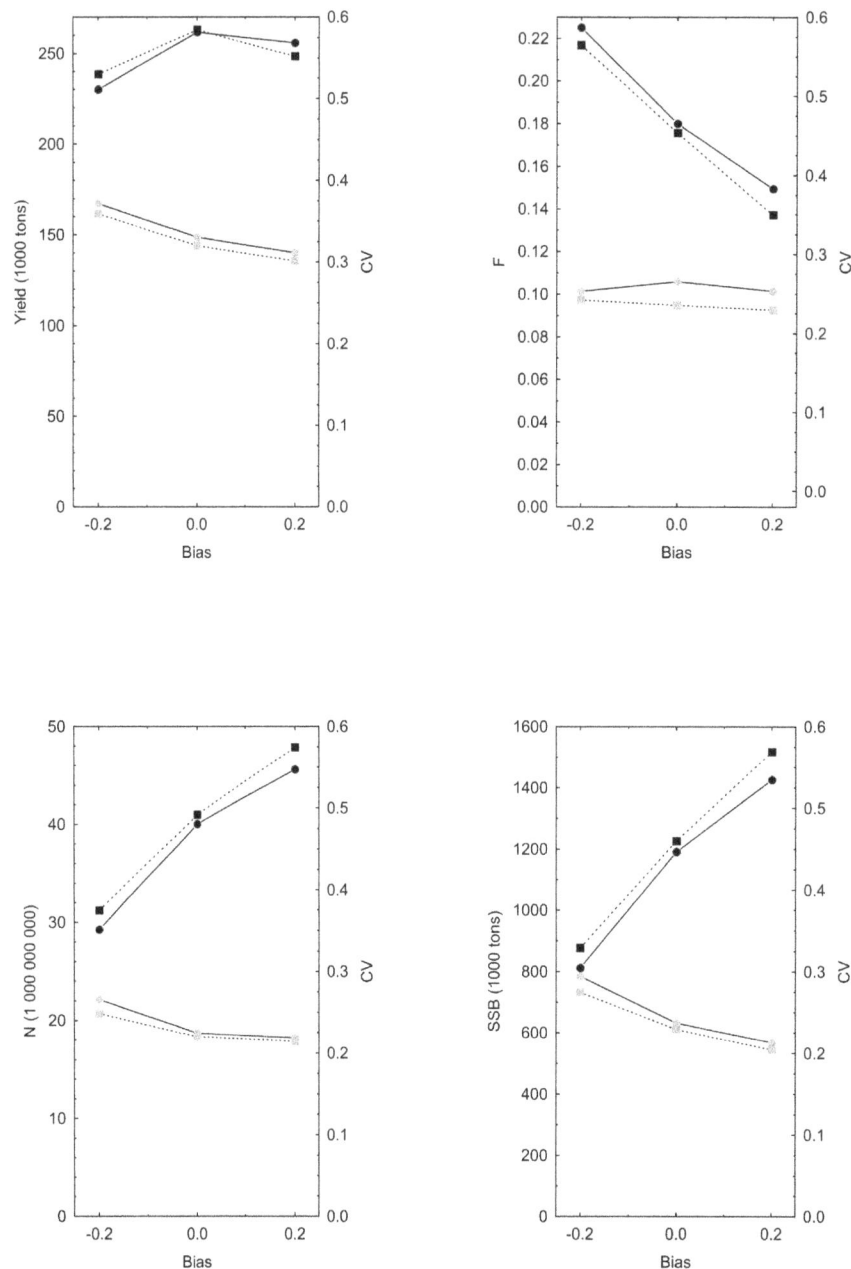

Figure 7. A comparison of yield, fishing mortality (*F*), post-harvest stock size (*N*) and spawning stock biomass (SSB) when random errors of the same CV are applied to both the stock size assessment (μ) and the estimated catchability parameter ($\hat{q}$). Relative positive and negative biases applied to $\hat{q}$ and μ are denoted on the x-axis. Circles denote results from *F*-targeted unsupervised BR and squares *F*-targeted supervised CR. Black symbols denote the value (left axis) and grey symbols the CV of the effect-variable (right axis).

Table 15. Coefficient of variation (CV) of pre- and post-harvest population numbers compared with the unexploited population.

	Pre-harvest N			Post-harvest N		
	CV	F-value	p	CV	F-value	p
CR_F^s	21.6	1.52	<0.001	22.0	1.42	<0.001
CR_N^s	24.5	1.18	<0.001	26.1	1.00	n.s.
BR_F^s	21.2	1.57	<0.001	21.2	1.53	<0.001
BR_N^s	14.8	3.25	<0.001	4.6	33.01	<0.001
BR_F^u	20.8	1.64	<0.001	20.7	1.60	<0.001
BR_N^u	15.1	3.12	<0.001	4.7	30.58	<0.001
Unexpl. Pop.	26.6			26.1		

CV is from simulations of 10,000 years. F-value is the ratio of squared CVs (i.e. comparison of standardized variances).

may depend on our implementation of errors in assessment, whereas earlier studies implemented perfect fishing mortality.

The comparison of Bayesian- and catch-regulation

We expected N-targeted BR to give the highest mean yields, because of its ability to maintain the population in the most productive state. It turned out to return the highest mean yield, but the differences between HCRs are small. The CV of fishing mortality and the population size is, however, much more affected using BR compared to CR. The advantage of F-targeted BR is seen in a reduced variation in fishing mortality and yield. Similarly, the N-targeted BR is associated with less variation in post-harvest population size and SSB. Under some conditions CR can be a blunt harvest control instrument; populations inevitably go extinct if not supervised, and it exhibits the largest CV in yield, F and SSB of all harvest controls when operated with N-target. For the better, F-target supervised CR is a commonly applied harvest control, e.g. within the European Union, and exhibits a smaller CV than N-targeted CR in all fisheries and stock variables investigated. Choosing BR, a manager has to consider the trade-off between the pros and cons of an F- versus an N-targeted fishery, since targeting one variable pushes the variability to the other [47]. From a fisheries perspective, F-targeted BR can appear as the most preferable HCR: low variability in effort (costs) and yield (income) enables a more stable economy, and an even process load is more manageable for the food industry [47]. At the same time, single-stock F-targeted BR can potentially yield a conflict of interest between fishers, and may be less desirable when taking into account the ecological interaction between different species of fish. N-targeted BR, on the other hand, gives low variation in population size, which means lower risks of extinction and smaller indirect effects of variability on prey and predator species, which may also be subject to commercial fishing. The stable post-harvest population size of N-targeted BR, in combination with the inherent variation in the productivity of the population causes a highly variable yield.

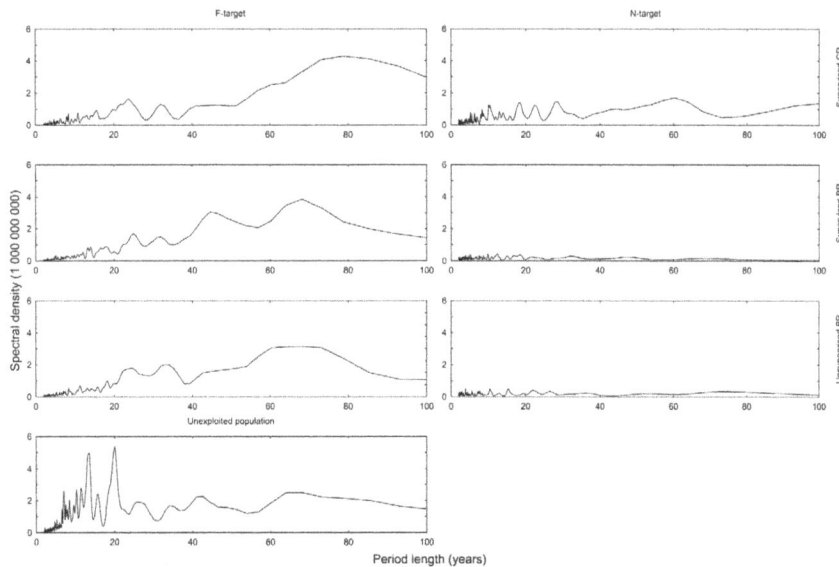

Figure 8. Spectral densities of Fourier analysis of 1,000 years of pre-harvest population size under various HCRs. Period lengths from 0 to 100 years are shown (depicted on the abscissa).

Table 16. Fourier analysis of pre-harvest population numbers in series of 1,000 years of fishing with alternative HCRs.

HCR	Period max	Density	K-S d	p
CR_F^s	78.769	4.29E+09	0.5013	<0.01
CR_N^s	60.235	1.71E+09	0.3152	<0.01
BR_F^s	68.267	3.84E+09	0.5370	<0.01
BR_N^s	12.488	4.62E+08	0.0929	<0.01
BR_F^u	68.267	3.13E+09	0.5368	<0.01
BR_N^u	15.515	5.10E+08	0.0716	<0.01
Unexpl. Pop.	20.078	5.42E+09	0.4995	<0.01

Period length with maximum spectral density is given in years along with density value. Kolmogorov-Smirnov d-value for deviation from white noise is given with its p-value.

Our results show that F-targeted BRs come close to resolving the trade-off between a high long-term yield and the stable fishing effort and yield acknowledged by some authors [20,35]. Although giving a slightly lower average yield than BRs with N-target, the yield is higher than with CR and the fluctuations in the yield are reduced considerably.

The ability of the BR to control the target variable depends on the reliability of the estimated fishery-dependent catchability. We have shown, as a rule of thumb, that if the CV of the error in catchability is as large as the error CV in the stock size assessment, the unsupervised BR performs similar to the supervised CR. Lack of knowledge makes it difficult to say in general which of the errors is the largest, in fishery-dependent catchability or in assessments. In the case of the Newfoundland cod, CPUE increased when stock abundance declined due to hyper-aggregation of the fish [40]. Such deceitful behavior of CPUE warrants close attention, and neither BRs nor CRs are immune to it. It is possible to estimate catchability accurately when we know the fishing gear, the fishers' behavior, and the fish dispersal pattern. One could say that a transfer from CR to BR would require a shift of focus by managers from surveys and assessments to gear efficiency, behavior of fish and fishing fleets.

HCRs' effects on period lengths of stock size variation

Not only the amplitude of population variation, but also the frequency of the variation has consequences for the population [48]. The induced periodic population variability in predator-prey interactions is well known in ecology [49], but the result of the dynamic interaction between the fish stock and the harvest control is much less studied [15]. It has been hypothesized that populations truncated by fishing have a shorter inherent periodicity, and track environmental fluctuations more closely. Given the assumption that unharvested populations fluctuate at longer frequencies than the environment, this could be the reason for increased amplitude of population variation [4]. Since populations have inherent dynamics controlled by age-specific demographic parameters, one can conclude that populations truncated by harvest should exhibit fluctuations of shorter period length than unexploited ones. In our study, this is true for the N-targeted BRs, but for the other harvest controls the period length of the fluctuations increase. For F-targeted fisheries, stock size deviations from equilibrium delays return times because of a positive feedback: enlarged populations face reduced mortality and reduced populations are subject to increased mortality. If stocks exhibit long-wave fluctuations with period lengths of 60–80 years due to fishing as indicated with F-targeted CR fisheries, the stress it imposes on ecosystems warrants more attention in future studies and consideration in the choice of method for harvest control.

Potential application of Bayesian regulation

The Bayesian regulation of fisheries modeled here is highly idealized. Real, well-informed management processes are far more complex, and may have sources of error not accounted for. We want to emphasize the potential of BR and stimulate ideas in how it might be implemented. The concept of BR can be developed further, for example to include a Bayesian estimate of the catchability parameter in the probability function (Equation 1). BR has the advantage that, within a year, it can give much higher precision in reaching the management target compared to CR. For instance, stock control with quantitative assessment and forecast can utilize a Bayesian HCR for the harvest year. New assessment methods that calculate a confidence interval for stock size estimate, such as SAM (Nielsen and Berg unpublished), is ideal as they provide priors for a Bayesian regulation.

Another useful property of BR is the ability to work without other assessments, since it is an assessment in itself. It can be an option for the more than 99% of fished species which lack assessments because of the costs and other requirements for collecting the necessary data [50]. An application of BR would still require daily reporting of catches, effort and gear used. But with such a system in place, Bayesian estimates of stock size and fishing mortality could be provided daily, with fishers and managers more up-to-date with current stock status.

Supporting Information

Appendix S1 Derivation of the generalized Bayesian estimate of remaining prey population.

Appendix S2 Description of the stochastic operating model of the Baltic Sea herring.

Code S1 R-code file "control.r". The file contains the highest levels of instructions like regulation type, target type etc. In addition, the targets and other constants are set here. Running the file in R with the three other files in the same folder will produce and plot the results of the specified setup.

Code S2 R-code file "MSE.r". The file contains most of the actual code of the management strategy evaluation procedure. Parameters like simulation length (number of generations), predator spawning stock biomass, and other parameters are set in this file.

Code S3 R-code file "BR.r". The file contains the algorithm to produce the fishing mortality that the BR suggests. The error and bias of A are set here, as is the tolerance of the algorithm. This code is used both for CR and BR where catch only regulation results are produced by having the variance set to zero.

Code S4 R-code file "OPMOD.r". The file contains the stochastic operating model. The parameters that are more directly linked to the population are set here, as are also the initial values of the population numbers-at-age and weights-at-age. The functions

that describe mortality, reproduction, growth, and the weight of one year olds are defined here.

Acknowledgments

Dr. Michael Errigo and an anonymous reviewer gave valuable comments on the manuscript. Oskar MacGregor provided improvements on the language.

Author Contributions

Conceived and designed the experiments: NMAH NN RA SK. Performed the experiments: NMAH NN. Analyzed the data: NMAH NN RA SK. Contributed reagents/materials/analysis tools: NMAH NN RA SK. Wrote the paper: NMAH NN RA SK.

References

1. Virtala M, Kuikka S, Arjas E (1998) Stochastic virtual population analysis. ICES J Mar Sci 55: 892–904.
2. Mäntyniemi S, Kuikka S, Rahikainen M, Kell LT, Kaitala V (2009) The value of information in fisheries management: North Sea herring as an example. ICES J Mar Sci 66: 2278–2283.
3. Polasky S, Carpenter SR, Folke C, Keeler B (2011) Decision-making under great uncertainty: environmental management in an era of global change. Trends Ecol Evol 26: 398–404.
4. Anderson CNK, Hsieh C, Sandin SA, Hewitt R, Hollowed A, et al. (2008) Why fishing magnifies fluctuations in fish abundance. Nature 452: 835–839.
5. Jonzén N, Ripa J, Lundberg P (2002) A Theory of Stochastic Harvesting in Stochastic Environments. Am Nat 159: 427–437.
6. Getz WM (1985) Optimal and Feedback Strategies for Managing Multicohort Populations. J Optimiz Theory App 46: 505–514.
7. Walters CJ, Ludwig D (1981) Effects of Measurement Errors on the Assessment of Stock-Recruitment Relationships. Can J Fish Aquat Sci 38: 704–710.
8. Hjerne O, Hansson S (2001) Constant catch or constant harvest rate? Baltic Sea cod (*Gadus morhua* L.) fishery as a modelling example. Fish Res 53: 57–70.
9. Aps R, Fetissov M, Holmgren N, Norrström N, Kuikka S (2011) Central Baltic Sea herring: effect of environmental trends and fishery management. In: Villacampa Y, Brebbia CA, editors. Ecosystems and Sustainable Development VIII. Boston: WIT Press Southampton. pp. 69–80.
10. Hightower JE, Grossman GD (1985) Comparison of constant effort harvest policies for fish stocks with variable recruitment. Can J Fish Aquat Sci 42: 982–988.
11. Koonce JF, Shuter BJ (1987) Influence of Various Sources of Error and Community Interactions on Quota Management of Fish Stocks. Can J Fish Aquat Sci 44: s61–s67.
12. Murawski SA, Idoine JS (1989) Yield Sustainability under Constant-Catch Policy and Stochastic Recruitment. T Am Fish Soc 118: 349–367.
13. Hsieh C, Reiss CS, Hunter JR, Beddington JR, May RM, et al. (2006) Fishing elevates variability in the abundance of exploited species. Nature 443: 859–862.
14. Shelton AO, Mangel M (2011) Fluctuations of fish populations and the magnifying effects of fishing. PNAS 108: 7075–7080.
15. Wikström A, Ripa J, Jonzén N (2012) The role of harvesting in age-structured populations: disentangling dynamic and age-truncation effects. Theor Popul Biol 82: 348–354.
16. Kell LT, Pilling GM, Kirkwood GP, Pastoors M, Mesnil B, et al. (2005) An evaluation of the implicit management procedure used for some ICES roundfish stocks. ICES J Mar Sci 62: 750–759.
17. Kraak SBM, Kelly CJ, Codling EA, Rogan E (2010) On scientists' discomfort in fisheries advisory science: the example of simulation-based fisheries management-strategy evaluations. Fish and Fisheries 11: 119–132.
18. Rochet M-J, Rice JC (2009) Simulation-based management strategy evaluation: ignorance disguised as mathematics? ICES J Mar Sci 66: 754–762.
19. Stephenson R, Peltonen H, Kuikka S, Pönni J, Rahikainen M, et al. (2001) Linking Biological and Industrial Aspects of the Finnish Commercial Herring Fishery in the Northern Baltic Sea. In: Funk F, Blackburn J, Hay D, Paul AJ, Stephenson R et al., editors. Herring: Expectations for a New Millennium. Fairbanks: University of Alaska Sea Grant. pp.741–760.
20. May RM, Beddington JR, Horwood JW, Shepherd JG (1978) Exploiting natural populations in an uncertain world. Mathematical Biosciences 42: 219–252.
21. Hilborn R, Mangel M (1997) The ecological detective. Princeton: Princeton University Press.
22. Dall SRX, Giraldeau L-A, Olsson O, McNamara JM, Stephens DW (2005) Information and its use by animals in evolutionary ecology. Trends Ecol Evol 20: 187–193.
23. McNamara JM, Green RF, Olsson O (2006) Bayes' theorem and its applications in animal behaviour. Oikos 112: 243–251.
24. Valone TJ (1991) Bayesian and prescient assessment: foraging with pre-harvest information. Anim Behav 41: 569–577.
25. Charnov EL (1976) Optimal foraging, the marginal value theorem. Theor Popul Biol 9: 129–136.
26. Olsson O, Holmgren NMA (1999) Gaining ecological information about Bayesian foragers through their behaviour. I. Models with predictions. Oikos 87: 251–263.
27. Olsson O, Wiktander U, Holmgren NMA, Nilsson SG (1999) Gaining ecological information about Bayesian foragers through their behaviour. II. A field test with woodpeckers. Oikos 87: 264–276.
28. Green RF (1980) Bayesian birds: a simple example of Oaten's stochastic model of optimal foraging. Theor Popul Biol 18: 244–256.
29. Iwasa Y, Higashi M, Yamamura N (1981) Prey distribution as a factor determining the choice of optimal foraging strategy. Am Nat 117: 710–723.
30. Olsson O, Holmgren NMA (1998) The survival-rate-maximizing policy of Bayesian foragers: wait for good news! Behav Ecol 9: 345–353.
31. Holmgren NMA, Norrström N, Aps R, Kuikka S (2012) MSY-orientated management of Baltic Sea herring (Clupea harengus) during different ecosystem regimes. ICES J Mar Sci 69: 257–266.
32. ICES (2011) Report of the Workshop on Implementing the ICES Fmsy Framework (WKFRAME-2), 10–14 February 2011, ICES, Denmark. ICES CM 2011/ACOM: 33. 110 pp.
33. European Commission (2002) On the conservation and sustainable exploitation of fisheries resources under the Common Fisheries Policy. Council regulation (EC) No 2371/2002.
34. Marchal P, Horwood J (1998) Increasing fisheries management options with a flexible cost function. ICES J Mar Sci 55: 213–227.
35. Pelletier D, Laurec A (1992) Management under uncertainty: defining strategies for reducing overexploitation. ICES J Mar Sci 49: 389–401.
36. ICES (2009) Report of the ICES Advisory Committee, 2009. ICES Advice, 2009. Book 8, 132 pp.
37. ICES (2012) Report of the Workshop 3 on Implementing the ICES Fmsy Framework, 9–13 January 2012, ICES, Headquarters. ICES CM 2012/ACOM 39: 1–33.
38. ICES (2012) Report of the ICES Advisory Committee 2012. ICES Advice, 2012. Book 8, 158 pp.
39. Valone TJ, Brown JS (1989) Measuring Patch Assessment Abilities of Desert Granivores. Ecology 70: 1800–1810.
40. Rose GA, Kulka DW (1999) Hyperaggregation of fish and fisheries: how catch-per-unit-effort increased as the northern cod (Gadus morhua) declined. Can J Fish Aquat Sci 56: 118–127.
41. Rohlf FJ, Sokal RR (1995) Statistical tables. New York: W.H. Freeman & Co.
42. Green RF (1984) Stopping rules for optimal foragers. Am Nat 123: 30–43.
43. Patterson KR (1999) Evaluating uncertainty in harvest control law catches using Bayesian Markov chain Monte Carlo virtual population analysis with adaptive rejection sampling and including structural uncertainty. Can J Fish Aquat Sci 56: 208–221.
44. Prager MH, Porch CE, Shertzer KW, Caddy JF (2003) Targets and limits for management of fisheries: a simple probability-based approach. N Am J Fish Manage 23: 349–361.
45. Shertzer KW, Prager MH, Williams EH (2008) A probability-based approach to setting annual catch levels. Fish Bull 106: 225–232.
46. Patterson K, Cook R, Darby C, Gavaris S, Kell L, et al. (2001) Estimating uncertainty in fish stock assessment and forecasting. Fish and Fisheries 2: 125–157.
47. Getz WM, Francis RC, Swartzman GL (1987) On Managing Variable Marine Fisheries. Can J Fish Aquat Sci 44: 1370–1375.
48. Ripa J, Lundberg P (1996) Noise colour and the risk of population extinctions. Proc R Soc London B 263: 1751–1753.
49. Yodzis P (1991) Introduction to theoretical ecology. New York: Harper & Row.
50. Costello C, Ovando D, Hilborn R, Gaines SD, Deschenes O, et al. (2012) Status and Solutions for the World's Unassessed Fisheries. Science 338: 517–520.

13

When Is Spillover from Marine Reserves Likely to Benefit Fisheries?

Colin D. Buxton[1]*, **Klaas Hartmann**[1], **Robert Kearney**[2], **Caleb Gardner**[1]

1 Fisheries Aquaculture and Coasts Centre, Institute for Marine and Antarctic Studies, University of Tasmania, Hobart, Tasmania, Australia, 2 Institute for Applied Ecology, University of Canberra, Canberra, Australian Capital Territory, Australia

Abstract

The net movement of individuals from marine reserves (also known as no-take marine protected areas) to the remaining fishing grounds is known as spillover and is frequently used to promote reserves to fishers on the grounds that it will benefit fisheries. Here we consider how mismanaged a fishery must be before spillover from a reserve is able to provide a net benefit for a fishery. For our model fishery, density of the species being harvested becomes higher in the reserve than in the fished area but the reduction in the density and yield of the fished area was such that the net effect of the closure was negative, except when the fishery was mismanaged. The extent to which effort had to exceed traditional management targets before reserves led to a spillover benefit varied with rates of growth and movement of the model species. In general, for well-managed fisheries, the loss of yield from the use of reserves was less for species with greater movement and slower growth. The spillover benefit became more pronounced with increasing mis-management of the stocks remaining available to the fishery. This model-based result is consistent with the literature of field-based research where a spillover benefit from reserves has only been detected when the fishery is highly depleted, often where traditional fisheries management controls are absent. We conclude that reserves in jurisdictions with well-managed fisheries are unlikely to provide a net spillover benefit.

Editor: Kerrie Swadling, University of Tasmania, Australia

Funding: This work was supported by the Fisheries Research and Development Corporation (FRDC 2010/226). http://frdc.com.au/Pages/home.aspx. The funders had no role in study design, data collection and analysis, decision to publish, or preparation of the manuscript.

Competing Interests: The authors have declared that no competing interests exist.

* Email: colin.buxton@utas.edu.au

Introduction

Marine reserves (MR), also known as no-take marine protected areas (MPA), are widely acknowledged as a conservation tool and their utility in a variety of situations is well established [1]. In particular over-exploited fish populations are shown to recover in the absence of fishing and generally become more abundant and attain a larger mean size in the reserve [2]. MPAs are also frequently promoted for the management of fisheries [3–7], even though compelling evidence in support of a net fisheries benefit is lacking [8]. Fisheries are proposed to benefit from reserves through increased production of eggs and larvae from the reserve (recruitment effect) and the net movement of adults into adjacent fishing grounds (spillover effect) [9].

In this study we focus on the spillover effect and, to avoid confusion over the use of terms, we define *spillover* as the net movement of fish across the boundary of a reserve into the fished ground, which would be expected to occur on the basis of fundamental physical principles of random movement. This is in contrast to *net spillover benefit* which involves spillover of sufficient magnitude to compensate for lost productivity due to the closure of fishing grounds, resulting in an overall benefit to the fishery through higher catch or economic yield.

Our review of the extensive literature reporting fisheries benefits reveals that there are surprisingly few empirical studies that attempt to quantify either the recruitment effect or a net spillover benefit. For example, Goni et al. [10] claims to be the first study to demonstrate a net spillover benefit in a fishery. Harrison et al. [11] make a similar claim with respect to the recruitment benefit of reserves in terms of larval export. Whilst spillover has been shown in several other studies, most do not accommodate the reduction in catch that results from reducing the area of the fishery, and consequently do not demonstrate a net spillover benefit.

Fishers are generally opposed to the introduction of reserves because they reduce the size of their fishing grounds, which is inferred to result in a loss of yield. Spillover is a common counter argument from reserve proponents, including Government agencies in the US, Europe and Australia, claiming that it will compensate for the lost fishing grounds to the extent that a net improvement in fisheries yield occurs [12–14].

The impact of the introduction of reserves on yield has been addressed in a number of theoretical studies (e.g., [15–17]), several of which progressively conclude that under broad assumptions well-managed fisheries should not benefit from the introduction of reserves [18–21]. Hart [22] quantifies this result to some degree by using an age-structured model, concluding that a benefit from spillover should not be anticipated unless open area fishing mortality considerably exceeds that which produces MSY.

The assumptions underlying these studies primarily concern the homogeneity of fish stocks and are reasonable for a large range of species. The obvious exception occurs in fish stocks with strong variability in spatial structure, for example where source-sink relationships exist or where reserves may result in the closure of disproportionately productive areas [23]. Such spatial heterogeneity is the basis of traditional spatial management of fisheries, and is a well-established and understood technique. Targeted spatial closures can be expected to benefit fisheries for selected species if the closed area is of disproportionate significance to the productivity of the species in question. Not surprisingly some models have shown that, at least under certain conditions, higher sustainable yields can be achieved with a marine reserve than without, e.g, [17], [23], [24]. But despite the common demonstration that special circumstances are required to achieve a spillover benefit from reserves, the implication of these findings have received limited attention and appear to have contributed little to the international public debate over fisheries benefits and to current management policy.

In this paper we use a widely applied fisheries population dynamics model which minimizes assumptions in order for the outputs to be applicable to a broad range of fisheries in non-structured environments ('normal' or 'average' fisheries). We modify this model to incorporate a MR and consider the management circumstances under which a non-specific reserve is likely to provide a benefit to the fishery. Our work highlights the effect that the degree of mismanagement under conventional fisheries management practices has on the ability of a reserve to provide a net fisheries benefit. It also investigates how this relationship changes with the rate that fish move between the reserve and the main population.

Methods

Population Dynamics

The population dynamics were modeled using a deterministic difference equation of the form:

$$N_{t+1} = f(N_t)N_t - C(N_t), \tag{1}$$

where N_t is the stock size at time t, $f(N_t)$ is the biological model that defines population growth and $C(N_t)$ is the catch. Common examples for the biological component of this model include the Ricker model:

$$f(N) = e^{r(1-N/K)}, \tag{2}$$

and logistic model:

$$f(N) = 1 + r(1 - N/K). \tag{3}$$

In both models r is the maximal growth rate and K the carrying capacity (maximum population size).

Throughout this analysis we assume that the population is homogenous - a small proportion, δ, of the population will behave identically in isolation to a larger proportion of the population. Mathematically, this implies that the carrying capacity can be reduced to δK. Alternatively we can consider the biological model to be a function of population density, in this case our model becomes:

$$N_{t+1} = f(N_t/\delta)N_t - C(N_t/\delta). \tag{4}$$

The divisor in the catch term indicates that catches are proportional to the population density (or constant).

Consider splitting a population into two areas: (i) a reserve occupying a proportion, α, of the original habitat size and (ii) the

remaining fishing grounds of size $1 - \alpha$. Denoting the two population sizes by R_t and M_t respectively, the model becomes:

$$R_{t+1} = f(R_t/\alpha)R_t - S_t$$
$$M_{t+1} = f(M_t/(1-\alpha))M_t - C(M_t/(1-\alpha), M_t) + S_t \tag{5}$$

where S_t denotes the spillover from the reserve into the fished population.

Spillover

We assume that a proportion, μ, of the population in the reserve moves into the fishing ground at each time step. As the population in the reserve is R_t, then μR_t will migrate out of the reserve. Similarly a proportion, v, of the population in the main fishing ground will migrate into the reserve. This results in the net movement from the reserve into the main fishing ground (the spillover) being:

$$S_t = \mu R_t - v M_t. \tag{6}$$

The values μ and v will depend on both the size and geometry of the reserve, however given the homogeneity of the population we also require that the net spillover is zero ($S_t = 0$) when the population density in the reserve and the fishing ground is equal (i.e. $R_t/\alpha = M_t/(1-\alpha)$). With this requirement and (6) we have:

$$S_t = \mu R_t - v M_t$$
$$0 = \mu \frac{\alpha}{1-\alpha} M_t - v M_t \tag{7}$$
$$v = \frac{\alpha}{1-\alpha} \mu.$$

As a direct result of the assumption of spatial homogeneity, a single parameter, μ, is sufficient to define the strength of the movement both in and out of the reserve. The net spillover from the reserve therefore becomes:

$$S_t = \mu \left(R_t - \frac{\alpha}{1-\alpha} M_t \right). \tag{8}$$

Note that we assume that μ (and v) are independent of the population density in and outside of the reserve. While there may be evidence to suggest that some individuals do follow a density gradient [25], [26] this does not substantially alter our findings, as it is akin to an increase in μ.

Fishing

We have specified the catch as a function of the population density and population size, $C_t = C(M_t/(1-\alpha), M_t)$. One common catch model is constant catch, as found, for example, in a subsistence fishery where a certain catch must be obtained each year to feed the population:

$$C(M_t/(1-\alpha), M_t) = P. \tag{9}$$

Well managed fisheries either have natural restrictions that prevent over-exploitation of the fish stock (e.g., limited demand of a niche product) or management controls to prevent over-exploitation. Management controls can be divided into two broad categories – input and output controls. Input controls limit the effort applied in the fishery. Denoting this by E we have:

$$C(M_t/(1-\alpha),M_t) = \frac{qEM_t}{1-\alpha}, \tag{10}$$

where q is a constant of proportionality. With this formulation, catch is directly proportional to the effort and population density (hence division of M by $1-\alpha$ to obtain a density). Other functional forms may be more appropriate for certain fisheries and fishing methods (e.g. purse seining of schooling fish). We considered all effort applied to the fishery to shift instantaneously from the reserve to the open area.

Output controls limit the catch that can be taken from a fishery and were not explored, as the existence of an effective output control (that does not cause a fishery collapse at equilibrium) implies effective fisheries management [27]. In reality there are many examples of ineffective output controls in fisheries that have not collapsed. These fisheries persist as the output controls are adjusted through time or, when the stock is in low abundance, effort controls (whether through management or limited numbers of participating fishers) restrict the fishery. Modeling such systems requires many assumptions; hence we have focused on input controlled fisheries in this analysis.

Net effect of the reserve on catch

We consider an effort-controlled fishery with a fish stock governed by the Logistic model. Stock size is measured in biomass, consequently growth encompasses both individual growth and recruitment. We assume that the population is homogenous and that introduction of the reserve will concentrate the effort in the remaining fishing grounds. The latter would be expected in a poorly managed fishery.

We assume that the population was at equilibrium prior to the introduction of a reserve and compare this with the post-reserve equilibrium. During the transient time between these two states spillover will be less. Since we are considering the equilibrium states we have $N_{t+1}=N_t$ which we simply denote by N, similarly for M, R and S.

Firstly, consider a fishery with a level of effort corresponding to near extinction, $E=E_E$. Introduction of a reserve will increase surplus production unless the population is beyond recovery.

At the other extreme, consider a pre-reserve fishery that is producing maximum sustainable yield (MSY) from the total area: $E=E_{MSY}$. By definition at this point, surplus sustainable production cannot increase. Therefore introduction of a reserve must decrease overall catch.

At E_{MSY} the spillover effect is less than the lost productivity and at E_E it exceeds the lost productivity. At some level of effort in between, the reserve must switch from having a net negative effect on the fishery to a net positive effect due to spillover. The level of effort at which this occurs is dependent on the model and its parameters. We now establish the point at which this occurs for a logistic model (equation (3)).

If spillover equals lost productivity in the fishing area, the pre-reserve and post-reserve catches must equal $qEN = qEM/(1-\alpha)$; hence $M=N(1-\alpha)$. Simply put, the population density in the fishing grounds must remain unchanged. Substitution in equation (5) yields:

$$N(1-\alpha)=f(N)N(1-\alpha)-qEN+S \tag{11}$$

subtracting equation (1) (at equilibrium) and solving for S gives:

$$S=\alpha qEN \tag{12}$$

Consequently, the spillover must equal the surplus production of the original fishing grounds that has now been encompassed in the reserve.

For a given level of effort, the pre-reserve fishery given by equation (4) will possess a solution, the nature of which depends on the population dynamics model. For example the non-zero solution for the logistic model is:

$$N=(r-qE)K/r \tag{13}$$

Using the full two area logistic model with effort controlled fishing (equations (5), (8) and (10)) and substituting equations (11) and (12) permits us to eliminate several of the unknowns. In this case we choose to eliminate N, M, R and S since conceptually we consider these to be determined by the remaining parameters. After algebraic manipulation (not shown here) we obtain the level of effort at which the introduction of the reserve does not change the overall catch:

$$qE=\left(r-2\mu+\sqrt{r^2+4\mu^2}\right)/2. \tag{14}$$

Note that $qE=0$ is also a solution (if no fishing is taking place, introduction of a reserve will not reduce the catch). A negative solution also exists but is of no further interest as the population would be extinct and negative densities are merely a mathematical curiosity. The same approach can be used for other population dynamics models, however for some models (e.g. the Ricker model) straight-forward analytic solutions do not exist. Qualitatively we would expect similar results for other population dynamics models and found this to be the case for numerical solutions to the Ricker model (results not shown here).

The optimal effort for this fishery without a reserve is $r=2q$. We divide equation (14) by this and subtract 1 to obtain the minimum excess effort (as a proportion) required for a reserve to be beneficial:

$$\hat{E}=\sqrt{1+(2\mu/r)^2}-2\mu/r. \tag{15}$$

This depends only on the ratio of the movement rate out of the reserve to the growth rate of the stock (μ/r), and not on the proportion of the area dedicated to the reserve (α). However, it should be noted that the movement rate out of the reserve, μ, is likely to depend on the reserve size. This link has not been explicitly explored here, however, for a given choice of μ, there is likely to be only a limited range of values of α that is possible.

Equations (14) and (15) are derived in more detail in Appendix S1.

Results

Figure 1 shows an example where a 10% reserve is introduced with 5% movement out of the reserve (μ) and a maximum growth rate (r) of 10%. This figure explores the effect of a reserve for different levels of initial effort applied to the fishery. The maximum sustainable yield (MSY) is obtained with an effort of 0.05 (E_{MSY}).

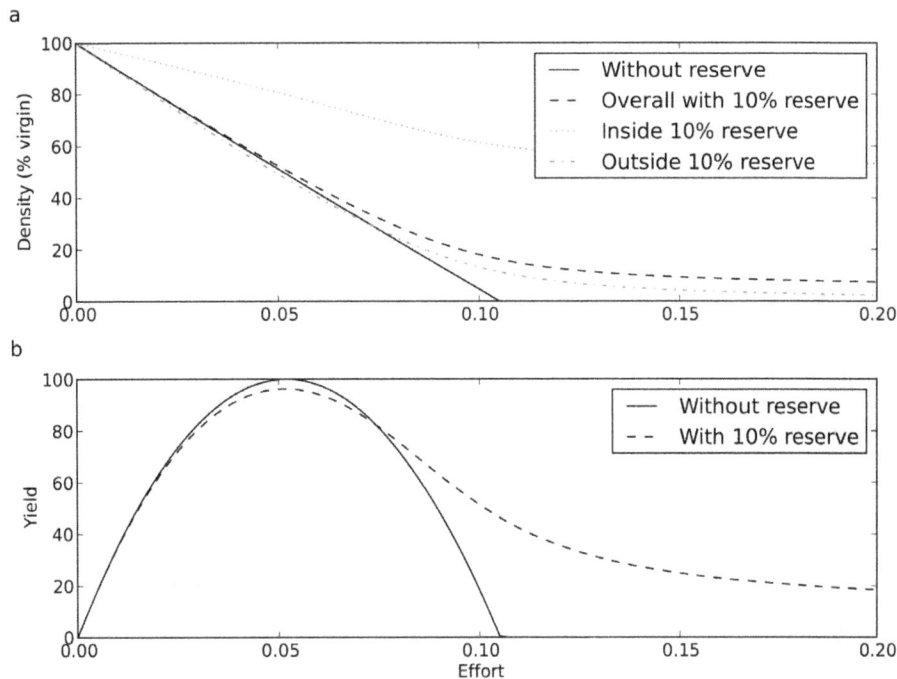

Figure 1. Changes in population and fishery dynamics resulting from the introduction of a reserve. (a) The equilibrium biomass density as a function of fishing effort. The density is shown for the whole stock without a reserve and with a 10% reserve. For the reserve scenario the density inside and outside of the reserve is also shown. (b) Yield as a function of fishing effort both with and without a reserve.

Introduction of the reserve decreases the yield at E_{MSY} and by definition there is no alternative effort that produces the same maximal yet sustainable yield. The point of intersection in the bottom panel corresponds to a level of effort, E_I, where the yield is the same with or without a reserve. At levels of effort above E_I, the introduction of a reserve increases yield. In this scenario, E_I is 150% of E_{MSY}, so a fishery would have to have 50% excess effort for the reserve to be beneficial in terms of the yield of the target species. At even higher levels of effort (>150% E_{MSY}) the MPA mitigates the impact of overfishing and permits sustainable (but substantially reduced) yield.

The level of excess effort at which a reserve has a neutral impact on fisheries yield depends only on the ratio of movement out of the reserve (μ) to the maximum growth rate (r) (equation (15)). This relationship is shown in Figure 2a, when the movement rate is high relative to the growth rate, a reserve is beneficial at low levels of excess effort. The extreme situation where μ/r approaches infinity corresponds for example to a miniscule reserve, which clearly will have negligible impact on a fishery. At the other extreme, $\mu/r = 0$, there is no movement out of the reserve, consequently it will always have a negative impact.

Alternatively we consider the excess effort required for a reserve to be beneficial as a function of the reserve density at equilibrium (Figure 2b). If the reserve is at 50% virgin biomass density it has neutral effect on the fishery. This is because 50% virgin biomass corresponds to MSY in this model and all surplus production is moved to the main population through spillover. At reserve densities above this, a fishery must have more excess effort to benefit from a reserve. In particular if reserves have a high percentage of virgin biomass (a common conservation goal for reserves) they will only benefit fisheries that have greater mismanagement. For example, at 80% virgin biomass a reserve will only benefit fisheries with more than 60% excess effort.

Discussion

Model outcomes

The model presented here examines the circumstances under which spillover from a reserve is sufficient to increase fishery yield (thus providing a net spillover benefit). As expected, density of exploited species was higher in the reserve than the fished area, which may be mistaken in itself as evidence that the reserve will create a net beneficial increase through larvae production [28]. However, it is important to consider the net effect, which in our model case was a decline in average density and a loss of yield except where effort exceeded E_{MSY}. While models are by necessity a simplification of ecological complexity, we show that the extent to which effort must exceed E_{MSY} for any yield benefit to occur from the reserve depends on the ratio of the rate of movement out of the reserve and the growth rate of the species concerned. Highly mobile/slow growing species received relatively less benefit from reserves where effort was above management targets compared to species with low movement/fast growth.

Our model is a relatively simple one chosen to illustrate a fundamental principle that is applicable across a broad range of fisheries. Different formulations for the biological model, $f(N_t)$, can be specified and similar results were obtained for the Ricker model (not shown here). Three major assumptions were made to maintain model simplicity: spatial homogeneity, density dependence and steady state dynamics.

Spatial homogeneity is an inappropriate assumption for some species. For example, where there are clear source-sink relationships protecting the source in a reserve is likely to provide an overall benefit [29]. The location of source areas can be consistent across different species and trophic levels, and in rare cases where these locations are known, it becomes possible to locate reserves that provide benefit to numerous, and theoretically all, species [30].

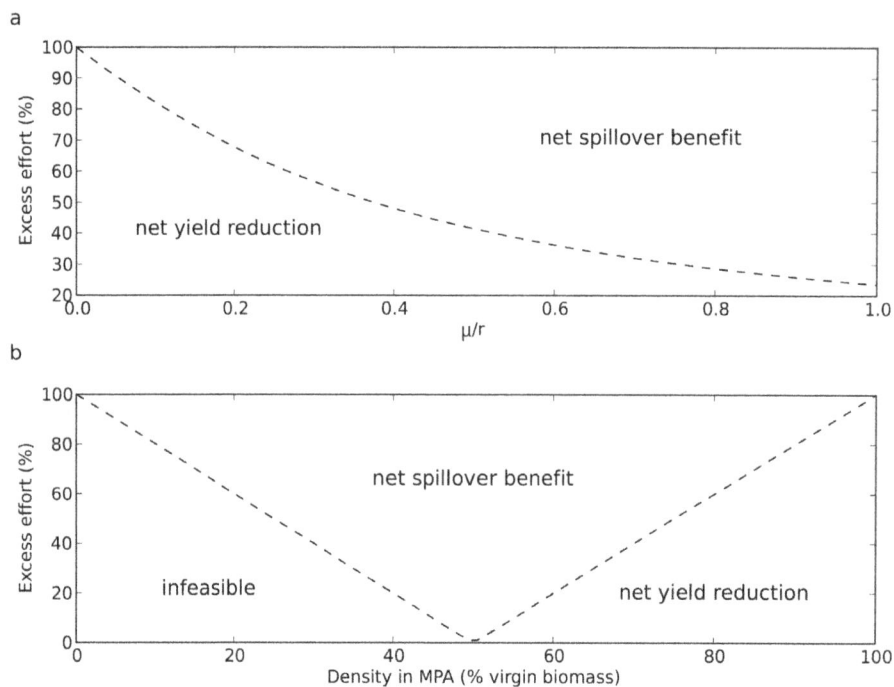

Figure 2. Characterisation the management and biological circumstances in which a reserve is beneficial. (a) The excess effort required for a reserve to improve fishery yield. For our simple model this was found to depend only on the ratio of the movement rate out of the reserve (and thus on reserve size) to the growth rate of the stock (μ/r). (b) The excess effort required for optimality as a function of the reserve density (at equilibrium). For example a reserve with 80% virgin biomass at equilibrium will provide a net economic benefit for a fishery that has more than 60% excess effort relative to optimal management. Combinations of excess effort and reserve density that fall in the bottom left region are infeasible; in these situations a reserve would have to decrease in population density after being formed (not possible in our model). Inside the "V" the reserve provides a net increase in fishery yield. In the right region the reserve decreases yield.

Density dependence in our model is a function of the total biomass in the local area (i.e. the fished population or the reserve population). This does not adequately capture the dynamics of species where density dependence varies substantially with age (e.g. density dependence occurring primarily during larval stages) and where different age classes have different movement rates across the reserve boundary. In such situations it could be possible for the reserve to provide a greater benefit by providing a recruitment increase to the fished region.

Steady state dynamics are widely used to explore fundamental fisheries principles. In the context of reserves, some models have shown that biological stochasticity may lead to theoretical net spillover benefits in fisheries where the biomass can be determined accurately on an annual basis and corresponding perfect catch limits set each year [31], [32]. Given the unrealistic nature of this assumption for most management situations there would be some value in further research that explored reserve benefits in a stochastic setting with realistic management. After the introduction of a reserve, it will take some time for the reserve population to build to the final density. Consequently it is expected that the reduction in yield will initially be much greater than predicted by our steady state model. With the concentration of effort the fished population would initially decrease before increasing some time later due to spillover from the reserve.

Our model did not consider that the introduction of a reserve may result in an effort reduction due, for example, to decreased accessibility or increased fishing costs. This would be beneficial for stock status and overall production in over-exploited fisheries, however, it would result in a reduction of production in well-managed fisheries.

Under our model there were no combinations of growth rate or movement where a net spillover benefit from reserves could occur unless effort exceeded E_{MSY}. Where effort is less than E_{MSY}, a loss of yield always occurs when reserves are implemented. The level of excess effort beyond E_{MSY} at which a reserve provides net spillover benefits was shown to depend only on the ratio of movement out of the reserve to the rate of growth of the population (μ/r). We also showed that reserve configurations that achieve higher densities of stock are only beneficial for mismanaged fisheries (Figure 2b). For example, a reserve that ultimately increases biomass density to 75% of unfished levels would benefit a fishery if the initial effort exceeds E_{MSY} by more than 50%. These results show that reserves will generally negatively impact yield for well managed fisheries. However reserves could minimize their impact on a well managed fishery by reducing the density increase of the fishery's target species in the reserve. For example, a reserve could be of a sufficient size to protect species with small home ranges whilst being small enough that individuals of the target species frequently move beyond reserve boundaries (a high movement rate, μ). This could also be achieved by having high reserve boundary length to total area ratios. The feasibility of this outcome will depend on the movement characteristics of the species involved.

Our finding that reserves cannot improve the yield of a well-managed fishery is consistent with several other theoretical studies [18–20]. The work here extends these findings by exploring the extent to which a fishery must be mismanaged before introduction of a reserve provides a benefit to the fishery in terms of yield.

Many fisheries have management objectives that constrain catch below the target of MSY assumed here, for example to manage risk from stochastic processes such as recruitment, or where there is an objective to target a maximum economic yield (MEY) that is variant to MSY. In these fisheries, effort and catch are lower than would occur with the MSY target [33], which reduces the negative impact of reserves on total yield, but also shifts the fishery further away from the level of depletion required for a net spillover benefit to occur.

Empirical context

The results from this study are consistent with other studies that have modeled the impact and/or benefits of reserves on fisheries in terms of improvements in yield. Following the publication of the early models on the potential net spillover benefits from reserves [16], [18], [34] there have been surprisingly few empirical studies that have attempted to demonstrate the effect. Most of the reserve literature has concentrated on the changes within reserves, showing an increase in size and abundance of resident fish and crustaceans, particularly of reef associated species (for a review see [35]). Despite the lack of empirical evidence the argument persists that reserves will confer a net spillover benefit to fisheries [5]. This view is actively promoted by government agencies [12–14]. However, the literature confirms that the evidence for such a benefit is far from conclusive. Several studies report a lack of evidence for spillover due to the low movement at the scale of the reserve [36–38], while others showed that spillover occurred but not that lost yield was compensated to produce a net benefit (e.g., [39–41]).

While density dependent export from reserves is considered to be a rational expectation [42], no studies have been able to conclusively demonstrate a net spillover benefit, and leakage from reserves is probably more related to random movement within species (e.g., [38], [41], [43], [44]). Several studies fail to provide conclusive evidence for net spillover benefits, yet argue that reserves are needed to provide fishery benefits (e.g., [39], [40]). Spillover has been inferred from observations of a density gradient between the reserve and adjacent fished area (e.g., [45], [46]) even though evidence was acknowledged to be equivocal (e.g., [42], [47], [48]), and where confounding factors such as a change in fishing practices (e.g., [49]) or changed fisheries management strategies over the study period were ignored (e.g., [4], [7]). Few of these studies consider whether the purported spillover to the fishery (as inferred from catch rates) has actually resulted in a net spillover benefit for the fishery. Even if CPUE goes up in a fished area it may be insufficient to result in a net production gain for the whole of the fishery.

Several studies have been able to demonstrate that spillover has contributed to an improvement in biomass and thus catch rate adjacent to the reserve [50–53]. These examples, all in the Mediterranean, were conducted in areas where the total fishery had been severely depleted. In this respect they are similar to several studies in other areas that, on multiple lines of evidence, infer a net spillover benefit to fisheries. Examples come from Africa [48] and Asia [7], [42], [47] where the fisheries in question were over-exploited and where there was limited application and/or enforcement of standard fisheries management controls. The result was that the proclamation of a reserve resulted in a recovery of the population in the reserve and a subsequent improvement in catches close to the reserve boundary. This is consistent with our conclusion that reserves can provide a net spillover benefit for severely depleted stocks. It does not, however, provide evidence that the declaration of the reserve was the most efficient means of achieving that benefit.

There are many possible variations on the biological assumptions made in our model. Aspects such as stock heterogeneity and variant density dependence assumptions will influence the impacts

of a reserve as well as the level of mismanagement, where a reserve switches from being beneficial to being detrimental for a fishery.

The model results presented here are for a general case, which is appropriate for consideration of reserves where a large number of species with variable life histories and spatial distributions are affected by change in management. Closed areas for traditional fishery management purposes are applied on a species by species basis and may have very different management outcomes to reserves because they can be designed and located to affect an individual stock. There are numerous cases where species with spatial heterogeneity, such as spawning aggregations or larval source-sink dynamics, benefit from fishery closures that target important source areas [54]. A total fishing closure would achieve the same result for those species, but can be expected to have less beneficial results for other exploited species.

Conclusions

We conclude that in fisheries where there is effective management, marine reserves are unlikely to produce a net spillover benefit for the total fishery, whereas they may be beneficial where the fishery has been mismanaged and stocks severely depleted. These results expand the implications of previous work by providing estimation and evaluation of the degree of mismanagement of fisheries that is necessary for non-specific closures to provide net benefits to fisheries.

The conclusions from the modeling presented here are supported by review of empirical studies, where spillover benefits have only been conclusively demonstrated in highly depleted areas. Together with the combined weight of earlier modeling work, they suggest that a net benefit from spillover should not be expected in areas already benefiting from quality traditional fisheries management.

These generalised findings in relation to reserves should not be confused with the use of targeted spatial closures for single fisheries, where it is possible to increase yield through closures by taking account of the spatial heterogeneity of life history traits.

While reserves may be proclaimed for a range of conservation objectives (including addressing impacts such as the effect of fishing on benthic environments, interactions with threatened species and catch of non-target species), we contend that it is misleading for governments to promote reserves on the basis of net spillover benefit in the context of well-managed fisheries. Reserves are only likely to be an effective strategy for fisheries management where effort is not or cannot be effectively controlled across the wider stock.

Acknowledgments

This work was supported by a grant from the Fisheries Research and Development Corporation (FRDC 2010/226). The authors would also like to thank Dr Zoe Doubleday for her assistance with the project and Prof Reg Watson for his valuable comments on an earlier draft of the manuscript, and two anonymous referees whose comments substantially improved the paper and made it accessible to a broader audience.

Author Contributions

Conceived and designed the experiments: CB KH RK CG. Analyzed the data: CB KH. Contributed reagents/materials/analysis tools: KH. Wrote the paper: CB KH RK CG.

References

1. Edgar GJ, Barrett NS, Stuart-Smith RD (2009) Exploited reefs protected from fishing transform over decades into conservation features otherwise absent from seascapes. Ecological Applications 19: 1967–1974.

2. Lester SE, Halpern BS, Grorud-Colvert K, Lubchenco j, Ruttenberg BI, et al. (2009) Biological effects within no-take marine reserves: a global synthesis. Marine Ecology Progress Series 384: 33–46.

3. Roberts CM (1997) Ecological advice for the global fisheries crisis. Trends in Ecology and Evolution 2(1): 35–38.

4. Roberts CM, Bohnsack JA, Gell FR, Hawkins JP, Goodridge R (2001) Effects of marine reserves on adjacent fisheries. Science 294: 1920–1923.

5. Gell FR, Roberts CM (2003) Benefits beyond boundaries: the fishery effects of marine reserves. Trends in Ecology & Evolution 18(9): 448–455.

6. Halpern BS, Lester SE, Kellner JB (2009) Spillover from marine reserves and the replenishment of fished stocks. Environmental Conservation 36(4): 268–276.

7. Russ GR, Alcala AC (2011) Enhanced biodiversity beyond marine reserve boundaries: The cup spillith over. Ecological Applications 21: 241–250. [doi:10.1890/09-1197.1].

8. Kerwath SE, Winker H, Gotz A, Attwood CG (2013) Marine protected area improves yield without disadvantaging fishers. Nature Communications 4: 2347 DOI:10.1038/ncomms3347.

9. Russ GR (2002) Yet another review of marine reserves as reef fishery management tools. In: Coral Reef Fishes (ed. Sale P) Academic Press, San Diego, CA. p 421–443.

10. Goni R, Hilborn R, Diaz D (2010) Net contribution of spillover from a marine reserve to fishery catches. Marine Ecology Progress Series. 400: 233–243. [doi:10.3354/meps08419].

11. Harrison HB, Wiiliamson DH, Evans RD, Almany GR, Thorrold SR, et al. (2012) Larval export from marine reserves and the recruitment benefit for fish and fisheries. Current Biology 22: 1023–1028. [doi:10.1016/j.cub.2012.04.008].

12. DEH (2003) The benefits of marine protected areas. Information Paper - The Commonwealth Department of Environment and Heritage, Australian Government. Available: http://www.environment.gov.au/coasts/mpa/publications/wpc-benefits.html.

13. Revenga S, Badalamenti F (2008) Management of marine protected areas for fisheries in the Mediterranean. Options Mediterraneennes Series B, 62: 107–111.

14. NOAA (2011) Benefits of a National System of Marine Protected Areas. Available: http://www.mpa.gov/resources/publications/factsheets/ Accessed 2012 May 31.

15. Sladek Nowlis J, Roberts CM (1999) Fishery benefits and optimal design of marine reserves. Fishery Bulletin 97: 604–616.

16. Sladek Nowlis J (2000) Short- and long-term effects of three fishery management tools on depleted fisheries. Bulletin of Marine Science 66(3): 651–662.

17. Steele JH, Beet AR (2003) Marine protected areas in 'nonlinear' ecosystems. Proceedings of the Royal Society London B (Supp) 270: S230–S233. [doi:1098/rsbl.2003.0074].

18. Polacheck T (1990) Year around closed areas as a management tool. Natural Resource Modeling 4: 327–354.

19. Hilborn R, Micheli F, De Leo GA (2006) Integrating marine protected areas with catch regulation. Canadian Journal of Fisheries and Aquatic Sciences 63: 642–649.

20. Le Quesne WJF, Codling EA (2009) Managing mobile species with MPAs: the effects of mobility, larval dispersal, and fishing mortality on closure size. ICES Journal of Marine Science 66: 122–131.

21. Barnes B, Sidhu H (2013) The impact of marine closed areas on fishing yield under a variety of management strategies and stock depletion levels. Ecological Modelling 269: 113–125.

22. Hart DR (2006) When do marine reserves increase fisheries yield? Canadian Journal of Fisheries and Aquatic Sciences 63: 1445–1449. [doi:10.1139/F06-071].

23. Apostolaki P, Milner-Gulland EJ, McAllister MK, Kirkwood GP (2002) Modelling the effects of establishing a marine reserve for mobile fish species. Canadian Journal of Fisheries and Aquatic Sciences 59: 405–415.

24. Ralston S, O'Farrell MR (2008) Spatial variation in fishing intensity and its effect on yield. Canadian Journal of Fisheries and Aquatic Sciences 65: 588–599.

25. Zeller D, Stoute SL, Russ GR (2003) Movements of reef fishes across marine reserve boundaries: effects of manipulating a density gradient. Marine Ecology-Progress Series 254: 269–280.

26. Abesamis RA, Russ GR, Alcala AC (2006) Gradients of abundance of fish across no-take marine reserve boundaries: evidence from Philippine coral reefs. Aquatic Conservation-Marine and Freshwater Ecosystems 16(4): 349–371.

27. Costello C, Gaines SD, Lynham J (2008) Can catch shares prevent fisheries collapse? Science (Washington), 321(5896): 1678–1681. doi:http://dx.doi.org/10.1126/science.1159478.

28. Bohnsack JA (1998) Application of marine reserves to reef fisheries management. Australian Journal of Ecology 23: 298–304.

29. Stockhausen WT, Lipcius RN, Hickey BM (2000) Joint effects of larval dispersal, population regulation, marine reserve design, and exploitation on production

and recruitment in the caribbean spiny lobster. Bulletin of Marine Science, 66(3): 957–990.

30. White JW, Samhouri JF (2011) Oceanographic coupling across three trophic levels shapes source-sink dynamics in marine metacommunities. Oikos 120(8): 1151–1164.

31. Yamazaki S, Grafton QR, Kompas T, Jennings S (2012) Biomass management targets and the conservation and economic benefits of marine reserves. Fish and Fisheries. [doi:10.1111/faf.12008].

32. Grafton RQ, Kompas T, Van Ha P (2006) The economic payoffs from marine reserves: resource rents in a stochastic environment. The Economic Record 82(259): 469–480.

33. Grafton RQ, Kompas T, Hilborn RW (2007) Economics of overexploitation revisited. Science 318(5856): 1601.

34. DeMartini EE (1993) Modelling the potential of fishery reserves for managing Pacific coral reef fishes. Fishery Bulletin 91: 414–427.

35. Halpern BS, Warner RR (2002) Marine reserves have rapid and lasting effects. Ecology Letters 5: 361–366.

36. Davidson RJ, Villouta E, Cole RG, Barrier RGF (2002) Effects of marine reserve protection on spiny lobster (Jasus edwardsii) abundance and size at Tonga Island Marine Reserve, New Zealand. Aquatic Conservation - Marine and Freshwater Ecosystems 12: 213–227. [DOI:10.1002/aqc.505].

37. Tewfik A, Bene C (2003) Effects of natural barriers on the spillover of a marine mollusc: implications for fisheries reserves. Aquatic Conservation Marine and Freshwater Ecosystems 13(6): 473–488.

38. Tupper MH (2007) Spillover of commercially valuable reef fishes from marine protected areas in Guam, Micronesia. Fishery Bulletin 105(4): 527–537.

39. Rowe S (2001) Movement and harvesting mortality of American lobsters (Homarus americanus) tagged inside and outside no-take reserves in Bonavista Bay, Newfoundland. Canadian Journal of Fisheries and Aquatic Sciences 58(7): 1336–1346.

40. Pillans S, Pillans RD, Johnstone RW, Fraft PG Haywood MDE, et al. (2005) Effects of marine reserve protection on the mud crab Scylla serrata in a sex-biased fishery in subtropical Australia. Marine Ecology Progress Series 295: 201–213.

41. Follesa MC, Cuccu D, Cannas R, Sabatini A, Deiana AM, et al. (2009) Movement patterns of the spiny lobster Palinurus elephas (Fabricius, 1787) from a central western Mediterranean protected area. Scientia Marina 73(3): 499–506.

42. Abesamis RA, Russ GR (2005) Density-dependent spillover from a marine reserve: Long-term evidence. Ecological Applications 15(5): 1798–1812.

43. Cole RG, Villouta E, Davidson RJ (2000) Direct evidence of limited dispersal of the reef fish Parapercis colias (Pinguipedidae) within a marine reserve and adjacent fished areas. Aquatic Conservation: Marine and Freshwater Ecosystems 10(6): 421–436.

44. Kelly S, Scott D, MacDiarmid AB (2002) The value of a spillover fishery for spiny lobsters around a marine reserve in Northern New Zealand. Coastal Management 30(2): 153–166.

45. Ashworth JS, Ormond RFG (2005) Effects of fishing pressure and trophic group on abundance and spillover across boundaries of a no-take zone. Biological Conservation 121(3): 333–344.

46. Abesamis RA, Alcala AC, Russ GR (2006) How much does the fishery at Apo Island benefit from spillover of adult fish from the adjacent marine reserve? Fishery Bulletin 104(3): 360–375.

47. Russ GR, Alcala AC (1996) Do marine reserves export adult fish biomass? Evidence from Apo Island, central Philippines. Marine Ecology Progress Series 132: 1–9.

48. McClanahan TR, Mangi S (2000) Spillover of exploitable fishes from a marine park and its effect on the adjacent fishery. Ecological Applications 10(6): 1792–1805.

49. McClanahan TR, Kaunda-Arara B (1996) Fishery recovery in a coral-reef marine park and its effect on the adjacent fisher. Conservation Biology 10(4): 1187–1199.

50. Stobart B, Warwick R, Gonzalez C, Mallol S, Diaz D, et al. (2009) Long-term and spillover effects of a marine protected area on an exploited fish community. Marine Ecology Progress Series 384: 47–60.

51. Forcada A, Valle C, Bonhomme P, Criquet G, Cadiou G, et al. (2009) Effects of habitat on spillover from marine protected areas to artisanal fisheries. Marine Ecology Progress Series 379: 197–211.

52. Goni R, Quetglas A, Renones O (2006) Spillover of spiny lobsters Palinurus elephas from a marine reserve to an adjoining fishery. Marine Ecology Progress Series 308: 207–219.

53. Vandeperre F, Higgins RM, Sanchez-Meca J, Maynou F, Goni R, et al. (2011) Effects of no-take area size and age of marine protected areas on fisheries yields: a meta-analytical approach Fish and Fisheries 12(4): 412–426.

54. Wakefield CB (2010) Annual, lunar and diel reproductive periodicity of a spawning aggregation of snapper Pagrus auratus (Sparidae) in a marine embayment on the lower west coast of Australian Journal of Fish Biology 77(6): 1359–1378.

14

Balancing Energy Budget in a Central-Place Forager: Which Habitat to Select in a Heterogeneous Environment?

Martin Patenaude-Monette[1], Marc Bélisle[2], Jean-François Giroux[1]*

1 Groupe de recherche en écologie comportementale et animale, Département des sciences biologiques, Université du Québec à Montréal, Montréal, Québec, Canada,
2 Département de biologie, Université de Sherbrooke, Sherbrooke, Québec, Canada

Abstract

Foraging animals are influenced by the distribution of food resources and predation risk that both vary in space and time. These constraints likely shape trade-offs involving time, energy, nutrition, and predator avoidance leading to a sequence of locations visited by individuals. According to the marginal-value theorem (MVT), a central-place forager must either increase load size or energy content when foraging farther from their central place. Although such a decision rule has the potential to shape movement and habitat selection patterns, few studies have addressed the mechanisms underlying habitat use at the landscape scale. Our objective was therefore to determine how Ring-billed gulls (*Larus delawarensis*) select their foraging habitats while nesting in a colony located in a heterogeneous landscape. Based on locations obtained by fine-scale GPS tracking, we used resource selection functions (RSFs) and residence time analyses to identify habitats selected by gulls for foraging during the incubation and brood rearing periods. We then combined this information to gull survey data, feeding rates, stomach contents, and calorimetric analyses to assess potential trade-offs. Throughout the breeding season, gulls selected landfills and transhipment sites that provided higher mean energy intake than agricultural lands or riparian habitats. They used landfills located farther from the colony where no deterrence program had been implemented but avoided those located closer where deterrence measures took place. On the other hand, gulls selected intensively cultured lands located relatively close to the colony during incubation. The number of gulls was then greater in fields covered by bare soil and peaked during soil preparation and seed sowing, which greatly increase food availability. Breeding Ring-billed gulls thus select habitats according to both their foraging profitability and distance from their nest while accounting for predation risk. This supports the predictions of the MVT for central-place foraging over large spatial scales.

Editor: Michael Sears, Clemson University, United States of America

Funding: The research was supported by grants to JFG and MB from the Natural Sciences and Engineering Research Council of Canada, the Canadian Wildlife Service, ICI Environnement, Falcon Environmental Services, Chamard et Associés, Waste Management, and BFI Canada. MPM was supported by a scholarship from the Natural Sciences and Engineering Research Council of Canada. The funders had no role in study design, data collection and analysis, decision to publish, or preparation of the manuscript.

Competing Interests: ICI Environnement, Falcon Environmental Services, Chamard et Associe's, Waste Management, and BFI Canada provided funding towards this study. There are no patents, products in development or marketed products to declare.

* Email: giroux.jean-francois@uqam.ca

Introduction

Animals face time and energy constraints leading to trade-offs in their activity budget, which can also be modulated by factors such as the spatio-temporal distribution of food resources, conspecifics, predation risk, and phenology. How animals respond to these constraints in order to maximize their fitness through foraging behaviour has been the main focus of optimal foraging theory [1], [2]. For instance, the marginal-value theorem (MVT) has been used to predict which resource patch an animal should exploit and how long it should stay before moving to another patch or return to its nest or shelter [3], [4]. Assuming that animals maximize their net energy gain, this model has provided relevant qualitative predictions [5]. However, it has been developed and used for small-scale systems in which animals are assumed to incur few or no travel costs and to be highly informed about their environment [1], [2].

This model may therefore be difficult to apply at the landscape level because of information uncertainty about the environment, which influences learning ability and because of the limited motion and navigation capacity of animals [6], [7], [8]. For example, classical central-place foraging models based on the MVT predict that prey load size should increase with the distance traveled by a forager from its central place [4], [9]. However, a forager moving across the landscape with a large load can incur increased travel costs due to greater energy expenditures or can encounter higher predation risks through increased exposure and reduced manoeuvrability [5]. Therefore, the impact of carrying a heavy load can influence the time and energy budget of a central-place forager in different ways, sometime far from the conclusions of the classical models [10].

The MVT predicts that a foraging path is the outcome of balancing trade-offs between energy expenditures and gains, especially within landscapes where resources are heterogeneously distributed. Although it is difficult to use the MVT to make precise predictions under relaxed assumptions, classical central-place foraging models nevertheless allow to predict that distant patches must provide higher energy "prey" than those found in nearby patches [9]. Hence, the profitability of a given load size may vary for a generalist forager traveling through a heterogeneous landscape. Also, habitats providing low energy food should only be used close to the central place whereas habitats with high-energy food may be exploited near or far from the central place. It remains that travel costs may increase the use of poor quality habitats when individuals must sample and learn the quality of their environment [11], [12]. Moreover, temporal variation in habitat availability and forager condition may alter the pattern of habitat use along a distance gradient [13].

Although assessing the costs and benefits of large spatio-temporal scale movements is difficult, analytical methods based on accurate location data (e.g., GPS) are now available to study movement behaviour. Combining these analytical methods with *in situ* observations of individual foraging strategies, patch quality, and environmental conditions while considering the individuals' characteristics has the capacity to provide insights into the cost-benefit trade-offs associated with foraging movements underlying habitat selection [13], [14]. For instance, resource selection functions (RSF) have been widely used to assess habitat selection. They are based on the comparison of relative habitat use (defined by presence-only data) and availability or on the presence/absence of individuals in habitat patches [15]. RSF are particularly informative if a distinction can be made between actively selected locations, such as foraging patches, and the incidentally selected locations visited during inter-patch movements [16], [17]. Bastille-Rousseau *et al.* [18] have advocated the use of a combination of RSF, residence time analysis, and ground surveys to study resource selection and foraging strategies at the landscape level. Considering the hierarchical aspect of the selection process, the difficulty of defining available habitats with presence-only data can be avoided by building RSF based on the habitats actually visited for foraging vs. those crossed when moving to a patch [19], [20]. Measuring the time spent by an animal within the surroundings of recorded locations (residence time) should allow discriminating between locations occurring within foraging patches and those found along movement paths [16], [21].

We used RSF and residence time analyses from GPS-tracking data, as well as survey data, diet characterization and calorimetric analyses to study the processes that determine habitat use by breeding Ring-billed gulls (*Larus delawarensis*). This species is a colonial central-place forager that feeds opportunistically upon a wide variety of prey items found in both aquatic and terrestrial habitats [22], [23]. We expected that gulls should be more likely to forage in a patch where the amount of habitats providing high-energy food increases and that such a relationship should be more pronounced far from the colony so that gulls reach a threshold of profitability. We also hypothesized that gulls should select habitats with a temporally variable food availability only when those habitats provide high food returns. For instance, agricultural lands and lawns should be selected on rainy days when annelids (earthworms) are more available to gulls [24]. By testing these predictions, our study sheds light on the process of habitat selection by animals from an energy trade-off perspective.

Materials and Methods

Ethics statement

Field methods to capture, mark, and collect Ring-billed gulls were approved by the Institutional Animal Protection Committee

of the Université du Québec à Montréal (No. 646). The capture and marking of gulls was conducted under Environment Canada scientific permit to capture and band migratory birds (No. 10546) while the collection of specimens was carried out under Environment Canada scientific research permit (No. SC-23)

Study area

We tracked the movements of Ring-billed gulls breeding on Deslauriers Island located in the St. Lawrence River 3 km downstream from Montreal, QC, Canada (45.717°N, 73.433°W). This colony covered 11.4 ha and supported 48,000 pairs at the time of the study. The surrounding foraging area encompassed approximately 6,000 km^2 and consisted of a mosaic of high and low density urban areas, agricultural lands of intensive (soybean, maize, and small cereals) and extensive cultures (hayfields and pastures), as well as riparian habitats along the River and its tributaries (Fig. 1). Four landfills and two waste material transhipment sites were located in the vicinity of the colony. Landfills attract gulls because of the anthropogenic food they supply but the implementation of deterrence programs may reduce their accessibility [25], [26]. During our study, the St-Thomas (41 km) and Lachute (63 km) landfills, as well as the two transhipment sites (12 and 27 km), had no deterrence program. On the other hand, the Ste-Sophie landfill (37 km) initiated a deterrence program in 2009 that combined pyrotechnics and selective culling. However, the program was limited to weekdays from 07:00 to 15:00 thereby leaving some feeding opportunities for gulls [26]. Lastly, the Terrebonne landfill (8 km) conducted a deterrence program since 1995 that included falconry, distress calls, and pyrotechnics. This program was in operation every day from sunrise to sunset, preventing all but few gulls to use the landfill (Thiériot, E., unpublished data).

Telemetry

Breeding Ring-billed gulls were fitted with 10–16-g GiPSy-2 data loggers (TechnoSmart, Italy) between April and June 2009–2010. The loggers represented (mean ± SD) 2.8±0.5% of the birds' body mass (485±49 g). Most gulls were captured and recaptured with nest traps or dip nets but some had to be recaptured by rifle shooting; carcasses were then kept for further analyses (see *Diet and calorimetric analyses*). Data loggers were attached on the two median rectrices with white TESA tape (no. 4651) and programmed to acquire locations at 4-min intervals. Tracking lasted 1 to 3 days depending on battery life. Half of the birds included in the analyses returned to their nest within 15 min after being released and 81% of them returned within 60 min. Birds that spent more than 1 h away from their nest took on average 4.5±3.9 h to return. Breeding stage upon capture was categorized as incubating or brood rearing. Gulls were sexed with genomic DNA isolation from chest feathers [27].

We recaptured 109 Ring-billed gulls (41 females, 68 males) with loggers that provided reliable data (Table S1). After removing locations within a 300-m buffer zone around the colony (see *Data analyses*), there were only 28 missing locations on a potential of 15,948. The remaining 15,920 locations had a low dilution of precision metric (DOP ≤6) and an estimated precision of ±5 m [28]. A total of 67 gulls were followed during incubation (164 foraging trips) and 42 during the brood rearing period (239 foraging trips).

Gull Surveys

We conducted weekly surveys from April to June alternating between three periods (05:00–10:00, 10:00–15:00, and 15:00–20:00) to determine the proportion of time Ring-billed gulls spent

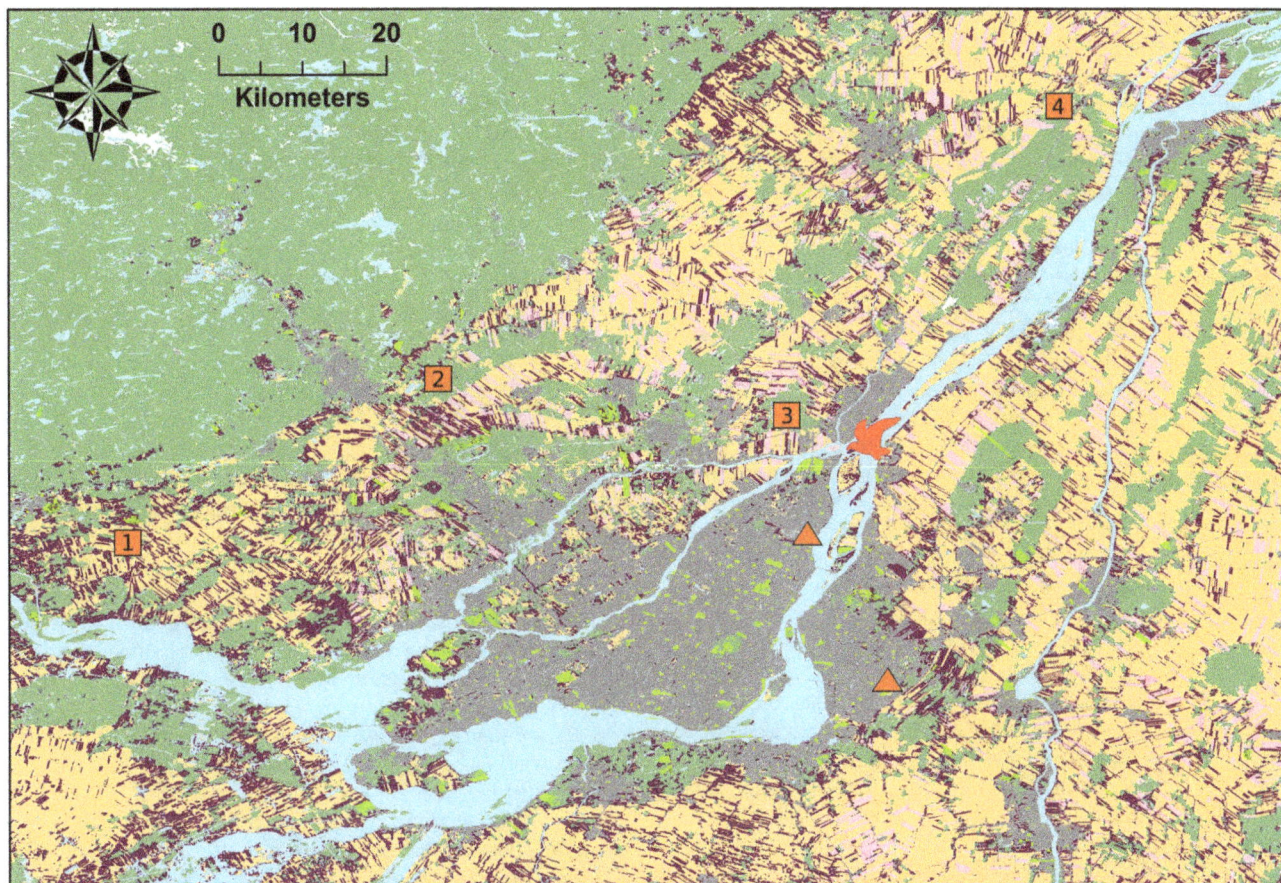

Figure 1. Map of the study area. Land cover types include water (blue), urban areas (gray), intensive cultures (mango), extensive cultures (purple), unidentified cultures (rose), lawns (olive green), and woodlots (dark green). Numbers in squares indicate landfill locations (1- Lachute, 2- Ste-Sophie, 3- Terrebonne, 4- St-Thomas), red triangles indicate transhipment site locations, and the bird pictogram indicates the location of the Deslauriers Island Ring-billed gull colony.

foraging in their main feeding habitats. In agricultural habitats, we surveyed a 50-km roadside transect on each shore of the St. Lawrence River ($N = 13$ and 21 surveys in 2009 and 2010, respectively). We tallied the number of birds in each flock and performed an instantaneous scan sampling to determine the proportion of birds foraging (head down below the horizontal or

Table 1. Cover percentage of eight habitat types available in the foraging range of 109 Ring-billed gulls breeding on Deslauriers Island established as the minimum convex polygon calculated with all gull locations and mean cover percentage (± 1 SD) in movement (residence time <100 s) and foraging (residence time ≥100 s) patches (200-m radius), 2009–2010.

| Habitat type | % cover | | |
| | Foraging range | Movement patches | Foraging patches |
	(5,565 km²)	(N = 2,599)	(N = 4,490)
Lawns (parks, golf courses, etc.)	1.2	1.8±8.3	2.1±9.4
Woodlots	20.6	13.7±28.8	4.8±16.0
Urban areas	16.8	27.8±38.5	23.1±36.6
Water bodies	5.3	18.7±34.8	22.4±37.9
Intensive cultures	39.5	24.3±34.1	31.4±38.7
Extensive cultures	11.7	8.4±16.2	8.0±15.6
Unidentified cultures	4.1	3.4±12.3	4.2±14.3
Landfills/Transhipment sites	0.1	1.3±11.3[a]	4.2±20.0[a]

[a]Percent occurrence.

probing into the soil) that we considered as the proportion of time spent foraging [29]. A flock was defined as a group of gulls using the same field type and not separated by more than 200 m from each other. Birds using different field types but closer than 200 m from each other were considered as different flocks. The total number of tractors and their activity (ploughing, harrowing or sowing) was also noted over the entire transect during each survey.

Observations in other habitats were conducted weekly in 2010 at fixed points located in urban ($N = 25$ points), suburban ($N = 53$ points), and riparian ($N = 10$ points) areas on the Montreal Island ($N = 16$ surveys) and along the North ($N = 18$ surveys) and the South shores ($N = 22$ surveys) of the St. Lawrence River. These sites were selected because they were susceptible to be visited by gulls while insuring that observers driving vehicles could stop safely. At each point, gulls using different habitat types (lawns, shores, water, grounds covered with concrete, asphalt or gravel, building roofs, and post lights) were counted and scanned to determine the proportion of birds foraging (erratic flight in emergent insect clouds above waterbodies, feeding on garbage, head down below the horizontal or probing into the soil or water).

Finally, we estimated the proportion of time gulls spent foraging at landfills by conducting 5-h observation periods once a week in 2009 ($N = 7$) and five days a week in 2010 ($N = 59$) at the Ste-Sophie landfill, again alternating among the three daily periods including periods with and without deterrence. Total bird counts and instantaneous scan sampling were conducted every half hour. The mean daily abundance of gulls was computed for each day as well as the proportion of birds that were actually foraging (flying less than 5 m above the active tipping area, head down below the horizontal or probing into refuse).

Diet and calorimetric analyses

We collected 496 boli from chicks of both sexes during weekly visits to the Deslauriers colony during the rearing period of 2009 and 2010. We selected chicks haphazardly and slightly pressed their proventriculus to make them regurgitate recently swallowed food. Spontaneous regurgitations of adults ($N = 13$) captured during banding operations throughout the breeding period were also collected. Samples were frozen until they were analysed. We also kept frozen the carcasses ($N = 51$) of adults fitted with data loggers and recaptured by shooting until the content of their oesophagus and proventriculus could be analysed. Similarly, we analysed stomach contents of birds collected by rifle shooting in agricultural lands ($N = 69$), riparian areas ($N = 54$), and at the Ste-

Sophie landfill ($N = 85$). We made sure that birds were actively feeding in these habitats before collecting them. For safety reasons, gulls could not be collected in urban areas. Each food item of a bolus or stomach was separated, identified, dried to constant weight and weighted (± 0.01 g). Food items were grouped into broad categories (e.g., arthropods, annelids, vertebrates, refuse, vegetation, other).

Food availability could not be assessed throughout the 6,000 km^2 of the foraging area to estimate the benefits obtained by gulls when feeding in different habitats. Instead, we relied on the relative area of each habitat and the food quality in these habitats based on energy content of the various food items. We therefore performed duplicate or triplicate calorimetric analyses of each food category using a bomb calorimeter (Parr, model 1108P).

Data analyses

We first created a 300-m buffer zone around Deslauriers Island (colony) to discriminate between foraging trips and short movements to the shore or surrounding shallow water where gulls rest and preen [30]. Our analyses were limited to locations outside this zone. The mean number of foraging trips per day, the mean direct (Euclidean) distance between the colony and the farthest location reached during a foraging trip (whether a stopover or not), the mean distance traveled on a foraging trip and the mean sinuosity of movement paths (traveled distance divided by round trip direct distance, [31]) were compared between breeding stages and sexes using linear mixed models with gull ID as a random factor.

For each foraging trip, we calculated the total amount of time spent at different locations on the landscape by estimating residence time without rediscretization [16]. Residence time was defined as the time spent in a circle of radius r centred on a given location along the foraging path. The circle, with its specific habitat composition and features, could then be viewed as a potential foraging patch. In the absence of precise information regarding the spatio-temporal distribution of resources, the hierarchy of spatial scales at which animals are likely to respond to landscape heterogeneity (i.e., patches, [32]) can only be identified through behaviour [16], [33]. For each trip, we thus computed the coefficient of variation (CV) of residence times for radii ranging between 200 and 2,000 m with 100-m increments. We averaged the CV across paths and plotted them against the circle radii (Fig. S1). The mean CVs of residence time across paths showed a plateau for radii of 200 to 400 m instead of a clear peak.

Table 2. Summary of a priori models based on resource selection functions that predict the probability that a breeding Ring-billed gull will forage in a patch (200-m radius) for a 100-s residence time threshold.

Model	Deviance	K	ΔAICc	w_i
H+D+B	7,891	26	0.00	0.813
H+D+R+B	7,886	30	2.94	0.187
H+D	7,940	19	35.06	0.000
H+D+R	7,936	23	38.63	0.000
H+B	8,466	18	558.98	0.000
H+R+B	8,459	22	560.26	0.000
H	8,533	11	611.33	0.000
H+R	8,527	15	613.67	0.000

H: habitat types; D: distance between a location and the colony; R: mean daily rainfall; B: breeding stage (egg incubation vs. chick rearing); K: number of parameters; w_i: Akaike weight.

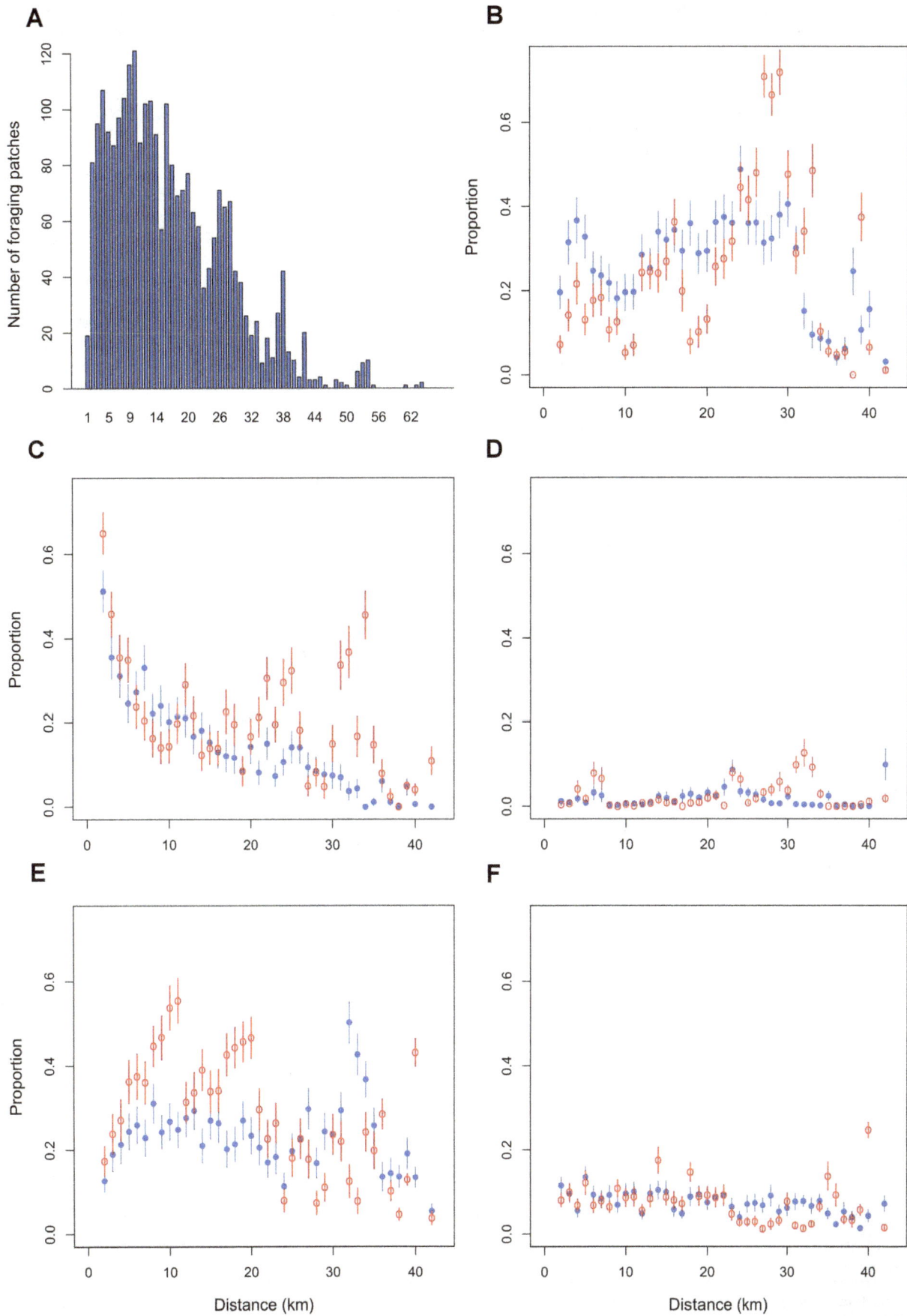

A

B

C

D

E

F

Figure 2. The effect of distance on habitat selection by foraging Ring-billed gulls. Number of foraging patches within 1-km concentric bands from the colony (a). Mean (±1 SD) proportion of urban areas (b), waterbodies (c), lawns (d), intensive cultures (e), and extensive cultures (f) in foraging (blue) and movement patches (red) in relation with the distance from the colony.

There was no significant difference in CVs distribution between males and females. We thus chose a 200-m radius to get a stronger contrast in habitat composition between foraging and movement patches. We finally retained locations distanced by at least two radii to limit spatial autocorrelation.

We calculated the landscape composition within each circle in which residence time was estimated based on a land cover map created in ArcGIS 9.3.1 [34] using both agricultural and topographic data ([35], [36], [37]; planimetric precision <30 m). Landscape composition was defined as the proportion of different habitats including lawns, woodlots, urban areas, and water as well as intensive, extensive and unidentified cultures. Because of their relatively small size, landfills and transhipment centres were noted as presence/absence in each circle. We also measured the distance between each location where a residence time was computed and the nest of the tracked gull. Finally, we calculated the mean daily

rainfall using data from 10 meteorological stations located throughout the entire foraging area [38].

We first described the habitats within the global home range of gulls breeding on Deslauriers Island by estimating the proportion of habitats within the 100% minimum convex polygon drawn using the foraging trip locations of all birds. Next, we built a RSF based on patches visited by gulls on their foraging trips. Considering that a foraging individual must reduce its flying speed and increase its turning rate, we used residence time to discriminate "foraging patches" from "movement patches". We assumed that if a gull spent more than 100 s in a 200-m radius circle, it was actively foraging. Otherwise, we considered that it was moving either between the colony and a foraging patch or between two foraging patches. Gulls observed foraging during surveys typically spent more than 100 s within 200 m from where they were first detected. Moreover, based on the flight speed of

Table 3. Mixed-effects averaged logit resource selection functions quantifying the probability that a breeding Ring-billed gull forage in a patch.

Variable	β	SE	95% CI	
Intercept	−0.612	0.491	−1.575	0.351
Distance*	0.089	0.013	0.063	0.115
Woodlots*	−2.963	0.448	−3.840	−2.086
Lawns	−0.212	0.834	−1.846	1.422
Urban areas*	−2.550	0.521	−3.570	−1.529
Landfills	0.992	0.510	−0.007	1.992
Water*	−1.130	0.516	−2.142	−0.118
Extensive cultures	−1.037	0.619	−2.251	0.177
Intensive cultures	−0.901	0.516	−1.913	0.111
Unidentified cultures	−0.319	0.688	−1.666	1.029
Lawns×Distance	0.017	0.033	−0.049	0.083
Urban areas×Distance*	0.067	0.016	0.036	0.098
Landfill×Distance	0.001	0.017	−0.033	0.035
Water×Distance	0.031	0.016	−0.001	0.063
Extensive cultures×Distance	−0.005	0.021	−0.045	0.035
Intensive cultures×Distance	−0.026	0.015	−0.056	0.004
Unidentified cultures×Distance	−0.036	0.026	−0.087	0.015
Lawns×Incubation	−0.050	0.671	−1.365	1.266
Urban areas×Incubation	0.210	0.244	−0.267	0.688
Landfill×Incubation	0.271	0.446	−0.603	1.145
Water×Incubation	0.205	0.254	−0.294	0.703
Extensive cultures×Incubation	−0.349	0.418	−1.168	0.469
Intensive cultures×Incubation*	1.429	0.237	0.965	1.893
Unidentified cultures×Incubation	0.911	0.474	−0.018	1.839
Lawns×Rainfall*	0.038	0.019	0.001	0.075
Extensive cultures×Rainfall	0.000	0.007	−0.013	0.013
Intensive cultures×Rainfall	−0.002	0.004	−0.009	0.005
Unidentified cultures×Rainfall	−0.003	0.008	−0.019	0.012

Model-averaged coefficients (β), unconditional standard errors (SE), and 95% confidence intervals (CI) are presented. Variables followed by an asterisk are significant (95% CI excluding 0).

Gulls

Tractors

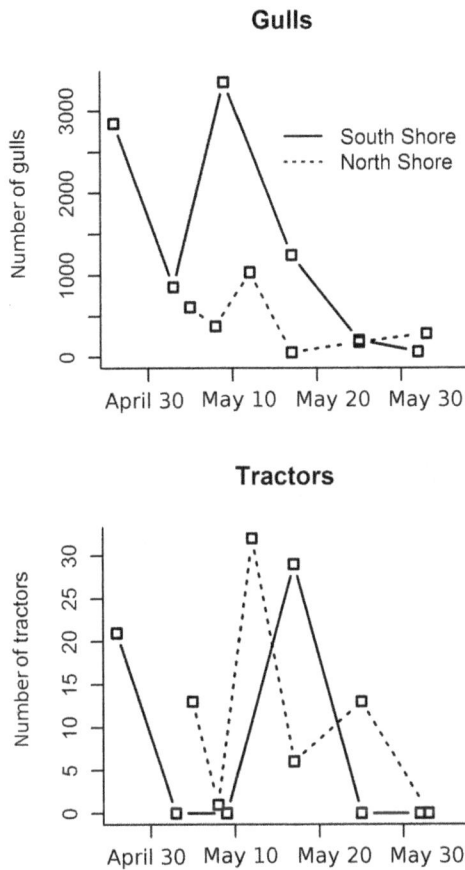

Figure 3. Use of agricultural lands by breeding Ring-billed gulls. Number of gulls and of tractors observed during surveys on the North and South shores of the St. Lawrence River, 2010.

Figure 4. Use of landfills and transhipment sites by Ring-billed gulls. Mean (± 1 SD) proportion of Ring-billed gull locations at landfills and transhipment sites in foraging patches within 1-km concentric bands located at different distances from the colony. All landfills and open transhipment sites were visited by at least one tagged individual. Some sites encompassed more than one band.

Black-headed gulls (*Chroicocephalus ridibundus*) and Lesser Black-backed gulls (*Larus fuscus*), which are respectively slightly smaller and larger than Ring-billed gulls (14.7–15.5 m/s, respectively; [39]), at least 26 s is required for a gull to cross a circle of 200-m radius. The remaining 74 s appears insufficient for a gull to forage significantly in such a circular patch. Although our tracking device did not allow to determine the precise activity of the birds while not moving, we consider justified to assume that gulls were actually foraging in patches where they spend more than 100 s. Indeed, during the breeding period, gulls must brood their eggs or feed their young and must therefore spend as much time as possible on the colony allowing the rest of their time to foraging.

We used mixed effects logistic regressions to quantify the influence of landscape composition on the probability that a gull foraged in a patch along its movement path. Gull ID and foraging trip ID (nested within gull ID) were treated as random factors. The addition of these terms dealt with the hierarchical structure of the data and allowed the estimation of the variability across individuals and foraging trips. Eight different models were built and compared based on the second-order Akaike information criterion (AIC$_c$, [40]). We included the proportion of each habitat type and the occurrence of landfills and transhipment sites in all eight models. We considered the interaction of rainfall with lawns as well as with each type of agricultural cover because annelids are more prevalent under wet conditions [24]. We also included the distance between the location of a gull while foraging and its nest as a proxy for foraging costs and accessibility [41]. We considered

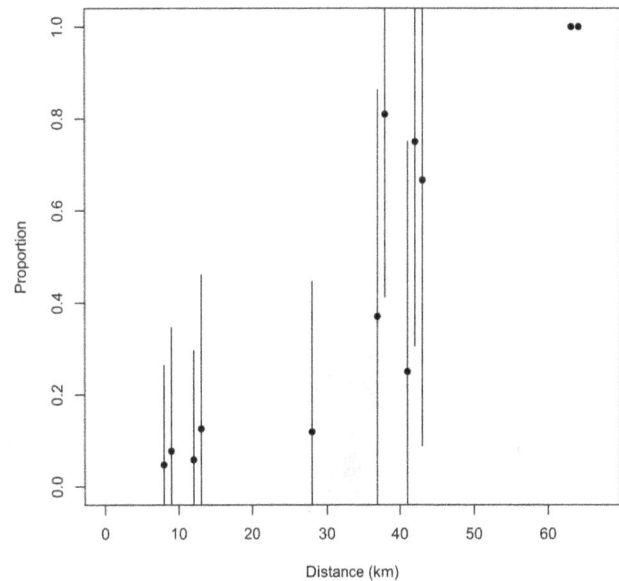

distance both as a main effect and in interaction with the relative amount of each habitat type (except woodlots) as well as with the occurrence of landfills or transhipment sites. We used this approach because we do not know which fitness currency gulls may be maximizing and because the profitability of the different habitats may not scale linearly with distance. Although woodlots are accessible, gulls avoid being under canopy and should thus avoid forest habitats whatever the distance from the colony. We included breeding stage in interactions with each habitat type to take into consideration the gulls' breeding phenology and their associated requirements as well as habitat phenology, particularly for agricultural cover types where farming practices and field conditions vary throughout the season. Finally, we built a second set of eight models, adding the sex of the birds in interaction with each habitat type and the distance of the patch from the colony as males and females differ in size (affecting travel costs and dominance on food patches) and provide different levels of parental care [23]. We fitted mixed effects logistic regressions using the Laplace approximation using the lme4 package (version 0.999375-39; [42]) run in the R statistical environment (version 2.12.2; [43]). AIC$_c$ were computed based on maximum log-likelihoods. Multi-model inference was performed following Burnham and Anderson [41] after testing that there was no problem of collinearity.

The effect of distance on habitat selection may be non-linear partly because central-place foragers often avoid habitats near their central-place by moving further away [44], [45]. Moreover, the relative abundance of different habitats varied with distance from the colony. To overcome this problem, we first draw 1-km wide circular bands up to 67 km from the colony, which corresponds to a few kilometers further than the farthest gull location (see *Results*). For each band with at least five foraging patches, we calculated the mean area covered by each habitat type within patches to estimate habitat use. We also calculated the mean area covered by the different habitats in movement patches

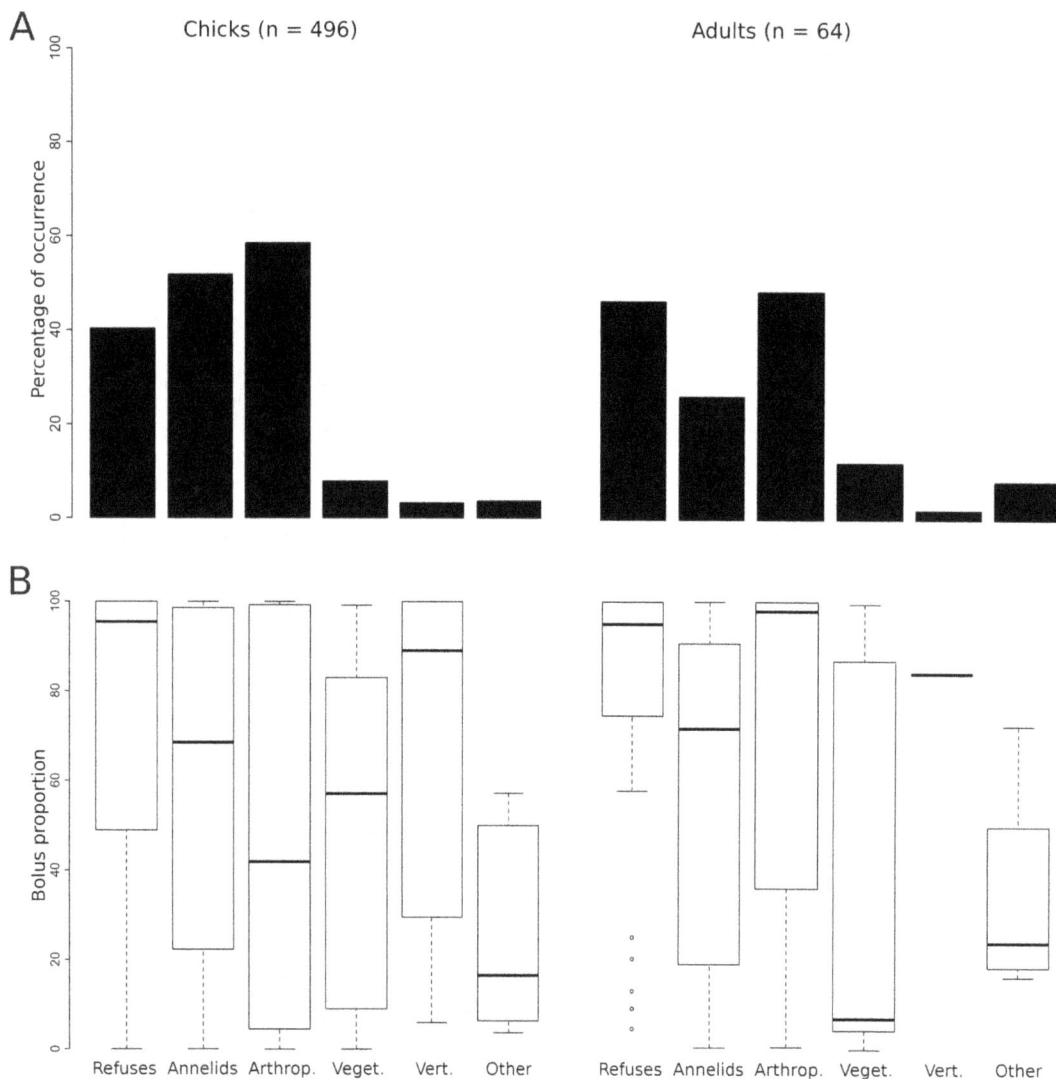

Figure 5. Diet of chicks and breeding adults of Ring-billed gulls. Diet of Ring-billed gull chicks (boli) and breeding adults (boli and stomach contents) at the Deslauriers Island colony, 2009–2010, expressed as the percentage of occurrence of each food category (a) and the proportion based on dry mass of each food category when present (b). Boxplots provide the first (bottom line), second (black midline) and third (top line) quartiles; whiskers extend to observations found up to 1.5 times the interquartile range; observations outside this range are indicated by empty dots.

(≥5 patches) within each circular band. We then plotted these two values for each band and each habitat to explore habitat selection as a function of distance.

The proportion of gulls observed foraging in a flock was considered as the time spent foraging in a given habitat [29]. This proportion was modeled as a two-column matrix, with the first column giving the number of gulls foraging and the second column giving the number of gulls involved in other activities for each flock, using a GLM with a binomial error distribution and logit link function (i.e., a logistic regression) using the stats package run in R. We also assessed whether the abundance of gulls in agricultural lands was related to the total number of tractors encountered along transects, which was considered an index of agricultural field work. This was done using a GLM with a Poisson error distribution and log link function (i.e., a Poisson regression) in R. This model included tractor number, transect location (South or North shore) and their interaction as explanatory variables.

Finally, we computed the proportion of boli containing at least one item of each food category for both chicks and adults. We then calculated the mean relative amount of each food item category when present in a bolus based on dry mass. The energy value of boli (kJ) was calculated for each gull collected at the Ste-Sophie landfill, in agricultural lands and at riparian sites by combining the dry mass of each item found in the stomach and their energy value. We compared the mean energy value of boli across habitats using an ANOVA.

Results

Characteristics of foraging trips

The mean number of foraging trips per day was greater during the rearing period (3.1 ± 1.0 trips/day, $\pm$SD) than during incubation (1.9 ± 0.8 trips/day; $t_{107} = -7.01$, $P<0.001$). The mean direct distance between the colony and the furthest location reached during a foraging trip (whether a stopover or not) was also greater during brood rearing (16.6 ± 12.4 km, maxi-

Table 4. Mean dry mass (g ± SD) and mean energy value (kJ) of nine food items gathered by sub-adult and adult Ring-billed Gulls in three habitat types (N = number of birds with food items present in their stomach).

| Food items | Mean dry mass (g) | | | |
| | Landfills | Agricultural lands | Riparian habitats | Energy |
	(N = 81)	(N = 54)	(N = 22)	(kJ)
Meat	1.84±4.71	-	0.87±2.81	30.1
Bread/rice	1.27±2.62	-	0.47±1.05	19.9
Potatoes/French fries	0.15±0.79	-	0.32±0.89	21.8
Miscellaneous refuse	0.79±2.69	-	-	22.6
Annelids	0.10±0.32	1.99±2.80	-	13.8
Arthropods	0.06±0.24	0.03±0.08	0.09±0.24	22.8
Corn/soybean grains	-	1.44±3.36	-	18.1
Vertebrates	0.01±0.11	-	0.46±0.61	19.4
Miscellaneous	0.10±0.32	0.08±0.35	-	21.3
TOTAL	4.47±6.37	3.55±3.68	2.36±3.39	-

mum = 63.5 km vs. 12.5±9.9 km, maximum = 42.4 km; t_{107} = −2.22, P = 0.03). Furthermore, the mean foraging distance traveled was greater during the rearing period compared to incubation (38.6±29.0 km, maximum = 156 km vs. 30.2±23.8 km, maximum 105 km; t_{107} = −2.55, P = 0.01). However, there was no difference in path sinuosity during the two periods (incubation: 1.2±0.2, maximum = 2.5; rearing: 1.2±0.2, maximum = 2.9; t_{107} = 1.38, P = 0.17). Finally, the mean trip duration was similar throughout the breeding period (incubation: 2.5±2.0 h, maximum = 9.6 h; rearing: 2.3±1.7 h, maximum = 12.4 h; t_{107} = 0.91, P = 0.36) but the trips lasted longer when a landfill was visited (3.5±1.8 h vs. 2.2±1.8 h; t_{107} = 5.95, P<0.0001). No significant effect of sex was found for the trip characteristics (all P>0.21).

Habitat selection

The composition of foraging and movement patches was highly variable (Table 1). Nevertheless, both movement and foraging patches were on average composed of smaller percentages of woodlots and of intensive and extensive cultures than what was found over the whole foraging range. An opposite trend was found for urban areas, waterbodies, landfills, and transhipment sites. While woodlots and urban areas covered a smaller proportion of foraging patches than movement patches, intensive cultures were relatively more important in foraging patches. Landfills and transhipment sites also occurred more often in foraging than movement patches. The distribution of residence time was strongly skewed to the right, with a peak under 100 s and a maximum reaching 19,377 s or 5.4 h (Fig. S2).

Model ranking based on AIC$_c$ remained similar when considering the sex of individuals and its interaction with habitat types or patch distance. Yet, we only show results for models without sex as they performed better with much less parameters (ΔAICc = 3.04). The best model (w_i = 0.813) included habitat types as well as the distance separating the foraging patch from the colony, the breeding stage and their two-way interactions with habitat types (Table 2). The model that also included rainfall and its interaction with habitat types scored as the second best model (w_i = 0.187), leaving barely any support from the data for the remaining models. Note that the same two models were selected with similar, strong levels of evidence for other residence time thresholds (60, 80, 120, and 140 s) and with patch radii of 200 and 400 m,

underlining the robustness of our results with respect to these two assumptions.

Ring-billed gulls had a greater probability of foraging in patches located farther from the colony (Table 3). The distribution of these patches with respect to their distance from the colony was skewed to the right and showed a noticeable mode at ~10 km notwithstanding habitat types (Fig. 2a). Not surprisingly, gulls strongly avoided foraging in patches that included large amounts of woodlots (Table 3). Patches containing urban areas were significantly avoided close to the colony but increasingly selected further away. In fact, gulls tended to forage in patches with more urban cover compared to movement patches when the birds were between 25 and 35 km from their nest site (Fig. 2b). Because the colony was surrounded by water and gulls foraged little near the colony, there was a significant overall avoidance of this habitat (Table 3). Nevertheless, there was nearly a significant positive interaction between waterbodies and distance. In fact, waterbodies were relatively more important in foraging patches compared to movement patches when the birds were at 12 km or more from the colony (Fig. 2c). As expected, Ring-billed gulls foraged to a greater extent in patches containing lawns on rainy days (Table 3). We observed a greater proportion of lawns in foraging than in movement patches at around 5 km and again between 27 and 35 km (Fig. 2d).

The probability that a gull foraged in a patch increased with the proportion of intensive cultures (i.e., cereal fields) during incubation but tended to decrease during chick rearing (Table 3). More specifically, there were more intensive cultures in foraging than in movement patches up to 23 km from the colony (Fig. 2e). Similarly, the likelihood that gulls foraged in patches with extensive cultures (i.e., hayfields and pastures) tended to decrease with increasing amounts of this habitat (Table 3). The effect of distance on the use of extensive cultures by foraging gulls was not important (Fig. 2f). Gull surveys conducted in agricultural landscapes support the above patterns as the presence of gulls in intensive agricultural lands was related to the occurrence of ploughing, harrowing, and sowing, which all took place during the incubation period (mid-April to mid-May; Fig. 3). Indeed, the number of gulls observed along transects in agricultural lands increased with cultivation activities as indexed by the number of operating tractors seen in the fields (Poisson regression: β ±

SE = 0.064±0.001, z = 53.9, P<0.01). Of 20,900 gulls counted along transects, 52% were observed on bare soil fields (ploughed or recently sown), 34% on cereal fields with short vegetation (< 10 cm), 8% on stubble cereal fields and the remaining 6% on recently mowed hayfields. Finally, Ring-billed gulls had a greater tendency to forage in patches where a landfill or transhipment site was present (Table 3) and this was especially true as distance from the colony increased (Fig. 4).

Foraging behaviour, diet, and energy

The mean proportion of time that Ring-billed gulls spent foraging varied among habitats (deviance = −1.5×10⁴; df = 1351, P<0.01). It was higher in agricultural lands (0.54±0.40) than in landfills and transhipment sites (0.17±0.20; z = −76.9, P<0.01), riparian habitats (0.12±0.25; z = −78.4, P<0.01), urban areas (0.15±0.32; z = −53.4, P<0.01) and on lawns (0.43±0.41; z = −3.1, P<0.01).

The four main food items (i.e., refuse, annelids, arthropods, and vegetation) were found in 40–60% of the boli collected from chicks reared on Deslauriers Island (Fig. 5a). The same items were found in the stomachs and boli of breeding adults, but in lower proportions (25–30%); it was compensated by a greater frequency of vertebrates and miscellaneous items. Yet, vertebrates occurred in less than 10% of the boli/stomachs in both chicks and adults. When refuse items were present, they contributed to a large proportion of the contents based on dry mass, unlike vegetation and miscellaneous items that usually represented a small proportion (Fig. 5b). The importance of annelids and arthropods was much more variable when present. Vertebrates were also quite variable in chick boli, whereas they clearly contributed to a very large proportion of the adult diet when they occurred.

Stomach contents from gulls collected in landfills were largely composed of fat meat typically found in refuse (Table 4). In agricultural lands, stomach contents were composed more or less equally of annelids and grains (soybean and corn). Stomach contents from riparian areas contained edible refuse, wild fishes, and arthropods. By pooling data on the relative importance of each food item and their respective energy content, we found that the mean energy value of stomach contents differed significantly among habitats ($F_{2,144}$ = 3.51, P = 0.03). It was significantly higher in landfills (112.8±169.8 kJ) than in agricultural lands (55.8±63.6 kJ) and riparian areas (56.5±97.5 kJ), which were not significantly different.

Discussion

By combining analyses of GPS-tracking data and information on the gulls' abundance, diet, and proportion of time spent foraging in different habitats, we found that the distance from the colony and habitat phenology had strong effects on the process of habitat selection by breeding Ring-billed gulls foraging in a heterogeneous environment. For instance, they positively selected areas managed intensively for agriculture at a distance up to about 23 km from the colony but only when fields were being ploughed, harrowed, or sown. Gulls also selected areas where landfills and transhipment sites were present, especially as the distance from the colony increased. The mean energy intake being significantly greater in landfills than in agricultural lands, these results clearly suggest a trade-off by Ring-billed gulls to balance their energy budget. The St. Lawrence River and its tributaries are often used as passageways when flying to and from the insular colony, which resulted in a general avoidance of this habitat as feeding site. Over 12 km, however, gulls may stop along the shores of the rivers and the lakes or feed on emergent insects over water resulting in a selection of this habitat.

Energy trade-offs in selected habitats

The spatial and temporal variation in food availability could not be measured across the 6,000-km² study area. Nevertheless, we believe that using energy as an index of food quality and the relative area covered by each habitat allowed us to assess the relative benefits of different habitats. The strong selection for intensive cultures during incubation corresponded to the period when fields were being cultivated and the new cereal shoots were still at a height that allowed the birds to feed without visual obstruction. This seems to be associated with the occurrence of short periods of high food availability. Although it is difficult to differentiate the confounding effects of the breeding stage from the timing of field work and food availability, the positive effect of soil preparation and seed sowing on the abundance of gulls in agricultural lands during the incubation period (vs. brood-rearing) supports the hypothesis that selection for a specific habitat is higher during the peak of food availability. During our surveys, most gulls foraged in bare soil fields as observed for Black-headed gulls [46]. Moreover, half of the gulls' diet in agricultural lands was made of annelids, which are more accessible when tractors are ploughing and harrowing. Sibly and McCleery [24] have shown a positive relationship between the abundance of Herring gulls (*Larus argentatus*) in agricultural lands and the biomass of earthworms near the ground surface. Yet, the averaged RSFs did not detect an effect of rainfall on the use of agricultural lands despite the positive effect of ground wetness on the availability of annelids and their use by gulls [24]. In agricultural fields, gulls rely on the presence of heavy machinery that cannot work on wet soils. This contrasts with the use of lawns by gulls that was strongly associated with rainfall. Although we could not sample birds using urban areas, the greater availability of annelids on rainy days on lawns and their use by gulls is well established [22]. The other half of the gulls' diet in agricultural lands was made of soybeans and corn, which availability increases when sowing takes place (e.g., seeds accidentally dropped along road and field edges when farmers fill their seeders and seeds sown in superficial ground; M. Patenaude-Monette, pers. obs.). Annelids, soybeans and corn composed a less energy-rich diet than the food gathered by gulls at landfills. Considering that gulls selected the intensive agricultural lands no further than 23 km, we suggest that the profitability of this habitat was limited by the travel costs associated with the distance from the colony and the relatively low energy value of the food.

Gulls selected areas comprising landfills or transhipment sites throughout the breeding season, a period during which food availability at these sites does not vary with time. Although the accessibility (distance from the colony and deterrence program effectiveness) and volume of refuse differed among sites, we could only account for variation in distance from the colony. The selection of landfills was stronger, but also more variable, as the distance increased. Thus, the selection of landfills was probably not constrained by their distance from the colony as was the selection of agricultural lands at the scale of the study area. Nevertheless, its high variability suggests that not all gulls used landfills and transhipment sites. Indeed, landfills and transhipment sites were present in less than 5% of foraging patches of all individuals. Moreover, when refuse food items occurred in boli, they accounted for a much larger proportion of the bolus than any other food items. Furthermore, both the mean bolus mass and the mean energy content of food were much higher in landfills than in any other habitats.

We can hypothesize that gulls incur higher travel costs when foraging in landfills, which are located farther from the colony than agricultural lands [47]. Habitat accessibility is indeed likely to be negatively correlated with the distance separating the foraging site from the nest as travel costs (time, energy) increase with distance [41], [48]. Accordingly, intensively managed agricultural lands may thus provide a profitable net energy gain to foraging gulls despite food items of lower energy value, at least during the incubation period. On the other hand, landfills with their more energy rich food may be valuable foraging sites despite their remoteness and are thereby selected by gulls. The stronger selection observed with increasing distance to the colony (up to 63 km) may result from the fact that the closest sites (<30 km from the colony) included two transhipment sites where refuse is less available than at landfills. Moreover, the Terrebonne landfill that received the largest tonnage of refuse and which is located the closest to the colony has a very effective deterrence program (É. Thiériot, unpublished data).

Time constraints in urban areas

Gulls are known to feed on refuse in commercial and residential areas and on handouts offered by citizens [49]. Nevertheless, we found that breeding Ring-billed Gulls avoided foraging in urban areas located <10 km from the colony, but showed the opposite trend at greater distances. This pattern may result from the profitability of urban areas as foraging sites, which likely depends on the type of development (e.g., residential, commercial, or industrial) and population density. The proportion of time foraging was indeed very low in urban areas where gulls adopted a sit-and-wait strategy to exploit spatially and temporally scattered feeding opportunities (e.g., people handouts and overfilled garbage bins). While the proportion of time foraging was comparable in urban areas and in landfills, foraging opportunities are probably much less predictable in the former habitat. Furthermore, commercial and residential areas of high population densities (i.e., with greater foraging opportunities) were located about 20 km from the colony, which is much further than the closest landfill or agricultural lands. Although urban refuse food may present high energy contents, the time to gather enough refuse is likely too long to make foraging trips to urban areas profitable, particularly during the rearing period when chicks are waiting to be fed at the colony [50], [51]. The situation may nevertheless be different during the post-breeding period when gulls are then actively using urban areas ([52], C. Girault and J.-F. Giroux, unpublished data).

Conclusion

Combining RSF to survey data, diet characterization, and calorimetric analyses allowed us to characterize habitat selection processes of a central-place forager from an energy trade-off perspective. It also shows that other factors such as predation risk associated to deterrence programs at landfills can also play a role in the process of habitat selection at large spatial scales as

suggested through the concept of landscape of fear [53]. This approach was applied to a species that had to move over a large area to find food in a heterogeneous environment where habitat profitability also varied in time. Despite the complexity brought up by travel costs and habitat sampling issues, we were able to show that classical optimal foraging theory can make qualitative predictions applicable at the landscape level. This adds to the few evidences that optimal foraging theory has the potential to be scaled-up to the landscape level as predicted by Lima and Zollner [6]. Moreover, once classical models will have been modified such that their constraints are adapted to large spatio-temporal scales (e.g., [11], [54], [55]), GPS data loggers will allow us to test these models by linking the foraging behaviour of individuals to their breeding performance [14]. Such progress would make significant strides toward understanding the links between movement behaviour, habitat selection, fitness, and population dynamics within heterogeneous landscapes. For instance, this approach could be applied to many gull populations around the world to link their dynamics to food availability through landfill, agriculture, and fishery management.

Supporting Information

Figure S1 Residence times of breeding Ring-billed gulls in relation with patch size. Mean coefficient of variation (CV) of residence times within circular patches of different radii centred on locations obtained by GPS data loggers ($N = 109$ birds).

Figure S2 Frequency distribution of residence times of breeding Ring-billed gulls. Residence times were established for 200-m radius circular patches centred on locations obtained by GPS data loggers ($N = 109$ birds).

Table S1 Characteristics of individual Ring-billed gulls tracked during the study.

Acknowledgments

This research was conducted in partnership with the municipalities of Terrebonne, Repentigny, Laval, Charlemagne, Mascouche, Saint-Hippolyte, Sainte-Sophie, Sainte-Anne-des-Plaines, and Saint-Lin-Laurentides. We thank F. St-Pierre and M. Tremblay for field and laboratory assistance and R. Zamojska for diet and calorimetric analyses. S. Benhamou provided valuable advice about residence time computation while A. Desrochers, P. Peres-Neto, and G. Bastille-Rousseau commented an earlier draft of this manuscript.

Author Contributions

Conceived and designed the experiments: MPM JFG MB. Performed the experiments: MPM JFG. Analyzed the data: MPM MB. Contributed reagents/materials/analysis tools: JFG MB. Wrote the paper: MPM. Provided methodological and editorial comments: JFG MB. Supervised the overall field study: JFG.

References

1. Stephen DW, Krebs JR (1986) Foraging theory. Princeton: Princeton University Press. 247 p.
2. Giraldeau L-A, Caraco T (2000) Social foraging theory. Princeton: Princeton University Press. 376 p.
3. Charnov EL (1976) Optimal foraging, the marginal-value theorem. Theor Popul Biol 9: 129–136.
4. Orians GH, Pearson NE (1979) On the theory of central place foraging. In: Horn DJ, Stairs GR, Mitchell DR, editors. Analysis of ecological systems. Columbus: Ohio State University Press. 155–177.
5. Nonacs P (2001) State dependent behavior and the marginal value theorem. Behav Ecol 12: 71–83.
6. Lima SL, Zollner PA (1996) Towards a behavioral ecology of ecological landscapes. Trends Ecol Evol 11: 131–135.
7. Zollner PA, Lima SL (1999) Search strategies for landscape-level interpatch movements. Ecology 80: 1019–1030.
8. Nathan R, Getz WM, Revilla E, Holyoak M, Kadmon R, et al. (2008) A movement ecology paradigm for unifying organismal movement research. P Natl Acad Sci-Biol 105: 19052–19059.

9. Schoener TW (1979) Generality of the size-distance relation in models of optimal feeding. Am Nat 114: 902–914.

10. Olsson O, Brown JS, Helf KL (2008) A guide to central place effects in foraging. Theor Popul Biol 74: 22–33.

11. Bernstein C, Kacelnik A, Krebs JR (1991) Individual decisions and the distribution of predators in a patchy environment. II. The influence of travel costs and structure of the environment. J Anim Ecol 60: 205–225.

12. Beauchamp G, Bélisle M, Giraldeau L-A (1997) Influence of conspecific attraction on the spatial distribution of learning foragers in a patchy habitat. J Anim Ecol 66: 671–682.

13. Owen-Smith N, Fryxell JM, Merrill EH (2010) Foraging theory upscaled: the behavioural ecology of herbivore movement. Philos T Roy Soc B 365: 2267–2278.

14. Gaillard J-M, Hebblewhite M, Loison A, Fuller M, Powell R, et al. (2010) Habitat-performance relationships: finding the right metric at a given spatial scale. Philos T Roy Soc B 365: 2255–2265.

15. Manly BFJ, McDonald LL, Thomas DL, McDonald TL, Erickson WP (2002) Resource Selection by Animals: Statistical Design and Analysis for Field Studies. Second Edition, London: Kluwer Academic, 221 p.

16. Barraquand F, Benhamou S (2008) Animal movements in heterogeneous landscapes: identifying profitable places and homogeneous movement bouts. Ecology 89: 3336–3348.

17. Beyer HL, Haydon DT, Morales JM, Frair JL, Hebblewhite M, et al. (2010) The interpretation of habitat preference metrics under use-availability designs. Philos T Roy Soc B 365: 2245–2254.

18. Bastille-Rousseau G, Fortin D, Dussault C (2010) Inference from habitat-selection analysis depends on foraging strategies. J Anim Ecol 79: 1157–1163.

19. Fauchald P, Tveraa T (2006) Hierarchical patch dynamics and animal movement pattern. Oecologia 149: 383–395.

20. Freitas C, Kovacs KM, Lydersen C, Ims RA (2008) A novel method for quantifying habitat selection and predicting habitat use. J Appl Ecol 45: 1213–1220.

21. Fauchald P, Tveraa T (2003) Using first-passage time in the analysis of area-restricted search and habitat selection. Ecology 84: 282–288.

22. Brousseau P, Lefebvre J, Giroux J-F (1996) Diet of ring-billed gull chicks in urban and non-urban colonies in Quebec. Colon Waterbird 19: 22–30.

23. Pollet IL, Shutler D, Chardine J, Ryder JP (2012) Ring-billed Gulls (*Larus delawarensis*). The Birds of North America Online. Ithaca: Cornell Lab of Ornithology. Available: http://bna.birds.cornell.edu.bnaproxy.birds.cornell.edu/bna/species/033doi:10.2173/bna.33. Accessed 5 July 2013.

24. Sibly RM, McCleery RH (1983) The distribution between feeding sites of herring gulls breeding at Walney island, U.K. J Anim Ecol 52: 51–68.

25. Belant JL, Ickes SK, Seamans TW (1998) Importance of landfills to urban-nesting herring and ring-billed gulls. Landscape Urban Plan 43: 11–19.

26. Thiériot E, Molina P, Giroux J-F (2012) Rubber shots not as effective as selective culling in deterring gulls from landfill sites. Appl Anim Behav Sci 142: 109–115.

27. Fridolfsson AK, Ellegren H (1999) A simple and universal method for molecular sexing of non-ratite birds. J Avian Biol 30: 116–121.

28. Frair JL, Fieberg J, Hebblewhite M, Cagnacci F, DeCesare NJ, et al. (2010) Resolving issues of imprecise and habitat-biased locations in ecological analyses using GPS telemetry data. Philos T Roy Soc B 365: 2187–2200.

29. Altmann J (1974) Observational study of behavior: Sampling methods. Behaviour 49: 227–267.

30. Racine F, Giraldeau L-A, Patenaude-Monette M, Giroux J-F (2012). Evidence of social information on food location in a ring-billed gull colony, but the birds do not use it. Anim Behav 84: 175–182.

31. Batschelet E (1981) Circular statistics in biology. London: Academic Press. 371p.

32. Kotliar NB, Wiens JA (1990) Multiple scale of patchiness and patch structure: a hierarchical framework for the study of heterogeneity. Oikos 59: 253–260.

33. Bellier E, Certain G, Planque B, Monestiez P, Bretagnolle V (2010) Modelling habitat selection at multiple scales with multivariate geostatistics: an application to seabirds in open sea. Oikos 119: 988–999.

34. ESRI (2009) ArcGIS version 9.3.1. Redlands, California.

35. FADQ (2010) Bases de données des cultures assurées 2009 et 2010. Direction des ressources informationnelles, la Financière agricole du Québec, Saint-Romuald, Canada. Available: http://www.fadq.qc.ca/geomatique/professionnels_en_geomatique/base_de_donnees_de_cultures_assurees.html. Accessed 5 December 2011.

36. Natural Resources Canada (2009) Land Cover, Circa 2000, Vector. NRCan, Earth Sciences Sector, Centre for Topographic Information, Sherbrooke, Canada. Available: http://geobase.ca/geobase/en/find.do?produit = csc2000v. Accessed 12 January 2012.

37. Natural Resources Canada (2010) CanVec version 1.1. NRCan, Earth Sciences Sector, Centre for Topographic Information, Sherbrooke, Canada. Available: ftp://ftp2.cits.rncan.gc.ca/pub/canvec/. Accessed 12 January 2012.

38. Environment Canada (2010) National Climate Data and Information Archive. Government of Canada website. Available: http://climate.weather.gc.ca. Accessed 10 January 2011.

39. Shamoun-Baranes J, van Loon E (2006) Energetic influence on gull flight strategy selection. J Exp Biol 209: 3489–3498.

40. Burnham KP, Anderson DR (2002) Model selection and multimodel Inference: a practical information-theoretic approach. Second Edition, New-York: Springer-Verlag. 448 p.

41. Matthiopoulos J (2003) The use of space by animals as a function of accessibility and preference. Ecol Model 159: 239–268.

42. Bates D, Maechler M, Bolker B (2011) Lme4: Linear mixed-effects models using S4 classes. R package version 0.999375-39. Available: http://CRAN.R-project.org/package = lme4. Accessed 20 June 2011.

43. R Development Core Team (2011) R: a language and environment for statistical computing. Version 2.12.2, Vienna. http://www.r-project.org/.

44. Gaston AJ, Ydenberg RC, Smith GEJ (2007) Ashmole's halo and population regulation in seabirds. Mar Ornithol 35: 119–126.

45. Elliot KE, Woo KJ, Gaston AJ, Benvenuti S, Dall'Antonia L, et al. (2009) Central-place foraging in an arctic seabird provides evidence for Storer-Ashmole's halo. Auk 126: 613–625.

46. Schwemmer P, Garthe S, Mundry R (2008) Area utilization of gulls in a coastal farmland landscape: habitat mosaic supports niche segregation of opportunistic species. Landscape Ecol 23: 355–367.

47. Wilson RP, Quintana F, Hobson VJ (2012). Construction of energy landscapes can clarify the movement and distribution of foraging animals. P Roy Soc Lond B Bio 279: 975–980.

48. Rosenberg DK, McKelvey KS (1999) Estimation of habitat selection for central place foraging animals. J Wildl Manage 63: 1028–1038.

49. Belant JL (1997) Gulls in urban environments: landscape-level management to reduce conflict. Landscape Urban Plan 38: 245–258.

50. Bukacinska M, Bukacinski D, Spaans AL (1996) Attendance and diet in relation to breeding success in Herring gulls (*Larus argentatus*). Auk 113: 300–309.

51. Shaffer SA, Costa DP, Weimerskirch H (2003) Foraging effort in relation to the constraints of reproduction in free-ranging albatrosses. Funct Ecol 17: 66–74.

52. Maciusik B, Lenda M, Skorka P (2010) Corridors, local food resources, and climatic conditions affect the utilization of the urban environment by the Black-headed Gull *Larus ridibundus* in winter. Ecol Res 25: 263–272.

53. Searle KR, Stokes CJ, Gordon IJ (2008) When foraging and fear meet: using foraging hierarchies to inform assessments of landscapes of fear. Behav Ecol 19: 475–482.

54. Amano T, Ushiyama K, Moriguchi S, Fujita G, Higuchi H (2006) Decision-making in group foragers with incomplete information: Test of individual-based model in Geese. Ecol Monogr 76: 601–616.

55. Mueller T, Fagan WF (2008) Search and navigation in dynamic environments – from individual behaviors to population distributions. Oikos 117: 654–664.

15

Eco-Label Conveys Reliable Information on Fish Stock Health to Seafood Consumers

Nicolás L. Gutiérrez[1]*, **Sarah R. Valencia**[2], **Trevor A. Branch**[3], **David J. Agnew**[1], **Julia K. Baum**[4], **Patricia L. Bianchi**[1], **Jorge Cornejo-Donoso**[5,6], **Christopher Costello**[2], **Omar Defeo**[7], **Timothy E. Essington**[3], **Ray Hilborn**[3], **Daniel D. Hoggarth**[1], **Ashley E. Larsen**[8], **Chris Ninnes**[1], **Keith Sainsbury**[9], **Rebecca L. Selden**[8], **Seeta Sistla**[8], **Anthony D. M. Smith**[10], **Amanda Stern-Pirlot**[1], **Sarah J. Teck**[8], **James T. Thorson**[3], **Nicholas E. Williams**[11]

1 Marine Stewardship Council, London, United Kingdom, 2 Bren School of Environmental Science and Management, University of California Santa Barbara, Santa Barbara, California, United States of America, 3 School of Aquatic and Fishery Sciences, University of Washington, Seattle, Washington, United States of America, 4 Department of Biology, University of Victoria, Victoria, British Columbia, Canada, 5 Interdepartmental Graduate Program in Marine Science, Marine Science Institute, University of California Santa Barbara, Santa Barbara, California, United States of America, 6 Universidad Austral de Chile, Centro Trapananda, Coyhaique, Chile, 7 UNDECIMAR, Facultad de Ciencias, Montevideo, Uruguay, 8 Department of Ecology, Evolution and Marine Biology, University of California Santa Barbara, Santa Barbara, California, United States of America, 9 University of Tasmania, Tasmanian Aquaculture & Fisheries Inst, Taroona, Tasmania, Australia, 10 Commonwealth Scientific and Industrial Research Organization, Wealth from Oceans Flagship, Hobart, Tasmania, Australia, 11 Department of Anthropology, University of California Santa Barbara, Santa Barbara, California, United States of America

Abstract

Concerns over fishing impacts on marine populations and ecosystems have intensified the need to improve ocean management. One increasingly popular market-based instrument for ecological stewardship is the use of certification and eco-labeling programs to highlight sustainable fisheries with low environmental impacts. The Marine Stewardship Council (MSC) is the most prominent of these programs. Despite widespread discussions about the rigor of the MSC standards, no comprehensive analysis of the performance of MSC-certified fish stocks has yet been conducted. We compared status and abundance trends of 45 certified stocks with those of 179 uncertified stocks, finding that 74% of certified fisheries were above biomass levels that would produce maximum sustainable yield, compared with only 44% of uncertified fisheries. On average, the biomass of certified stocks increased by 46% over the past 10 years, whereas uncertified fisheries increased by just 9%. As part of the MSC process, fisheries initially go through a confidential pre-assessment process. When certified fisheries are compared with those that decline to pursue full certification after pre-assessment, certified stocks had much lower mean exploitation rates (67% of the rate producing maximum sustainable yield vs. 92% for those declining to pursue certification), allowing for more sustainable harvesting and in many cases biomass rebuilding. From a consumer's point of view this means that MSC-certified seafood is 3–5 times less likely to be subject to harmful fishing than uncertified seafood. Thus, MSC-certification accurately identifies healthy fish stocks and conveys reliable information on stock status to seafood consumers.

Editor: Myron Peck, University of Hamburg, Germany

Funding: SRV, JCD, AEL, RLS, SS, SJT, and NEW thank the Henry Luce Foundation and the National Center for Ecological Analysis and Synthesis, which is funded by National Science Foundation (NSF) Grant EF-0553768, the University of California, Santa Barbara, and the State of California. TAB was funded by NSF grant 1041570. The funders had no role in study design, data collection and analysis, decision to publish, or preparation of the manuscript.

Competing Interests: The authors have declared that no competing interests exist.

* E-mail: nicolas.gutierrez@msc.org

Introduction

The global per-capita demand for seafood has reached an all-time high, and is likely to continue to increase [1]. Wild capture seafood harvest has also peaked, and while management measures have led to rebuilding in some fish stocks, one-third of the world's well-studied fisheries are overfished [1–3]. To maintain or enhance wild fish supplies on a sustainable basis management agencies and governments must rebuild fisheries whose stocks are at low biomass and maintain healthy stocks at or above sustainable levels. Fisheries and conservation objectives can be attained by redundancy in management actions, including catch controls, gear modifications, closed areas, and community-based management,

depending on local context and specific features [2,4]. One way of influencing these fishery practices is through market-based approaches such as "eco-labeling" which aim to harness consumer preferences to increase market demand, and often prices, for well-managed fisheries and diminish demand for others [5,6]. To further this aim, national and global schemes designed to allow consumers to make informed choices when purchasing seafood have proliferated [5]. These efforts include awareness campaigns such as consumer guides produced by Monterey Bay Aquarium, World Wildlife Fund and Greenpeace, the risk of extinction Red List categories of the IUCN, and certification and eco-labeling programs such as the Marine Stewardship Council (MSC) and Friend of the Sea [7].

Unlike some consumer awareness campaigns, certification programs such as the MSC consider a fishery or fish stock, rather than a species, to be the primary unit of certification. This acknowledges variation in harvest practices among fleets and recognizes those that adopt environmentally sound activities [7]. In theory, eco-labels convey information about these improved fishing practices to consumers, who then make choices about what seafood to buy based on this information. Consumer preference can result in increased prices [8] and indirect non-economic benefits for fishers [9,10] and access to markets looking to exclusively source certified fish products [7,11]. Moreover, leading supermarkets and restaurant chains recognize that consumers increasingly expect retailers to make responsible purchasing decisions as part of their corporate social responsibility, and may require third-party certification in the products they source. These act as incentives for improvement in uncertified fisheries and for continued stewardship in certified fisheries. The conservation value of eco-labels, however, relies on their ability to convey accurate information to consumers about the sustainability of fisheries.

The Marine Stewardship Council is the most prominent global fisheries eco-label program. It arose from a partnership between the World Wildlife Fund and Unilever in 1996, and has operated as an independent non-profit since 1999. There are currently 132 MSC-certified fisheries and 141 more at different stages of the certification process. MSC-certified seafood covers 10% of the annual global harvest of wild capture fisheries and more than 13,000 products, by far the highest representation of eco-labeled seafood in global markets [7,8]. The MSC's rapid growth has stoked debate about what constitutes a sustainable fishery, and the decisions to certify certain fisheries as sustainable have been scrutinized. Recent criticisms have questioned the rigor of the MSC's certification standards for ecosystem impacts of fisheries that damage habitats or result in high levels of bycatch [12–16]. There have been calls to focus certification on small-scale fisheries [12,16–18] based on the perception that they have a lower environmental impact than industrial fisheries [16], despite a paucity of data with which to assess the sustainability of small-scale fisheries. The strongest criticism of the MSC, however, argues that its certification standards fail to accurately identify healthy stocks, with several case studies cited by critics as not being sustainably managed, including Pacific Hake (*Merluccius productus*) and Eastern Bering Sea (EBS) walleye pollock (*Theragra chalcogramma*) [12,16].

The MSC certifies fisheries as sustainable only if they score highly on each of 3 principles [19]: (1) fishing should be conducted in a way that prevents overfishing (depletion of exploited populations beyond biological limits) through the use of target reference points that should maintain the stock at or above the biomass that produces the maximum sustainable yield (MSY), and overexploited stocks must be demonstrably on a path to recovery; (2) fishing operations must maintain the structure, diversity, function, and productivity of associated ecosystems; and (3) the management system must respect national and international regulations. Fisheries applying for MSC certification first undergo a confidential pre-assessment stage to evaluate their potential for meeting the certification standard. Based on this evaluation, fishing industry groups decide whether to undergo a public full assessment by an independent third party. Fisheries meeting the above standards are certified for five years and undergo annual surveillance audits. Those that meet the standard but are weak in certain areas can be certified if they commit to and demonstrate progress toward meeting agreed conditions on improvement. Thus, fisheries must demonstrate continued adherence to, and

improvement in, a variety of aspects of sustainability to maintain their certification status.

The term "sustainable" is difficult to define because it encompasses ecological, social, and economic components. At a basic level, however, a renewable resource must be extracted no faster than the level at which it can replace itself for it to be considered sustainable. If certified fisheries are no better at identifying and responding to low biomass levels than uncertified fisheries, then eco-labeling is unlikely to catalyze widespread improvements in fisheries management [14]. Therefore the decline of some certified stocks has cast doubt on the validity of the information conveyed by the MSC label.

Here we assess the performance of fish stocks against Principle 1 (targeted population status), specifically evaluating the status and harvest levels of fish stocks targeted by MSC-certified fisheries, because recent criticisms of MSC have questioned whether these fisheries actually target healthy stocks [12,16]. While all three principles are equally weighted criteria in the MSC's assessment process, Principles 2 and 3 address effects of fishing whose impacts can be difficult to measure directly. Thus assessing the performance of certified vs. uncertified fish stocks is a critical first step in evaluating the effectiveness of the MSC's certification standards.

Methods

Data

We compiled time series of catch data and model estimates of biomass and fishing mortality rates for all stocks where the above information was available (45 certified stocks, Table S1, and 179 uncertified stocks, Table S2). Certified stocks managed under different schemes than single-species MSY (e.g., salmon and invertebrates) or without biomass time series and thus qualitatively assessed under a risk-based framework [19] (e.g., small-scale and data-deficient fisheries) were excluded from our analysis. However, given that some stocks are targeted by multiple certified fisheries, analyzed stocks represented 62% ($n = 82$) of the total certified fisheries ($n = 133$) and 85% of the certified landings (4.5 million tons).

The majority of data was sourced from the RAM Legacy Stock Assessment Database (20), which represents the largest global stock assessment database currently available, but status was updated if newer stock assessments were available (Table S1 and Table S2). For each stock we recorded the biomass (B_{MSY}) and exploitation rate (u_{MSY}) or fishing mortality (F_{MSY}) that results in MSY from published stock assessments. When estimates of one or both reference points were not available, they were estimated by fitting Schaefer surplus production models [2,20] to time series of biomass estimates using a maximum likelihood approach in AD Model Builder [21]. Table S1 includes the method used for each certified stock, and Table S2 lists the method for each uncertified stock. Finally, we collected time series of biomass estimates from stock assessments in order to compare long-term trends in biomass relative to target B_{MSY} for certified and uncertified stocks.

The confidential MSC pre-assessment process screens out many applicants that are unlikely to achieve full certification. Since 1997, 447 fisheries have applied for pre-assessment. Of these fisheries, 55% were not recommended to enter full assessment due either to major management weaknesses (35% of the 447 fisheries) or because they had low biomass, high exploitation rates, or insufficient information to judge stock status (20% of the 447 fisheries) [22]. Of the fisheries not recommended to enter full assessment, 25 stocks for which information was available were included in our analysis (Text S1). Thus our analysis compared certified stocks with both uncertified stocks and the 25 stocks (counted among the 179) that a pre-assessment had suggested would fail the Principle 1 standard. Due to confidentiality, details about the identities of these stocks cannot be released.

Definitions and Analysis

According to international agreements and many national laws, fish stocks should be maintained at or rebuilt to a size that can support MSY [23,24]. This biomass, denoted B_{MSY}, is typically 20–50% of the average population biomass in the absence of fishing [25]. The corresponding annual exploitation rate (u, catch divided by total biomass) that stabilizes the stock around B_{MSY} is u_{MSY}. We follow international convention by using MSY-based reference points, which remain the most widely used method of assessing whether stocks are overfished or not. For example, the UN Convention on the Law of the Sea [23] requires countries to rebuild to MSY levels, and the United Nations Fish Stock Agreement [24] specifies F_{MSY} as a reference point. We acknowledge that there is some debate about whether MSY reference points are an appropriate target for sustainability, and that alternative methods may result in differing estimates of B_{MSY} and F_{MSY}, but these are larger issues than can be addressed in this paper.

For a fishery to be MSC-certified, biomass should be at or fluctuating around B_{MSY}, or if consistently below B_{MSY}, should be under a rebuilding plan (i.e., $F<F_{MSY}$) that will lead to recovery of the stock in the near future [19]. There is also a minimum level of biomass, or limit reference point, below which stocks are considered overfished and certification cannot be obtained irrespective of any rebuilding plan. Limit reference points must be set such that if a stock is maintained above these, there is a very low risk of impaired recruitment. These MSY reference points are currently the most informative benchmark with which to assess status across a global sample of fish stocks [2,20].

We compared the biomass status and exploitation rate in relation to MSY targets of certified, uncertified, and non-recommended fish stocks by plotting B/B_{MSY} vs. u/u_{MSY} or F/F_{MSY} and using kernel density smoothing functions to describe the probability of occurrence in each quadrant. To determine whether B/B_{MSY} and u/u_{MSY} were significantly different between groups we used re-sampling inference (100,000 times without replacement), which allows us to assess how often a difference of the observed magnitude or larger would arise by chance. We also estimated the proportion of stocks in each group that met or exceeded biomass and harvest targets and that were below $0.5B_{MSY}$ (as a proxy for the U.S. legal definition of overfished, or the point below which recruitment can be impaired for certain stocks; [26]) and above a more conservative target of $1.3B_{MSY}$ [27]. The long-term performance of certified and uncertified stocks in relation to B_{MSY} was assessed using time series data from 1970 to the present (available for 165 uncertified stocks and 31 certified stocks). Using an autoregressive model we tested for differences in the conditional mean of B/B_{MSY} between certified and uncertified stocks over time. The model structure was selected through Akaike's Information Criteria (AIC) and model parameters were estimated using Ordinary Least Squares.

Results and Discussion

Statuses of Certified and Uncertified Stocks

Our analysis indicates that the ratio of current biomass to the biomass at MSY ($B_{current}/B_{MSY}$) is significantly different between certified and uncertified fisheries (1.25 vs. 0.87; $P<0.005$), and between certified and non-recommended fisheries (1.25 vs. 0.48; $P<0.005$; Table S3 and Table S4). We found 74% of the certified stocks to be currently above sustainable target biomass levels (i.e., $B_{current}>B_{MSY}$; Fig. 1A) (Text S1), compared with 44% of

uncertified stocks, and 16% of non-recommended fisheries (Fig. 1B; Table S3). Given that certification assumes harvesting at MSY levels, which will result in biomass fluctuating around B_{MSY}, we would expect 50% of stocks above B_{MSY} and 50% below B_{MSY}. Thus, our finding that three quarters of stocks targeted by MSC-certified fisheries are above B_{MSY} suggests that managers of these stocks are often aiming to ensure biomass is not kept near B_{MSY} but above B_{MSY}. Additionally, 82% of certified stocks had current exploitation rates that are expected to maintain the stocks at B_{MSY} or allow for rebuilding to B_{MSY} (i.e., $u_{current}<u_{MSY}$) compared with 65% of uncertified stocks and 52% of non-recommended stocks.

Non-recommended Stocks

An analysis of non-recommended stocks revealed that 52% were below $0.5B_{MSY}$ and 48% had exploitation rates higher than u_{MSY} (Fig. 1C). These levels are significantly worse than those for certified stocks ($P<0.005$; Table S3), providing further evidence that the pre-certification process screens out stocks that do not meet internationally recognized standards for stock health. It may seem puzzling that such fisheries would apply for certification in the first place, but this may reflect differing perceptions of what sustainability means, as well as different reasons for pursuing certification. Non-recommended fisheries also show poorer stock status when compared with the global sample of uncertified fisheries. Reasons for such differences may be due to a lack of market incentives to pursue certification for some fisheries targeting healthy stocks [10], or due to weaknesses in other aspects of uncertified fisheries such as management systems or bycatch considerations [22]. Moreover, fisheries with known poor stock status may use pre-assessment as an endorsement to conduct formal stock assessments or as benchmark for amendment of management plans towards MSC certification [22].

Comparing Rates of Poor Performance

Fisheries of most concern are those with biomass below B_{MSY} and continued high harvest rates, which hinder stock rebuilding to sustainable levels ($u>u_{MSY}$; top left sector of Fig. 1A). Four certified stocks (9% of those analyzed) fell into this category, including North Sea saithe (*Pollachius virens*), North Sea sole (*Solea solea*), Atlantic Iberian sardine (*Sardina pilchardus*) and deep-water Cape hake (*Merluccius paradoxus*) (Table S1), compared with 51 (28%) uncertified stocks (Table S3) and 11 (44%) non-recommended stocks. In other words, certified fisheries are 3–5 times less likely to be subject to harmful overfishing than uncertified fisheries. At the time of the original MSC assessment, all three European stocks met certification requirements, as they were above the limit reference points defined by the International Council for the Exploration of the Seas (ICES). However, for the Iberian sardine, the mandatory annual audit revealed poor stock condition (i.e., below the limit reference point where recruitment to the population could be impaired) causing the suspension of the MSC certificate in January 2012, which will remain in place until the stock recovers. The other 3 stocks (North Sea saithe, North Sea sole and deep-water Cape hake) are currently well above their respective limit reference points as defined by the relevant advisory bodies [14,28].

The MSC Standard requires stocks to be above a limit reference point, which itself is above the point at which recruitment is impaired, and requires that if this point is not empirically determined, a biomass level of $0.5B_{MSY}$ could be used as an acceptable proxy. The U.S. also uses $0.5B_{MSY}$ as a default Minimum Stock Size Threshold (MSST), below which stocks are classified as "overfished" [26]. In our analysis, 46 (27%) un-

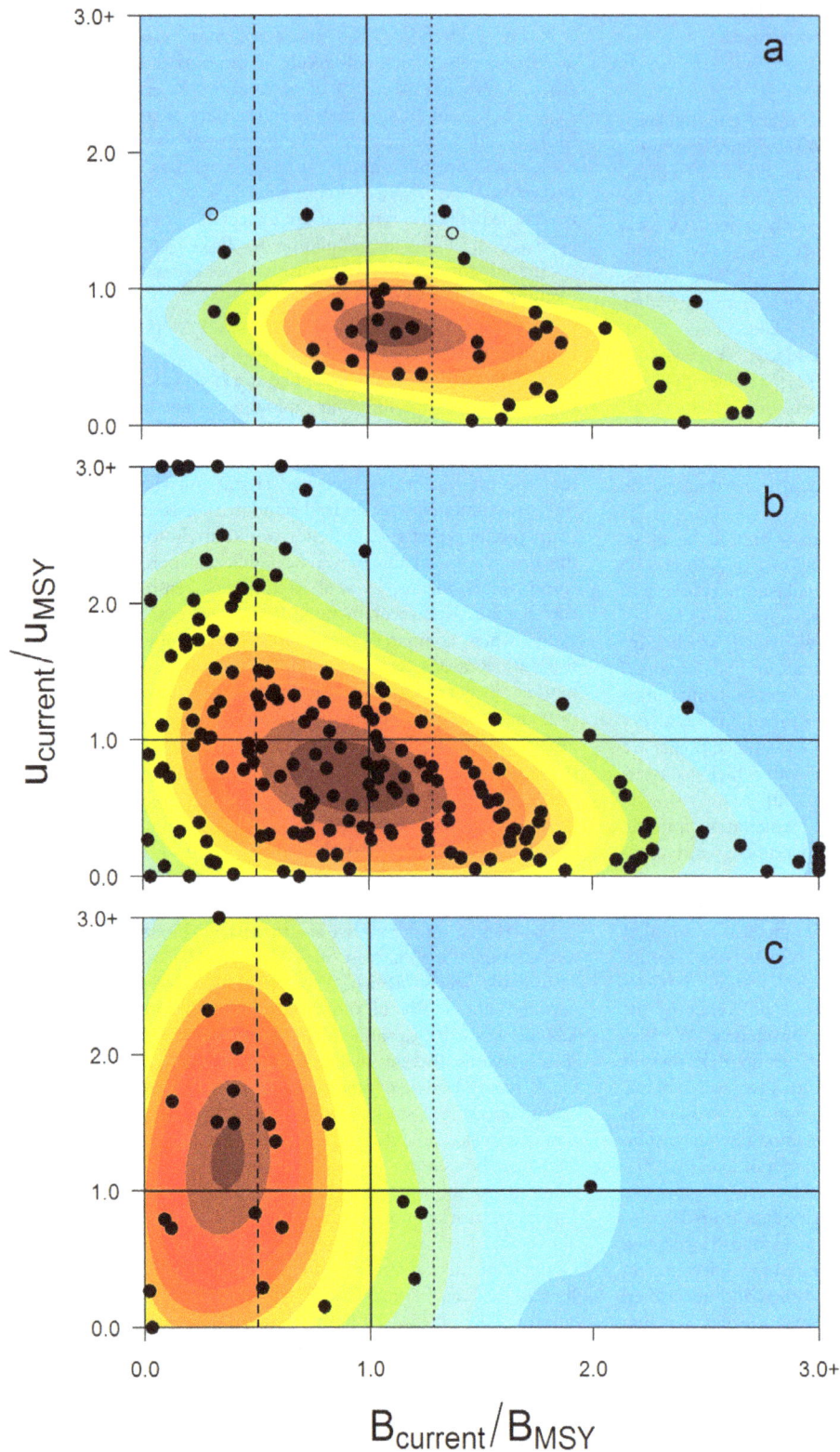

Figure 1. Sustainability of certified and uncertified seafood. Current (i.e., most recent year with available information) biomass and exploitation rate for (**A**) individual certified ($n = 45$); (**B**) all uncertified ($n = 179$); and (**C**) uncertified stocks that went through pre-assessment and were not recommended for certification ($n = 25$). Data are scaled relative to BMSY and uMSY or FMSY (the biomass and exploitation rates or fishing mortality rates that produce maximum sustainable yield). Contour colors show probability of occurrence (red indicates the highest probability and blue the lowest). Vertical and horizontal solid lines represent reference points common to all fisheries ($B/BMSY = 1$ and $u/uMSY$ or $F/FMSY = 1$). Dotted lined represents $B = 1.3BMSY$ and dashed line $B = 0.5BMSY$. Footnote: New assessments for some fish stocks were released while this paper was in press, but this figure was not updated to maintain consistency in year of release.

certified stocks had biomass levels below $0.5B_{MSY}$, compared with 4 (9%) certified stocks: North Sea saithe and Atlantic Iberian sardine as previously described, Eastern Baltic cod (*Gadus morhua*) and North Sea haddock (*Melanogrammus aeglefinus*). With the exception of the Atlantic Iberian sardine, which has had its MSC certificate removed, all 3 stocks are above the limit reference points defined by ICES without signs of recruitment overfishing [29]. Moreover, Eastern Baltic cod is under a strict rebuilding plan, which has resulted in an 80% reduction in exploitation rate and a three-fold increase in spawning biomass over the last five years [30] and North Sea haddock has experienced a reduction of the exploitation rate of 55% in the last 5 years [31].

As a target, B_{MSY} is often conditionally defined as the biomass that produces the maximum sustainable yield "under existing environmental conditions" [32]. Thus, even under an MSY control rule where B_{MSY} is the target stock biomass level, natural variation in productivity will result in stock fluctuations, being half of the time below B_{MSY} and half of the time above B_{MSY} [33]. Similarly, harvest rates might exceed u_{MSY} in some years. This natural population variability (typically driven by recruitment fluctuations in marine fishes) precludes the possibility of keeping stocks at constant levels. Successful management therefore must include continuous monitoring and an ability to adjust and enforce harvest levels [33,34]. It is for these reasons that stocks can be certified, and retain their certification even when they drop below B_{MSY}, provided they include proper management feedbacks and precautionary limits. Specifically, fisheries must have a management system in place that will detect decreases in biomass and respond by reducing the exploitation rate to a level that should enable recovery (i.e., harvest control rules). These feedback mechanisms present in certified fisheries result in well-managed fisheries fluctuating around their target reference points. Fisheries must meet these and other criteria in terms of ecosystem impacts and compliance with local and international laws to achieve and maintain certification [19].

Weighting Results by Stock Size

By using individual fish stocks as our unit of analysis we weight all stocks equally. However, this does not account for the fact that their total biomass can differ by several orders of magnitude. Recent annual landings of certified stocks range from 7 metric tons for the North-eastern inshore sea bass (*Dicentrarchus labrax*) fishery to 1.7 million tons for the Pacific skipjack (*Katsuwonus pelamis*) fishery (Table S1). While 179 uncertified fisheries are considered in the present analysis, their combined landings are lower than the total landings of the 45 certified fisheries analyzed (6.8 vs. 8.0 million metric tons; Tables S1 and S2). When we compared the biomass and exploitation rates in relation to MSY reference points for the 10 largest certified and uncertified stocks we found that 8 of the certified stocks, including the largest which represent almost 6 million metric tons of landed seafood, are at or above B_{MSY}, and are harvested at rates that should maintain the stocks above or fluctuating around their reference points (i.e., $u_{current} \leq u_{MSY}$; Table 1). Notably, EBS walleye pollock, which has been the target of many criticisms due to declines in biomass since certification [11,15,17], currently has a biomass level 25% higher than the target level (B_{MSY}) and an exploitation rate that is less than half of u_{MSY} (Table 1). In contrast, most of the 10 largest uncertified stocks have biomass levels substantially lower than B_{MSY} and exploitation rates considerably higher than u_{MSY} (Table 1).

Evaluating Performance against Conservation Targets

The sustainable yield in an ecosystem context depends on both the trophic level of the species [35] and the structure of the ecosystem [36]. As a result, the target biomass associated with an ecologically sustainable yield is unknown for each stock, and may be higher or lower than B_{MSY} depending on the species. To account for this, we also considered the more conservative target biomass of $1.3 B_{MSY}$ [37]. We found that 49% of the certified stocks were above $1.3 B_{MSY}$, compared with 29% of the uncertified fisheries and 4% of non-recommended fisheries (Table S3). However, such conservative target biomass reference points should be evaluated on a case by case basis taking into consideration biological, economic and social aspects of fisheries [38].

Long-term Stock Performance

Given the relatively young age of the MSC, with 40% of stocks and 65% of fisheries certified in the last two years, trends in population biomass for individual fisheries after certification could only be analyzed for 10 certified stocks (23% of those analyzed) that had more than 5 years of available data after certification (Fig. 2; Table S1). For 7 out of 10 stocks, a combination of favorable environmental conditions, improved compliance to catch quotas (total allowable catch) as part of a rebuilding plan, and multiple management regulations (e.g., spatial and temporal closures, minimum landing sizes) have contributed to observed biomass increases and in some cases a rapid recovery in abundance to sustainable levels [30,31] (Fig. 2). The rest of the analyzed stocks (3 out of 10) have shown slight declines in biomass since certification but remain above B_{MSY}.

This fluctuation of stock size over time is evident in the time series data of biomass in relation to B_{MSY} after certification (Fig. 2). However, whether the changes over time seen in the 10 stocks with sufficient post-certification data are the product of good management depends on the initial biomass in relation to the target at the time the fishery was certified. South Georgia Patagonian toothfish (*Dissostichus eleginoides*), for example, declined in biomass since certification, but was still 22% larger than B_{MSY} in 2009. Conversely, New Zealand's eastern and western hoki (*Macruronus novaezelandiae*) stocks were below B_{MSY} and certified under a rebuilding plan, and have since increased 300% in biomass. EBS walleye pollock and Northeast Atlantic mackerel (*Scomber scombrus*) both dropped below B_{MSY} after being certified but have improved in recent years and are now above B_{MSY}. Certification of EBS walleye pollock was criticized [16] because the biomass fell 64% between 2004 and 2009. However, climate regime shifts in this region in the 1970s [39] greatly increased pollock productivity leading to a marked surge in biomass. For example, pollock spawning biomass in the 2000s has averaged 3.3 times that in the 1960s and is currently 5 times higher than when the fishery developed in 1964 [40]. Finally, for those stocks currently below B_{MSY}, biomass has increased since the time of certification (Fig. 2). Although determining a causal connection between certification and increase in biomass would require an evaluation of reference uncertified fisheries and longer time series, the observed patterns are consistent with conservative harvest levels and responsive management systems required under MSC certification standards.

Natural variability, changes in fisheries management systems, and adjustments in the MSC's standards would likely affect the performance of certified fisheries against the standards in a particular year. When we examined four decades (1970-present) of data to characterize long-term trends in biomass relative to B_{MSY} we found that certified stocks on average performed better over the long-term than uncertified fisheries (Fig. 3). Biomass of uncertified fish stocks globally has been below B_{MSY} since the 1970s but shows signs of recovery towards B_{MSY} since 2000, while certified stocks have on average been consistently above B_{MSY} since 1980 (biomass long-term average $= 1.3B_{MSY}$). It is possible that differences early in the time series may reflect differences in

Table 1. Stock status by landings.

Stock name	Species	Large Marine Ecosystem	Landings (MT)	Most recent year with data	$B_{current}/B_{MSY}$	$u_{current}/u_{MSY}$
Certified						
Skipjack tuna	*Katsuwonus pelamis*	Pacific High Seas	1,700,000	2010	2.67	0.34
Herring	*Clupea harengus*	North East Atlantic	1,687,371	2010	1.24	1.05
Bering Sea walleye pollock	*Theragra chalcogramma*	East Bering Sea	813,000	2011	1.25	0.46
North East Atlantic mackerel*	*Scomber scombrus*	Celtic-Biscay Shelf	734,889	2010	1.37	1.27
Barents Sea Atlantic cod	*Gadus morhua*	Barents Sea	523,430	2010	1.02	0.58
Barents Sea saithe	*Pollachius virens*	Barents Sea	520,529	2010	1.08	0.99
Pacific hake	*Merluccius productus*	California Current	216,910	2010	1.75	0.82
Barents Sea haddock	*Melanogrammus aeglefinus*	Barents Sea	200,512	2010	1.20	0.71
North Sea herring	*Clupea harengus*	North Sea	168,443	2010	0.93	0.47
North Sea saithe	*Pollachius virens*	North Sea	161,462	2010	0.37	1.27
Median					*1.22*	*0.77*
Uncertified						
Chilean Jack Mackerel	*Trachurus murphyi*	Humboldt Current	744,495	2010	0.09	3.66
Blue Whiting Northeast Atlantic	*Micromesistius poutassou*	Iceland Shelf	634,978	2010	0.29	1.01
Yellowfin tuna Central Western Pacific	*Thunnus albacares*	Pacific High Seas	413,418	2005	1.29	0.80
Capelin Iceland	*Mallotus villosus*	Iceland Shelf	391,000	2010	0.40	0.01
Yellowfin tuna Indian Ocean	*Thunnus albacares*	Indian Ocean	325,854	2009	1.02	1.15
Capelin Barents Sea	*Mallotus villosus*	Barents Sea	323,000	2010	1.01	0.27
Sandeel North Sea Dogger Bank SA1	*Ammodytes marinus*	North Sea	285,794	2010	1.86	0.28
Yellowfin tuna Eastern Pacific	*Thunnus albacares*	Pacific High Seas	255,923	2010	0.71	1.13
Sardine South Africa	*Sardinops sagax*	Benguela Current	217,138	2006	0.75	0.55
Argentine hake Southern Argentina	*Merluccius hubbsi*	Patagonian Shelf	212,618	2008	0.40	1.49
Median					*0.73*	*0.91*

Biomass status and exploitation rate in relation to MSY reference points for the 10 largest certified and uncertified stocks by landings (metric tons, in 2010). Rows in **bold italics** represent median values for certified and uncertified stocks.
*MSC certificate currently suspended.

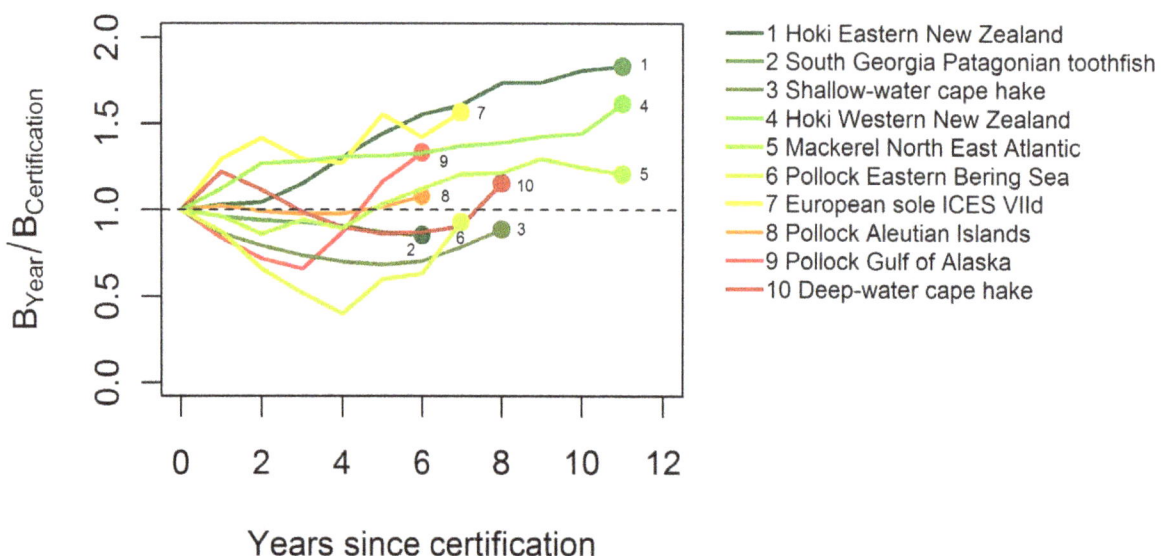

Figure 2. Time trends of current biomass (total or spawning) relative to biomass levels at the time of MSC assessment for stocks with available information and more than 5 years of certification. Colors represent current stock status (green to yellow: $B > BMSY$; orange to red: $B < BMSY$) and dots represent most recent year of available information (as per Table S1).

how and when fisheries developed. However, MSC began certifying fisheries in 1999, while certified and uncertified stocks diverged from each other in the 1980s. This suggests that the stocks certified by MSC were performing well prior to certification. This improved performance has continued over the last 10 years, with certified stocks experiencing an average 45% increase in biomass compared with a 9% increase for uncertified stocks (Fig. 3; Table S5).

Conclusions

Successful single-species management is only one part of fostering sustainable fisheries. There has been increasing support to move away from the single-species paradigm and towards an ecosystem-based approach [41,42] which recognizes that fishing has both direct and indirect effects on marine systems [43]. Conserving ecosystem diversity and structure will play an important role in helping to retain ecosystem function in the face of climate change [44,45]. While the impacts of fishing on ecosystems are difficult to measure, a few studies have attempted to quantify damage done not only by different types of gear but also by the volume of gear in the water [46,47]. Through their certification standards the MSC has the opportunity to recognize fisheries that limit the collateral impacts of fishing on food webs, habitats, and ecosystems structure. However, more research is needed to quantify the performance of certified fisheries in these areas in comparison to uncertified fisheries [22].

Most certified fisheries come from developed countries with strong central governments, sophisticated fisheries management and data-rich situations. The analyzed time series (Fig. 3) suggests that these fisheries were already well managed by their agencies before certification. Given the increase in the number of fisheries seeking MSC certification in the last three years, future analyses will be able to examine the effect of MSC certification on initially less well managed fisheries, particularly small-scale and data-limited fisheries, which are critically important to developing world economies in terms of employment, national food security, and foreign exchange earnings [1]. An open question is whether certification and eco-label programs should raise the bar of sustainability, which may result in decreased market opportunities

for small-scale and data-limited fisheries, or attempt to catalyze positive changes in those fisheries with informal and traditional management that are characteristic of many parts of the developing world, and in fisheries with current poor performance. These fisheries must be part of the solution in the pursuit of sustainable fisheries on a global scale.

Our study reveals that MSC-certified stocks are on average more likely to meet or exceed MSY-based target reference points, with higher biomass and lower exploitation rates than uncertified stocks. Certified stocks are also more likely to meet more conservative targets than uncertified stocks. Further, for those stocks with lower biomass levels, rebuilding plans are in place to improve stock health. While our time series analysis indicates that the observed difference in performance between certified and uncertified stocks existed prior to certification, the MSC eco-label is a reliable indicator of target stock health to consumers. It is important to note that MSC-certified fisheries not only must pass the sustainability criterion for the target stock but also must minimize ecosystem impact and have robust management systems. As more agencies attempt to implement ecosystem-based management, certified fisheries will need to demonstrate enhanced performance in these other areas to meet evolving definitions of sustainability and maintain the integrity of the MSC eco-label. A key part of MSC certification is the chain of custody to ensure that seafood labeled as MSC-certified indeed comes from the certified fishery and is not mislabeled catch from uncertified fisheries [48,49]. Nevertheless, the current study shows that certification and eco-labeling can effectively recognize healthy stocks and fisheries that are achieving internationally accepted management targets. This is a critical first step in providing a mechanism for consumers to effectively influence change in fishing practices and ensure future ocean health and productivity.

Supporting Information

Table S1 Summary information on certified stocks and their estimated current biomass and exploitation rates relative to MSY reference points ($B_{current}/B_{MSY}$ and $u_{current}/u_{MSY}$ or $F_{current}/F_{MSY}$). Rows in grey indicate certified stocks without available reference points or biomass estimates. These stocks were not used in the analysis (Fig. 1A). "Method used" indicates the method used to estimate B_{MSY} and F_{MSY} or u_{MSY}: 1 = stock assessment model, 2 = surplus production model, and 3 = combination.

Table S2 Summary information on uncertified stocks and their estimated current biomass and exploitation rates relative to MSY reference points (B/B_{MSY} and u/u_{MSY} or F/F_{MSY}). "Method used" indicates the method used to estimate B_{MSY} and u_{MSY} or F_{MSY}: 1 = stock assessment model, 2 = surplus production model, and 3 = combination.

Table S3 Median ($\pm$SE) biomass and exploitation rates relative to their targets and differences among certified, uncertified and not-recommended stocks. *denotes statistical significance (*$P<0.05$; **$P<0.005$).

Table S4 Median biomass and exploitation rates relative to their targets and differences among certified, uncertified and not-recommended stocks for those fisheries with both B_{MSY} and F_{MSY} available from stock assessments

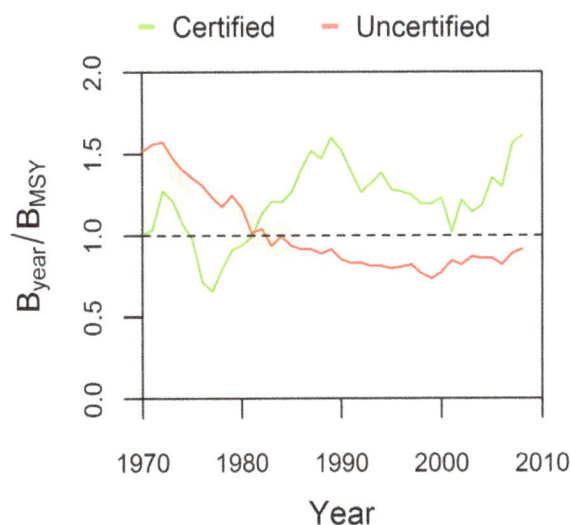

Figure 3. Performance of MSC-certified and uncertified fisheries. Long term trends (1970–2009) of biomass relative to their targets levels (i.e., estimated biomass at which the maximum sustainable yield should be obtained: B_{MSY}; median $\pm$S.E.). B_{MSY} is set to 1 (broken line).

Table S5 Results of test for difference in mean B/B_{MSY} over time between certified and uncertified stocks. *denotes significance ($*P<0.05$; $**P<0.005$).

Text S1 This supporting information file includes expanded descriptions of methods used, additional results and model outputs and supporting references.

Acknowledgments

We thank the Certification and Accreditation Bodies that provided confidential information on fisheries pre-assessments. The data reported in this paper are tabulated in the Supporting Information.

Author Contributions

Conceived and designed the experiments: NLG SRV TAB. Analyzed the data: NLG SRV. Wrote the paper: NLG SRV. Discussed the results and contributed to the manuscript: NLG SRV TAB DJA JKB PLB JCD CC OD TEE RH DDH AEL CN KS RLS SS ADMS ASP SJT JTT NEW.

References

1. FAO (2010) *The State of World Fisheries and Aquaculture 2010* (Food and Agriculture Organization of the United Nations, Rome) Available at: http://www.fao.org/docrep/013/i1820e/i1820e.pdf.
2. Worm B, Hilborn R, Baum JK, Branch TA, Collie JS, et al. (2009) Rebuilding global fisheries. Science 325: 578–585.
3. Branch TA, Jensen OP, Ricard D, Ye Y, Hilborn R (2011) Contrasting global trends in marine fishery status obtained from catches and from stock assessments. Conservation Biology 25: 777–786.
4. Gutierrez NL, Hilborn R, Defeo O. (2011) Leadership, social capital and incentives promote successful fisheries. Nature 470: 386–389.
5. Ward T, Phillips B eds. (2008) *Seafood Ecolabelling: Principles and Practice* Ward T, Phillips Beds (Wiley-Blackwell).
6. Hilborn R, Cowan JH (2010) Marine stewardship: high bar for seafood. Nature 467: 531–531.
7. Parkes G, Young JA, Walmsley SF, Abel R, Harman J, et al. (2010) Behind the signs–a global review of fish sustainability information schemes. Reviews in Fishery Sciences 18: 344–356.
8. Roheim CA, Asche F, Santos JI (2011) The elusive price premium for ecolabelled products: evidence from seafood in the UK market. Journal of Agronomic Economy 62: 655–668.
9. Perez-Ramirez M, Phillips B, Lluch-Belda D, Lluch-Cota S (2012) Perspectives for implementing fisheries certification in developing countries. Marine Policy 36: 297–302.
10. Perez-Ramirez M, Ponce-Diaz G, Lluch-Cota S (2012) The role of MSC certification in the empowerment of fishing cooperatives in Mexico: The case of red rock lobster co-managed fishery. Ocean and Coastal Management 63: 24–29.
11. Gulbrandsen L (2009) The emergence and effectiveness of the Marine Stewardship Council. Marine Policy 33: 654–660.
12. Jacquet J, Pauly D (2007) The rise of seafood awareness campaigns in an era of collapsing fisheries. Marine Policy 31: 308–313.
13. Ponte S (2008) in *Seafood labelling: principles and practice*, Ward T, Phillips Beds (Wiley-Blackwell, Oxford, UK), 287–306.
14. Ward TJ (2008) Barriers to biodiversity conservation in marine fishery certification. Fish and Fisheries 9: 169–177.
15. Potts T, Haward M (2007) International trade, eco-labelling, and sustainable fisheries – recent issues, concepts and practices. Environmental Development Sustainability 9: 91–106.
16. Jacquet J, Pauly D, Ainley D, Holt S, Dayton P (2010) Seafood stewardship in crisis. Nature 467: 28–29.
17. Ponte S (2006) Ecolabels and fish trade: Marine Stewardship Council certification and the South African hake industry. TRALAC Working Paper, DIIS, Denmark. 66.
18. Jacquet J, Pauly D (2008) Funding priorities: big barriers to small-scale fisheries. Conservation Biology 22: 832–835.
19. Marine Stewardship Council (2011) *MSC Certification Requirements v1* (London, UK) Available at: http://www.msc.org/documents/scheme-documents/msc-scheme-requirements/msc-certification-requirement-v1.1/view.
20. Ricard D, Minto C, Jensen OP, Baum JK (2011) Examining the knowledge base and status of commercially exploited marine species with the RAM Legacy Stock Assessment Database. Fish and Fisheries Online early: DOI: 10.1111/j.1467-2979.2011.00435.x.
21. Fournier DA, Skaug HJ, Ancheta J, Ianelli J, Magnusson A, et al. (2012) AD Model Builder: using automatic differentiation for statistical inference of highly parameterized complex nonlinear models. Optimization Methods Software 27: 233–249.
22. Martin S, Cambridge T, Grieve C, Nimmo F, Agnew DA (2012) Environmental impacts of the Marine Stewardship Council certification scheme. Reviews Fisheries Sciences 20: 61–69.
23. UNCLOS (1982) United Nations Convention on the Law of the Sea. 1833: 1–186. Available at: http://treaties.un.org/doc/Publication/UNTS/Volume%201833/volume-1833-A-31363-English.pdf.
24. UNFSA (1995) Agreement for the Implementation of the Provisions of the United Nations Convention on the Law of the Sea of 10 December 1982. 1–40. Available at: http://daccess-dds-ny.un.org/doc/UNDOC/GEN/N95/274/67/PDF/N9527467.pdf.

25. Hilborn R (2010) Pretty Good Yield and exploited fishes. Marine Policy 34: 193–196.
26. Rosenberg AA, Swasey JH, Bowman M (2006) Rebuilding US fisheries: progress and problems. Frontiers in the Ecology and Environment 4: 303–308.
27. Froese R, Branch TA, Proelß A, Quaas M, Sainsbury K, et al. (2011). Generic harvest control rules for European fisheries. Fish and Fisheries 12: 340–351.
28. Rademeyer RA (2012) Routine Update of the South African Hake Base Reference Case Assessment. FISHERIES/2012/AUG/SWG-DEM/39. 7.
29. ICES (2011) Report of the ICES Advisory Committee, 2011. ICES Advice, 2011. Books 11 (Copenhagen).
30. Eero M, Köster FW, Vinther M (2012) Why is the Eastern Baltic cod recovering? Marine Policy 36: 235–240.
31. ICES (2011) Report of the ICES Advisory Committee, 2011. ICES Advice, 2011. Books 6 (Copenhagen) Available at: http://www.ices.dk/committe/acom/comwork/report/2011/2011/had-34.pdf.
32. Mangel M, Marinovic B, Pomeroy C, Croll D (2002) Requiem for Ricker: Unpacking MSY. Bulletin of Marine Sciences 70: 763–781.
33. Hilborn R, Stokes K (2010) Defining overfished stocks: have we lost the plot? Fisheries 35: 113–120.
34. Punt AE, Smith ADM (2001) The Gospel of maximum sustainable yield in fisheries management: birth, crucifixion and reincarnation in *Conservation of Exploited Species*, Reynolds JD, Mace GM, Redford KH, Robinson JG eds (Cambridge University Press, Cambridge, UK), 41–66.
35. Walters CJ, Christensen V, Martell SJ (2005) Possible ecosystem impacts of applying MSY policies from single-species assessment. ICES Journal of Marine Sciences 62: 558–568.
36. Smith ADM, Brown CJ, Bulman CM, Fulton EA, Johnson P, et al. (2011) Impacts of fishing low-trophic level species on marine ecosystems. Science 333: 1147–1150.
37. Froese R, Branch TA, Proelß A, Quaas M, Sainsbury K, et al. (2011). Generic harvest control rules for European fisheries. Fish and Fisheries 12: 340–351.
38. Dichmont CM, Pascoe S, Kompas T, Punt AE, Deng R (2012) On implementing maximum economic yield in commercial fisheries. Proceedings of the National Academy of Sciences of the USA 107: 16–21.
39. Mantua NJ, Hare SR, Zhang Y, Wallace JM, Francis RC (1997). A Pacific interdecadal climate oscillation with impacts on salmon production. Bulletin of the American Meteorology Society 78: 1069–1079.
40. Ianelli J, Barbeaux S, Honkalehto T, Kotwicki S, Aydin K, et al. (2010) Assessment of the walleye pollock stock in the Eastern Bering Sea. Available at: http://www.afsc.noaa.gov/refm/docs/2010/EBSpollock.pdf.
41. Hilborn R (2011) Future directions in ecosystem based fisheries management: A personal perspective. Fisheries Research 108: 235–239.
42. Pikitch EK, Santora C, Babcock EA, Bakun A, Bonfil R, et al. (2004) Ecology: ecosystem-based fishery management. Science 305: 346–7.
43. Crowder LB, Hazen E, Avissar N, Bjorkland R, Latanich C, et al. (2008) The impacts of fisheries on marine ecosystems and the transition to ecosystem-based management. Annual Reviews Ecology and Evolution Systems 39: 259–278.
44. Hooper DU, Chapin FS, Ewel JJ, Hector A, Inchausti P, et al. (2005) Effects of biodiversity on ecosystem functioning: a consensus of current knowledge. Ecological Monographs 75: 3–35.
45. Hughes TP, Bellwood DR, Folke C, Steneck RS, Wilson J (2005) New paradigms for supporting the resilience of marine ecosystems. Trends in Ecology and Evolution 20: 380–386.
46. Hixon MA, Tissot BN (2007) Comparison of trawled vs untrawled mud seafloor assemblages of fishes and macroinvertebrates at Coquille Bank, Oregon. Journal of Experimental Marine Biology and Ecology 344: 23–34.
47. Zhou S, Smith ADM, Punt AE, Richardson AJ, Gibbs M, et al. (2010) Ecosystem-based fisheries management requires a change to the selective fishing philosophy. Proceedings of the National Academy of Sciences of the USA 107: 9485–9489.
48. Marko PB, Nance HA, Guynn KD (2011) Genetic detection of mislabeled fish from a certified sustainable fishery. Current Biology 21: 621–622.
49. Martinsohn JT (2011). Tracing fish and fish products from ocean to fork using advanced molecular technologies in *Food Chain Integrity: a holistic approach to food traceability, safety, quality and authenticity* (Cambridge, UK), p. 259.

16

Outlook on a Worldwide Forest Transition

Chris Pagnutti[1,2], Chris T. Bauch[2,3], Madhur Anand[1,3]*

1 School of Environmental Sciences, University of Guelph, Guelph, Ontario, Canada, **2** Department of Mathematics and Statistics, University of Guelph, Guelph, Ontario, Canada, **3** Department of Ecology and Evolutionary Biology, Princeton University, Princeton, New Jersey, United States of America

Abstract

It is not clear whether a worldwide "forest transition" to net reforestation will ever occur, and the need to address the main driver–agriculture–is compelling. We present a mathematical model of land use dynamics based on the world food equation that explains historical trends in global land use on the millennial scale. The model predicts that a global forest transition only occurs under a small and very specific range of parameter values (and hence seems unlikely) but if it does occur, it would have to occur within the next 70 years. In our baseline scenario, global forest cover continues to decline until it stabilizes within the next two centuries at 22% of global land cover, and wild pasture at 1.4%. Under other scenarios the model predicts unanticipated dynamics wherein a forest transition may relapse, heralding a second era of deforestation; this brings into question national-level forest transitions observed in recent decades, and suggests we need to expand our lexicon of possibilities beyond the simple "forest transition/no forest transition" dichotomy. This research also underscores that the challenge of feeding a growing population while conserving natural habitat will likely continue for decades to come.

Editor: Bruno Hérault, Cirad, France

Funding: This work was funded by the Natural Sciences and Engineering Council of Canada (Discovery grants program) to MA and CB. The funders had no role in study design, data collection and analysis, decision to publish, or preparation of the manuscript.

Competing Interests: The authors have declared that no competing interests exist.

* E-mail: manand@uoguelph.ca

Introduction

According to the Food and Agriculture Organization (FAO) of the United Nations, global forest cover was reduced by more than 70 Mha since 1990, an area larger than France and roughly 0.5% of the global land area. The main driver of deforestation is agricultural expansion [1], [2]. It currently takes about 0.8 ha of cropland and pasture and 0.06 ha of urban land per person per year to feed and shelter the global population [3], [4]. At this rate, if population stabilizes at around 10 billion, then agricultural and urban land will cover over 67% of the Earth's land area. Since around 15% of the global land area is classified as arid, [5] there will be little land area remaining for other purposes such as forest and wild pasture conservation. Despite the apparent demise of the world's forests, over the last two centuries many countries, particularly in the industrialized world, have experienced a forest transition; that is, a transition from declining to expanding forested area [6], [7].

In the classical formulation of the forest transition the dynamics are simple: deforestation proceeds until the onset of the forest transition, after which time forest cover increases and eventually stabilizes. Some alternate approaches have been proposed to give the theory the flexibility needed to account for some real-world scenarios [8]. To this end, the possibility of a time lag between the end of the deforestation period and the start of the reforestation period has been suggested. However, there has been little focus on possible alternative forest cover dynamics that may ensue *after* the onset of the forest transition. For example, under what conditions may a forest transition be followed by a subsequent period of deforestation? In such a case, the original forest transition may be regarded as spurious.

The situation wherein a forest transition is followed by a second period of deforestation has been documented in the case of France where two forest transitions are believed to have occurred. The first was due to a decline of the French population during the time of the Black Death. The second was due to agricultural intensification among other factors [9]. Here we are interested in forest transitions of the second type, which we regard to be more important because they occur despite increasing population.

Land use models (e.g. IMAGE and GLOBIOM) generally assume that local parcels of land can be in one of several states and thus imply spatial localization [10]. These models are often quite complicated, and may account for a wide variety of ecological, biochemical, economic and political factors. Most studies of forest transitions and forest decline focus on spatial scales at the national and sub-national levels, and on temporal scales of a few decades. There are some important notable exceptions that have focused on policies and drivers that could potentially either trigger or inhibit a global forest transition in the future. [6], [7], [9], [11] However, few data-driven mathematical models have been developed to predict the timing of the forest transition and the ultimate (global) scale of deforestation [6].

In contrast, the model we present here is interpreted at the global level (i.e., is not spatially explicit) and the transitions between abundances of various land types do not correspond directly to transitions from one land type to another within a given patch. This removes confounding factors related to international trade, and other spatial processes that affect land use [12]. The model is capable of capturing historical land-use dynamics, and sheds some light on potential non-classical forest transition scenarios. The model is primarily based on the world food equation, is parameterized almost exclusively by historical data, and makes relatively few assumptions. Despite this simplicity, the model is capable of capturing estimates of historical land use dynamics at the global level going back several centuries. This is, to our knowledge, the first dynamic mathematical model based on the world food equation [10].

Model

Model Description

The model is illustrated conceptually in Fig. 1, and the three key inputs to the model are the time-dependent functions of world population, agricultural yield and per capita food consumption. The model captures how transitions between five possible land states–forest, agricultural fields, pasture, abandoned lands, and urban areas– are driven by these three inputs.

Our model is formally written as

$$F_{t+1} = F_t - \alpha R_t \Theta(R_t) + \beta B_t^{(F)} - \zeta(U_{t+1} - U_t) \quad (1)$$

$$P_{t+1} = P_t - (1-\alpha) R_t \Theta(R_t) + \delta B_t^{(F)} - (1-\zeta)(U_{t+1} - U_t) \quad (2)$$

$$A_{t+1}^{(F)} = A_t^{(F)} + \alpha R_t \Theta(R_t) + \gamma R_t \Theta(-R_t) \quad (3)$$

$$A_{t+1}^{(P)} = A_t^{(P)} + (1-\alpha) R_t \Theta(R_t) + (1-\gamma) R_t \Theta(-R_t) \quad (4)$$

$$B_{t+1}^{(F)} = B_t^{(F)} - \gamma R_t \Theta(-R_t) - \beta B_t^{(F)} \quad (5)$$

$$B_{t+1}^{(P)} = B_t^{(P)} - (1-\gamma) R_t \Theta(-R_t) - \delta B_t^{(P)} \quad (6)$$

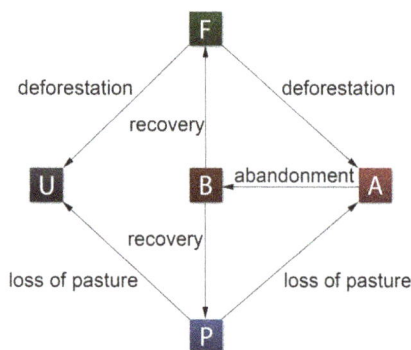

Figure 1. Conceptual depiction of the model. Our model assumes that non-barren land may be in one of five states: forested (F), agricultural (A), wild pasture (P), urban (U) and abandoned (B). The proportions of these land types may change over time in the following ways: forested land and wild pasture may be converted into either agricultural or urban land, agricultural land may become abandoned land, and abandoned land may recover to either forested or agricultural land.

$$U_{t+1} = U_t + s(p_{t+1} - p_t) \quad (7)$$

where

$$R_t = \frac{c_{t+1} p_{t+1}}{y_{t+1}} - A_t^{(F)} - A_t^{(P)} \quad (8)$$

The symbols F, P, A and U represent the global area of forested, wild pasture (i.e that not used for agriculture/domestic grazing), agricultural and urban land respectively. The superscripts (F) and (P) on A and B are used to keep track of whether the agricultural land was derived from either forested land or pasture. The heaviside functions $\Theta(\cdot)$ restrict the directions of land use conversion; for example, agricultural land cannot be converted to forested land without first going through a period of abandonment. The parameters and their values are summarized in Table 1. The model's processes can be explained in terms of the equations as follows: The α and $(1-\alpha)$ terms correspond to conversion of forest and pasture (respectively) to agricultural land. The ζ and $(1-\zeta)$ terms correspond to conversion of forest and pasture (respectively) to urban land. The γ and $(1-\gamma)$ terms correspond to the abandonment of agricultural land. The β and δ terms correspond to the conversion of abandoned land to their natural state. The code that was used to generate our results can be downloaded at https://github.com/Pacopag/fpau.

Formally, the model contains 19 parameters; however, all but four of these parameters can be fixed using physical and historical data and three of these three are insensitive with respect to model output (Table 1). To achieve a good fit with both FAO and independent data, we tuned only one parameter (i.e. the initial yield, $y_0 = 150$ kg/ha/year). A detailed description of how the model was calibrated from the available data is given in the following section. The basic processes of our model are similar to those commonly used in models of deforestation and forest transitions [6], [13]. The mechanism of our model is the following: as the population grows and requires more food, agricultural land is created, and as surplus food is produced, agricultural land is abandoned and left to recover to its natural state.

The main driving term in our model is the world food equation; that is, equation (8) which states that agricultural land area A is related to population p, per capita consumption c and agricultural yield y by the equation $cp/y - A = 0$ [7]. Parameters c, p and y all grow logistically to reflect a paradigm shift from a pre-industrialized world to a maximally industrialized one. The mechanistic basis for using a logistic function is the Levins metapopulation model; in a population divided into N patches (countries), where each patch can be either "high yield" or "low yield", and where high yield technology disperses from a "high yield" patch to a "low yield" patch at some rate (thereby converting it to a "high yield" patch), the growth of total population yield is logistic. Since the onset of the Green Revolution such technologies have been transferred to developing countries over time, [14] not unlike metapopulation dynamics, and this transfer explains the majority of global yield gains. A similar argument can be made for per capita consumption trends; as the economies of developing countries grow, their diets tend to become more similar to that of developed countries [15].

The data and justifications we used to fix the parameters and initial conditions are described in detail below. In general, we used historical estimates to fix the initial conditions, and FAO data to fit the growth rates and carrying capacities. The specific details and justifications of the fitting procedure are given in. Logistic behavior

Table 1. Summary of model parameters.

Symbol	Description	Value	Source
	Parameters fitted from data		
T	Total land area (excluding barren land)	11.26×10^9 ha	FAO
α	Fraction of agricultural land derived from forest	0.4	24
ζ	Fraction of urban land derived from forest	0.9	25,26
s	Urban area per person	0.06 ha	4
K_p	Maximum un-translated population	10.5×10^9 people	FAOF
K_c	Maximum un-translated per capita consumption	1940 kg/person/year	FAOF
K_y	Maximum un-translated annual yield	3391 kg/ha/year	FAOF
r_p	Growth rate of population	0.032	FAOF
r_c	Growth rate of per capita consumption	0.019	FAOF
r_y	Growth rate of the annual yield	0.039	FAOF
p_0	Initial population	310×10^6 people	16
c_0	Initial per capita consumption	571 kg/person/year	FAO
t_{Ip}	Population inflection time	1998.3 years	FAOF
t_{Ic}	Per-capita consumption inflection time	1995.8 years	FAOF
t_{Iy}	Annual yield inflection time	1995.7 years	FAOF
	Free (fitting) parameters		
y_0	Initial annual yield	150 kg/ha/year	FP
γ	Fraction of abandoned land that is naturally forest	0.4	NS
β	Recovery rate of forests	0.01	NS
δ	Recovery rate of pastures	1.0	NS

FAO and FAOF indicate data extracted or fitted, respectively, from FAO data. FP indicates a fitting parameter. NS indicates that the model is not sensitive to this parameter in the absence of a forest transition.

of population growth has been well documented, and the world population is expected to stabilize at around 10 billion [16]. Both yield per hectare and consumption per capita have been growing almost linearly over the past few decades. The growth in yield is primarily the result of the Green Revolution of the 1960s [14]. Prior to this revolution yield increases were likely very slow becoming negligible for times before the Agricultural Revolution that took place in Europe in the late 1800s. As we push the limits of technological advancements in the future, we expect yield increases to slow down and eventually stabilize close to a biophysical maximum dictated by energy available in incident sunlight per unit area. Thus, a maximum long-term yield is appropriate. The increase in per capita consumption during the past few decades is greatly attributable to the westernization of the diets people in developing countries [15]. It is reasonable to expect that consumption will also eventually level off as large developing countries complete their economic and dietary transitions. We further assume that urban expansion is simply proportional to population size.

Model Calibration

The inputs to the model are population p, per capita consumption c and agricultural yield y, which are all time-dependent functions [2], [6]. We model these as logistic functions of the form

$$x = K_x (1 + \exp[-r_x(t - t_{Ix})])^{-1} + x_0 \qquad (9)$$

where x indicates either p, c or y. The vertical translations of the curves are used to fix the initial conditions. The values of K_x and r_x were fit using a least-squares approach [17]. The parameters and initial conditions are summarized in Table 1. The curves for p, c and y are summarized in Fig. 2.

Population. To model the time-dependence of world population p, we fitted a logistic curve to data released by the United Nations and the FAO. [16] Our fit is consistent with the UN projection that world population is expected to stabilize in the 23rd century at around 10 billion people due to a demographic fertility transition. There is a discrepancy between the fit and the data between the years 1700 and 1900 because the logistic growth cannot keep up with the anomalous population explosion catalyzed by the industrial revolution. However, this does not affect our conclusions. Urban area was estimated to be at about 3% of the global land area, or roughly 390 Mha, in 2004 [4]. The population was estimated to be 6.43 billion. Together this gives roughly 0.061 hectares of urban area per person, which we use to fix s. Although the required urban area per person s may actually be time-dependent, we assume that it is changing slowly enough to be regarded as constant over our time scale, especially in comparison to the other input variables c, p, and y.

Yield. The yield function y is a logistic curve fitted to data extracted from the FAO database. Yield was calculated as total crop, meat and milk production divided by total agricultural land. These data account for both food and non-food agricultural products. FAO data from 1985 to 1993 contain anomalies for some individual fodder crops. These anomalies are not present in the sum of data for all of the agricultural item groups. We removed

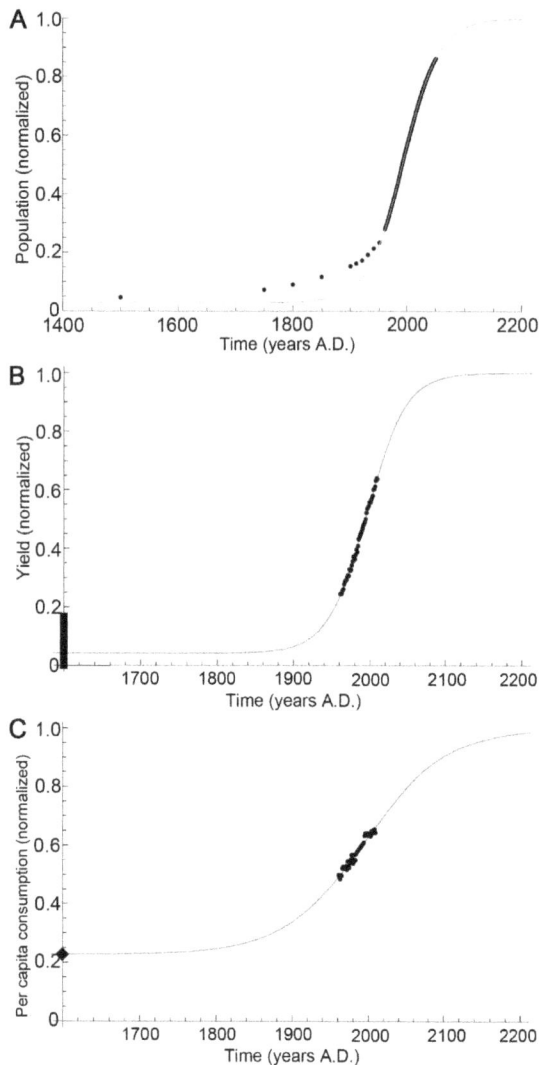

Figure 2. Logistic input curves. (A) Population $p(t)/p(\infty)$, (B) yield $y(t)/y(\infty)$, and (C) consumption $c(t)/c(\infty)$ as functions of time t in years A.D. Black dots represent historical data. Grey curves represent the logistic fit (see eq. (9) and Table 1). In (B), the vertical black line indicates the constraint on the initial yield. In (C), the black diamond represents the per capita consumption in least developed countries in 2009 (FAO). The R^2 values are 0.992, 0.997 and 0.956 respectively.

the anomalous data points and replaced them using linear interpolation. Many models of agricultural yields treat the time-dependence as linear [18], [19]. Indeed, from 1960 until 2009, yields of almost all major crop types and animal products (including our derived average) grew very quickly with a very high degree of linearity. Of course, linear growth cannot be sustained indefinitely, so it is not appropriate for large time scales. Rather, the rapid linear rise in yields that has been observed in recent decades is more likely to level off as humans push the limits of physical bio-energetic constraints on yield through genetic improvements on important crops, and as these technologies are implemented across the globe [2], [14].

Despite some advancements made to farming technology during the middle ages, agricultural yields remained fairly constant until the agricultural revolution, beginning in 17th century Britain, and further accelerated during the industrial revolution [14], [20-23].

However, notable yield increases did not make a global impact until the Green Revolution in the middle of the 20th century, when both mechanization and modern varieties of high-yielding crops were beginning to be introduced in many developing countries [14]. The timing of this revolution corresponds very closely with the left endpoint of the available yield data, so the observed linear growth of agricultural yields is largely attributed to the spreading of existing technologies to developing countries, and may not have persisted for very long before the 1960s. Estimates of medieval grain yields suggest that 700 kg/ha was typical of a harvest in those times [20], [22]. Estimates of grain yields in the 19th and 20th centuries suggest a higher, but constant 2000–3000 kg/ha before rapidly increasing in the late 1960s [21], [23]. Less is known about the yield of animal products, but estimates of the number of livestock (equine and bovine) per hectare are given to be in the order of 1 ha^{-1} [21]. However, not all livestock were harvested for food each year, so the annual yield of livestock food products would have been much lower than that for grain (as it is today). From 1960 to 2009, the ratio of the total global yield to that of cereal crops was roughly 0.6. Applying this ratio to estimates of medieval grain yield we constrained the initial yield in our model to be less than 420 kg/ha. On the other hand, the maximum attainable yield of around 3500 kg/ha is determined our fitting algorithm.

Consumption. The third driving factor in the world food equation is annual per capita consumption c. Global per capita food consumption has been steadily rising over the past five decades. The tendency is for the diets of developing countries to approach that of the United States and other industrialized western countries [15]. We fitted a logistic function to the FAO data for global per capita food consumption (or more accurately, per capita food supply) for the years from 1961 to 2007. As an initial condition we took the value of 571 kg/capita/yr. This value corresponds to the average of per capita consumption in least developed countries, which was found to be almost constant over the period from 1961–1995. The assumption here is that medieval consumption patterns were similar to those presently observed in lesser developed countries. The maximum value $c(\infty) = 2512$ kg/capita/year compares well with the present-day value of about 2300 kg/capita/year for the United Kingdom, which might be expected in a completely industrialized world.

Other parameters. In the absence of a forest transition, the model is only sensitive to two external parameters (i.e. not intrinsic to the input function), α and ζ. We fixed $\alpha = 0.4$ from historical estimates [24]. We did not find any data allowing us to fix ζ, but comparing population density maps to biome maps, it is clear that $0 << \zeta < 1$ [25], [26]. We coarsely tuned ζ to 0.9, but the model is not very sensitive to its variations which can be compensated by small changes in y_0.

Initial conditions. To select initial conditions for the state variables F_0, P_0 and $A_0^{(F,P)}$, we began with a pre-agricultural landscape estimated previously [24]. Then we assumed that agricultural and urban land existed as per the world food equation and equation (8). We took these areas from both forested and pasture land in the proportions dictated by the parameters α and ζ.

Model Fitting Procedure

There are 19 parameters in our model. We were able to fix 15 of these parameters using data and other estimates found in the literature (Table 1) as described in the previous section. For the logistic fits (Fig. 2) we used a least-square method on data translated vertically downward by an amount dictated by the initial conditions, and translated horizontally to center the curve approximately about the vertical axis. Three of the remaining four unspecified parameters (i.e. γ, β and δ) had no effect on the model's output in the absence of a forest transition. The only remaining unspecified parameter is the initial yield y_0, which is constrained as described in section 2.2. We varied this parameter until we achieved a good fit with both the recent FAO data as well as with estimates of land-type cover in the more distant past.

Exploring the Model's Phase Space

In order to probe our model for interesting post-forest transition dynamics, we analyzed the phase diagram of our model using maximum yield K_y and maximum consumption K_c as our control parameters. To do this we smoothly fit new logistic curves to the endpoints of the FAO data on the baseline curves. For any given value of K_y one can fix the value and slope of $y(t)$ at the endpoint of the data $t_e = 2009$ by setting

$$r_y = \frac{y'(t_e)K_y}{y(t_e)[K_y - y(t_e)]} \tag{10}$$

and

$$t_{Iy} = t_e - \frac{(1 - y(t_e)^2)}{y'(t_e)K_y} \ln\left(1 - \frac{K_y}{y(t_e)}\right) \tag{11}$$

and likewise for $y \rightarrow c$ to smoothly fit new logistic curves to the historical ones in Fig. 2, beginning at the endpoint of the FAO data corresponding to the year $t_e = 2009$. In this way, we scanned across a range of values to compute a section of the phase diagram of the our model.

We also did a sensitivity analysis against all model parameters. The results are summarized in Figure S1. We varied each parameter individually from -10% to 10% of the baseline values while holding all other parameters fixed. We found that the model is quite insensitive with respect to changes in most parameters. In all cases, a relative change of any parameter by 10% resulted in less than 10% a change in land cover fractions. In most cases, the change was much less than 10%.

Results

Baseline Scenario: Fitting to Historical Data

Using historical data to fix all but one parameter, the model is able to fit FAO data as well as independent estimates of pre-industrial land cover (Fig. 3) [24]. Projections of land use change for the years 2000 to 2030 have been made in previous studies. For example, 125–416 Mha of new agricultural land, 104–345 Mha of deforestation, and 48–100 Mha of new urban land are expected. [11], [27] These estimates are based on a combination of statistical extrapolation of FAO data as well as by assuming the goals set by the UNFCCC will be met (i.e. to cut the 2010–2020 deforestation rate in half as compared to that from 2000–2010). In comparison, our mathematical model predicts 168, 338 and 132 Mha respectively for new agricultural land, deforested land and new urban land, without the need to assume we can meet UNFCCC goals (Fig. 3). Furthermore, our baseline results for agricultural land cover are in perfect agreement with the "pessimistic" value of 5820 Mha for 2050 derived in an in-depth review of other dynamical models of land use change [10]. Thus, with a few simple processes that are calibrated almost entirely with data, our model simultaneously fits estimates of land use change over several centuries.

Extrapolating far enough in time so that the model's dynamics stabilize, our baseline scenario (Fig. 3) predicts that a global forest

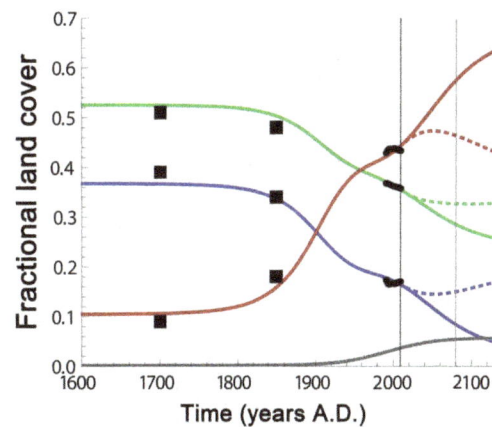

Figure 3. Land cover change over time. Forest area (green), pasture area (blue), agricultural area (red), and urban area (grey) versus time in years A.D. The vertical axis represents the fraction of the total area of non-barren land (1.13×10^{10} ha). The black vertical line corresponds to the year 2009 A.D. The solid black circles represent the FAO data. The solid black squares represent historical estimates. [24] The solid lines correspond to our baseline scenario with the logistic inputs given in Fig. 2. The dashed lines correspond to the case where consumption c and yield y continue to grow linearly beyond the domain of the FAO data.

transition is not likely to occur, and that forest and wild pasture cover will stabilize at roughly 25% and 2.5%, respectively, of the non-barren land area. This corresponds to roughly 22% and 1.4% of total land area excluding Antarctica.

Sustained Linear Growth in Yield and Consumption

Although it is reasonable to expect that both agricultural yield and per capita food consumption will eventually stabilize due to biophysical constraints, we have actually seen almost linear growth in these quantities over the past five decades (Fig. 2). Using a linear extrapolation on these data in our model, we find that a forest transition is likely to occur within the next century if such spectacular growth in yield can be sustained (Fig. 3, dashed lines). In cases in which a forest transition occurs, the model develops sensitivity to the parameter β, which determines the rate at which abandoned land becomes re-forested. Although the value of β affects the rate at which the forest cover expands, it has little effect on the timing of the forest transition. The reason for this is that the forest transition is driven primarily by the abandonment of agricultural land, which is independent of the re-forestation rate.

Future Maximal Yield and Consumption Dependence

The phase diagram (Fig. 4a) shows five distinct phases: No Forest Remaining (NFR), No Forest Transition (NFT), Classical Forest Transition (CFT), Overshot Forest Transition (OFT), and False Forest Transition (FFT). The phases can be understood in terms of the number of turning points in the function $c(t)p(t)/y(t)$. The OFT and FFT (Fig. 4(a) and 4(b)) are examples of non-classical forest transitions. Yield and consumption corresponding to the 2009 values are represented approximately by the axes' origin. Under the assumption of logistic-like stabilization of yield and consumption, our model predicts that in cases where a global forest transition is likely to occur, it will likely occur within the next century and that the forest cover would be between 33 and 35 percent (Fig. 4(c) and 4(d)). Estimating the scale of reforestation following the transition is encumbered by unknown factors that

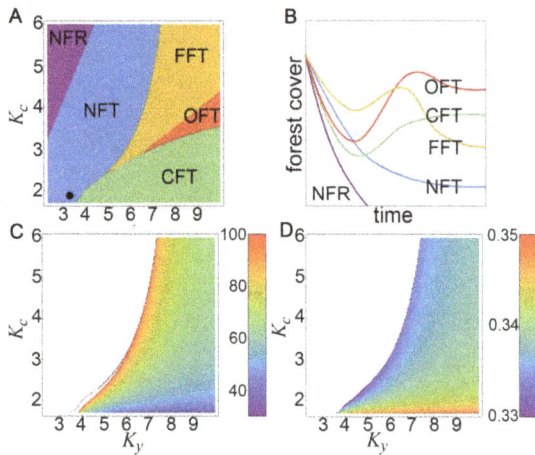

Figure 4. Analysis of $K_y K_c$ space. (A) The phase diagram indicates five distinct phases in which forest cover dynamics have distinct qualitative behaviors illustrated in (B): No Forest Remaining (NFR), No Forest Transition (NFT), Classical Forest Transition (CFT), Overshot Forest Transition (OFT), and False Forest Transition (FFT). The solid black circle in (A) corresponds to our baseline scenario. (C) The timing of the forest transition is indicated by the color scale measured in number of years after 2009. (D) The forest cover measured at the time of the forest transition indicated in (C).

determine the amount of time for which such increases in yield and consumption will occur. Our model suggests that an ultimate forest cover of less than 40% is typical even of extremely optimistic scenarios.

The phase structure of our model reveals two particularly interesting scenarios in which a forest transition gives way to a second wave of deforestation (Fig. 4 "False Forest Transition"). To our knowledge this is the first model of land use change to capture these dynamics despite strictly increasing population.

Although our model is a global-level model, the potential for a false forest transition brings into question the sustainability of national-level forest transitions observed in recent decades. For example, we looked at data for land use over the past two decades in Finland and found that the 1990s was a decade in which the forest area expanded significantly in that country, whereas the following decade was marked by a period of subsequent deforestation. Although the scale of deforestation is rather slight, it appears to be coupled with a corresponding decrease of agricultural area in the 1990s followed by an increase in the early 2000s (Fig. 5). Detailed data on forest area for long time scales are rare, so it is currently difficult to do an exhaustive study of the possibility of false forest transitions in other countries.

Agricultural Land-sparing Strategies

Next we considered the question as to which strategy would spare the most land: increasing yield, or inhibiting increases in consumption. We found that independently increasing K_y or decreasing K_c from the baseline value indicated in Fig. 2 by a given percentage had roughly the same effect on the ultimate forest cover. Yield increases depend on the continued success of genetic improvements of crops and are ultimately limited by biophysical constraints, and much of the recent increase in yield has occurred through transmission of existing technologies to developing countries, not necessarily through development of new technologies [14]. On the other hand, inhibiting consumption increases may seem to require a change in the global conscience about food

intake, but there are other more controllable ways to spare land through reduced consumption. For example, an estimated 32% of all food is lost or wasted every year [27]. If this could be reduced to just 29%, then our model predicts that almost 230 Mha of land would ultimately be spared with respect to our baseline scenario in Fig. 3. This is in good agreement with estimates of land sparing for similar reductions in food wastage by 2030 [27]. Moreover, a 5% decrease in per capita food intake could spare as much as 158 Mha of forest and wild pasture.

Although the situation for wild pastures appears rather dire in our baseline scenario, it should be noted that the FAO reports a considerable amount of wild pasture as agricultural land. So the 2% pasture cover should be interpreted as pasture that is untouched by domestic grazing livestock. Over the past 50 years, a fairly constant 69% of agricultural land was categorized as permanent pasture. Thus, the actual pasture cover (wild and cultivated) is more like 47%. This implies that human agricultural activities will have *created* pasture in addition to the natural global pasture cover, although very little of it will be completely natural. It should also be mentioned that agricultural land includes crops coming from trees (e.g. fruits and nuts), which could add to the global forest cover. However, these crops constitute less than 2% of all of the agricultural area harvested from 1961–2009, so the significance is small. On the other hand, the area occupied by these crops has more than doubled over this period. If this trend continues, land area covered by tree crops may account for a significant portion of the global forest cover in the future.

Discussion

Here, we showed that a mathematical model based on the world food equation explains past land use patterns at the global scale, and over the past millennium. The model requires only one free parameter. The same model predicts that a global-level forest transition is unlikely at baseline parameter values, and would require very significant changes in technology and/or consumption patterns. While this is not the first time these predictions have been made, mathematical models of this type are underutilized at the global level, and the model has shown how our lexicon regarding the future of forest cover must expand beyond the simple forest transition/no forest transition dichotomy to include possibilities such as false forest transitions and overshot forest transitions (Figure 4).

It should be noted that our model implicitly assumes a "business as usual" scenario as it extrapolates on contemporary trends in consumption and yield. While efforts to improve agricultural yield will likely continue, there are also possibilities for increased widespread demand for food that is not particularly land-intensive. For example, increased dependence on aquaculture and insect protein could have an enormous impact on reducing pressure on the world's remaining natural land habitats. There are other technologies that we do not consider here that could potentially increase yield to a level that would almost certainly induce a forest transition. For example, *in vitro*-cultured meat or the widespread adoption of alternative protein sources could change the face of the planet [28].

Our predictions rely on the assumption that the inputs to the world food equation, p, c and y, will all stabilize in a future where the world has become maximally industrialized. If this will be the case, and we are already observing national-level forest transitions in industrialized countries across the world, then we should consider the reasons why a global-level forest transition may not occur. A national-scale forest transition is often heralded as a great success in forest management and conservation; however, some

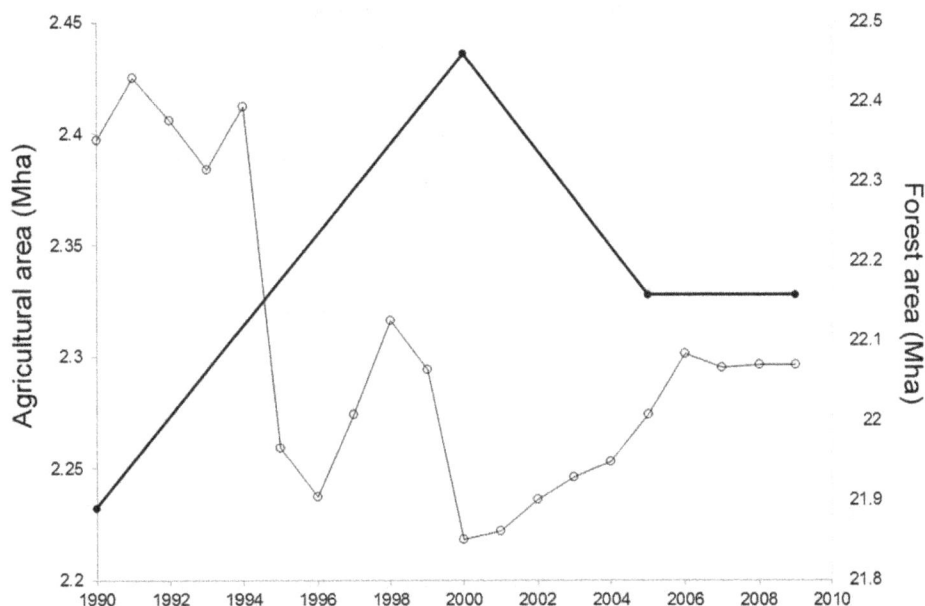

Figure 5. Recent land use dynamics in Finland. The thick line and solid circles represent forest area. The thin line and open circles represent agricultural area. Here we see evidence of a possible "False Forest Transition" where the reforestation following the forest transition (likely to have occurred in Finland's past) subsequently gives way to a second period of deforestation.

countries achieve this simply by importing more food and forest products [11]. By evaluating dynamics at the global level, we avoid this confounding factor. Our model finds that a forest transition occurs when yields do not increase fast enough to keep up with growing consumption. In this light, a forest transition may be viewed as a mismanagement of forest and agricultural resources, or a crisis in population growth until a certain point in time when policy, increasing yields, imports, or other factors halt further deforestation [11]. In light of the suggestion that regenerating forests may have lesser or different ecological quality than the original forest, [12] perhaps the best-case scenario is not a global forest transition, but for forest cover to settle at its natural equilibrium corresponding to an ultimate stable population.

Some researchers have suggested that the factors p, c and y may not have the effects on deforestation and land use that might be expected by conventional wisdom. For example, the inverse relationship between population and deforestation has been shown to be weakening in some countries in recent decades, which is likely due to increasing yields. [1] Yield gains on commodities that have elastic demand may actually promote agricultural expansion and thus deforestation [1], [29]. We account for these effects with time varying consumption and yield terms. Also, local decreases in yield may lead to agricultural abandonment and decreased pressure on local forests, but so-called "leakage" effects simply transfer the pressure to other localities, thereby resulting in net deforestation [7], [11]. Here, in contrast, we avoid such confounding factors by focusing on global-scale land use changes and the prospects of a global forest transition.

Efforts to conserve forest ecosystems are often directed at setting aside tracts of land in countries that can import whatever they need. However, the long-term explanatory power of the world food equation as we have demonstrated here, together with the observation that many national-level forest transitions may essentially be luxury imports, suggest that equal effort should be directed toward finding ways to boost agricultural yield, disseminate those technologies to developing countries, and decrease per capita consumption, thus reducing land use pressures [7].

Supporting Information

Figure S1 Sensitivity analysis. The horizontal axis is the percentage change of the corresponding parameter from its baseline value. The vertical axis is the absolute change in the fractional land cover. Varying the excluded parameters has a negligible effect.

Acknowledgments

The authors thank Simon A. Levin for stimulating discussions and Thomas Rudel for discussions and the encouragement to scale up our ideas to the global level.

Author Contributions

Conceived and designed the experiments: MA CP CB. Performed the experiments: CP. Analyzed the data: CP. Contributed reagents/materials/analysis tools: MA CP CB. Wrote the paper: MA CP CB.

References

1. Angelsen A, Kaimowitz D. (1999) Rethinking the Causes of Deforestation: Lessons from Economic Models. World Bank Research Observer 14: 73–98.
2. Mather AS, Needle CL. (2000) The relationships of population and forest trends. The Geographical 166: 2–13.
3. de Fraiture C, Wichelns D, Kemp-Benedict E, Rockstrom J. (2007) Looking ahead to 2050:scenarios of alternative investment approaches. In: Molden D, editor. Water for food, water for life: A comprehensive assessment of water

management in agriculture. London: Earthscan and Colombo: International Water Management Institute. 91–145.
4. Balk D, Pozzi F, Yetman G, Deichmann U, Nelson A. (2005) The distribution of people and the dimension of place: Methodologies to improve the global estimation of urban extents. Proceedings of the Urban Remote Sensing Conference; Tempe, AZ, USA.

5. Walker AS. (1996) Deserts - Geology and resources. U.S. Geological Survey general interest publication. Available: http://pubs.usgs.gov/gip/deserts. Accessed 13 January 2012.

6. Wagonner PE, Ausubel JH. (2001) How Much Will Feeding More and Wealthier People Encroach on Forests? Population and Development Review 27: 239–257.

7. Lambin EF, Meyfroidt P. (2011) Global land use change, economic globalization, and the looming land scarcity. Proceedings to the National Academy of Sciences 108: 3465–3472.

8. Grainger A. (1995) The forest transition: An alternative approach. Area 27: 242–251.

9. Mather AS, Fairbairn J, Needle CL. (1999) The Course and Drivers of the Forest Transition: the case of France. Journal of Rural Studies 15: 65–90.

10. Smith P, Gregory PJ, van Vuuren D, Obersteiner M, Havlik P, et al. (2010) Competition for land. Phil. Trans. R. Soc. B. 365: 2941–2957.

11. Meyfroidt P, Lambin EF. (2011) Global Forest Transition: Prospects for an End to Deforestation. Annu. Rev. Environ. Resourc. 36: 343–37.

12. Lambin EF, Meyfroidt P. (2010) Land use transitions: Socio-ecological feedback versus socio-economic change. Land Use Policy 27: 108–118.

13. Satake A, Rudel TK. (2007) Modeling the forest transition: Forest scarcity and ecosystem service hypotheses. Ecological Applications 1: 2024–2036.

14. Evenson RE, Gollin D. (1991) Assessing the Impact of the Green Revolution, 1960 to 2000. Science 300: 758–762.

15. Kearney J. (2010) Food consumption trends and drivers. Phil. Trans. R. Soc. B. 365: 2793–2807.

16. The World at Six Billion. Population Division Department of Economic and Social Affairs United Nations Secretariat (1999). Available: http://www.un.org/esa/population/publications/sixbillion/sixbillion.htm. Accessed 9 January 2012.

17. Cavallini F. (1993) Fitting a logistic curve to data. College Mathematics Journal 24: 247–253.

18. Chavas J, Bromley DW (2005) Modelling Population and Resource Scarcity in Fourteenth-century England. Journal of Agricultural Economics 56: 217–237.

19. Norwood B, Roberts MC, Lusk JL. (2004) Ranking Crop Yield Models Using Out-of-Sample Likelihood Functions. American Journal of Agricultural Economics 86: 1032–1043.

20. Farmer DL. (1977) Grain Yields on the Winchester Manors in the Later Middle Ages. The Economic History Review 30: 555–566.

21. Campbell BMS. (1983) Agricultural Progress in Medieval England: Some Evidence from Eastern Norfolk. The Economic History Review 36: 26–46.

22. Clark G. (1991) Yields Per Acre in English Agriculture, 1250–1860: Evidence from Labour Inputs. The Economic History Review 44: 445–460.

23. Miflin B. (2000) Crop improvement in the 21st century. Journal of Experimental Botany 51: 1–8.

24. Goldewijk KK, Ramankutty N. (2004) Land use changes during the past 300 years. In: Land use, land cover and soil sciences. In: Verheye WH, editor. Encyclopedia of life support systems. vol. 1. Oxford: EOLSS.

25. Tobler W, Deichmann V, Gottsegen J, Maloy K. (1995) The global demography project. Univ. Santa Barbara, CA. National Center for Geographic Information Analysis. Technical Report TR-95-6. Available: http://soils.usda.gov/use/worldsoils/mapindex/popden.html. Accessed 12 February 12 2012.

26. Olson DM, Dinerstein E, Wikramanayake ED, Burgess ND, Powell GVN, et al. (2001) Terrestrial Ecoregions of the World: A New Map of Life on Earth. BioScience 51: 933–938.

27. Wirsenius S, Azar C, Berndes G. (2010) How much land is needed for global food production under scenarios of dietary changes and livestock productivity increases in 2030? Agricultural Systems 103: 621–638.

28. Edelman PD, McFarland DC, Mironov VA, Matheny JG. (2005) In vitro-cultured meat production. Tissue Engineering 11: 659–662.

29. Angelsen A. (2010) Policies for reduced deforestation and their impact on agricultural production. Proceedings to the National Academy of Sciences 107: 19639–19644.

17

Colonization of Abandoned Land by *Juniperus thurifera* Is Mediated by the Interaction of a Diverse Dispersal Assemblage and Environmental Heterogeneity

Gema Escribano-Avila[1]*, **Virginia Sanz-Pérez**[2], **Beatriz Pías**[3], **Emilio Virgós**[1], **Adrián Escudero**[1], **Fernando Valladares**[4,1]

1 Departamento de Biología y Geología, Universidad Rey Juan Carlos, Madrid, Spain, 2 Facultad de Ciencias Ambientales, Universidad de Castilla-La Mancha, Toledo, Spain, 3 Departamento de Biología Vegetal I, Universidad Complutense de Madrid, Madrid, Spain, 4 Museo Nacional de Ciencias Naturales, Consejo Superior de Investigaciones Científicas, Madrid, Spain

Abstract

Land abandonment is one of the most powerful global change drivers in developed countries where recent rural exodus has been the norm. Abandonment of traditional land use practices has permitted the colonization of these areas by shrub and tree species. For fleshy fruited species the colonization of new areas is determined by the dispersal assemblage composition and abundance. In this study we showed how the relative contribution to the dispersal process by each animal species is modulated by the environmental heterogeneity and ecosystem structure. This complex interaction caused differential patterns on the seed dispersal in both, landscape patches in which the process of colonization is acting nowadays and mature woodlands of *Juniperus thurifera*, a relict tree distributed in the western Mediterranean Basin. Thrushes (*Turdus* spp) and carnivores (red fox and stone marten) dispersed a high amount of seeds while rabbits and sheeps only a tiny fraction. Thrushes dispersed a significant amount of seeds in new colonization areas, however they were limited by the presence of high perches with big crop size. While carnivores dispersed seeds to all studied habitats, even in those patches where no trees of *J. thurifera* were present, turning out to be critical for primary colonization. The presence of *Pinus* and *Quercus* was related to a reduced consumption of *J. thurifera* seeds while the presence of fleshy fruited shrubs was related with higher content of *J. thurifera* seeds in dispersers' faeces. Therefore environmental heterogeneity and ecosystem structure had a great influence on dispersers feeding behaviour, and should be considered in order to accurately describe the role of seed dispersal in ecological process, such as regeneration and colonization. *J. thurifera* expansion is not seed limited thanks to its diverse dispersal community, hence the conservation of all dispersers in an ecosystem enhance ecosystems services and resilience.

Editor: Anna Traveset, Institut Mediterrani d'Estudis Avançats (CSIC/UIB), SPAIN

Funding: Gema Escribano-Ávila was supported by a FPU-MEC doctoral grant from the Spanish Ministry for Education (http://www.educacion.gob.es/portada.html). Funding was provided by the Spanish Ministry for Innovation and Science (http://www.idi.mineco.gob.es/) with the grants CGL2010-16388/BOS, Consolider Montes (CSD2008_00040), VULGLO (CGL2010-22180-C03-03) and CALCOFIS (CGL2009-13013), and by the Community of Madrid grant (http://www.madrimasd.org/) REMEDINAL 2 (CM-S2009/AMB-1783). The funders had no role in the study design, data collection and analysis, decision to publish, or preparation of the manuscript.

Competing Interests: The authors have declared that no competing interests exist.

* E-mail: gema.escribano@urjc.es

Introduction

Ecosystems are changing at an unprecedented rate in response to global change [1]. One of its most powerful and probably least studied drivers is land use change. During the last century two opposing forces have coexisted in well-developed regions, such as at the northern fringe of the Mediterranean basin [2], regarding land use: either intensification or abandonment [3,4]. Abandonment is currently occurring in low productive areas (e.g. difficult accessible slopes, steep mountain areas) where the rural exodus has been very significant [5]. Current vegetation dynamics in abandoned fields is modulated by factors such as, past use and management history, soil characteristics, climate and propagules availability [6]. The arrival of seeds to non-forested areas is a key stage in the process of colonization of abandoned agricultural

lands [7] as seed dispersal decreases with the distance to the forest edge [8].

For fleshy fruited species the arrival of seeds to non-forested areas is a function of the abundance, composition and behaviour of the members of the dispersers' community [9]. The service provided by each disperser to a given plant species varies according to differences in both, the quantitative and qualitative components of the dispersal process [10]. The quantitative components are related to differences in the number of visits to a feeding plant, fruits dispersed per visit and local abundance of dispersers. While qualitative component would be mediated by differences in seed retention time, gut treatment and movement behaviours such as home *versus* foraging range and daily movement patterns (e.g. scent marking, anti-predator behaviour) [11]. Despite dispersers differ in these dispersal components [12,13,14], most of the research on seed dispersal mutualisms has been focused on single species, or at the best, in single functional groups. However the few works in which the complete assemblage was considered have reinforced the idea that each species differentially contributed to the quantitative and qualitative terms of the dispersal process [12,15,16,17]. In order to assess how seed dispersal contributes to critical ecological processes such as forest maintenance, regeneration and colonization, the complete community of potential dispersers and their differential behaviour need to be taken into account [16,18,19]. Composition of the dispersers' community and their behavior could vary according to the environmental heterogeneity and ecosystem structure (e.g. woody and shrub cover, fruiting environment) [19,20]. This knowledge seems critical in order to unveil how woodland expansion due to land abandonment operates and to develop adequate management strategies.

Woodlands of *Juniperus thurifera* have been subjected to a traditional management (e.g. logging, grazing and destruction for crop cultivation), however since the middle of the XIX century these activities have drastically decreased allowing the species to increase in density and currently to colonize abandoned fields [21,22] which is provoking a spectacular shift into new colonization areas [23]. *J. thurifera* is a fleshy fruited relict tree with a diverse assemblage of legitimate seed dispersers, such as thrushes *Turdus* spp. [24,25]; carnivores, such as red foxes (*Vulpes vulpes*) and stone martens (*Martes foina*) [12,15]; herbivores, as rabbits (*Oryctolagus cuniculus*) [12,26] and even domestic sheeps (*Ovis aries*) [27]. Since all these animals profoundly differ in their quantitative and qualitative efficiency on the dispersal process [10], their relative contribution to the expansion process of *J. thurifera* must be markedly different.

Dispersal season of *J. thurifera* occurs from the middle autumn until the end of the winter, during this period carnivores, thrushes and rabbits are active frugivores when other food resources are scare [12,24,25,26,27,28] hence we expect all of them will contribute to increase the density of mature woodlands by dispersing high quantity of seeds. Our expectations are that in new colonization areas, where isolated junipers remained, carnivores and thrushes should be the main dispersers, as they can transport a relatively large quantity of seeds [16]. However, the contribution of thrushes, as specialist feeders, will be conditioned to the presence of fruiting trees and crop size. In addition, we expect that the arrival of seeds to remaining active agricultural lands, where *J. thurifera* is absent, will be carried out mainly by carnivores according to their generalist diet and wide home range [19,28]. We also expect that sheeps disperse a low amount of seeds being their contribution to the dispersal process marginal [12]. Dispersers' post-feeding behaviour that conditioned the deposition pattern is also critical. In this sense we hypothesized that thrushes will disperse more seeds beneath the canopy of adult junipers or other fleshy fruited species. Rabbits will disperse more seeds on open pasture where they feed and under the canopy of fruiting trees where warrens used to be located. Carnivores, according to their territoriality and scent marking, will disperse more seeds to visible and conspicuous non canopied microhabitats.

In order to test our hypothesis we evaluate the role of seed dispersal in the expansion process of *J. thurifera* woodlands into new colonization areas by considering the whole dispersal assemblage community, their feeding behaviour and dispersal deposition pattern and how environmental heterogeneity occurring in the ecosystems influence dispersers behaviour. Specifically, we evaluate the following questions: a) what is the quantitative contribution of each member of the dispersal assemblage community to the seed dispersal in the different habitats: mature woodland, new colonization areas and active agricultural lands? b)

How the deposition patterns of each disperser condition the arrival or seeds to different microhabitats? c) Are the quantitative contribution of each disperser and their seed deposition patterns consistent across different sites?

Materials and Methods

Ethics Statement

All necessary permits were obtained for the described field studies from the Dirección General de Montes y Espacios Naturales de Castilla-La Mancha. All animal work was conducted according to relevant Spanish and international guidelines.

Study Area

The study was conducted during *J. thurifera* dispersal season 2008–2009 in *Alto Tajo* and *Parameras de Maranchón, Hoz de Mesa y Arangocillo*, both of them Special Areas of Conservation of the Natura 2000 network located in Guadalajara province, central Spain. The study area covers 40 km^2 (centroid 40° 55¢ N, 2° 10¢ W) (Figure 1A). The climate is Mediterranean continental with rainfall around 500 mm per year with a pronounced summer drought. Mean annual temperature is 10.2°C, with January being the coldest month (mean temperature: 2.4°C) and July the warmest (mean temperature: 19.5°C). Snowfalls occur from November to April (www.aemet.es). The mean elevation of the area is 1200 m where the vegetation is mainly composed by open woodlands dominated by *J. thurifera*.

Sampling Design

The complexity of the territory was classified in three habitats which describe the ongoing process of expansion, mature woodland remnants (MW), new colonization areas (NCA) and ongoing active agricultural lands (AL). *J. thurifera* cover on MWs was over 30% with a high abundance of adult trees. NCAs were abandoned agricultural fields or livestock pastures patches where *J. thurifera* cover was under 15%, being most of the individuals newcomers. In order to evaluate if the expansion process was limited to areas with some *J. thurifera* remanent or contrary if patches without any individual of the species could receive seeds, AL without any *J. thurifera* tree were also included in the study. These habitats did not suffer any management tasks during the survey process. We selected different sites in which the habitats MW, NCA and AL were represented: Maranchón, Torremocha, Cobeta, Riba and Huertahernando (Figure 1A). All habitats were represented in Maranchón and Torremocha, in Cobeta only MW was found while in Riba and Huertahernando only NCA and AL habitats were studied (Figure 1B). In each MW, we selected a total area of 2000×50 m in which seed dispersal process was studied independently for each functional group of dispersers (see Faeces collection below). In NCA and AL seed dispersal was studied in plots of 100×50 meters. A total of 16 NCA plots were selected which were unevenly distributed among Maranchón (6 plots), Torremocha (5), Riba (3) and Huertahernando (2) as a function of the available habitat fragments. For AL a total of 12 plots were selected distributed among Maranchón (4 plots), Torremocha (4), Riba (2) and Huertahernando (2) (Figure 1B).

Faeces Collection

Thrush pellets, carnivore scats and rabbit and sheep droppings (discrete clumps containing from 15 to 20 pellets) were considered as individual and comparable faecal deposits (hereafter faeces). Faeces collection was conducted during the dispersal season (December-March). Each faeces was collected and packed in a paper bag and the microhabitat (*J. thurifera*, shrub or open area) in

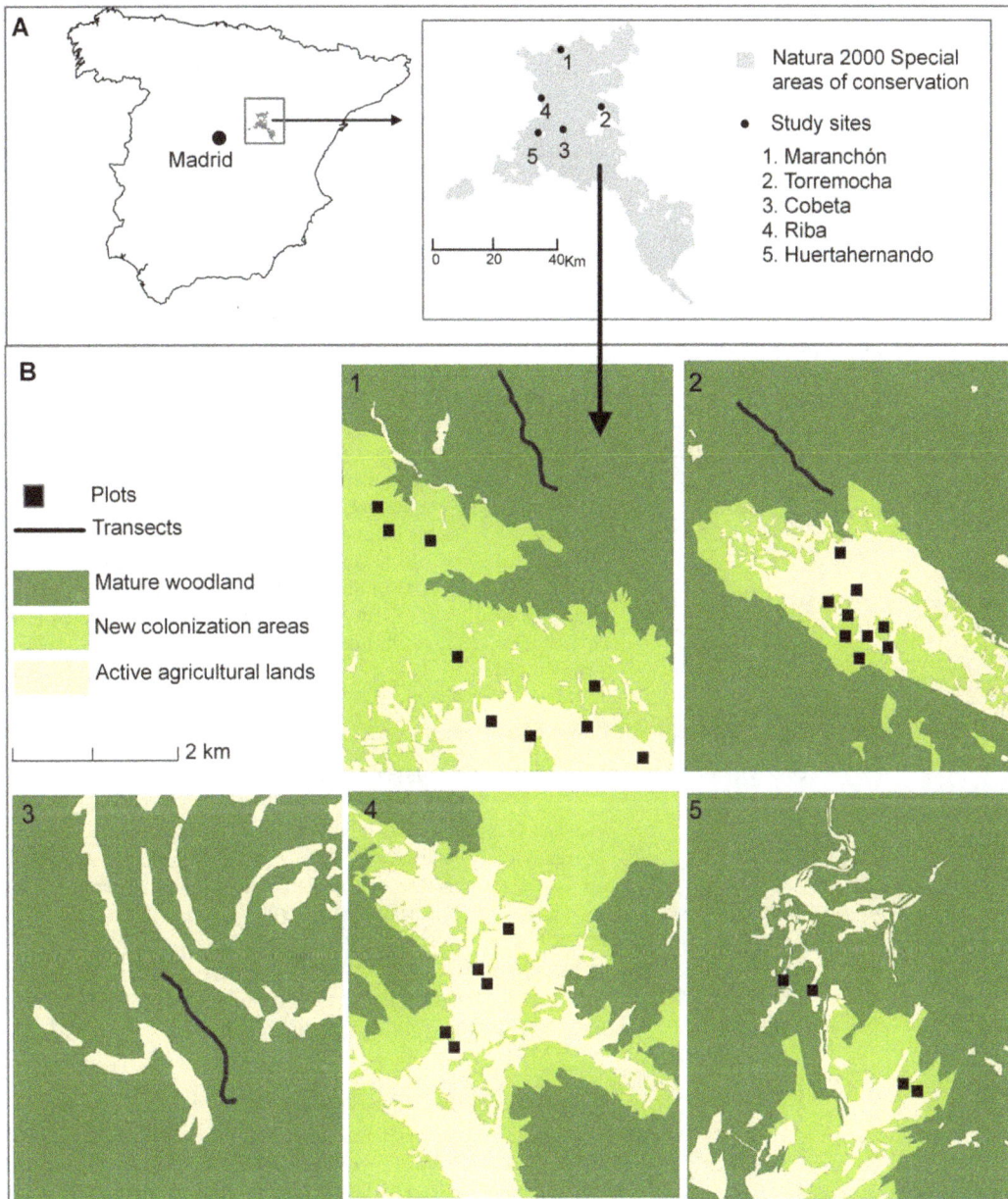

Figure 1. Location, study area and sampling design. Fig. 1A, on the top left of the figure we present the location of the study area within the Iberian Peninsula. On the top right, a zoom to the protected areas, Alto Tajo and Parameras de Maranchón, Hoz de Mesa y Arangocillo (Natural 2000 Network) where the study sites are located. Fig.1B, a graphical description of the different study sites and the amount and distribution of the habitats describing the ongoing process of *J. thurifera* expansion.

which each faeces was found recorded. We considered the influence area of *J. thurifera* and shrub microhabitats of 6 and 4 meters respectively away from the trunk. When a faeces was found within a further distance of these microhabitats it was considered as an open microhabitat.

The survey procedure differs between the studied habitats according to the detectability of the dispersers in each habitat and species behaviour. MWs consisted on homogeneous and continuous areas in the territory with a high percentage of tree cover and fruiting trees, therefore dispersers will highly occupied these habitats [12]. Whereas NCAs-ALs consisted on discrete and small fragments with a reduced tree cover and fruits availability. Thus, a reduced occupancy in relation MW is expected. A reduced

occupancy is related with a lower detectability [29,30,31,32], therefore in order to obtain reliable data about the occupancy of all studied habitats the sampling effort cannot be identical in areas of large occupancy (MW) and those with sporadic or lower intensity of occupancy (NCA and AL). Hence, more sampling effort was needed in those areas where the occupancy and thus detectability is lower. According to this, and attending to different behaviour and movement patterns of dispersers we performed a different sampling scheme for thrushes and mammals. Thrushes dispersal was assessed twice along the study period, coinciding with the moment of thrushes' censuses (see thrushes' abundance section). Thrushes perform a non-random use of the habitat being quite focused to trees with large fruit availability and high size for perching [12,24]. Therefore we performed a stratified sampling focused on trees in mature woodland where the cover of trees was high. We sampled 15 sub-plots within the area selected in each MW. We sampled 10 transects of 1×10 m located at random compass direction away from the microhabitats *J. thurifera tree* and shrub [12] in each sub-plot. However, in NCA very few trees for perching were available and trees with cones were usually one while in AL they were totally absent. In order to avoid overestimation of seed dispersal in NCA and AL patches the total surface of the plot of those habitats was sampled. Mammal's dispersal was assessed fortnightly during the whole dispersal season. Faeces in each MW were collected in a 2000×3 m transect within the selected area, this methodology is optimum for areas with high occupancy and detectability [31]. However, according to the lower detectability on NCA and AL, we needed to increase the sampling effort, therefore we sampled the whole surface of the plot of the mentioned habitats. All fresh faeces were collected and the microhabitat in which they occurred, according to the criteria defined above, recorded.

Juniperus Thurifera Crop Size

We randomly selected 20 adult fruiting trees in each MW and 1 in each NCA (rarely more than one was present). In each tree we counted all the arcestides (organs equivalent to fleshy fruits) inside 4 quadrates of 15×20 cm randomly located in the crown at different heights and compass directions. Since we used a density estimate of arcestides as surrogate of crop size, the sum of the counted arcestides in the four samples was divided by the total sampled area for each tree.

Tree and Shrub Description

We identified all tree and woody shrub species and their covers in all sites. In each MW we walked along 2 km lineal transect. Every 100 meters we established a 15 meter meters radius circumference in which the percentage cover of each tree species and woody shrubs was estimated. The final cover was the result of add up the twenty partial percentage covers for each species. In the case of NCA plots all trees and woody shrubs present were identified and their percentage cover established.

Thrushes' Abundance

We conducted two thrushes' censuses during the study period, in November 2008 (early winter) and in February 2009 (late winter). Within the area selected in MW we established a 2 km length transect with a main belt 50 m wide [12]. In NCA and AL the census were undertaken from a watching point in the centre of each plot. All thrushes seen or heard, walking along the transects or in the watching points were recorded. Total observation effort was 20 hours, 6 in MW, 8 in NCA and 6 in AL. We started census at sunrise and stopped at 11:00 hours. Both early and late winter censuses were conducted during three consecutive days with favourable weather by the same two observers.

Data Analyses

Variation in crop size was analysed using a general mixed model with the number of arcestides (log transformed for obtaining normal error distributions) as response variable with, habitat (MW, NCA) as fixed factor and site as random factor. In order to evaluate the possible effects of environmental heterogeneity (total cover of tree species different to *J. thurifera* and total cover of fleshy fruited species) in dispersers feeding behaviour we performed two Generalized Linear Mixed models (GLMMs). To test if environmental heterogeneity influenced the choice of

dispersers to consume or not consume *J. thurifera* seeds we used the presence/absence of *J. thurifera* seeds in each faeces as response variable with Binomial error distributions and logit as link function. To test if the environmental heterogeneity influenced the quantity of *J. thurifera seeds* consumed by disperses we used the total number of *J. thurifera* seeds dispersed in each faeces as response variable with Poisson error distribution and log as link function. In both models habitat and disperser were used as fixed factors and the variables defining environmental heterogeneity: total cover of tree species different to *J. thurifera* and total cover of fleshy fruited species different of *J. thurifera*, were used as additional fixed factors maintaining site as random factor. In order to test for differences in seed dispersal according to site, habitat, microhabitat and disperser we performed a new GLMM with the density of dispersed seeds per hectare as response variable and habitat, microhabitat and disperser as fixed factors together with site as a random factor being the corresponding error distribution Gaussian and the link function identity. Active agricultural lands habitats (AL) were analysed separately according to the lack of covered microhabitats. In this case a GLMM was performed with density of dispersed seeds per hectare as response variable, disperser as fixed factor and site as random factor with a Gaussian error type and the link function identity. All analyses were conducted in R environment [33] with additional packages "nlme" [34] and "lme4" [35].

Results

Tree and Shrub Description

Maranchón MW had the lowest tree cover among juniper woodlands having Cobeta the highest. In NCAs, Huerta and Torremocha were more open than Maranchón and Riba, which presented the highest cover (72%). The most common fleshy fruited species apart from *J. thurifera* was *J. communis* except in Riba where this rank position was occupied by *Rosa* spp (11.7%) and *J. oxycedrus* (4%). On the other hand *J. phoenicea* (2.3%) was present only in Cobeta which was the MW with the highest number of tree species with more than a tenth percentage covered by pines and oaks (Table 1).

Juniperus Thurifera Crop Size

Arcestides density differed significantly between habitats MWs had a higher density of arcestides than NCA. The estimates were 2.02, 1.37 for MW and NCA respectively which were significantly different from 0 (P value <0.001) in both cases. Arcestides density was similar for the MWs of Torremocha, Maranchón and Cobeta (Figure 2).

Thrushes' Abundance

A total of five species of thrushes were recorded during the censuses: *Turdus viscivorus* (57.1% of thrushes), *Turdus philomelos* (35.8%), *Turdus merula* (4.3%), *Turdus iliacus* (2.1%), and *Turdus pilaris* (0.7%). In general thrushes' abundance was higher in MW than in NCA and AL. However in Torremocha and Riba thrushes' abundance in NCA was similar to MW during the late winter. A similar abundance was found for the three MWs in the two dates, though a decreased was found in Cobeta in late winter. In general we did not observe thrushes in AL with the exception of Torremocha (Table 2).

Environmental Heterogeneity, Faeces Abundance and Number of *J. thurifera* Seeds Per Faeces

A total of 1627 faeces were collected during the study period, of which the highest number were thrushes' pellets (1192), then

Table 1. Tree and shrub species cover.

Site	Habitat	Cover	JT	JC	JO	JP	R	GS	PH	QI	QF
Maranchón	MW	42	30.6	10.2			0.8	1.2			
	NCA	28	10	8.1			0.9	9.3			0.4
Torremocha	MW	60	46	6.6			0.1	9.7		2.5	
	NCA	21	8	0.8			1	10.6		0.1	0.9
Cobeta	MW	71	47	5.7	2.1	2.3	0.3	1.8	17.6	10.8	0.7
Riba	NCA	72	4	0.2	4		11.7	60			0.4
Huertahernando	NCA	10	10				1	1		1	

MW: mature woodland; NCA: new colonization areas; Cover: Total percentage cover (%); Percentage cover of the species (%): JT: *Juniperus thurifera*; JC: *Juniperus communis*; JO: *Juniperus oxycedrus*; JP: *Juniperus phoenicea*; R: *Rosa spp*; GS: *Genista scorpius*; PH: *Pinus halepensis*, QI: *Quercus ilex*; QF: *Quercus faginea*. Blank space indicates the species was not present.

herbivores droppings (224) and lastly carnivores (211). One hundred faeces were collected in AL and therefore were not used for the analysis of dispersers feeding behaviour and environmental heterogeneity analyses due to the lack of natural vegetation in these plots.

Stone marten was the species which dispersed the higher number of faeces with presence of *J. thurifera* and sheeps the lowest. Tree cover excluding *J. thurifera* was negatively related to the presence of *J. thurifera* seeds on dispersers faeces but not to the cover of fleshy fruited species different to *J. thurifera* (Table 3). The number of seeds contained in each faeces was opposite to the rank of number of faeces: carnivores dispersed the highest number of seeds per faeces (36 in average), then herbivores (4) and lastly thrushes (0.7). In general both variables, the number of faeces and *J. thurifera* seed per faeces decreased from the MW to AL for all dispersers (Table 4 and Table S1). Tree cover excluding *J. thurifera* was also negatively related to the number of seeds of *J. thurifera* dispersed per faeces while total cover of fleshy fruited species apart from *J. thurifera* was positively related (Table 4).

In relation to deposition patterns in the MWs, the role of thrushes was similar in the three sites and they usually preferred *J. thurifera* canopied microhabitats. Herbivores, especially rabbits, deposited more faeces and a higher number of seeds per faeces in Torremocha than in the other sites and mainly in open microhabitats. Most carnivores faeces collected contained *J.*

thurifera seeds in the MW of Maranchón (84%) and Torremocha (94%) with a high number of seed per faeces (67 and 96 respectively). By contrast, in the MW of Cobeta most carnivores faeces contained small-mammals remnants and seeds of two coexisting congeners, *Juniperus oxycedrus* and *Juniperus phoenicea* being the average number of *J. thurifera* seeds per faeces really low (5). The microhabitat preferred by carnivores in MW was shrub (Table S1).

Deposition patterns in NCA patches of Maranchón and Torremocha were quite similar to those on the MW but with a lower abundance of faeces and seeds per faeces. In Riba we found that carnivores, mainly stone martens, disperse a high number of faeces in comparison with the rest of the sites. It is worth noting that faeces did not contain any seed of *J. thurifera* in this last case, while congener *J. oxycedrus* seeds were abundant. A similar pattern in the number of deposited faeces *versus* dispersed seeds per faeces was found for thrushes which deposited the highest number of faeces of all NCAs in Riba although only 15% of them contained *J. thurifera* seeds. They had a low number of seed per faeces (0.21), being *Rosa* spp seeds abundant. We did not found any carnivores' faeces in Huerta and only thrushes generated some seed dispersion preferring the *J. thurifera* canopied microhabitat (Table S1).

In AL the total number of faeces and seed number per faeces

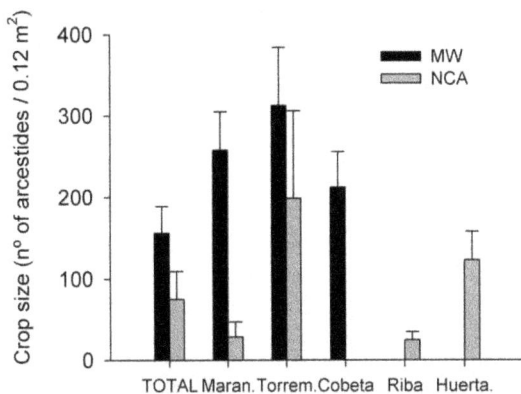

Figure 2. Crop size. Crop size (mean ± SE number of arcesti-des*0.12m^{-2}) on mature woodlands (MW) and new colonization areas (NCA) for all studied sites. TOTAL (mean ± SE number of arcestides/0.12m^2) for each habitat type is represented on the first two columns.

Table 2. Thrushes abundance.

Site	Habitat	Early Winter	Late Winter
Maranchón	MW	9.1	13.8
	NCA	1.7	0.33
	AL	0	0
Torremocha	MW	11.4	9.1
	NCA	0	12.8
	AL	0.5	7.5
Cobeta	MW	12.3	6.4
Riba	NCA	2	12
	AL	0	0
Huertahernando	NCA	0	0
	AL	0	0

Thrushes' abundance (*Turdus spp* Ha^{-1}) in the three habitats, MW: mature woodland; NCA, new colonization areas; AL, active agricultural lands.

Table 3. General linear model result for *J. thurifera* seeds presence/absence in each faeces.

Fixed effects		Parameter value	SE	z value	Pr(>|z|)
Intercept		0.31	1.56	0.20	0.842
Disperser	**Stone marten**	**1.46**	**0.56**	**2.61**	**0.009**
	Red fox	0.69	0.50	1.39	0.166
	Thrushes	0.46	0.45	1.02	0.309
	Sheep	**−1.64**	**0.49**	**−3.32**	**0.001**
Habitat	**New colonization areas**	**−0.67**	**0.26**	**−2.58**	**0.010**
Fruits cover		0.03	0.05	0.66	0.510
Tree cover		**−0.45**	**0.10**	**−4.26**	0.000
Random effects					
	Intercept	Residual			
SD	9.67	3.11			

Significant effects (P<0.05) are indicated in bold. When P value was smaller than 0.001, <0.001 was indicated. D.F: degrees of freedom. SE: Standard Error. SD: Standard Deviation. Missing levels of factors (disperser: rabbit; habitat: woodland) are included on the intercept.

sharply decreased in relation to MW and NCA. The dispersal pattern generated by dispersers was site depended. Thus in Maranchón only foxes produced dispersal patterns, in Torremocha both species of carnivores, sheeps and thrushes deposited faeces and seeds while in Huerta only foxes and thrushes did. Finally we did not found any dispersed seed by any disperser in the Riba AL (Table S1).

Seed Dispersal

Seed dispersal was one hundred fold times higher in MW than in NCA. The disperser which produced the highest seed dispersal was the red fox followed by the stone marten, thrushes and lastly herbivores (Table 5, Figure 3). However their relative efficiency was habitat-depend (Figure 3, Figure 4). Carnivores played a more important role on the seed dispersal process in MW while in NCA the relative importance of all dispersers was similar (Figure 3, Figure 4). Seed dispersal did not vary across microhabitats, thus all microhabitats receive a similar amount of seeds as a result of the different deposition patterns of dispersers (Figure 4). Both species of carnivores presented a clear preference for shrub and open

microhabitats in MW and in NCA whereas thrushes and herbivores changed their deposition patterns with the habitat. Thrushes dispersed similar quantities of seeds beneath the crown of *J. thurifera* trees and open areas in MW while in NCA most of the seeds dispersed by them where on *J. thurifera* microhabitat. On the other hand herbivores in MW dispersed small number of seeds and mostly in open microhabitats while in NCA they dispersed more seeds and mostly beneath the crown of *J. thurifera* trees (Figure 4). For AL, carnivores were the main dispersers especially foxes (figure 5), which were the only ones with an estimate significantly different from 0 (Estimate 29.37, P = 0.0031).

Discussion

The main functional groups of dispersers of *J. thurifera* were thrushes and carnivores since herbivores, especially sheeps, dispersed significantly fewer seeds. According to our expectations thrushes and carnivores had a critical role in the process of woodland expansion and colonization of abandoned fields. Both thrushes and carnivores differed in their efficiency according to

Table 4. General linear model result for *J. thurifera* seeds abundance per faeces.

Fixed effects		Parameter value	SE	z value	Pr(>|z|)
Intercept		**2.62**	**1.18**	**2.23**	**0.03**
Disperser	**Red fox**	**1.05**	**0.06**	**18.21**	**<0.001**
	Stone marten	**0.94**	**0.06**	**15.95**	**<0.001**
	Sheep	**−2.33**	**0.10**	**−24.01**	**<0.001**
	Thrushes	**−3.13**	**0.07**	**−47.15**	**<0.001**
Habitat	**New colonization areas**	**−1.81**	**0.09**	**−20.16**	**<0.001**
Fruits cover		**0.06**	**0.02**	**2.64**	**0.008**
Tree cover		**−0.40**	**0.05**	**−8.68**	**<0.001**
Random effects					
	Intercept	Residual			
SD	6.61	2.57			

Significant effects (P<0.05) are indicated in bold. When P value was smaller than 0.001, <0.001 was indicated. D.F: degrees of freedom. SE: Standard Error. SD: Standard Deviation. Missing levels of factors (disperser: rabbit; habitat: woodland) are included on the intercept.

Table 5. General mixed model result for seed dispersal analyses for mature woodland (MW) and new colonization areas (NCA).

Fixed effects		Parameter value	SE	DF	t-value	p-value
Intercept		−1088.47	686.86	93	−1.58	0.116
Habitat	**Woodland**	**2139.98**	**502.88**	**93**	**4.26**	**<0.001**
Disperser	Stone marten	983.49	676.23	93	1.45	0.149
	Sheep	−41.71	676.23	93	−0.06	0.951
	Red fox	**2160.13**	**676.23**	**93**	**3.19**	**0.002**
	Thrushes	514.63	676.23	93	0.76	0.449
Microhabitat	Shrub	704.36	523.81	93	1.34	0.182
	J. thurifera tree	−305.48	523.81	93	−0.58	0.561
Random effects						
	Intercept	Residual				
SD	726.70	2191.24				

Significant effects (P<0.05) are indicated in bold. When P value was smaller than 0.001, <0.001 was indicated. D.F: degrees of freedom. Missing levels of factors (disperser: rabbit; habitat: new colonization areas; microhabitat: open) are included on the intercept.

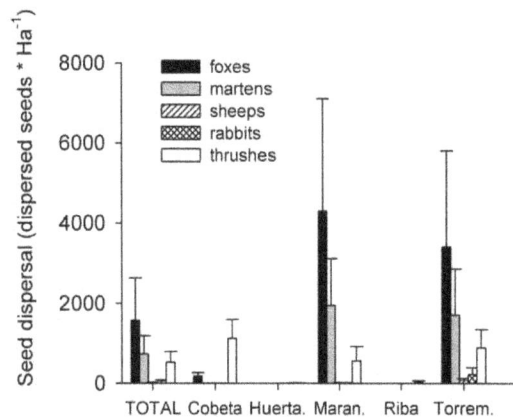

Figure 3. Seed dispersal site and dispersers. Seed dispersal (mean ± SE of dispersed seeds*Ha^{-1}) generated by the different dispersers in the five study sites. TOTAL (mean ± SE) for each disperser is represented by the columns on the left.

their feeding behavior and post dispersal pattern, which was modulated by habitat type and particularly by environmental heterogeneity and ecosystem structure. This complex interaction between dispersers and ecosystem heterogeneity surely is conditioning the spatial and genetic structure of the expanding and former woodlands. Our results also shed light on the poorly understood process of seed dispersal in heterogeneous habitats at the landscape scale, especially complex in the case of very fragmented or highly disturbed landscapes such as abandoned fields [36].

Differential Role of the Disperser Community on the Process of Woodland Expansion

It is well known the role of frugivores birds on plant dispersal far from parental trees [37,38], since large and medium-sized birds can fly intermediate and long distances (>200 m), even across open areas located between forest or shrubland remnants [16,39,40]. In the case of thrushes, some species, as *T. viscivorus*, *T. pilaris* and *T. iliacus*, showed flight distances ranging from 50 m to 300 m after feeding [16,17], or even longer (>500 m, personal observation). They perform high-height exploratory flights in large flocks [41] which may contribute to long dispersal events outside mature plant populations on woodland remnants, even in agricultural or abandoned lands as also found in our paper (see [42] for similar north-American landscapes). However the importance of thrushes as disperses in new colonization areas is conditioned to the presence of attractive perches as high trees [17,43] with big crop size [41,44]. According to our results a higher percentage cover of other fleshy fruited species was positively related with a higher number of seeds dispersed per faeces. This could be the explanation to the highly different dispersal patterns found for thrushes in two of our study sites, Riba and Huerta. It seems that a more abundant fruiting neighbourhood could produce an attraction effect on thrushes and enhance seed dispersal for the target species even if this one is not very abundant. A similar pattern has previously been shown in other

ecological contexts [44,45]. It seems a general rule that for fleshy fruited species is very positive in dispersal terms to have a heterospecific fruiting neighbourhood. Obviously, depending on total availability of different fruiting species and their spatial location this positive effect on seed dispersal could turn into a competition effect [46].

Regarding carnivores many studies in Mediterranean ecosystems point out their relevance in long-distance seed dispersal due to their wide home range, generalist diet and high retention time in the gut [15,16,47,48]. Seed dispersal produced by carnivores is independent on the current presence of mature fruiting trees [19] as shown also in our study system where we found a higher incidence of red foxes and stone martens as dispersers in active agricultural lands. This means that for primary colonization where no trees of the species are present, carnivores, especially red foxes, could be essential members of the dispersal community and may promote natural restoration of degrade lands as recently proposed [49].

Differential Role of the Disperser Community on Mature Woodlands

Numerous studies point out that thrushes are the main dispersers of *Juniperus* while the contribution of mammals is occasional and less relevant [24,25,50,51,52]. Our results do not support such statement, as carnivores and specially red foxes, were by far the main dispersers in two of the tree studied mature woodlands. Changes in tree species cover and structure could produce variations in the availability of different food resources for carnivores. According to our results a higher diversity in tree cover was negatively related with the consumption of *J. thurifera* fruits and with the total number of seeds dispersed per faeces, both variables were notably lower in the mature woodland of Cobeta compared with the other MWs studied. This result could be explained by the existence of a high diversity in the tree layer with the presence of pines and oak species. Higher tree diversity and the presence of some deciduous species promote a more abundant litter cover which has been related with a more abundant and diverse insects and rodents communities [53,54,55,56]. Both, insects and small rodents are consumed and frequently preferred by carnivores [57,58,59] due to higher protein content. Therefore we speculate that a higher diversity on the tree layer may had produced a higher availability of trophic resources different from fruits provoking a shift in carnivores diets, More specifically this

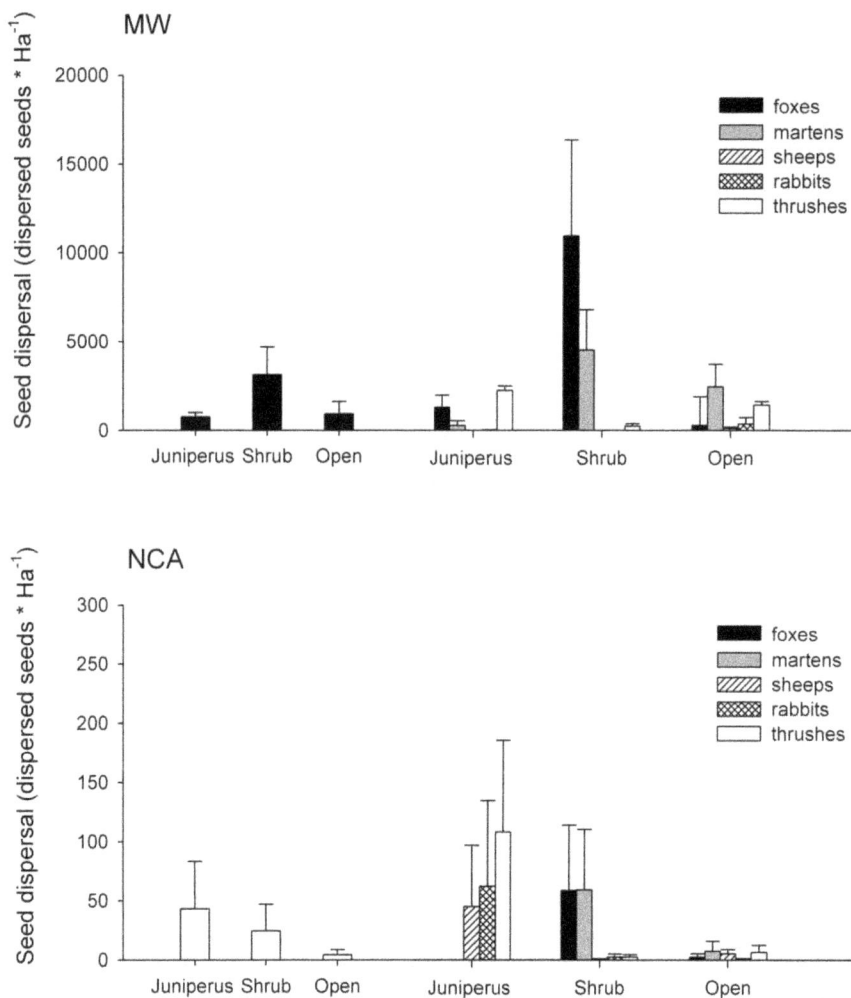

Figure 4. Seed dispersal, habitats, microhabitats and dispersers. Seed dispersal (mean $\pm$ SE of dispersed seeds* Ha^{-1}) generated by the different dispersers in each microhabitat in Mature Woodland (MW) on the upper panel and New colonization Areas (NCA) on the lower panel. For both panels the first three columns correspond to the mean $\pm$ SE total dispersed seeds by all dispersers on each microhabitat. Black columns correspond to Mature Woodland and white columns correspond to New colonization areas.

hypothesis should help to explain the low seed dispersal generated by carnivores in Cobeta [12]. In support to this hypothesis carnivores' faeces found in the MW of Cobeta contained mainly small-mammals remains (personal observation).

Thrushes' population present in the study area were mainly wintering migrants. In their arrival and during their stay they make prospect flights searching for good patches for feeding and avoiding predators [12,24,60]. According to our results, thrushes profited fruit resources according to their abundance. In the studied sites where *J. thurifera* cover and crop were abundant, thrushes dispersed a high and similar quantity of seeds. As a result mature woodlands which offered enough crop size, independently of their spatial structure, and presence of other tree species, would result appropriate for wintering thrushes. Thus, as long as migratory period of thrushes and fruiting moment will be accomplished we could assure that thrushes would be constant and faithful dispersers of *J. thurifera*.

Seed Dispersal and Deposition Pattern Between Microhabitats

As a result of dispersers' deposition pattern, shrubs and open microhabitat will receive more seeds dispersed by carnivores while *J. thurifera* canopies will receive more seeds dispersed by thrushes. Thrushes and carnivores had a different clumping pattern (1 seed/faeces for thrushes *versus* 50 for carnivores) and we detected higher seeds weight for carnivores than thrushes (unpublished data) which suggests the existence of a playground for evolution to operate. Therefore as a result of dispersers behaviour seeds dispersed on open or shrub microhabitats are heavier than those dispersed beneath the crown of *J. thurifera* tree. Whether or not heavier seeds in open microhabitats will increase or decrease the probability to be predated, secondary dispersed or germinated and finally established remains unknown. Any case recruitment, and species traits' evolution, would be highly influenced by the interaction among the quantity of seeds do arrive to a microhabitat, the traits selected by dispersers (e.g. seed size) and the environmental characteristics of each microhabitat [61].

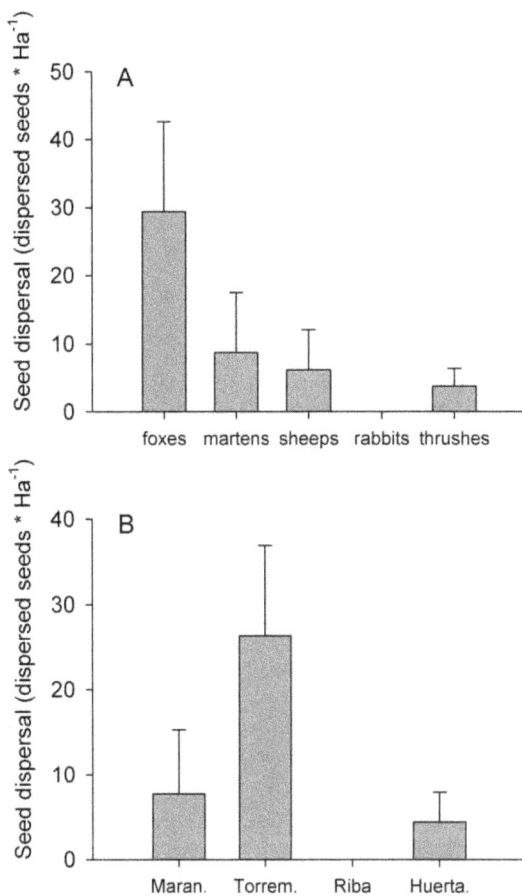

Figure 5. Seed dispersal in agricultural lands. Seed dispersal (mean ± SE of dispersed seeds* Ha^{-1}) generated by each disperser species on panel A and on each study site at the panel B.

Conclusions

Abandonment of agricultural activities has promoted the colonization of many fields by shrub and tree species. For ecosystems dominated by fleshy fruited species, the arrival of seeds as the first step in the process of colonization is mediated by the feeding behaviour and post dispersal deposition pattern of the community of dispersers as shown here. Our results are congruent with our expectations that the role of carnivores is critical for moving seeds into agricultural lands where isolated trees and perches are absent, therefore this functional group of dispersers are a critical member of the dispersal assemblage for promoting the colonization of abandoned fields. Although, in order to describe the importance of seed dispersal in ecological processes, it is essential to take into account the whole dispersal community together with the environmental heterogeneity occurring at the landscape level, (e.g. vegetation cover and composition), as these variables significantly influence dispersers' behaviour. Our results showed a decrease in seed dispersal when tree species apart from *J. thurifera* are present, therefore in ecosystems where *J. thurifera* is not the dominant species its dispersal and therefore regeneration and colonization of abandoned fields could be constrained. This finding together with the results described in [62] suggest that *J. thurifera* open woodlands with a reduced grazing pressure, as occurred in our study sties, could produce a shift of their typical open monospecific woodlands towards a closed and more diverse canopy forest. In this scenario seed dispersal of *J. thurifera* could result limited. Therefore, having into account that these formations have conservation concerns the diversity of the dispersal community is an important ecosystem feature that should be preserved and even managed (e.g. avoiding hunting thrushes and predators control of medium carnivores such as red foxes and stone martens) in order to promote the colonization of abandoned fields, the regeneration of former woodlands [63], and the ecosystem services provided by them, such as net gain of value habitat, water and nutrient cycling and carbon sink capacity [23].

Supporting Information

Table S1 Faeces abundance and number of *J. thurifera* seeds per faeces. MW: mature woodland. NCA: new colonization areas. AL: active agricultural lands. Mh: Microhabitat. N: Number of faeces* ha^{-1}. Nj: Percentage of faeces with *J. thurifera* seeds respect the total number of faeces. S/N: average number of *J. thurifera* seeds*faece^{-1}.

Author Contributions

Conceived and designed the experiments: BP EV AE FV. Performed the experiments: VSP GEA BP EV. Analyzed the data: GEA BP EV. Wrote the paper: GEA VSP BP EV. Critically reviewed the paper: AE FV.

References

1. Sala OE, Chapin FS, Armesto JJ, Berlow E, Bloomfield J, et al. (2000) Biodiversity - Global biodiversity scenarios for the year 2100. Science 287: 1770–1774. 10.1126/science.287.5459.1770.
2. Lavorel S, Canadell J, Rambal S, Terradas J (1998) Mediterranean terrestrial ecosystems: research priorities on global change effects. Global Ecol. Biogeogr. 7: 157–166. 10.1046/j.1466-822X.1998.00277.x.
3. Mottet A, Ladet S, Coque N, Gibon A (2006) Agricultural land-use change and its drivers in mountain landscapes: A case study in the Pyrenees. Agric. Ecosyst. Environ. 114: 296–310. 10.1016/j.agee.2005.11.017.
4. Rey Benayas JM, Martins A, Nicolau JM, Schulz JJ (2007) Abandonment of agricultural land: an overview of drivers and consequences. CAB Reviews 057: 1–14.
5. Westhoek HJ, van den Berg M, Bakkes JA (2006) Scenario development to explore the future of Europe's rural areas. Agric. Ecosyst. Environ. 114: 7–20. 10.1016/j.agee.2005.11.005.
6. Tasser E, Walde J, Tappeiner U, Teutsch A, Noggler W (2007) Land-use changes and natural reforestation in the Eastern Central Alps. Agric. Ecosyst. Environ. 118: 115-129. 10.1016/j.agee.2006.05.004.
7. Duncan RS (2006) Tree recruitment from on-site versus off-site propagule sources during tropical forest succession. New For. 31: 131–150. 10.1007/s11056-004-5395-7.
8. Aide T, Cavelier J (1994) Barriers to Lowland Tropical Forest Restoration in the Sierra Nevada de Santa Marta, Colombia. Restor. Ecol. 2: 219–229. 10.1111/j.1526-100X.1994.tb00054.x.
9. Nathan R, Muller-Landau HC (2000) Spatial patterns of seed dispersal, their determinants and consequences for recruitment. Trends Ecol. Evol. 15: 278–285. 10.1016/s0169-5347(00)01874-7.
10. Schupp EW, Jordano P, Gomez JM (2010) Seed dispersal effectiveness revisited: a conceptual review. New Phytol. 188: 333–353. 10.1111/j.1469-8137.2010.03402.x.
11. Fragoso JMV (1997) Tapir-generated seed shadows: scale-dependent patchiness in the Amazon rain forest. J. Ecol. 85: 519–529. 10.2307/2960574.
12. Santos T, Tellería JL, Virgós E (1999) Dispersal of Spanish juniper *Juniperus thurifera* by birds and mammals in a fragmented landscape. Ecography 22: 193–204.
13. Traveset A, Verdú M (2002) A Meta-analysis of the Effect of Gut Treatment on Seed Germination. CAB International 2002 Seed Dispersal and Frugivory: Ecology, Evolution and Conservation Eds DJ Levey, WR Silva and M Galetti: 339–350.
14. Russo SE (2003) Responses of dispersal agents to tree and fruit traits in *Virolacalophylla* (*Myristicaceae*): implications for selection. Oecologia 136: 80–87. 10.1007/s00442-003-1239-y.

15. Herrera CM (1989) Frugivory and seed dispersal by carnivorous mammals and associated fruit characteristics, in undisturbed mediterranean habitats. Oikos 55: 250–262. 10.2307/3565429.

16. Jordano P, García C, Godoy JA, García-Castano JL (2007) Differential contribution of frugivores to complex seed dispersal patterns. Proc. Natl. Acad. Sci. U S A. 104: 3278–3282. 10.1073/pnas.0606793104.

17. Martinez I, Garcia D, Obeso JR (2008) Differential seed dispersal patterns generated by a common assemblage of vertebrate frugivores in three fleshy-fruited trees. Ecoscience 15: 189–199. 10.2980/15-2-3096.

18. Calviño-Cancela M, Martín-Herrero J (2009) Effectiveness of a varied assemblage of seed dispersers of a fleshy-fruited plant. Ecology 90: 3503–3515. 10.1890/08-1629.1.

19. López-Bao JV, González-Varo JP (2011) Frugivory and Spatial Patterns of Seed Deposition by Carnivorous Mammals in Anthropogenic Landscapes: A Multi-Scale Approach. PLoS One 6.e14569.

20. García D, Zamora R, Amico GC (2010) Birds as Suppliers of Seed Dispersal in Temperate Ecosystems: Conservation Guidelines from Real-World Landscapes. Conserv. Biol. 24: 1070–1079. 10.1111/j.1523-1739.2009.01440.x.

21. Blanco E, Casado M, Costa M, Escribano R, García M, et al. (2005) Los bosques ibéricos: una interpretación geobotánica. 4ª ed rev Barcelona, Planeta.

22. Olano JM, Rozas V, Bartolome D, Sanz D (2008) Effects of changes in traditional management on height and radial growth patterns in a Juniperus thurifera L. woodland. For. Ecol. Manage. 255: 506–512. 10.1016/j.foreco.2007.09.015.

23. Gimeno T, Pías B, Martínez-Fernández J, Quiroga D, Escudero A, et al. (2012) The decreased competition in expanding versus mature juniper woodlands is counteracted by adverse climatic effects on growth. Eur. J. For. Res. 131: 977–987. 10.1007/s10342-011-0569-2.

24. Jordano P (1993) Geographical ecology and variation of plant-seed disperser interactions-Southern spanish junipers and frugivorous thrushes Vegetatio 108: 85–104.

25. Santos T, Tellería JL (1994) Influence of forest fragmentation on seed consumption and dispersal of spanish juniper Juniperus thurifera Biol. Conserv. 70: 129–134. 10.1016/0006-3207(94)90280-1.

26. Muñoz J (1993) Consumo de gálbulos de sabina (Juniperus sp. turbinata GUSS,1891) y dispersión de semillas por el conejo (Oryctolaguscuniculus L.) en el Parque Nacional de Doñana. Donaña, Acta Vertebrata 20: 49–58.

27. Heras J, Ruiz M, Aguilera E, Herranz J (1994) Determinación del potencial reproductivo de Juniperus thurifera e influencia antrópica sobre el mismo. Albasit 34: 161–181.

28. Rosalino LM, Rosa S, Santos-Reis M (2010) The role of carnivores as Mediterranean seed dispersers. Ann. Zool. Fenn. 47: 195–205.

29. Kerr JT, Sugar A, Packer L (2000) Indicator taxa and biodiversity assessment and nestedness in an endangered ecosystem. Conserv. Biol. 14: 1726–1734.

30. MacKenzie DI, Nichols JD, Lachman GB, Droege S, Royle JA, et al. (2002) Estimating site occupancy rates when detection probabilities are less than one. Ecology 83: 2248–2255.

31. Virgós E, Tellería JL, Santos TA (2002) A comparison on the response to forest fragmentation by medium-sized Iberian carnivores in central Spain. Biodivers. Conserv. 11: 1063–1079.

32. Gu W, Swihart RK (2004) Absent or undetected? Effects of non-detection of species occurrence on wildlife-habitat models. Biol. Conserv. 116: 195–203.

33. R Development Core Team, (2011) R: A language and environment for statistical computing. R foundation for Statistical Computing, Vienna, Austria.Available: http://www.R-project.org.

34. Pinheiro J, Bates D, DebRoy S, Sarkar D, the R Development Core Team (2012). nlme: Linear and Nonlinear Mixed Effects Models. R package version 3.1–104.

35. Bates D, Maechler M, Bolker B (2012). lme4: Linear mixed-effects models using S4 classes. R package version 0.999999–0. Available: http://CRAN.R-project. org/package = lme4.

36. McConkey K, Prasad S, Corlett RT, Campos-Arceiz A, Brodie J, et al. (2012) Seed dispersal in changing landscapes. Biol. Conserv. 10.1016/j.biocon.2011.09.018.

37. Deckers B, Verheyen K, Hermy M, Muys B (2005) Effects of landscape structure on the invasive spread of black cherry Prunus serotina in an agricultural landscape in Flanders, Belgium. Ecography 28: 99–109. 10.1111/j.0906-7590.2005.04054.x.

38. Holbrook KM, Smith TB, Hardesty BD (2002) Implications of long-distance movements of frugivorous rain forest hornbills. Ecography 25: 745–749. 10.1034/j.1600-0587.2002.250610.x.

39. Breitbach N, Laube I, Steffan-Dewenter I, Bohning-Gaese K (2010) Bird diversity and seed dispersal along a human land-use gradient: high seed removal in structurally simple farmland. Oecologia 162: 965–976. 10.1007/s00442-009-1547-y.

40. Weir JES, Corlett RT (2007) How far do birds disperse seeds in the degraded tropical landscape of Hong Kong, China? Landsc. Ecol. 22: 131–140. 10.1007/s10980-006-9002-5.

41. García D, Ortiz-Pulido R (2004) Patterns of resource tracking by avian frugivores at multiple spatial scales: two case studies on discordance among scales. Ecography 27: 187–196. 10.1111/j.0906-7590.2004.03751.x.

42. Livingston R (1972) Influence of birds, stones and soil on establishment of pasture juniper, Juniperus communis and Red cedar, Juniperus virginiana in New England pastures. Ecology 53: 1141–1147.

43. Debussche M, Escarre J, Lepart J (1982) Ornithochory and plant succession in Mediterranean abanoned orchards. Vegetatio 48: 255–266.

44. García D, Zamora R, Gomez JM, Hodar JA (2001) Frugivory at Junipepus communis depends more on population characteristics than on individual attributes. Journal of Ecology 89: 639–647. 10.1046/j.1365–2745.2001.00577.x

45. Herrera CM (1984) Seed dispersal and fitness determinants in wild rose: Combined effects of hawthorn, birds, mice, and browsing ungulates. Oecologia 63: 386–393.

46. Carlo TA, Aukema JE, Morales JM (2007) Plant-Frugivory Interactions as Spatially Explicit Networks: Integrating Frugivore Foraging with Fruiting Plant Spatial Patterns. In Dennis AJ, Schupp EW, Green RJ, Westcott DA, editors. Seed Dispersal: Theory and its Application in a Changing World. CAB International. 369–390.

47. Calisti M, Ciampalini B, Lovari S, Lucherini M (1990) Food-habits and trophic niche variation of the red fox Vulpes vulpes (L, 1758) in a mediterranean coastal area. Revue D Ecologie-La Terre Et La Vie 45: 309–320.

48. Virgós E, Llorente M, Cortes Y (1999) Geographical variation in genet (Genetta genetta L.) diet: a literatura review. Mammal Rev. 29: 117–126. 10.1046/j.1365–2907.1999.00041.x.

49. Matías L, Zamora R, Mendoza I, Hodar JA (2010) Seed Dispersal Patterns by Large Frugivorous Mammals in a Degraded Mosaic Landscape. Restor. Ecol. 18: 619–627. 10.1111/j.1526–100X.2008.00475.x.

50. Barlow J, Silveira JM, Mestre LAM, Andrade RB, Camacho D'Andrea G, et al. (2012) Wildfires in Bamboo-Dominated Amazonian Forests: Impacts on Above-Ground Biomass and Biodiversity. PLoS One 10.1371/journal.pone.0033373.

51. Vasconcelos HL, Pacheco R, Silva RC, Vasconcelos PB, Lopes CT, et al.(2009) Dynamics of the Leaf-Litter Arthropod Fauna Following Fire in a Neotropical Woodland Savanna. PLoS One 10.1371/Journal.pone.0007762.

52. Camacho J, Moreno S (1989) Datos sobre la distribución espacial de micromamíferos en el Parque Nacional de Doñana. Donaña, Acta Vertebrata 16: 239–245.

53. Díaz M, Gonzalez E, Muñoz-Pulido R, Naveso M (1993) Effects of food abundance and habitat structure on seed-eating rodents in Spain wintering in man-made habitats. Zeitschriftfür Säugetierkunde 58: 302–311.

54. Pandolfi M, DeMarinis AM, Petrov I (1996) Fruit as a winter feeding resource in the diet of stone marten (Martes foina) in east-central Italy. Z Saugetierkd-Int J Mamm. Biol. 61: 215–220.

55. Barrientos R, Virgós E (2006) Reduction of potential food interference in two sympatric carnivores by sequential use of shared resources. Acta Oecol.-Int. J. Ecol. 30: 107–116. 10.1016/j.actao.2006.02.006.

56. Padial JM, Avila E, Gil-Sanchez JM (2002) Feeding habits and overlap among red fox (Vulpes vulpes) and stone marten (Martes foina) in two Mediterranean mountain habitats. Mamm. Biol. 67: 137–146. 10.1078/1616-5047-00021.

57. Herrera CM (1995) Plant-vertebrate seed dispersal systems in the mediterranean-ecological, evolutionary and historical determinants. Annu. Rev. Ecol. Syst. 26: 705–727. 10.1146/annurev.ecolsys.26.1.705.

58. Alcántara JM, Rey PJ (2003) Conflicting selection pressures on seed size: evolutionary ecology of fruit size in a bird-dispersed tree, Olea europaea. J. Evol. Biol. 16: 1168–1176. 10.1046/j.1420–9101.2003.00618.x.

59. De Soto L, Olano JF, Rozas V, De la Cruz M (2009) Release of Juniperus thurifera woodlands from herbivore mediated arrested succession in Spain. Appl. Veg. Sci. 10.1111/j.1654–109X.2009.01045.x.

60. García D, Martínez D (2012) Species richness matters for the quality of ecosystem services: a test using seed dispersal by frugivorous birds. Proceedings of the Royal Society of London. 10.1098/rspb.2012.0175.

18

A Multi-Criteria Index for Ecological Evaluation of Tropical Agriculture in Southeastern Mexico

Esperanza Huerta[1]*, Christian Kampichler[2,3], Susana Ochoa-Gaona[4], Ben De Jong[4], Salvador Hernandez-Daumas[1], Violette Geissen[5,6]

1 El Colegio de la Frontera Sur, Unidad Campeche, Dpto. Agroecología, Campeche, México, 2 Universidad Juárez Autónoma de Tabasco, División de Ciencias Biológicas, Villahermosa, Tabasco, México, 3 Sovon Dutch Centre for Field Ornithology, Natuurplaza (Mercator 3), Nijmegen, The Netherlands, 4 El Colegio de la Frontera Sur, Unidad Campeche, Dpto. Sustainability Sciences, Campeche, México, 5 University of Bonn - INRES, Bonn, Germany, 6 Wageningen University and Research Center – Alterra, Wageningen, Gelderland, Netherlands

Abstract

The aim of this study was to generate an easy to use index to evaluate the ecological state of agricultural land from a sustainability perspective. We selected environmental indicators, such as the use of organic soil amendments (green manure) versus chemical fertilizers, plant biodiversity (including crop associations), variables which characterize soil conservation of conventional agricultural systems, pesticide use, method and frequency of tillage. We monitored the ecological state of 52 agricultural plots to test the performance of the index. The variables were hierarchically aggregated with simple mathematical algorithms, if-then rules, and rule-based fuzzy models, yielding the final multi-criteria index with values from 0 (worst) to 1 (best conditions). We validated the model through independent evaluation by experts, and we obtained a linear regression with an $r^2 = 0.61$ ($p = 2.4e-06$, $d.f. = 49$) between index output and the experts' evaluation.

Editor: Yong Deng, Southwest University, China

Funding: Financial support was obtained from the Conacyt-SEMARNAT Project "Uso sustentable de los recursos naturales en la frontera sur de México" (Sustainable use of natural resources on the southern border of Mexico) (code SEMARNAT-2002-C01-1109). El Colegio de la Frontera Sur provided infrastructural resources. The funders had no role in study design, data collection and analysis, decision to publish, or preparation of the manuscript.

Competing Interests: The authors have declared that no competing interests exist.

* Email: ehuerta@ecosur.mx

Introduction

In the past 60 years, degradation and deforestation of tropical forests worldwide have occurred much faster and more extensively than in any other period in history [1], [2]. Furthermore, countries like Mexico have been undergoing drastic land use changes. In the tropics of Mexico large parts of its lowland rainforest areas have been converted into pasture and cropland. In the state of Tabasco, for example, only 3.4% of the state is covered with original forest [3], whereas 76.4% of the surface was used for cattle production in 2000 [4] and 15.6% was used for agriculture, principally sugarcane and fruit plantations [3]. The ecological consequences of these land-use changes in Tabasco are well documented. Soil losses in hilly regions are very high up to 200 t ha^{-1} year^{-1} [5]; high pesticide and fertilizer inputs to crops that have replaced forests have caused considerable environmental contamination [6], [7], and soil fertility is decreasing [8], [5], [9].

It is of the utmost importance to identify sustainable land use strategies which are economically attractive for the region's farmers and which may also reconcile the need for food production with that of soil conservation. In order to assist a variety of stakeholders at the local and regional level in making land use decisions, simple evaluation tools are needed. This is even more needed since a high percentage of the population consists of immigrants from other Mexican states, who are unfamiliar with the conditions of the humid tropics and use intensive techniques

for farming the land. This is mostly due to large-scale agricultural development projects, such as "Plan Balancán-Tenosique," named after the two municipalities involved, in which, in the 1970 s, over 1100 km^2 of lowland rain forest was destroyed and converted into crop and pasture land. Ecological values of the 1970 s were very different from those of the present, and government representatives were willing to deforest in order to grant land to farmers [10].

The direct impact of farming is difficult to measure due to methodological difficulties (impossibility of measurement, complexity of the system) or practical reasons (time, costs) [11]. Therefore, the use of indicators appears to be an alternative way of guiding land use decisions [12], [13]. However, the "indicator explosion" [14], that is, the use of an exaggerated number of indicators aimed at assessing environmental impacts of agricultural activities, has been of little use to local decision-makers. Particularly in the tropics, land use decisions are still based on the informal opinion of local experts rather than on implementation of Decision Support Systems for environmentally sound resource management [15]. On the one hand, this is due to farmers' restricted access to modern communications and information technologies; on the other, application of indicators is often beyond the capability of local farmers. For example, the agricultural sustainability index proposed by Nambiar et al. [16], which aims to measure agricultural sustainability as a function of biophysical, chemical, economic, and social indicators, would require considerable training of government stakeholders and farmers in order to be applied, and such training is rarely available.

Thus, any tool which local farmers or regional decision makers may use to support their decision making must be as simple as possible. Agroecosystems (like any other ecosystem) are too complex to be precisely measured and evaluated [17]. We agree with the view of Darnhofer et al. [18] in favour of developing less precise rules of thumb which may be used by farmers as well as to guide local land-use decisions toward a more environmentally friendly system of agriculture.

This index may provide farmers and others involved with a tool to evaluate the sustainability of their management of crop land which is oriented toward diminishing soil damage and conserving soil fertility. The index, exclusively based on terms which describe the environmental conditions of the crop system, is accessible to most farmers, and may be calculated using a simple internet application. In this paper we present a simple, easy-to-use index in order to evaluate the ecological state of farms in south-eastern Mexico. We applied the indicator system to 52 crop production systems in south-eastern Mexico and compared the results of the indicator system with expert opinions.

Human knowledge of how to efficiently and sustainably manage complex systems (including agricultural systems) is incomplete and much of what is thought to be known about this topic is actually incorrect. Yet, decisions must be made by policy makers, agricultural extension agents, and farmers despite uncertainty and knowledge gaps [19]. Therefore, tools to support local decision-makers must be flexible, should not enter into too much detail or precision, and should allow for an adaptive strategy which promotes "learning through management" [19]. Consequently, our rationale for developing an index which aids farmers in making environmentally friendly land-use decisions is based on basic, simplified ecological concepts, i.e. the presence of trees, since trees within an agroecosystem enhance soil microclimate in terms of radiation partitioning (shading), evapotranspiration partitioning, and rain interception/redistribution [20]. These factors all help to retain soil moisture. Branches, bark, roots, and living and dead leaf surfaces provide shelter [21] for soil micro-, meso-, and macro-invertebrates. Tree cover for instance, enhances above- and below-ground diversity, serving to support agricultural sustainability [22].

Materials and Methods

1. Rationale of index composition

We define conventional agriculture as a cropping system, typically promoted by government development programs, that is "capital-intensive, large-scale, highly mechanized agriculture with monocropping and extensive use of synthetic fertilizers, and pesticides" [23], [24]. Furthermore, we acknowledge that farming systems are sustainable only if "they minimize the use of external inputs and maximize the use of internal inputs already present on the farm" [25], [26]. The strategy most frequently linked to sustainability is reduction or elimination of agrochemicals, particularly chemical fertilizers and pesticides [25], [27], [28], [29], [30], [31]. Another key to sustained productivity of agricultural systems is the maintenance of soil functions, such as organic matter and nutrient cycling [32], based on organic inputs [33], above-and below-ground biodiversity [22], and diversifying crop systems with nitrogen-fixing legumes [34]. The principal role of the index we propose is to characterize methods of tillage, external inputs, and crop structure.

2. Primary indicators

We chose 12 field variables as primary indicators related to the above mentioned aspects of ecologically sound agricultural land use based on farmer's practices. These are easy to evaluate in the field and characterize plot structure (primary indicators: tree cover, tree density, tree diversity), crop structure and crop conditions (primary indicators: crop type, crop rotation, crop density, crop colour), tillage (primary indicators: type of tillage, timing of tillage), the use of fertilizers (chemical versus organic) and pesticide application.

2.1 Tree cover. Tree cover is defined as the canopy of trees, measured in the field, and recorded as percentage classes of tree cover in three height classes (trees >15 m, 10–15 m, <10 m). Thus, this variable characterized one aspect of agroecosystem management: the farmers' decision to maintain the canopy of the trees in his or her agroecosystem.

2.2 Tree density. Tree density is defined as the number of trees per area. To measure this, we distinguished three categories of tree density: high density (abundant), medium density, and low density (isolated or no trees). A high number of trees per area guarantee carbon sequestration [35] while a stable microclimate is maintained. This variable, measured in the field, is one the variables that characterize the effect of trees in the agroecosystem.

2.3 Tree diversity. This variable was measured in the field by counting the number of trees species within the plot. In agroecosystems, biodiversity may; (i) contribute to constant biomass production and reduce the risk of crop failure in unpredictable environments, (ii) restore disturbed ecosystem services such as water and nutrient cycling, and (iii) reduce risks of pests and diseases through enhanced biological control or direct pest control [36], [20], [22].

2.4 Crop type. This variable indicates whether the crop is annual, seasonal, or perennial. We obtained this information by observing the type of crop. Annual crops in general have higher environmental impacts, ie: greenhouse emissions, and nutrient leaching, than perennial crops [37].

2.5 Crop rotation. In sustainable farm systems leguminous crops are increasingly used in crop rotations as a source of nutrients, particularly nitrogen for crop growth [38], [39], nitrogen-fixing legumes, contribute to maintaining biodiversity above and in the soil, contribute nitrogen to the soil/plant system, and help avoid the build-up of pest populations [34]. In this study, we asked farmers whether they planted another crop before planting the main crop and whether they practice crop rotation, as crop rotation may assist with weed and pest control [40], [34]. According to Bellon [41], an activity which leads into the maintenance or increase of renewable resources in agroecosystems, is considered as an ecological technology. This variable helped us to characterize the technology used in the agroecosystem.

2.6 Crop density. Crop density is defined as the number of plants (individuals) per area. Three categories were recorded in the field: abundant (high density: 3,000 plants/ha), medium density (1000–1600 plants/ha), and sparse (<1000 plants/ha). This variable also indicated the level of technology applied to the crop, as less intensive techniques typically yield lower densities [42].

2.7 Crop colour. The colour of a crop indicates the nutritional status of the plants; green plants generally have sufficient nutrients, while yellow plants lack nitrogen [43].

2.8 Type of tillage. Type of tillage was categorized into no tillage, manual tillage, and mechanical tillage using machinery, the latter of which generally indicates high disturbance of the soil surface and rapid loss of soil organic carbon and other nutrients

[44]. In the field, we asked the farmers how they prepared the land for crop planting.

2.9 Frequency of tillage. This variable indicates frequency of tillage - every year or every 2 years. With this information, it was possible to estimate the frequency of soil disturbance due to tillage.

2.10 Chemical input. Searching for sustainable productions, it is recommended no or low or use of inorganic fertilizers and pesticides [45], [46]. Long term use of some pesticides, as glifosate can decrease earthworm species number, density and biomass [47], [48]. This information allowed us to estimate the amount of chemical fertilizers and pesticides applied within a given area. These variables (chemical fertilizers and pesticides) are included in the indicator for chemical disturbance.

2.11 Green manure. Is defined as the presence or absence of leguminous crops mixed with the principal crop, generally used to increase total soil nitrogen content. Green manures should always be intercropped, as it has been proven that growing legumes with cereal crops decreases N_2O emissions [49], and therefore is a sustainable, environmentally friendly practice. Examples of green manure use have been observed in traditional Mesoamerican cultures, for example, intercropping beans, as well as other edible plants, within the *milpa* or traditional maize cropping system [50].

3. Index development

Primary indicators were hierarchically aggregated into higher levels, forming intermediate variables, which in turn are structured into a single index that evaluates the ecological condition of a given plot on a scale from 0 (worst) to 1 (best). An index close to zero either would mean that a more environmentally friendly farming techniques need to be implemented or that the plot should be subjected to a fundamental change in land-use, e.g., reforestation, in order to return to a more ecologically sound state. An index close to one, on the other hand, would indicate an ecologically sound land-use. Methods applied in indicator aggregation were (i) simple mathematical operations, ii) sets of if-then rules, and (iii) sets of if -then rules combined with fuzzy logic.

3.1. Aggregation through mathematical operations. Mathematical operations include calculating averages, weighed averages, minimum values, and maximum values. For example, if primary indicator A has the value a and primary indicator B has the value b, then the value x of the intermediate variable X is determined as $x = (a+b)/2$, $x = (w_1*a+w_2*b)$, where $w1$ and $w2$ are weights, or $x = min(a, b)$ or $max(a, b)$.

3.2. Aggregation by IF-THEN rules. If primary indicator A can have the discrete values a_1, a_2, and a_3, and primary indicator B can have the discrete values b_1, b_2, and b_3, then the value x of the intermediate variable X is determined by a set of nine (number of levels of A x number of levels of B) rules. For example, IF $A = a_1$ and $B = b_1$ THEN $X = x_1$.

3.3. Aggregation by IF-THEN rules and fuzzy logic. We used small fuzzy rule-based models for aggregation in the case of non-linear interactions among indicators using a continuous numerical scale or an ordinal scale with a large number of possible values. In classic set theory, an object can either be a member (membership = 1) or not (membership = 0) of a given set. The central idea of fuzzy set theory is that an object may have a partial membership of a set, which consequently may possess all possible values between 0 and 1. The closer an element is to 1, the more it belongs to the set; the closer the element is to 0, the less it belongs to the set. To apply the fuzzy set theory, three steps are involved in calculating the model's output. First, for any observed value of the primary indicators, its corresponding membership value in the fuzzy set domain is calculated (*fuzzification*); second,

the memberships of the intermediate variable X are calculated, applying the rules in the fuzzy set theory (*fuzzy inference*); third, the fuzzy results are converted into a discrete numerical output (*defuzzification*; see Wieland 2008 for an introduction to fuzzy models). Fuzzy rule-based models have become popular in ecological modelling [51], [52], and several examples exist of its usefulness in the context of ecosystem evaluation, bioindication, and sustainable management [15], [53], [54]. Here, if both primary indicators A and B can have numerical values from 0 to 1, then the value x of the intermediate variable X is determined by a series of fuzzy set rules representing the linguistic variables "low A", "medium A", "high A", and "low B", "medium B", and "high B", as well as the output "low X", "medium X", and "high X". The value of the intermediate variable X is determined by nine levels of A x the number of levels of B; for example: If A = low and B = low Then X = [low, medium, or high].

To maintain the number of rules as well as their complexity as low as possible, we aggregated only two variables at a time. A simple example shows the reasoning behind this decision; if there are three primary variables, A, B, and C with three categories for each variable, a single rule node requires $3*3*3 = 27$ If-Then rules of the type "If $A = a_1$, a_2, or a_3 and If $B = b_1$, b_2, or b_3 and if $C = c_1$, c_2, or c_3 Then...", whereas if an additional intermediate variable X is introduced to the model, only 18 rules are needed: $3*3 = 9$ rules to aggregate A and B to X (If $A = a_1$, a_2, or a_3 and If $B = b_1$, b_2, or b_3), and $3*3 = 9$ rules to aggregate X and C (If $X = x_1$, x_2, and If $C = c_1$, c_2, or c_3).

4. Study area and application of the index

The state of Tabasco in south-eastern Mexico is characterized by a humid tropical climate with a mean annual rainfall between 1200 and 4000 mm and a mean annual temperature of 27°C [55]. Predominant soils are Gleysols and Fluvisols over alluvial sediments in the plains, Vertisols, Cambisols, Luvisols, and Acrisols over Miocene or Oligocene sediments, and Leptosols and Regosols over limestone mountains [9], [56]. We chose the municipalities Balancán and Tenosique in western Tabasco (17°81′50″–18°81′00″ N, 91°80′10″–91°84′60″ W) as a study area (Figure 1), we worked with private, *ejidal* (multipurpose land, where owners can or cannot sell the property according to their legal status in the National Agrarian File), and communal lands (coordinates of each plot are shown in Table 1). The region is mainly a plain towards the North (67% of area has an elevation <20 m. a. s. l.) with hills (29%, 20–200 m. a. s. l.), and mountains (4%, max. 640 m. a. s. l.) in the South, comprising a total area of 5474 km^2. These municipalities have undergone a high degree of land use change over the past 40 years. Until the early 1970 s, this region was still covered by lowland rain forest. The principal form of land use is pastureland, and covers 60% of the land [57]. An additional 30% is cropland, mainly cultivated under small-medium holder systems with seasonal conventional agriculture using high levels of agrochemical inputs (Table 2 & 3). Common crops are maize (*Zea mays*), a variety of hot peppers (*Capsicum* sp), cucumbers (*Cucurbita argyrosperma*), watermelon (*Citrullus lanatus*), perennial fruit crops such as papaya (*Carica papaya*), and biannual crops such as sugar cane (*Saccharum officinarum*) [58].

We chose 52 farms in the study area (Table 1), and selected those farms whose main economic activity (100%) is agriculture and chose one agricultural plot from each farm for evaluation. There were annual and biannual crops with or without trees (Table 2). Average plot size was 32.4±55.1 ha, and the average time that the plot had been used for a given crop system was 2.5±3.0 years.

Table 1. Characterization of plots.

Plot #	Municip	Lat	Long	Altitud (mls)	Plot size (ha)	Type of property	Original-vegetation
1	Bal	2007880	652445	43	18	Ejidal	A
2	Bal	1964368	642505	10	0.5	Private	TRF
3	Bal	1970936	647701	11	160	Private	TRF
4	Bal	1957459	667496	14	2	Ejidal	TRF
5	Bal	1964107	660656	14	7	Private	PT
6	Bal	1960297	658989	22	15	Private	TRF
7	Bal	1976300	676424	40	30	Private	TRF
8	Bal	1990326	665947	40	28	Private	TRF
9	Bal	1964590	667264	11	3	Private	RV
10	Bal	1972260	672382	28	2.5	Ejidal	TRF
11	Bal	1950339	666409	49	40	Private	TRF
12	Bal	1966112	682994	40	80	Private	PT
13	Ten	1931754	704474	65	14	Private	TRF
14	Ten	1912951	690381	130	2	Communal	TRF
15	Ten	1932644	665521	15	3	Ejidal	TRF
16	Ten	1926163	678608	56	68	Private	TRF
17	Ten	1933663	674227	59	35	Private	A
18	Ten	1926570	669486	35	8.5	Ejidal	TRF
19	Ten	1942431	663885	21	1.5	Ejidal	TRF
20	Ten	1942068	671156	31	7.25	Ejidal	TRF
21	Ten	1943167	650736	82	14	Ejidal	A
22	Ten	1923838	669841	118	0. 25	Private	TRF
23	Bal	1973301	695548	105	40	Ejidal	TRF
24	Ten	1982392	701383	74	20	Ejidal	A
25	Bal	1965354	709053	44	11	Ejidal	A
26	Bal	1960496	706464	51	44	Ejidal	TRF
27	Bal	1942484	709731	66	10	Ejidal	A
28	Ten	1944148	673673	16	320	Ejidal	TRF
29	Ten	1930350	659650	29	20	Private	TRF
30	Ten	1924549	656910	41	53	Ejidal	TRF
31	Ten	1926226	652978	65	3	Ejidal	TRF
32	Bal	1955194	688843	45	44	Private	TRF
33	Bal	1960980	679265	49	29	Ejidal	TRF
34	Bal	1942552	696222	48	10	Private	TRF
35	Ten	1938120	686470	49	5	Private	TRF

Table 1. Cont.

Plot #	Municip	Lat	Long	Altitud (mls)	Plot size (ha)	Type of property	Original-vegetation
36	Bal	1968750	695595	60	2	Private	A
37	Bal	1977546	699417	53	200	Ejidal	TRF
38	Bal	1949770	708854	53	23	Ejidal	TRF
39	Bal	1951216	697890	46	2	Ejidal	TRF
40	Ten	1931041	667544	31	17	Private	TRF
41	Ten	1927447	664223	3	3	Ejidal	TRF
42	Ten	1928226	664551	31	3	Ejidal	TRF
43	Ten	1930579	691785	54	15	Communal	TRF
44	Ten	1931219	677895	59	63	Private	TRF
45	Ten	1931525	676500	60	6.75	Ejidal	TRF
46	Ten	1908445	675040	140	40	Ejidal	TRF
47	Ten	1910875	669750	222	45	Ejidal	TRF
48	Ten	1924222	704702	42	30	Ejidal	A
49	Ten	1935076	712227	88	20	Ejidal	TRF
50	Ten	1914876	686285	113	4	Ejidal	TRF
51	Ten	1911365	684878	389	22	Ejidal	TRF
52	Ten	1911503	698606	139	10	Ejidal	TRF

Municip: municipality, Bal: Balancan, Ten: Tenosique. TRF: tropical rain forest, PA: pasture, A: acahual (secondary vegetation). Ejidal: multipurpose land, where owners can or cannot sell the property according to their legal status in the National Agrarian File.

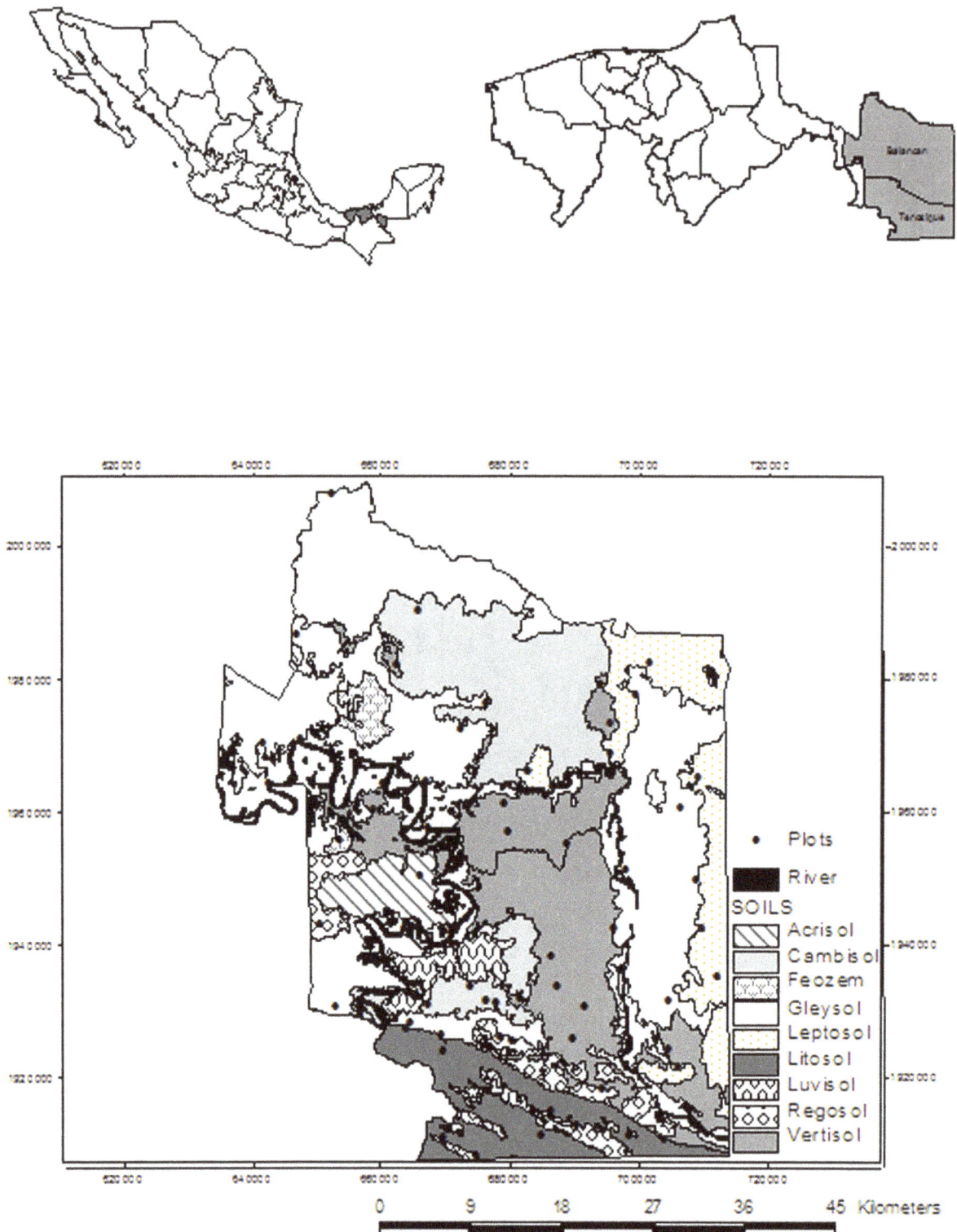

Figure 1. Distribution of 52 evaluated agricultural plots in tropical South-East Mexico.

We questioned farmers as to frequency, amount and type of chemical fertilizers and pesticides applied per area of cropland. After their verbal consent, farmers gave us their complete name and signed next to their name on the record sheet, validating all the information. Due to the fact that we only asked about the use of fertilizers and management of the land, the procedure approval from the Ethics Committee was not required.

Between March and October 2004, for each plot, the values of the primary indicators were determined and the index of ecological condition was calculated. Prior to index calculation, the plots were also evaluated by experts (2 scientists, each with

Table 2. Crop characterization.

Plot	Cycle	Main seasonal or perennial crop	Trees
1	s-s	Zm, Pv, Cs, Cl	Tr, Pa, Cpa
2	s-s	Zm, Pv, Cs, Cl	Sh, Dg
3	w-s	Sv	Sa
4	a-w	Zm, Pv, Cpa	0
5	Y	Zm, Pv, Cl, Cm	0
6	Y	Zm	Cpa, Eg, Tr
7	s-a	Ca	Tr
8	Y	Cp	Cp
9	Y	Zm	0
10	Y	Zm	0
11	w-s	Zm, Pv, Cl	Sm, Cpa
12	Y	Co	Co
13	s-s	Zm	0
14	s-s	Le, Se	0
		Cpa, Zm, Pv	
15	s-s	Cl, Ta, Mp,	Fruit trees
16	Y	Jj	Gs, Hc, Cc, Tr
17	Y	So	0
18	Y	So	0
19	Y	So	0
20	Y	Zm	Mi
21	s-s	Zm	0
22	Y	Csi, Cli, Js	Co, Pa, Bc, Cn, Mi
23	s-s	Zm, Cpa	Cpe
24	s-s	Zm	Cpa
25	s-s	Zm	0
26	s-s	Zm, Cpa	0
27	s-s	Zm, Ca	Sm
28	Y	So	0
29	s-s	Zm, Pv, Cs, Cl	0
30	Y	Zm, Pv, Cs, Cl, Ta	0
31	Y	Zm, Pv, Cs	0
		Cl, Os, Le	
32	s-s	Zm, Pv, Cpa, Cl, Os	Tr
33	Y	Zm, Pv	Sm, Bg
34	s-s	Zm, Pv, Os	Cpa
			Tr
35	Y	Zm, Cl	Tr
36	Y	Zm, Pv, Cpa, Cl, Ca	0
37	s-s	Zm, Pv, Ca	Sm, Pa, Sh
38	Y	Zm, Pv, Cpa, Cl, Ta	Co, Sh
39	Y	Zm, Pv, Cpa, Cl, Ca	
		Cs, Cm	
40	Y	So	Tr, Cpe
			Co, D a
41	Y	So	Tr, Cpe
			Co, D a
42	Y	So	Tr, Cpe
			Co, D a

Table 2. Cont.

Plot	Cycle	Main seasonal or perennial crop	Trees
43	Y	So	Tr, Cpe
			Co, Da
44	Y	So	Tr, Cpe
			Co, Da
45	Y	So	Tr, Cpe
			Co, D a
46	Y	Zm, Pv, Cpa, Me	Sm
47	Y	Zm, Pv, Me, Ib	Bc, Mi, Dg, Cs, Ll
48	Y	Zm, Pv, Cpa, Cl	0
49	Y	Zm, Pv, Ca	Tr, Gs
50	Y	Zm, Pv, Cpa, Cl	Sm, Co, Ma
		Cs, Ib, Me	0
51	Y	Zm, Pv, Cpa, Cs	Fruit trees
52	Y	Zm, Pv, Cpa, Cl, Cs	Co, Sh, Bc
		Mp, Ca, Me	0

Cycle: s–s: spring-summer, w–s: winter-summer, s-a: summer-autumn, Y: all the year. **Main seasonal or perennial crop**: Ca: *Cucurbita argyrosperma* (cushaw pumpkin); Cp: *Carica papaya* (papaya); Cl: *Citrullus lanatus* (watermelon); Cli: *Citrus limon* (lemon); Cm: *Cucumis melo* (muskmelon); Cpa: *Cucurbita pepo* (squash); Cs: *Capsicum* sp. (pepper); Csi: *Citrus sinensis* (orange tree); Ib: *Ipomoea batatas* (sweet potato); Jj: *Jarthropha jurcas* (oil palm); Js: *Jobo spondia* (plum); Le: *Lycopersicon esculentum* (tomato); Me: *Manihot esculenta* (cassava); Mp: *Musa paradisiaca* (banana); Os: *Oryza sativa* (rice); Pv: *Phaseolus vulgaris* (bean); So: *Saccharum officinarum* (sugarcane); Se: *Sechium edule* (pear squash); Sv: *Sorghum vulgare* (milo); Ta: *Triticum aestivum* (wheat); Zm: *Zea mays* (maize). **Trees**: Bc: *Byrsonima crassifolia* (nanche); Cc: *Crescentia cujete* (calabash tree); Co: *Cedrela odorata* (Mexican cedar); Cpa: *Carludovica palmate* (toquilla palm); Cpe: *Ceiba pentandra* (kapok); Cn: *Cocos nucifera* (Coconut); Dg: *Dialium guianense* (wild tamarind); Eg: *Eucalyptus grandis* (Eucalyptus); Gs: *Gliricidia sepium* (Cocoite, gliricidia, cacao de nance); Ll: *Leucaena leucocephala* (white lead tree, jumbay); Ma: *Mammea americana* (mamey apple); Mi: *Mangifera indica* (mango); Pa: *Persea americana* (avocado); Sa: *Sterculia apetala* (camoruco, manduvi tree); Sh: *Swietenia humilis* (small mahogini); Sm: *Spondias mombin* (Yellow plum, bai makok); Tr: *Tabebuia rosea* (savannah oak, macuilis).

over 15 years of experience in agroecology) on a scale from 0 (poor condition) to 1 (good condition) in order to test the correlation between the experts' opinions and the index.

Data was normalized for carrying out multiple regression with plot size, altitude of plot site, or previous vegetation cover and the ecological index.

Finally we located the index in this public web address: http://201.116.78.102/~modelo/Index.html, where farmers in the near future can introduce independent data sets and evaluate or monitor their own agro ecosystems.

Results

1. Index structure

Ten nodes were used to aggregate primary indicators to the final index of ecological conditions of agricultural systems (Figure 2). Six of these used simple mathematical operations; two were based on rule sets, and two on fuzzy rule-based models (Table 4).

2. Index application

2.1. General characterization of sampled plots. A total of 67% of the plots were cultivated after cutting of primary lowland forest, 24% after cutting secondary forests of various ages, and 1.5% after cutting riparian vegetation, the rest conversion from pastureland (livestock production) to crop system. On 63% of the farms, maize, beans and pumpkins were cultivated, 13% sugar cane, 12% watermelon, 9% rice and 3% pepper. On 68.5% of the plots, trees were scattered among the crops. Conventional tillage was used on 55.5% of the plots, and pesticides were used on 79.4%, green manure was used on 21% of the plots and chemical

fertilizers on 67%. All farmers understood the meaning of the variables evaluated for their plots (for a complete list of plots, see Table 5).

2.2. Plot evaluation with the index. Ecological condition of the plots ranges from 0.0 to 0.8125 (Figure 3; Table 5). One plot which was intercropped with timber and fruit trees presented the highest value. This site was characterized by an absence of agrochemical use, use of green manure, presence of annual crop rotation, and high tree diversity. We found 18 plots with index values of 0–0.25, 7 plots with values of 0.3–0.5, and 26 plots with values of 0.5–0.7. Thus, the majority of the plots evaluated in this study had an intermediate index value. 50% of the plots with this intermediate index are *ejidal* property, 38% private land, and 12% is under another type of ownership (smallholder or communal). 80% of these intermediate plots were lowland rain forest before being converted into agricultural land, 12% were secondary vegetation, and 8% were pastureland (Table 1). One might believe that a prior forest condition implies a more environmentally friendly agroecosystem that allows for preserving a more diverse system. However, plots with low index values were also lowland rain forest before being turned into agricultural land (Table 5). In this case, the land managers or owners decided to deforest the land to subsequently plant annual crops. 13 plots were larger than 40 ha (Table 1). All of these were previously covered with lowland rain forest; the ecological index ranges from 0.5–0.56 for 46% of these plots, and another 23% have an ecological index of 0.37–0.43. 10 plots fell under the smallest size category (0.25–3 ha) and had an ecological index of 0.5–0.68 (4 plots), 0 to 0.18 (5 plots), and 0.39 (1 plot).

Carrying out multiple regression with normalized data, we did not observe significant correlations between plot size, altitude of

Table 3. Crop land preparation and inputs characterization.

Plot	Tillage	Chemical Fertilization	Pesticides Use	Gm	Oa
1	Ma & Me	NPK 17:17:17 & U	Endosulfan**	0	0
2	Ma	0	Chlorpyrifos*	0	0
3	Me	Pholiar	0	0	c s m
4	Ma	0	0	0	0
5	Ma & Me	NPK 17:17:17	Chlorpyrifos**	Y	Y
6	Ma & Me	NPK 17:17:17	Chlorpyrifos*	0	0
7	Me	0	ID	0	0
8	Me	ID	ID	Y	0
9	Ma & Me	0	Carbofuran*	0	0
10	Me	NPK 18-46-0	ID	0	0
11	Me	NPK 18- 46- 0, U & P	ID	Y	0
12	Me	NPK 18-46-0	ID	0	0
13	Me	NPK 17:17:17 & U	Chlorothalonil** Chlorpyrifos*	0	0
14	Ma	0	Endosulfan*	0	0
15	Me	0	0	Y	0
16	Ma	Urea	0	0	0
17	Me	NPK 19-19-19	ID	0	0
18	Me	NPK 19-19-19	ID	0	0
19	Me	Urea	ID	0	bg
20	Me	0	Methilic*	0	0
21	Me	NPK 17:17:17 & U	Zeta**	0	0
22	Ma	0	0	0	0
23	Me	NPK 17:17:17 & U	Chlorpyrifos**	0	lcf
24	Me	Urea & P	Zeta*	Y	lcf
25	Me	Urea	(2,4-D AMINA)*	Y	0
26	Me	NPK 17:17:17 & U	(Z)-(1R,3R)**	0	0
27	Me	0	Zeta*	N	0
28	Me	NPK 17:17:17	(RS)*	0	0
29	Me	NPK 17:17:17 & U	Chlorpyrifos**	0	0
30	Ma	0	Chlorpyrifos**	Y	0
31	Me	NPK 17:17:17 & U	Chlorpyrifos*	0	0
32	Me	0	0	Y	0
33	Me	0	0	0	0
34	Me	0	0	0	0
35	Me	NPK 17:17:17 & U	0	0	0
36	Me	NPK 17:17:17 & U	Chlorpyrifos*	Y	0
37	Me	0	Urea	0	0
38	Me	NPK 17:17:17 & U	Carbofuran*	Y	0
39	Me	NPK 17:17:17 & U	Chlorpyrifos**	0	0
40	Me	NPK 17:17:17	(Z)-(1R,3R)*	0	0
41	Me	NPK 17:17:17 & U	(Z)-(1R,3R)*	0	0
42	Me	NPK 17:17:17 & U	(Z)-(1R,3R)*	0	0
43	Me	NPK 17:17:17 & U	(Z)-(1R,3R)*	0	0
44	Me	NPK 17:17:17 & U	(Z)-(1R,3R)*	0	0
45	Me	NPK 17:17:17 & U	(Z)-(1R,3R)*	0	0
46	Ma	0	0	Y	0
47	Me	U	0	0	0
48	Ma &Sb	Urea	Chlorpyrifos**	0	0

Table 3. Cont.

Plot	Tillage	Chemical Fertilization	Pesticides Use	Gm	Oa
49	Me	Urea & P	Zeta*	0	0
50	Ma	Urea	Endosulfan*	0	0
51	Ma	0	ID	0	0
52	Me	NPK 17:17:17 & U	Chlorpyrifos*	0	0

Tillage: Ma: Manual; Me: Mechanized. **Pesticides**: Endosulfan: 6,7,8,9,10,10-Hexachlor- 1,5,5a,6,9,9a-Hexahidro-6,9-metane-2,4,3-benzodioxatiepin-3-oxide; Chlorpyrifos: 0,0-dimetil 0-(3,5,6-trichlore-2-piridinil) fosforotioate (33.8%), Permetrine: 3-fenoxibenzil (1RS)-cis, trans-3-(2,2 diclorovinil)-2,2 dimetil ciclopropane-carboxilate (4.8%); Carbofuran: 2,3 Dihidro-2,2-dimetil-7-benzofuranil metil carbamate; Chlorothalonil: Tetrachloroisoftalonitrile; Zeta: Zeta-cipermetrine a-ciano-3-(fenoxifenil) metil (±) cis-trans; (Z)-(1R,3R): (Z)-(1R,3R)-3-(2-chloro-3,3,3-trifluoroprop-1-enil)-2,2-dimetilciclopropanecarboxilate de (S)-α-ciano-3-fenoxibencile & (Z)-(1S,3S)-3-(2-chloro-3,3,3-trifluoroprop-1-enil)-2,2-dimetilciclopropanecarboxilate de (R)-α-ciano-3-fenoxibencile; (RS): (RS)- alfa- ciano-3-fenoxybencil(1RS)-cis-trans-3-(2,2-dichlorvinil)- 1,1-dimetilcichlopropanecasrboxilate. * Once per year, ** 2 per year. **Chemical Fertilization**: U: Urea, P: Phosphorus. **Gm**: Green manure, Y: Yes, 0: null application. **Oa**: Organic amendments: csm: cow, sheep manure, bg: burned grass, lcf: last crop fallow, 0: null application.

plot site, or previous vegetation cover and the ecological index (Kendall's Tau, T = 0.016 p = 0.98, T = 0.11 p = 0.90, T = −0.31 p = 0.75, respectively). Therefore, in this study, it seems that neither size nor location of the plot determines the type of plot management; rather, plot management is likely determined by government development programs and traditional farming techniques.

3. Correlation between index and expert opinion

We obtained a Pearson correlation coefficient of $r^2 = 0.61$ ($p = 2.4\text{e-}06$, $d.f. = 49$) between the index and the values determined by independent experts, indicating a satisfactory correspondence (Figure 4). However, the experts systematically awarded higher scores to the plots than did the index. Moreover, they suggested to include additional variables to the index which would yield better information regarding (i) type of organic inputs to the crop system, (ii) types of pest and disease control used, (iii) number of native plant species among the crop, (iv) origin of crop seeds, (v) vegetation surrounding the crop, (vi) presence of vertebrate fauna, and (vii) diversity of soil macroinvertebrates.

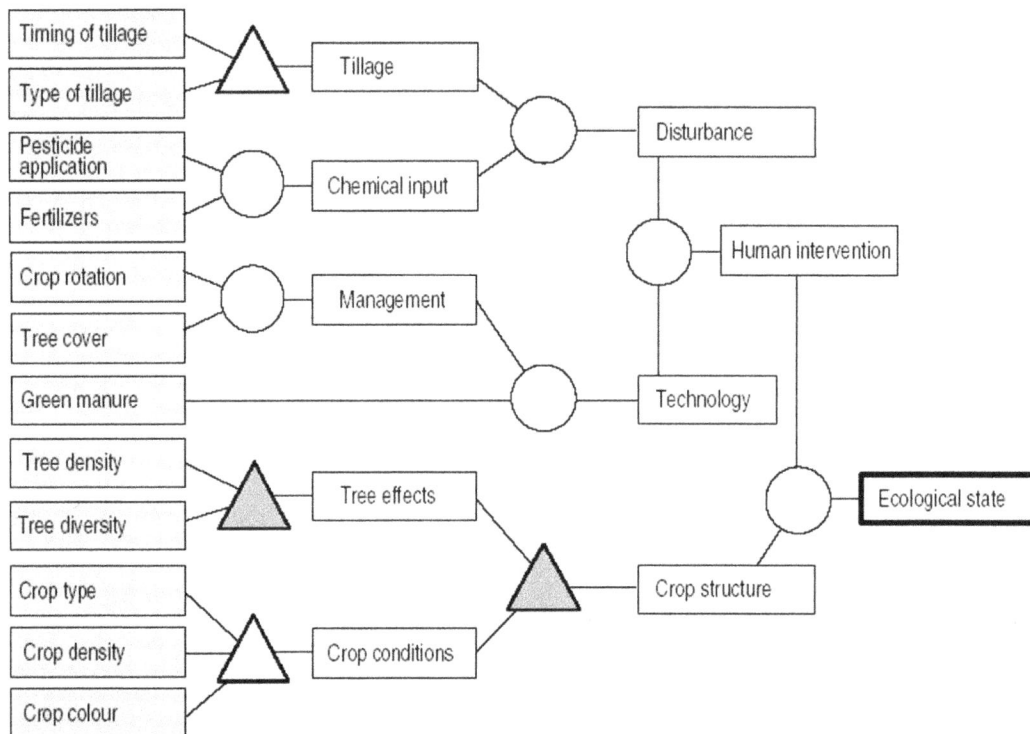

Figure 2. Structure of the Index of ecological condition of tropical agroecosystems. Primary indicators are shaded in grey. Circles represent simple mathematic algorithms, white triangles represent rule sets, and grey triangles represent rule sets, and grey triangles represent rule sets based on fuzzy logic. This index was presented together with other indexes within a frame of Indicators of environmentally sound land use in the humid tropics [15].

Table 4. Description of the measurement levels of the variables.

Groupal variables	Field data variables	Best	Intermediate			Worst
		1	**0.75** (5 levels)	**0.5**	**0.25**	**0**
Management	Tree cover	all year	presence polyculture	present	rare	null
Tree effect	Tree density	high	medium	low	isolated	null
		1	4 levels / 0.66	0.33		0
	Tree diversity	>4 species	4 species	2 species		1 species
Crop structure — Crop conditions	Crop type	perennial polyculture	perennial monoculture	biannual		annual
		1	3 levels / 0.33			0
Plough (tillage)	Tillage time	null	frequent			constant
	Tillage form	null	manual			technical
		1	0.33			0
Disturbance — Chemical inputs	Pesticides	null	frequent			constant
	Fertilizers	null	frequent			constant
Management	Crop rotation	constant	frequent			null
Technology — Green manure	Green manure	constant	frequent			null
		1	2 levels			0
Crop conditions	Crop density	abundant				disperse
	Crop coleur	green				yellow

Table 5. Normalized Ecological Index for plots evaluated (52).

Plot	TT	TF	P	F	GM	RC	CA	DA	DivA	CT	CD	Ccol	Ev	Index
1	n	m	u	u	n	n	>1	l	>4	pm	a	g	0.24	0.56
2	n	n	u	n	n	n	>1	i	>4	a	a	g	0.35	0.5
3	u	m	n	u	u	n	>1	i	>4	a	a	g	0.26	0.56
4	n	n	n	u	n	n	n	n	1	a	a	g	0.01	0.37
5	n	n	u	n	u	u	>1	i	>4	a	a	g	0.09	0.62
6	n	m	u	u	n	n	>1	l	>4	a	a	g	0.22	0.5
7	u	t	u	n	n	n	>1	i	>4	a	d	b	0	0.15
8	u	t	u	u	n	u	n	n	1	pp	d	g	0.1	0.12
9	n	t	u	n	n	n	n	n	1	a	d	g	0.09	0.25
10	u	t	u	u	n	n	n	n	1	a	d	b	0	0
11	u	t	u	u	u	u	>1	i	>4	pm	ID	ID	0.03	0.21
12	u	t	u	n	n	n	>1	h	>4	ba	a	g	0.2	0.5
13	n	t	u	u	n	n	>1	i	>4	a	a	g	0.03	0.5
14	n	n	u	n	n	n	n	n	1	pp	a	g	0.29	0.37
15	n	t	n	u	n	u	>1	h	>4	pp	a	g	0.2	0.81
16	n	n	u	u	n	n	>1	l	>4	ba	a	g	0.25	0.5
17	u	t	u	u	n	n	n	n	1	a	a	g	0	0.25
18	u	t	u	u	n	n	n	n	1	a	d	g	0	0
19	u	t	u	u	u	n	>1	i	1	a	d	g	0	0.06
20	u	t	u	n	n	n	n	n	>4	a	a	g	0.1	0.5
21	u	t	u	n	n	n	n	n	1	a	a	g	0	0.25
22	n	n	u	u	n	n	>1	i	>4	pp	d	g	0.33	0.62
23	n	t	u	u	u	n	>1	i	>4	a	a	g	0.15	0.56
24	n	n	u	n	n	u	>1	l	>4	a	a	g	0.2	0.62
25	n	n	u	u	n	u	n	n	1	a	d	g	0.01	0.06
26	n	m	u	n	n	n	n	n	1	a	a	g	0.25	0.25
27	n	t	u	n	n	n	>1	i	>4	a	a	g	0.01	0.5
28	n	t	u	u	n	n	n	n	1	a	a	g	0	0.25
29	u	t	u	u	n	n	n	n	1	pm	a	g	0.01	0.37
30	n	m	u	u	n	n	n	i	1	pm	a	g	0.1	0.43
31	u	t	n	u	n	n	n	n	1	a	a	g	0.01	0.25
32	u	t	n	u	n	u	>1	i	>4	a	d	y	0.09	0.34
33	u	t	n	n	n	n	>1	h	>4	a	a	g	0.15	0.62
34	u	t	n	u	n	n	>1	i	>4	ba	a	g	0.13	0.62
35	u	t	n	n	n	n	>1	l	>4	a	a	g	0.01	0.5

Table 5. Cont.

Plot	TT	TF	P	F	GM	RC	CA	DA	DivA	CT	CD	Ccol	Ev	Index
36	c	t	c	c	n	c	n	n	1	a	a	g	0	0.31
37	c	t	c	n	n	n	>1	i	>4	pm	a	y	0.01	0.15
38	c	t	c	c	n	c	>1	h	>4	pm	a	g	0.2	0.68
39	c	t	c	c	n	n	>1	l	>4	a	a	b	0.06	0.18
40	c	t	c	c	n	c	>1		>4	a	a	g	0.19	0.5
41	c	t	c	c	n	c	>1		>4	a	a	g	0.19	0.5
42	c	t	c	c	n	c	>1		>4	a	a	g	0.17	0.5
43	c	t	c	c	n	c	>1	l	>4	a	a	g	0.2	0.5
44	c	t	c	c	n	c	>1	l	>4	a	a	g	0.2	0.5
45	c	t	c	c	n	c	>1	l	>4	a	a	g	0.2	0.5
46	n	n	n	n	n	c	>1	l	>4	pp	a	b	0.6	0.37
47	n	t	c	c	n	n	>1	l	>4	pp	d	g	0.18	0.5
48	c	n	c	c	n	n	n	n	1	a	a	b	0	0
49	c	t	c	c	n	n	>1	l	>4	a	a	b	0.07	0.18
50	c	n	c	c	n	n	>1	l	>4	pm	a	g	0.2	0.56
51	c	n	c	n	n	n	>1	l	>4	pm	a	g	0.15	0.56
52	c	t	c	c	n	n	>1	l	>4	pm	a	ID	0.24	0.18

TT: Tillage Time: c: >once per year; n: once per year; TF: Tillage Form: n: null, m: manual, t: technical; P. Pesticide use: c: constant, n: null; GM: Green Manure, c: constant, n: null; RC: Crop Rotation: c: constant, n: null; CA: tree cover: c: constant, n: null; F: Use of chemical Fertilizers: c: constant, n: null; CA: tree cover: >1 year, n: null; DA: tree density: h: high, l: low, i: isolated, n:null; DivA: tree diversity: >4 species, 1 species, CT: crop type: a: annual, ba: biannual, pp: perennial polyculture, pm: perennial monoculture; CD: crop density: a: abundant, d: disperse; Ccol: crop colour: g: green, b: brown, y: yellow; EV: evaluation.

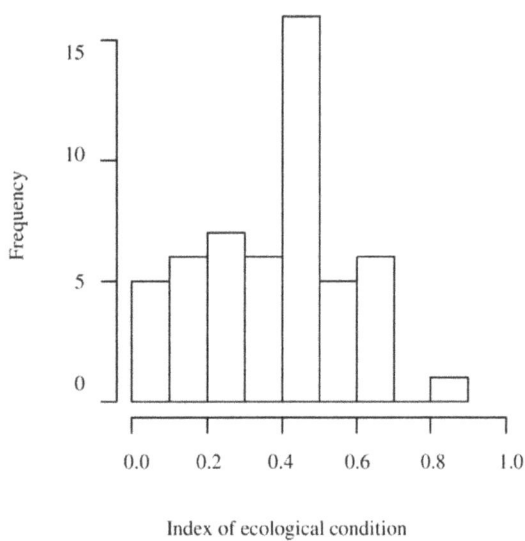

Figure 3. Frequency histogram of values of the index of ecological condition applied to 52 plots in South-East Mexico.

Discussion

Since the Rio Earth Summit, there has been a concerted effort to construct indicators to monitor progress toward sustainable development [59]. Most of these indicators have been developed in Europe (10) and Asia (2) [60]. In Latin America, sustainable indicators have been developed by Astier et al. [61]; this index mainly focuses on subsistence level agriculture, and evaluates the sustainability of a system.

Our index is geared toward small and medium scale producers who principally grow for market and have a fairly large crop area (32.4 ha on average). According to Bockstaller and Girardin [62], indicators must be elaborated according to a scientific approach, and one of the important steps in this elaboration is validation.

Our index was developed according to the consensus of a group of scientists, with knowledge in agroecology, and was validated independently by 2 scientists, each with over 15 years of experience in agroecology. The evaluation included 3 important steps: design validation, output validation, and end use validation [62].

In previous studies, only seven indicators have been used to evaluate farm systems: crop diversity, crop succession, pesticide use, nitrogen level, phosphorus level, soil organic matter, and irrigation methods [63]. All of these practices depend on the farmer's decision, and to a large extent they impact the environment.

Our index is based on qualitative and quantitative concrete data and includes most of these 7 indicators, except nitrogen and phosphorus, both of which are observed indirectly via plant health, through the crop colour indicator (according to whether plants have a greenish or yellowish colour); organic matter, which is indirectly characterized by the technology applied in the system (green manure indicator, Figure 2); and irrigation, which in this study was not evaluated, given that all plots evaluated were only used for seasonal rainfed agriculture, according to local rainfall patterns.

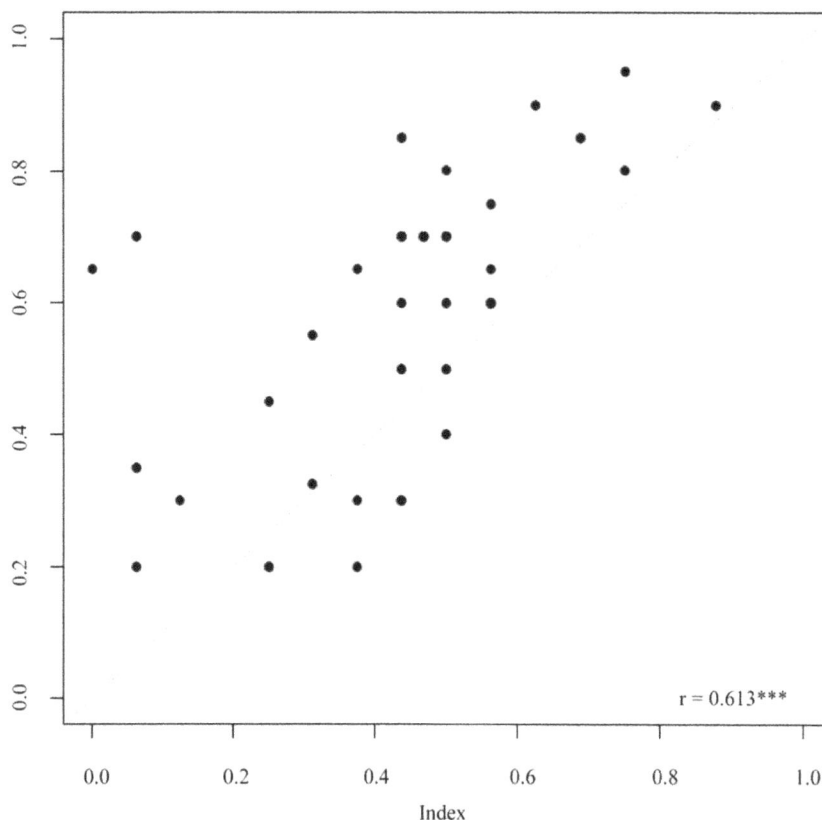

Figure 4. Scatterplot of the values of the index of ecological condition applied to 52 plots in South-East Mexico (x-axis) versus quality values between 0 (worst) and 1 (best) assigned by experts to the same plots (y-axis).

At the farm level there are indicators that evaluate the environmental impact of the agricultural practices and indicators that evaluate the effect of those practices at the local and global level [60].

Our index includes both types of indicators; evaluating agricultural practices: those variables taken in the field: type and frequency of tillage, pesticide application, fertilizer use, crop rotation, tree cover and density, green manure, crop density (see Table 2 & 3). Effect indicators used were disturbance (mainly soil disturbance) measured by tillage (frequency and type) and chemical inputs, technology used, and crop structure (see index, Figure 2). Some existing indexes focus on evaluation or evolution of environmental performance, thus encouraging environmentally sound practices [60], such as crop rotation, organic fertilizers, and no-tillage. Meanwhile, our index identifies the indicator that has the highest environmental impacts in each plot evaluated, with the idea that the farmer could potentially improve these with a given practice, ie. to monitor the soil ecological condition via the use of a soil macroinvertebrates index [64], where the lack of macroinvertebrates informs of a severe negative activity as pollution or conventional tillage.

In developing the variables to be included in our index, we reviewed bibliographic studies and carried out field work obtaining data which we hoped would reflect the negative and environmentally friendly practices commonly used in agro ecosystems of southeastern Mexico. The index can be used by farmers, using the following web address: http://201.116.78.102/~modelo/Index.html.

However, we did not evaluate certain indicators such as water use, and water quality, as did - for example - the index (monitoring tool) of ecological indicators used on a Flemish dairy farm [65]. Nor did we evaluate environmental impacts due to energy consumption [66]. Within a tropical framework, in south-eastern Mexico, the priorities were to identify those practices that were soil perturbing and environment polluting, practices that can be modified by the farmers, by an attitude changing. Nambiar et al. [16], proposed an agricultural sustainability index (ASI) to measure sustainability as a function of biophysical, chemical, economic, and social indicators, our index only measures the ecological state of the agroecosystem, and provides easy to use tools for improving those practices which negatively impact the environment. Van der Werf and Petit [60], state that indicators based on farmer practices cost less in data collection but do not allow for an actual evaluation of environmental impact. In the case of our study, the experts' evaluation correlated satisfactorily with the index, although the index rendered more penalizing scores than did the experts. Some improvements must be made to our index, relating quality and quantity of the applied inputs for

instance; the index should specify the kind of manures used and then to evaluate their effect when added to the systems. The consulted experts found important to integrate this information into the index, they also found that the possible relations among different crops and environmental effects of using cow manure, vermicompost, or traditional compost have to be considered. Another variable which the experts suggested should be added to the index is the presence of natural vegetation surrounding the crop. Farmers see advantages of having crops surrounded by secondary forest, diversity in the agricultural area can be increased, ie when different pollinators arrive.

The advantages of using this index is that the common agricultural practices (mechanized land preparation and use of common pesticides ie. Carbofuran, Chlorpyrifos), evaluated in this study as indicators are used throughout the world; over time, through practice, the index may be improved. Farmers and other land owners may realize which of the practices they use are disturbing the environment, due to the fact that these practices generate a value in each of the evaluated indicators. The variables obtained in the field contribute to the information of each of the indicators, and the index is the compendium of all the indicators. Our index doesn't give a sustainability measure, because it does not include socioeconomic indicators of the farms. Further studies are required in order to observe the acceptance of this index by farmers in a regional scale.

Acknowledgments

We thank Lorena Soto and Guillermo Jimenez for participating in the evaluation of the plots. We thank Simon Hernandez de la Cruz, Alejandra Sepulveda Lozada, Lauritania Ibarra Hernandez, and Marcelo Rodriguez Ricardes, who helped to collect data in the field. We are also thankful to Lorena Reyes for bibliographic support. El Colegio de la Frontera Sur provided infrastructural resources. Language revision was done by Ann Greenberg.

Author Contributions

Conceived and designed the experiments: EH CK SOG BDJ SHD VG. Performed the experiments: EH CK. Analyzed the data: EH CK. Contributed reagents/materials/analysis tools: EH CK SOG BDJ SHD VG. Contributed to the writing of the manuscript: EH CK SOG BDJ VG. Figures elaboration: CK SOG.

References

1. Houghton RA (1994) The world wide extent of land-use change. BioScience 44: 305–313.
2. Tilman D, Lehman C (2001) Human-cause environmental change: impacts on plant diversity and evolution. Proceedings of the National Academy of Sciences 98: 5433–5440.
3. INEGI (2008) Cuaderno Estadístico Municipal de Tenosique. Villahermosa, Tabasco: Gobierno del Estado de Tabasco.
4. Grande D, de Leon F, Nahed J, Perez-Gil F (2010) Importance and Function of Scattered Trees in Pastures in the Sierra Region of Tabasco, Mexico. Research Journal of Biological Sciences 5: 75–87.
5. Geissen V, Morales-Guzman G (2006) Fertility of tropical soils under different land use systems-a case study of soils in Tabasco, Mexico. Appl Soil Ecol 841: 1–10.
6. Melgar C, Geissen V, Cram S, Sokolov M, Bastidas P, et al. (2008) Pollutants in drainage channels following long-term application of Mancozeb to banana plantations in southeastern Mexico. Journal of Plant Nutrition and Soil Science 171: 597–604.

7. Aryal DR, Geissen V, Ponce-Mendoza A, Ramos-Reyes RR, Becker M (2012) Water quality under intensive banana production and extensive pastureland in tropical Mexico. Journal of Plant Nutrition and Soil Science 175(4): 553–559.
8. Ortiz SM, Anaya G, Estrada BW (1994) Evaluación, Cartografía y Políticas Preventivas de la Degradación de la Tierra. Chapingo, Mexico.
9. Geissen V, Sanchez-Hernandez R, Kampichler C, Ramos-Reyes R, Sepulveda-Lozada A, et al. (2009) Effects of land-use change on some properties of tropical soils – An example from Southeast Mexico. Geoderma 151: 87–97.
10. Moreno-Unda AA (2011) Environmental effects of the National Tree Clearing Program, Mexico, 1972–1982 Cologne: Cologne University of Applied Sciences 119 p.
11. Bockstaller CGL, Keichinger O, Girardin P, Galan MB, Gaillard G (2009) Comparison of methods to assess the sustainability of agricultural systems. A review. Agron Sustain Dev 29: 223–235.
12. Mitchell GMA, McDonald A (1995) PICABUE: a methodological framework for the development of indicators of sustainable development. Int J Sust Dev World 104–123.

13. Bockstaller C, Guichard L, Makowski D, Aveline A, Girardin P, et al. (2008) Agri-Environmental Indicators to Assess Cropping and Farming Systems: A Review Sustainable Agriculture. In: Lichtfouse E, Navarrete M, Debaeke P, Véronique S, Alberola C, editors. Springer Netherlands. 725–738.

14. Riley J (2001) The indicator explosion: local needs and International challenges. Agr Ecosyst Environ 87: 119–120.

15. Kampichler C, Hernández-Daumás S, Ochoa-Gaona S, Geissen V, Huerta-Lwanga E, et al. (2010) Indicators of environmentally sound land use in the humid tropics: The potential roles of expert opinion, knowledge engineering and knowledge discovery. Ecological Indicators 10: 320–329.

16. Nambiar KK, Gupta AP, Fu Qinglin, Li S (2001) Biophysical, chemical and socio-economic indicators for assessing agricultural sustainability in the Chinese coastal zone Agriculture Ecosystems and Environment 87: 209–214.

17. Cabell JF, Oelofse M (2012) An indicator framework for assessing agroecosystem resilience. Ecology and Society 17: 18 http://dx.doi.org/10.5751/ES-04666-170118.

18. Darnhofer I, Bellon S, Dedieu B, Milestad R (2010) Adaptiveness to enhance the sustainability of farming systems. A review. Agronomy for Sustainable Development 30: 545–555.

19. Allen CR, Fontaine JJ, Pope KL, Garmestani AS (2011) Adaptive management for a turbulent future. Journal of Environmental Management 92: 1339–1345.

20. Malezieux E, Crozat Y, Dupraz C, Laurans M, Makowski D, et al. (2009) Mixing plant species in cropping systems: concepts, tools and models. A review. Agron Sustain Dev 29: 43–62.

21. Jones CG, Lawton J, Shachak M (1997) Positive and negative effects of organisms as physical ecosystem engineers. Ecology and Society 78: 1946–1957.

22. Brussaard L, Caron P, Campbell B, Lipper L, Mainka S, et al. (2010) Reconciling biodiversity conservation and food security: scientific challenges for a new agriculture. Current opinion in Environmental sustainability 2: 34–42.

23. Knorr D, Watkins TR (1984) Alterations in Food Production Knorr DW, Watkins, T.R., editor. New York: Van Nostrand Reinhold.

24. Seufert V, Ramankutty N, Foley JA (2012) Comparing the yields of organic and conventional agriculture. Nature 485: 229–232.

25. Carter H (1989) Agricultural sustainability: an overview and research assessment. Calif Agric 43: 1618–1637.

26. Tellarini V, Caporali F (2000) An input/output methodology to evaluate farms as sustainable agroecosystems: an application of indicators to farms in central Italy. Agriculture, Ecosystems and Environment 77: 111–123.

27. Stinner BR, House GJ (1987) Role of ecology in lower-input, sustainable agriculture: an introduction. Am J Alternative Agric 2: 146–147.

28. Lockeretz W (1988) Open questions in sustainable agriculture. Am J Alternative Agric 3: 174–181.

29. Hauptli H, Katz D, Thomas BR, Goodman RM (1990) Biotechnology and crop breeding for sustainable agriculture. In: Edwards CA, Lal R, Madden P, Miller RH, House G, editors. Sustainable Agricultural Systems Soil and Water Conservation Society: Ankeny, Iowa. 141–156.

30. Madden P (1990) The economics of sustainable low-input farming systems. In: Francis CA, Flora CB, King LD, editors. Sustainable Agriculture in Temperate Zones. New York: John Wiley & Sons. 315–341.

31. Dobbs TL, Becker DL, Taylor DC (1991) Sustainable agriculture policy analyses: South Dakota on-farm case studies. J Farming Systems ResExt 2: 109–124.

32. Blair GJ, Lefroy RD, Lisle L (1995) Soil carbon fractions based on their degree of oxidation, and the development of a carbon management index for agricultural systems. Aust J Soil Res 46: 1459–1466.

33. Ouédraogo E, Mando A, Zombré NP (2001) Use of compost to improve soil properties and crop productivity under low input agricultural system in West Africa. Agriculture, Ecosystems and Environment 84: 259–266.

34. Pretty J, Toulmin C, Williams S (2011) Sustainable intensification in African agriculture. International Journal of Agricultural Sustainability 9: 5–24.

35. Lal R (2004) Soil carbon sequestration impacts on global climate change and food security. Science 304: 1623–1627.

36. Gurr GM, Wratten SD, Luna JM (2003) Multi-function agricultural biodiversity: pest management and other benefits. Basic Appl Ecol 4: 107–116.

37. Börjesson P (1999) Environmental effects of energy crop cultivation in Sweden I: Identification and quantification. Biomass and Bioenergy 16: 137–154.

38. Rommelse R (2001) Economic assessment of biomass transfer and improved fallow trials in western Kenya In: ICRAF Natural Resource Problems PaPP, editor. Natural Resource Problems. Nairobi (Kenya): International Centre for Research in Agroforestry (ICRAF).

39. Nyende P, Delve R (2004) Farmer participatory evaluation of legume cover crop and biomass transfer technologies for soil fertility improvement using farmer criteria, preference ranking and logit regression analysis. Exp Agric 40: 77–88.

40. Koocheki A, Nassiri M, Alimoradi L, Ghorbani R (2009) Effect of cropping systems and crop rotations on weeds. Agronomy for Sustainable Development 29: 401–408.

41. Bellon M (1995) Farmers 'Knowledge and Sustainable Agroecosystem Management: An Operational Definition and an Example from Chiapas, Mexico. Human Organization 54: 263–272.

42. Belalcázar S, Espinosa J (2000) Effect of Plant Density and Nutrient Management on Plantain Yield. Better Crops International 14: 12–15.

43. Shaahan MM, El-Sayed AA, Abou El-Nour EAA (1999) Predicting nitrogen, magnesium and iron nutritional status in some perennial crops using a portable chlorophyll meter. Scientia Horticulturae 82: 339–348.

44. Agbede TM (2008) Nutrient availability and cocoyam yield under different tillage practices. Soil Tillage Research 99: 49–57.

45. Edwards CA (1989) The Importance of Integration in Sustainable Agricultural Systems. Agriculture, Ecosystems and Environment 27: 25–35.

46. Edwards CA, Grove TL, Harwood RR, Pierce Colfer CJ (1993) The role of agroecology and integrated farming systems in agricultural sustainability. Agriculture, Ecosystems and Environment 46: 99–121.

47. García-Pérez JA, Alarcón-Gutiérrez E, Perroni Y, Barois I (2014) Earthworm communities and soil properties in shaded coffee plantations with and without application of glyphosate. Applied Soil Ecology 83: 230–237.

48. Correia FV, Moreira JC (2010) Effects of Glyphosate and 2,4-D on Earthworms (Eisenia foetida) in Laboratory Tests. Bull Environ Contam Toxicol 85: 264–268.

49. Dick J, Kaya B, Soutoura M, Skiba U, Smith R, et al. (2008) The contribution of agricultural practices to nitrous oxide emissions in semi-arid Mali Soil. Use and Management 24: 292–301.

50. Morales H, Perfecto I (2000) Traditional knowledge and pest management in the Guatemalan highlands. Agriculture and Human Values 17: 49–63.

51. Li BL, Rykiel EJ (1996) Introduction. Ecological Modelling 90: 109–110.

52. Salski A (1996) Introduction Fuzzy Logic in Ecological Modelling. Ecological Modelling 85: 1–2.

53. Mendoza GA, Prabhu R (2003) Fuzzy methods for assessing criteria and indicators of sustainable forest management. Ecological Indicators 3: 227–236.

54. Kampichler C, Platen R (2004) Ground beetle occurrence and moor degradation: modelling a bioindication system by automated decision-tree induction and fuzzy logic. Ecological Indicators 4: 99–109.

55. INEGI (2000) Cuaderno Estadístico Municipal de Tenosique. Gobierno del Estado de Tabasco; INEGI, editor. Villahermosa, Tabasco.

56. INEGI (1985) Carta Edafológica, Villahermosa In: E15-8, editor. 1: 250,000. Aguascalientes, México: Instituto Nacional de Estadística, Geografía e Informática.

57. Manjarrez-Muñoz B (2008) Ordenamiento territorial de la ganadería bovina en Balancán y Tenosique, Tabasco. Villahermosa Tabasco: El Colegio de la Frontera Sur. 105 p.

58. Isaac-Márquez R (2008) Análisis del Cambio de Uso y Cobertura del Suelo en los Municipios de Balancán y Tenosique, Tabasco, México. Villahermosa, Tabasco, México: El Colegio de la Frontera Sur.

59. Rigby D, Woodhouse P, Young T, Burton M (2001) Constructing a farm level indicator of sustainable agricultural practice. Ecological Economics 39: 463–478.

60. van der Werf HMG, Petit J (2002) Evaluation of the environmental impact of agriculture at the farm level: a comparison and analysis of 12 indicator-based methods. Agriculture, Ecosystems and Environment 93: 131–145.

61. Astier M, Speelman S, López-Ridaura S, Masera O, Gonzalez-Esquivel CE (2011) Sustainability indicators, alternative strategies and trade-offs in peasant agroecosystems: analysing 15 case studies from Latin America. International Journal of Agricultural Sustainability 9: 409–422.

62. Bockstaller C, Girardin P (2003) How to validate environmental indicators. Agr Syst 76: 639–653.

63. Bockstaller C, Girardin P, Van der Werf HGM (1997) Use of agroecological indicators for the evaluation of farming systems. Eur J Agron 7: 261–270.

64. Huerta E, Kampichler C, Geissen V, Ochoa-Gaona S, de Jong B, et al. (2009) Towards an ecological index for tropical soil quality based on soil macrofauna. Pesq agropec bras 44 (8): 1056–1062.

65. Meul M, Nevens F, Reheul D (2009) Validating sustainability indicators: Focus on ecological aspects of Flemish dairy farms. Ecological Indicators 9: 284–295.

66. Pervanchon F, Bockstaller C, Girardin P (2002) Assessment of energy use in arable farming systems by means of an agro-ecological indicator: the energy indicator. Agricultural Systems 72: 149–172.

19

Two Challenges for U.S. Irrigation Due to Climate Change: Increasing Irrigated Area in Wet States and Increasing Irrigation Rates in Dry States

Robert I. McDonald[1]*, **Evan H. Girvetz**[2]

1 Central Science Program, The Nature Conservancy, Arlington, Virginia, United States of America, **2** Global Climate Change Program, The Nature Conservancy, Seattle, Washington, United States of America

Abstract

Agricultural irrigation practices will likely be affected by climate change. In this paper, we use a statistical model relating observed water use by U.S. producers to the moisture deficit, and then use this statistical model to project climate changes impact on both the fraction of agricultural land irrigated and the irrigation rate ($m^3 ha^{-1}$). Data on water withdrawals for US states (1985–2005) show that both quantities are highly positively correlated with moisture deficit (precipitation – PET). If current trends hold, climate change would increase agricultural demand for irrigation in 2090 by 4.5–21.9 million ha (B1 scenario demand: 4.5–8.7 million ha, A2 scenario demand: 9.1–21.9 million ha). Much of this new irrigated area would occur in states that currently have a wet climate and a small fraction of their agricultural land currently irrigated, posing a challenge to policymakers in states with less experience with strict regulation of agriculture water use. Moreover, most of this expansion will occur in states where current agricultural production has relatively low market value per hectare, which may make installation of irrigation uneconomical without significant changes in crops or practices by producers. Without significant increases in irrigation efficiency, climate change would also increase the average irrigation rate from 7,963 to 8,400–10,415 $m^3 ha^{-1}$ (B1 rate: 8,400–9,145 $m^3 ha^{-1}$, A2 rate: 9,380–10,415 $m^3 ha^{-1}$). The irrigation rate will increase the most in states that already have dry climates and large irrigation rates, posing a challenge for water supply systems in these states. Accounting for both the increase in irrigated area and irrigation rate, total withdrawals might increase by 47.7–283.4 billion m^3 (B1 withdrawal: 47.7–106.0 billion m^3, A2 withdrawal: 117.4–283.4 billion m^3). Increases in irrigation water-use efficiency, particularly by reducing the prevalence of surface irrigation, could eliminate the increase in total irrigation withdrawals in many states.

Editor: A. Mark Ibekwe, U. S. Salinity Lab, United States of America

Funding: The authors are supported by The Nature Conservancy and its members. The funders had no role in study design, data collection and analysis, decision to publish, or preparation of the manuscript.

Competing Interests: The authors have declared that no competing interests exist.

* E-mail: rob_mcdonald@tnc.org

Introduction

Anthropogenic emissions of greenhouse gases are very likely increasing average temperatures and are more likely than not altering the amount and timing of precipitation [1]. The effects of climate change on agriculture will likely be multifaceted [2–5]. Changes in temperature, precipitation [6] and CO_2 concentration effects will affect plant growth. Moreover, the indirect effects of changes in weed and pest abundance and disease outbreaks may also significantly affect agriculture [7,8]. This paper focuses on irrigation, the largest human water use, which will likely be affected by climate change [9].

Climate change will likely affect the supply of water available from surface water or groundwater, decreasing it in many cases, and may also increase evapotranspiration from crops and hence demand for irrigation [10–16]. While several papers have quantified how changes in climate will affect irrigation water use by agricultural producers, results have been mixed depending on the time horizon of the analysis, the particular general circulation models (GCMs) consulted, and assumptions about how producers' irrigation practices respond to climate change [17–20].

The decision of whether or not to irrigate an agricultural field, as well as the amount of irrigation water applied, is a complex decision for producers, and depends on a number of factors [18,21]. Water supply is constrained by water policy and law as well as the infrastructure available to bring water to farmers. The potential economic return from production depends on the crops grown, the productivity of the soil, and the other inputs farmers add such as fertilizer. Demand for irrigation also depends on the crop grown, as well as the prevailing climate. Climate change might increase moisture deficit, or decrease it, depending on the relative change in temperatures and precipitation. Accounting for all these factors is difficulty, which makes forecasting irrigation water use challenging.

This paper uses an empirical approach, showing for the United States that changes in climate and irrigation equipment over the past 25 years have resulted in predictable changes in both the fraction of agricultural land irrigated and the irrigation rate ($m^3 ha^{-1}$). We then use a simple statistical model to make state-level projections of how both quantities will be affected by climate change. We incorporate information on historical changes over time in irrigation equipment into our statistical model, and project the potential scope of future irrigation efficiency improvements to serve as an adaptation to climate change. We use this analysis to estimate the how irrigation rate and area irrigated will need to change in the future–based on the assumptions of this model–to keep up with the change in water demand of crops due to climate change.

Materials and Methods

Data

Data on historical withdrawals for the period 1985–2005 were obtained from the United States Geological Survey (USGS), which conducts county-level surveys every five years [22]. For this study, we use total freshwater withdrawals (surface water plus groundwater) by the agriculture sector. For our analysis, we lump county-level data to states, since the county-level data vary widely in withdrawals when the county in which an irrigation system withdraws water is different than the county where the water is applied.

Withdrawal information was supplemented by data on area irrigated, as defined in the United States Department of Agriculture (USDA) Farm and Ranch Irrigation Survey [23], and total agricultural area and its market value, as defined in the USDA Census of Agriculture [24]. In this study, our definition of "agricultural area" follows the USDA definition of "total cropland", which includes harvested cropland, cropland used for pasture or grazing, and other miscellaneous cropland (e.g., areas where crops failed, fields were left intentionally left fallow, or with cover crops). Note that the subcategory of "cropland used for pasture and grazing" includes only land currently used for pasture or grazing that could be immediately used for crops without any additional improvement; it does not include the much larger land area of woodland and rangeland that is used for grazing or pasturing.

Data on historical temperature and precipitation were taken from the Parameter-elevation Regressions on Independent Slopes Model (PRISM) dataset (~4 km resolution) and used to calculate historic observed moisture deficit [25]. Future climate scenarios are based on the ensemble mean of a panel of 16 general circulation models (GCM) [26,27] for each of the A2, A1B, and B1 scenarios of the Intergovernmental Panel on Climate Change [28]. The 16 GCMs used in this study are BCCR-BCM2.0, CGCM3.1(T47), CNRM-CM3, CSIRO-Mk3.0, GFDL-CM2.0, GFDL-CM2.1, GISS-ER, INM-CM3.0, IPSL-CM4, MIROC3.2(medres), ECHO-G, ECHAM5/MPI-OM, CCSM3, PCM, and UKMO-HadCM3. For more information on the structure of each GCM, see Meehl et al. [26]. We examined three time periods: 2020–2039 (hereafter "2030"), 2040–2059 (hereafter "2050"), 2060–2079 (hereafter "2070"), and 2080–2099 (hereafter "2090"). To capture uncertainty in the GCM predictions of climate under different emissions scenarios, we also include the 20% and 80% quintiles of the ensemble's prediction of moisture deficit.

By using an ensemble of 16 GCMs, and considering the variation among the ensemble predictions for GCMs, we hope to improve on previously published work by considering a broad set of GCMs. Other papers that have quantified how changes in climate will affect irrigation water use have only considered a few GCMs, and some of the variation in the results among and within the papers seems to be driven by the particular GCMs chosen [17–20]. For instance, Fisher et al. [17] looked at the climate change impacts of the a2r emissions scenario with the HadCM3 and CISRO GCMs, over the time period 1990–2080, projecting a 45% increase in irrigated area globally. Their predictions for North America varied between the two GCMs, because the CSIRO MK3 predicts a decrease in moisture deficit for much of the US, while the Hadley CM3 predicts an increase in moisture for much of the U.S. Similarly, Thomson et al. [19] looked at the climate change impacts of different average temperature increases with two GCMs, the HadCm3 and CGCM, but incorporated in their model both demand for irrigation and supply of available water. They predict that irrigated area will decline in the continental United States under both GCMs, despite strong growth in demand, because of a decline in the available supply of irrigation water.

Calculation of Moisture Deficit

We calculated the soil moisture deficit during the growing season by subtracting precipitation from potential evapotranspiration (PET), as measured with the modified Thornthwaite (Hamon) method [29], which is based on temperature and number of daylight hours, and widely used in global and regional hydrologic models [30]. This metric should theoretically be related to irrigation water required. Moreover, it integrates intra-annual variability in water demand into a cumulative number for the growing season.

GCM predictions of moisture deficit were first resampled to average county values, and then average state values were calculated, weighting by the agricultural area (for the analysis of fraction of area irrigated) or irrigated area (for the analysis of irrigated area).

There has been considerable debate about the merits of different methods of estimating PET, and Kingston et al. [31] showed that the relative magnitude of the effect of climate change on PET depends on which method is used. The Hamon method appears to be one of the more sensitive metrics, increasing with climate change greatly, whereas other methods like Blaney-Criddle and Priestley-Taylor increase less [31]. To ensure that our results were not influence by our choice of metrics, we also calculated PET using the Blaney-Criddle method, a simple method that is known to be relatively less sensitive to climate change than most metrics [31]. The Blaney-Criddle method is a function solely of the mean daily temperature and the mean daily percentage of daytime hours.

Preliminary results suggest no major change in our findings. To see why this is likely to be so, note that we build a statistical model that relates irrigation use by farmers to historical values of PET. Since Hamon and Blaney-Criddle give highly correlated estimates of PET (Figure 1), and changes in Hamon and changes in Blaney-Criddle are highly correlated (Figure 2), they both give similar results in the linear regression we use as our statistical model. While change in PET as estimated by Hamon is systematically higher than change in PET by Blaney-Criddle, the fitted slopes in our regression correct for this fact and given similar estimates of irrigation use. Because we are fitting either estimate of PET to empirical data on irrigation use, as long as the estimates are strongly linearly correlated with one another, the projections of the statistical model will be very similar.

The spatial patterns of change in Hamon and Blaney-Criddle are quite similar (Figure 3). However, there are slight differences among states. For example, the Hamon metric predicts higher increases in PET in arid parts of Arizona than does Blaney-Criddle.

Statistical Analysis

Total withdrawals were modeled as a function of total agricultural land (A) times the fraction of agricultural land that is irrigated (F) times the irrigation rate (R):

California

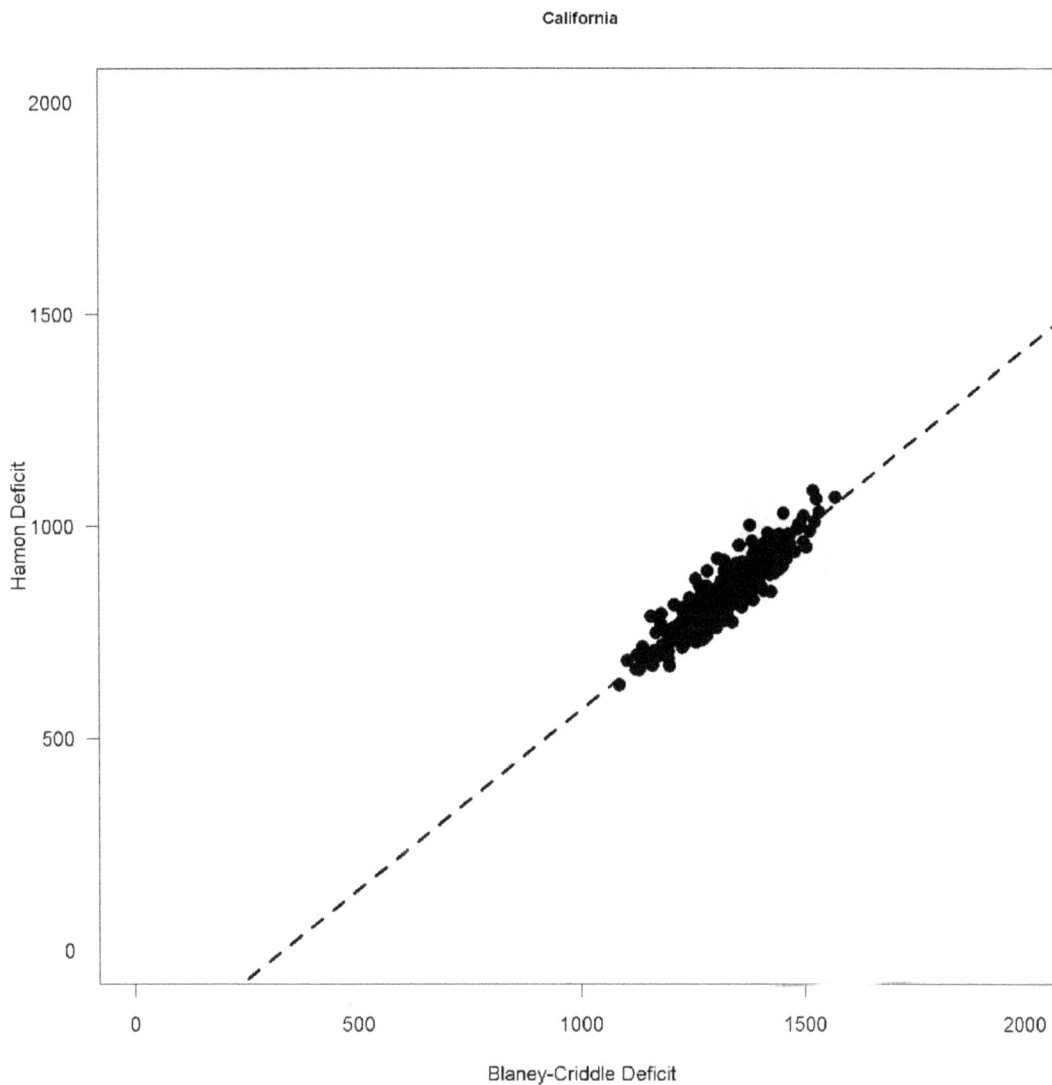

Figure 1. Scatterplot of estimated PET for California in 2005, using the Hamon and Blaney-Criddle metrics. Each dot represents one year in a particular combination of GCM and greenhouse gas emissions Scenario. Note that the strong linear correlation between the two means that when either of these two metrics are statistically related to irrigation use, the quantitative predictions of the effect of climate change are quite similar. Other states also show a linear correlation, with R^2 ranging from 0.75 to 0.93.

$$W_t = A_{tot,t} F_t R_t$$

$$\pi_2 = \mu_2 + \alpha_{t,2} + \beta_2 Flood_\Delta$$

Historical information is taken from the USDA survey. F and R were both modeled as functions of moisture deficit in 1985 and the change in moisture deficit ($Deficit_t - Deficit_{1985}$):

$$\pi_0 + \pi_1 Deficit_{1985} + \pi_2 Deficit_\Delta$$

where

$$\pi_0 = \mu_0 + \alpha_{t,0} + \beta_0 Flood_t$$

$$\pi_1 = \mu_1 + \alpha_{t,1} + \beta_1 Flood_t$$

The first two terms (π_0 and π_1) can be interpreted as, respectively, the slope and intercept of the relationship between moisture deficit and the response variable at a point in time. This relationship is the long-term average relationship with climate, and is the product of decades of economic decisions and irrigation policies. The third term (π_2) is the effect on the response variable of an inter-annual change in moisture deficit. It is the short term response of the agricultural sector to an interannual change in moisture deficit. All three terms are a function of a grand mean (μ), a time-specific shock (α_t), and the percent of surface irrigation (*Flood*). Surface irrigation (irrigation techniques where water is applied to the soil surface using gravity) was selected because it was believed to be causally negatively related to irrigation water-use efficiency. In this dataset proportion surface plus proportion

California

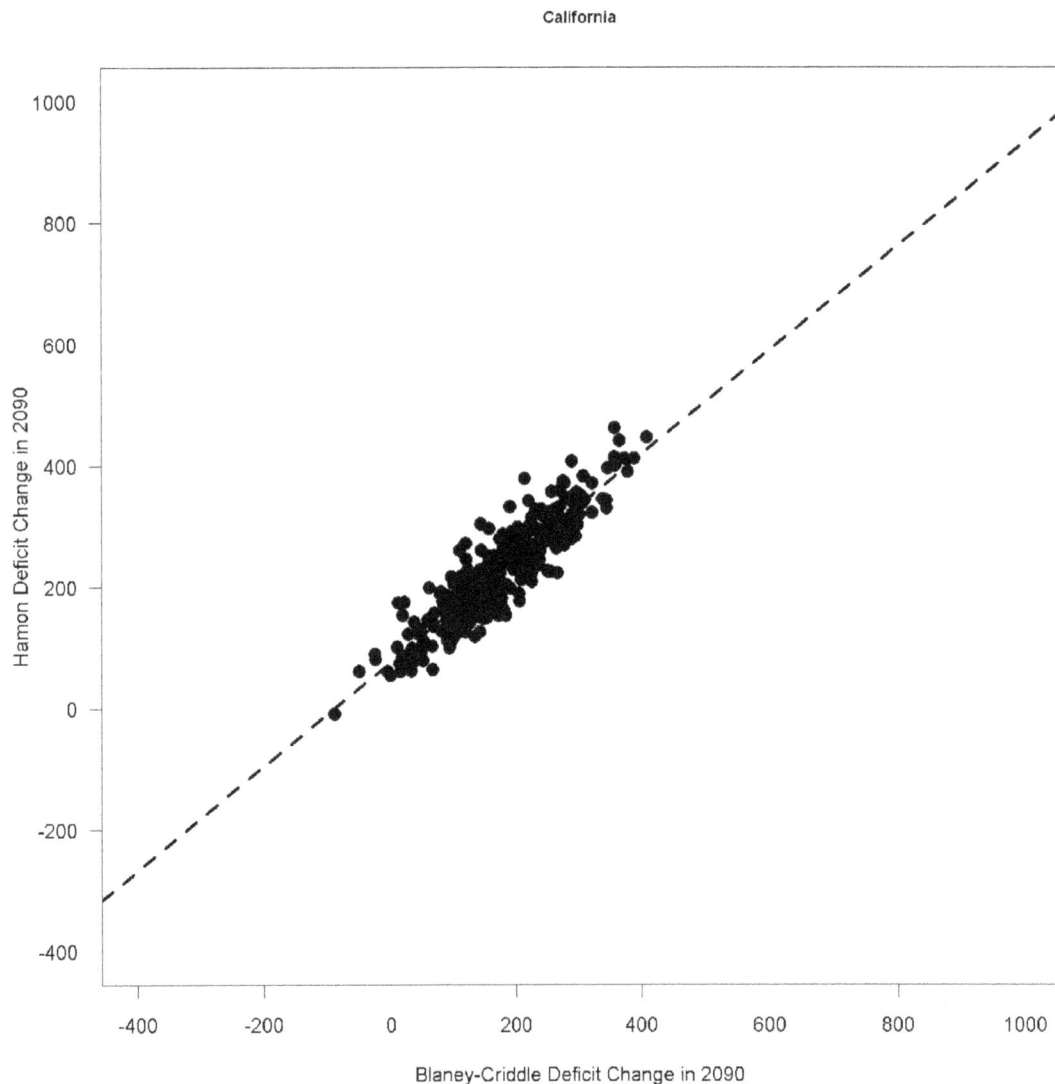

Figure 2. Scatterplot of change in PET for California between 2005 and 2090, using the Hamon and Blaney-Criddle metrics. Each dot represents one year in a particular combination of GCM and greenhouse gas emissions scenario. Note that the strong linear correlation between the two means that when either of these two metrics are statistically related to irrigation use, the quantitative predictions of the effect of climate change are quite similar. Other states also show a linear correlation, with R^2 ranging from 0.74 to 0.93.

sprinkler is approximately 1, since microirrigation is relatively uncommon. Note that the time-specific shock terms (α) will also incorporate other factors that varied over time but are not explicitly modeled here. For instance, if the steadily rising ambient CO_2 concentrations increased crop water-use efficiency over the observed 20 year time period, then one would expect the $\alpha_{t,1}$ to decline over time, all else being equal.

To account for autocorrelated errors by state, we used a repeated measured design for the error term, using Proc Mixed in SAS, version 9.2. We added terms one by one to the model, stopping when the added variables did not significantly improve the negative log-likelihood. The functional form of the regression of F was a logistic regression, although model fits using other functional forms that are bounded between 0 and 1 (e.g., probit) yielded similar results. For the regression of R, irrigation rates were log transformed to improve normality and to bound our results to the positive domain.

Future projections were calculated by assuming the fitted parameters in the regression equations, based on historical data, continue into the future. This implicitly assumes that some adaptation by producers takes place: regions that dry out, for example, adapt the agricultural practices of current comparable dry regions. Of course, hypothetically much larger water savings (or waste) are possible with more (or less) drastic change in agricultural practices, but we do not consider these more extreme possibilities in this analysis. Note that our analysis ignores the potential effect of increased CO_2 concentrations increasing crop water-use efficiency, although we highlight the importance of this issue in the Discussion section.

Results

Analysis of Observed Historic Irrigation Patterns

During 1985–2005 the proportion of irrigated agricultural land was greater in areas with greater moisture deficit ($p<0.001$, Table 1; and Figure 4). In any year, states with higher moisture

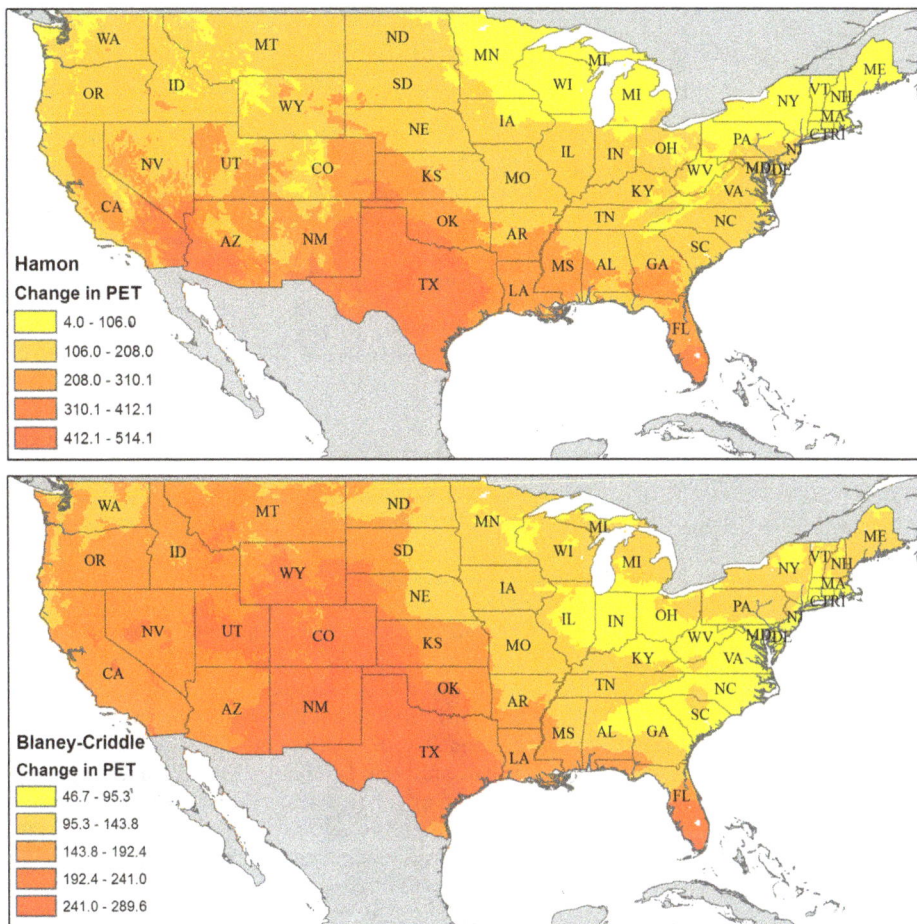

Figure 3. The effect of climate change on the Hamon estimate of PET (top) and the Blaney-Criddle estimate of PET (bottom). Both panels are colored with 5 equal interval categories that linearly span the range of the pixel values, with areas of less increase in PET being yellow and areas of greater increase in PET being red.

deficits have greater proportions of agricultural land under irrigation. This relationship represents the long-term outcome of individual farmers' decisions as a function of climate. Note that this long-term historical relationship shows adaptation to historical climate, in that dry areas are growing different crops and using different practices than wet areas. On the shorter year-to-year time scale, inter-annual changes in moisture deficit have resulted in changes to the proportion of agricultural area irrigated: higher moister deficit resulted in higher proportion of area irrigated (p = 0.03, Table 1).

Similarly, the irrigation rate (m^3 of water per hectare of irrigated land) is greater in sites with greater moisture deficit (p<0.001, Table 2 and Figure 4). At any point in time, states with higher moisture deficits have higher irrigation rates. Again, this relationship among different sites is the long-term outcome of a historical process of agricultural development that was influenced by the prevailing climate, and the gradient occurs despite some presumed adaptation of agricultural crops and practices to historical climate. Interestingly, the historical shift away from surface irrigation and toward sprinkler and drip irrigation has resulted in decreased irrigation rates over time (p<0.001, Table 2), all else being equal. In the short run on the year-to-year time scale, increases in moisture deficit are associated with increased irrigation rate, although this result is not statistically significant (p = 0.11, Table 2).

Projected Impact of Climate Change on Irrigation Area

Moisture deficit is projected to increase on average for US states under all three climate change scenarios (Figure 5A). After 2060, the increase in moisture deficit is largest in the A2 scenario, and smallest in the B1 scenario. Regardless of emissions scenario, climate change increases moisture deficit in 2090 throughout most of the contiguous US for 13 of the 16 GCMs considered [27], with the exception of the NCAR CCSM3, CSIRO MK3, and NCAR PCM models. Generally, the largest increase in moisture deficit is in the South followed by the Southwest, while the Northeast has

Table 1. Regression coefficients that predict the proportion of agricultural area that is irrigated, logit transformed.

Effect	Estimate	SE	P
Intercept (μ_0)	−4.4049	0.1798	<.0001
Deficit$_{1985}$ (μ_1)	0.000807	0.000067	<.0001
Deficit$_\Delta$ (μ_2)	0.000371	0.000164	0.0252

Only the final, best fit model is shown. Greek letters correspond to the regression parameters discussed in section 2.3.

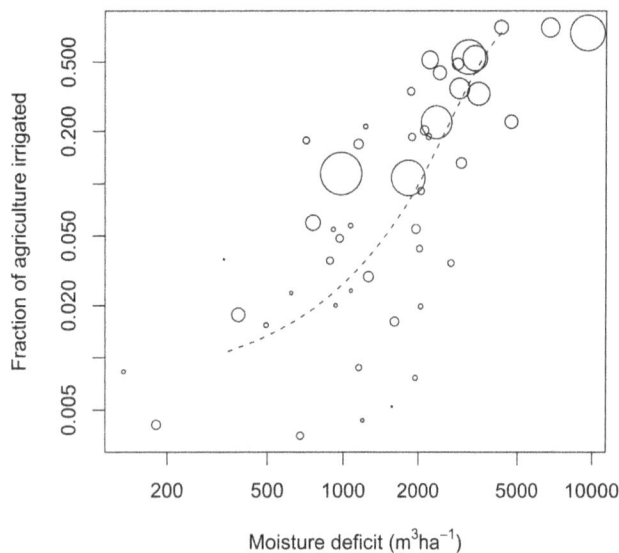

Figure 4. The fraction of agricultural land irrigated in U.S. states in 2005 as a function of moisture deficit. Note the logarithmic scale on both axes. The size of circle is proportional to the irrigation rate (m³ha⁻¹); states with high moisture deficit have higher irrigation rates. The fitted regression line is shown for the middle 90% of the data range.

Table 2. Regression coefficients that predict the irrigation rate (m^3ha^{-1}), log transformed.

Effect	Estimate	SE	P
Intercept (μ_0)	7.7495	0.07894	<.0001
Deficit$_{1985}$ (μ_1)	0.000132	0.000037	0.0009
Deficit$_\Delta$ (μ_2)	0.000118	0.000073	0.11
Fraction Flood$_{1985}$ (β_1)	2.1475	0.4675	<.0001
Fraction Flood$_\Delta$ (β_2)	1.3409	0.1988	<.0001

Only the final, best fit model is shown. Greek letters correspond to the regression parameters discussed in section 2.3.

small increases or decreases in moisture deficit. The most disagreement among climate models in the direction of moisture deficit change is in the Southwest, where some GCMs project a decrease and some an increase in moisture deficit.

Given the projected increases in moisture deficit with climate change, we project that the proportion of agricultural land irrigated in the U.S. will increase (Figure 5B). Under all emissions scenarios, farmers will increasingly need to irrigate their fields as it gets drier. The total area irrigated nationally will increase by 4.5–21.9 million ha, with the uncertainty due both to differences among emissions scenarios and to differences among GCMs. The B1 emissions scenario has the smallest increase (4.5–8.7 million ha) and the A2 emissions scenario has the greatest (9.1–21.9 million ha). Note that our projection estimates new irrigated area based on observed trends, after a moderate degree of adaptation: our methodology implicitly assumes farmers in a site made drier by climate change adopt the crops and practices of current farmers in comparable climates. Depending on the state, laws governing access to water or a shortage of available water might prevent this much expansion of irrigated area from occurring [18]. Alternatively, failure of farmers to adapt to climate change as much as is assumed in our model would imply an even larger increase in irrigated area.

The states with the greatest demand for new irrigated area, in ha, are those in the South-Central U.S. (Figure 6A) and to a lesser extent the Northwest and California. There is projected to be less of an increase in the Northeast, where climate impacts on moisture deficit are projected to be smaller. The extent of expansion of irrigation is a function of the amount of greenhouse gas emissions, with the A2 scenario requiring the most expansion and the B1 scenario the least. The difference in projected new irrigation area demanded between the A2 scenario and the B1 scenario, at the ensemble median climate projections, is 6.4 million ha. While exact cumulative emissions depend on the GCM implementation of the scenarios, every 1 GtC increase in cumulative emissions

increases the demand for new irrigated land in the US by roughly 7000 ha.

This expansion of agricultural area due to climate change will pose an adaptation challenge to some states that have traditionally had a very small percentage of their agricultural land irrigation. One way to see the magnitude of this challenge is to look at the fraction of agricultural water withdrawals in 2090 that would go to fields that were not irrigated in 2005 (Figure 7). In some states with currently limited irrigation systems, more than half of all irrigation withdrawals in 2090 will be for fields that were not irrigated in 2005. Changes in Kansas (KS), Texas (TX), and Montana (MT) are particularly notable, since these states have a lot of agricultural area that is currently mostly rain-fed but will have significant expansion of their irrigation systems. Conversely, Arizona (AZ) and California (CA) already have a large fraction of their agricultural area irrigated and there is little potential for expansion to increase withdrawals in 2090.

Installing irrigation in new fields costs money, and producers will only install irrigation if they find the increased revenue from crops grown on irrigated fields worth more than the cost of irrigation. While a full analysis of the costs and benefits of installing irrigation is beyond the scope of the paper, some insights can be gained by looking at state-level averages (Figure 8). In some states that will have a large increase in proportion of agricultural area irrigated, current market value per ha of cropland is fairly high, and may be enough to support the costs of installing irrigation equipment. Conversely, producers in states with lower market value per ha may find it difficult to support the installation of irrigation equipment unless new crops or markets are found. For instance, both Arizona (AZ) and Texas (TX) are projected to have increased irrigated area in response to climate change. However, Arizona has a relatively high average market value for cropland of $3700 ha⁻¹, driven by production of high-value crops like vegetables and horticultural products, while Texas has a relatively low average market value for cropland of $420 ha⁻¹, primarily from production of grains and cotton.

Projected Impact of Climate Change on Irrigation Rate

Increases in moisture deficit caused by climate change will likely increase the irrigation rate (Figure 5C). Under all emissions scenarios, irrigation rates will increase with climate change from 7,963 to 8,400–10,415 m³ha⁻¹ in 2090 if the current mix of irrigation technologies stays in place. The B1 emissions scenario has the smallest rate (8,400–9,145 m³ha⁻¹), while the A2 emissions scenario has the largest rate (9,380–10,415 m³ha⁻¹). However, if the current trend away from surface irrigation continues at the same pace into the future, needed irrigation rates would decline from current levels to 4,911–6,205 m³ha⁻¹ in 2090 (B1 rate: 4,911–5,355 m³ha⁻¹, A2 rate: 5,475–6,205 m³ha⁻¹). Note that

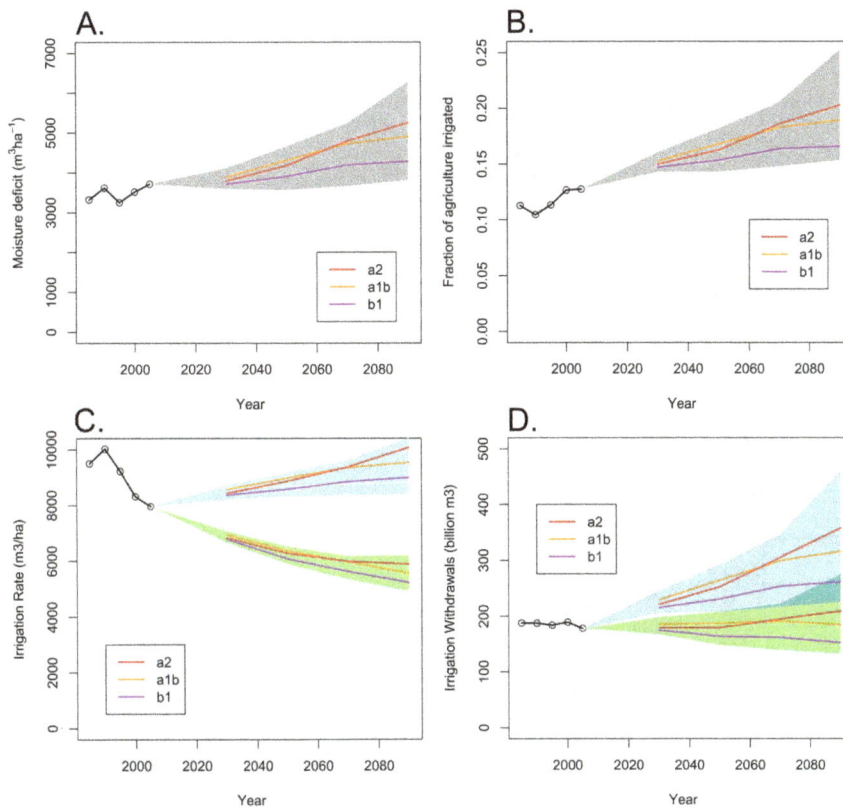

Figure 5. The effect of climate change on irrigation water use by United States agriculture. A.) Historical and future trends in U.S. mean moisture deficit. Displayed is the irrigated area weighted average, which is most directly relevant to irrigation rate. For each climate change scenario, the projected moisture deficit at the climate of the ensemble median is shown. The grey area shows the range of fraction irrigated for climate of the 20th and 80th quintiles of the ensemble. B.) Historical and future trends in the fraction of agriculture irrigated. C.) Historical and future trends in the irrigation rate. The blue area shows the effect of climate change on irrigation rates if the current mix of irrigation equipment persists over time; the green area shows the effect of climate change if the observed (1985–2005) trend away from relatively inefficient surface irrigation continues over time. D.) Historical and future trends in total irrigation withdrawals. Note the confidence intervals of the blue (current mix of irrigation equipment) and green (decreased use of surface irrigation) areas overlap after 2030.

our projection only estimates irrigation rates if current observed trends continue, after a moderate degree of adaptation: our methodology implicitly assumes farmers in newly dry climates will irrigate at a rate like that of comparable dry climates. Water policy or water availability might constrain increases in the irrigation rate. Alternatively, failure of farmers to adapt to climate change as much as is assumed in our model would imply an even larger increase in the irrigation rate.

Note also that higher CO_2 levels might increase the water-use efficiency of crop plants, which would further lower the curves in Figure 5C. However, the degree to which this water-use efficiency increase from higher CO_2 concentrations will occur in real systems is unclear, depending on the water balance of the site and the plant species involved [32–38]. For one important crop species, wheat, increases in water-use efficiency due to higher CO_2 concentrations may exceed 20% in some circumstances [37,39,40]. Interestingly, the $\alpha_{t,1}$ terms in our model were not statistically significant and are not included in our final model, so there is no clear historical effect of rising ambient CO_2 concentrations on crop water-use efficiency.

The second challenge that climate change poses for irrigation in the US will be this increase in irrigation rates, particularly important in already arid climates. If the current mix of irrigation technologies continues into the future, the biggest increase in irrigation rate is in the Western US (Figure 6B). In places like Arizona, California, Montana, Nevada, and Idaho, farmers will

have incentive to apply more water to irrigated lands. However, in some of these states (Arizona, California, and Nevada) there is high uncertainty in the predictions of the ensemble of the GCMs (Figure 6B).

Moving away from surface irrigation could generate substantial irrigation water-use efficiency gains, offsetting demand for an increase in irrigation rate in these states (Figure 9). States with the greatest potential for decrease in irrigation rate with a move away from surface irrigation are Arkansas (AR), Louisiana (LA), Arizona (AZ), Wyoming (WY), and Colorado (CO). In these states, the potential for decreased irrigation rate can more than offset the potential for irrigated area expansion due to climate change. On the other hand, some states such as Iowa and Illinois currently use very little surface irrigation, and there is little potential for decreasing irrigation rates with a move away from surface irrigation. Note also that there is considerable variability among GCMs for the Southwest, and while the median projection under all emissions scenarios is a large increase in irrigation rates in these states, a few GCMs actually project a decrease in moisture deficit and hence a decrease in irrigation rates in these states.

In the absence of gains in irrigation water-use efficiency, total water withdrawals for irrigation are likely to increase substantially (Figure 5D). This is true over all emissions pathways, but the increase in particularly striking for the A2 emissions pathway. Thus, irrigation water-use efficiency gains from switching away

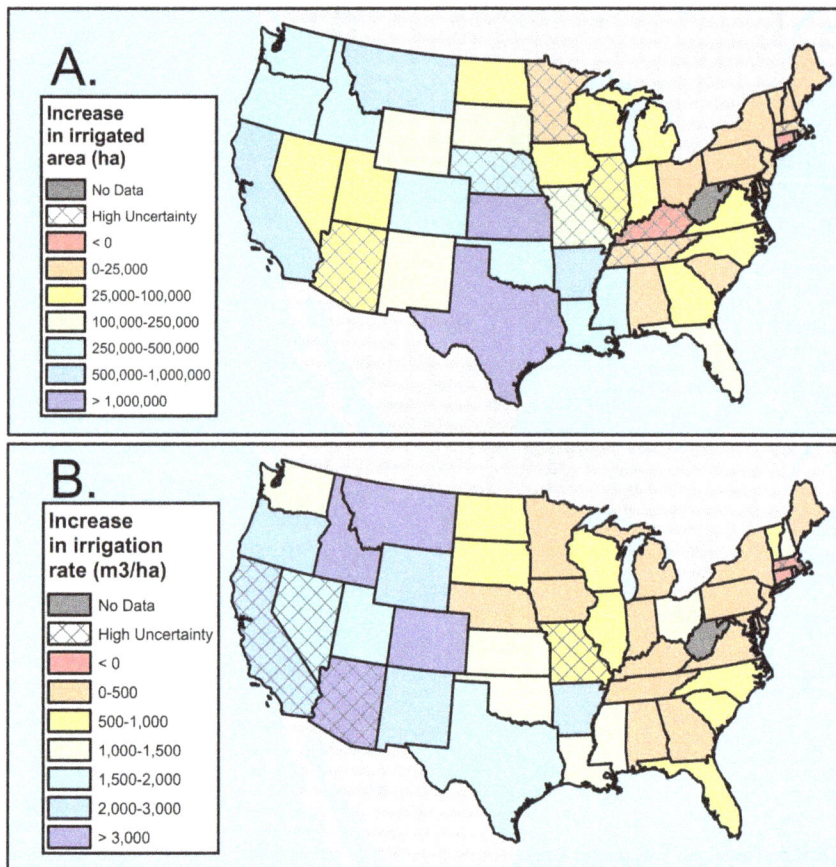

Figure 6. The effect of climate change on irrigation by states. A.) Projected increase in irrigated area by 2090 under the A1B scenario, ensemble median. Most states have an increase in irrigated area under all emission scenarios in more than 80% of the GCMs in the ensemble; those that have a decrease in some cases are marked high uncertainty. B.) Projected increase in irrigation rate by 2090 under the A1B scenario, ensemble median. Most states have an increase in irrigation rate under all emission scenarios in more than 80% of the GCMs in the ensemble; those that have a decrease in some cases are marked high uncertainty.

from surface irrigation can be an effective adaptation under moderate climate change, but cannot be wholly effective if climate change is severe. Moreover, increases in irrigation efficiency can only serve as adaptation effectively in some states (Figure 9).

Discussion

Our results suggest that there are two different challenges that climate change will pose to agricultural irrigation in the US. In states in currently relatively wet climates, expansion of irrigated area is likely to be an important driver in increases in overall agricultural withdrawals. In some cases, these states have relatively little experience managing water use under scarcity, at least compared to states in the western US. For instance, many states in the eastern United States use a riparian water rights system, where all land owners along a waterbody have a right to make reasonable use of water at any time. In these states, there may currently be little regulation on the installation of a new irrigation system. Our results suggest that a key goal for these states should be developing appropriate regulations that make sure new irrigation systems have access to sufficient water without affect other water users in the area.

In contrast, in states in currently dry climate that already have a large fraction of agricultural area irrigated, our results project the predominant driver of increases in water withdrawal will be the

tendency of producers to apply more irrigation water in response to droughts. Fully using technologies to increase irrigation efficiency will be a key goal for these states. The switch away from surface irrigation is included in our model, but many other forms of adaptation are possible [2,41]. Changes in agricultural practices such as tilling and plant spacing can affect water demand, as can changes in crop varieties beyond what is implicitly assumed by our model. Many of these currently dry states are in the western US and follow a prior appropriation water rights system, where older uses of water have priority over new uses of water. New users of water may be limited by their water right in how much they can increase their irrigation volume. However, currently many states lack mechanisms for water shortages to affect the price or distribution of existing water rights to farmers, leaving them little incentive to consider low water availability in their individual decision-making [9,42–45]. Farmers would have more incentive to switch to more water-efficient crops if the price of irrigation water more accurately reflects its scarcity.

Our results also highlight the importance of greenhouse gas mitigation for reducing impacts on freshwater demand. The median difference in extra demand for withdrawals in the A2 and B1 scenarios is 57 billion m^3 per year in 2090. This means that avoiding the release of 1 GtC will result in the U.S. needing roughly 65 million m^3 less irrigation water annually. Our results

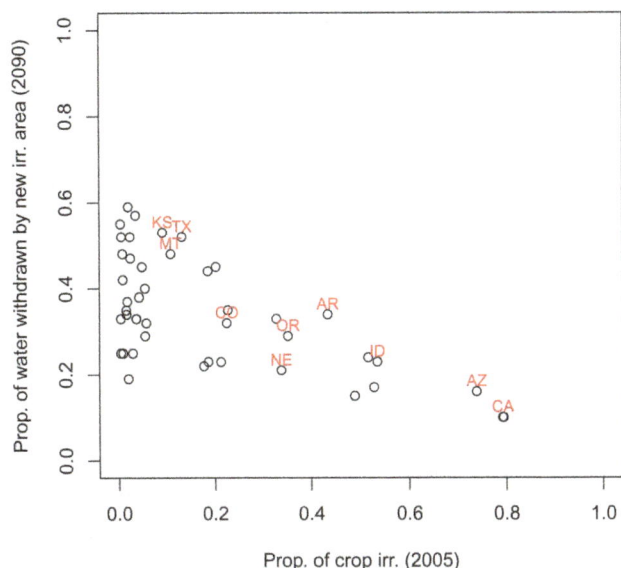

Figure 7. Currently wet states will have significant increases in irrigated area. The relationship between the proportion of agricultural land irrigation in 2005 and the predicted proportion of water withdrawals in 2090 (median A1B scenario of the GCM ensemble) that will come from fields not currently irrigated. A few states with significant agricultural area are labeled, using standard two-digit abbreviation for US states (see Figure 3). Three states are excluded from this graph, because climate change will have a net decrease on irrigated area there (Massachusetts, Connecticut, and Rhode Island).

demonstrate that, in addition to reducing climate change's many other impacts, climate change mitigation saves irrigation water.

Our results stress that climate change will likely increase irrigation water use in the US, consistent with some previously published research [17,18,21], but in contrast with the results of other studies that showed mixed results depending on the GCM examined [19,20]. At the same time, other research suggests climate change will decrease water supply, at least in some regions, either because of changes in annual supply or because of intra-annual changes in the hydrologic cycle such as changes in snowpack [19]. The exact response of agricultural producers depends on irrigation policy and law, the price of irrigation and water, and the income from crop production, among other things.

The central assumption when making climate change projections with such a statistical model is that past is prologue: how agricultural irrigation changed in response to past climate events is indicative of how it will respond in the future. While this is a reasonable assumption used by many modeling studies, readers should remember that more radical changes in water governance or water availability in the U.S. are theoretically possible, and would imply different forecasts than the ones we present.

Further research into the mechanisms that control the decisions of agricultural producers is key to evaluating the extent to which observed trends in the past two decades are indicative of the response of the agricultural system climate change. For instance, additional field-based studies of the impact of increased CO_2 on crop water use efficiency are needed to incorporate these feedbacks into mechanistic models of responses to climate change. Other factors known to affect irrigation withdrawals include the price of irrigation and income from commodity production.

Our simple statistical model provides a useful complement to more complex, mechanistic models of agricultural irrigation use. This model points out on a national scale how climate change is

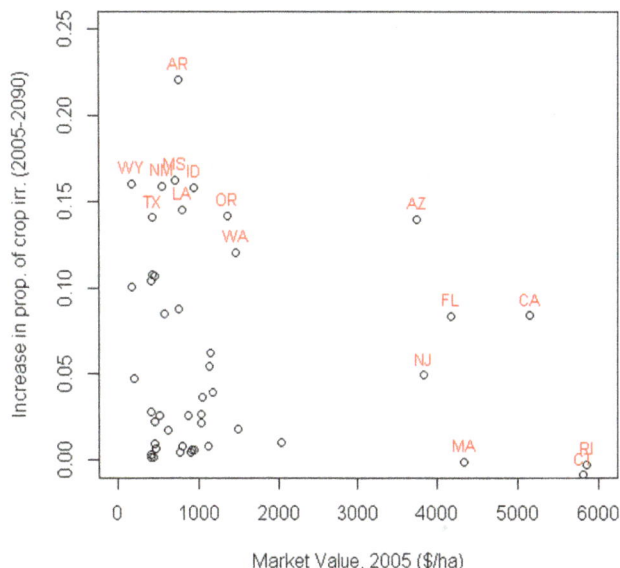

Figure 8. Market value and increase in proportion irrigated. The relationship between the average per hectare market value of cropland in 2005 and the predicted change in proportion of agricultural area irrigated (median A1B scenario of the GCM ensemble). A few states are labeled, using standard two-digit abbreviation for US states (see Figure 3).

projected to impact agricultural water demand in different states, and how those states will likely be driven to change their agricultural practices in response to these impacts–either through increasing the area of agricultural land irrigated or through increasing the irrigation rate on already irrigated land. This information is useful to agricultural and water resource managers for informing their long-term planning, and provides insight more broadly into how changes to agricultural water demand will likely

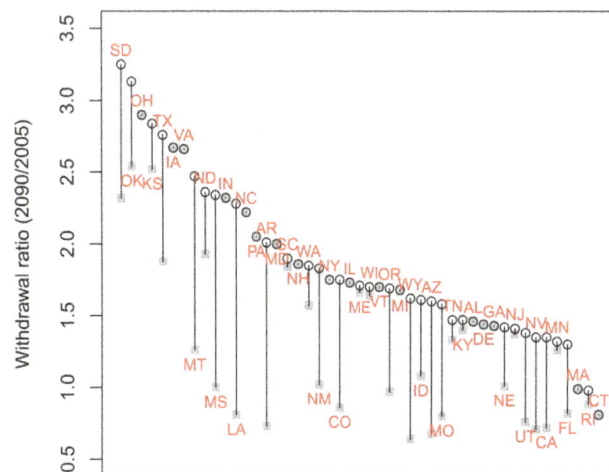

Figure 9. Withdrawal ratios by state. The ratio of irrigation withdrawals in 2090 to irrigation withdrawals in 2005 if the mix of irrigation technologies stays the same as today (open circles) or if the current trend away from surface irrigation continues into the future (grey squares). A ratio of more than one indicates withdrawals will increase, while a value of less than one indicates withdrawals will decrease. Data points are labeled using the standard two-digit abbreviation for US states (see Figure 3), staggered so that labels do not overlap.

complicate water management in the future, among many other demands from other sectors.

Acknowledgments

RIM is grateful for an appointment as an affiliate of the CONS Program, University of Maryland at College Park, and EHG is grateful for his affiliate appointment in the University of Washington, School of Forest Resources.

Author Contributions

Conceived and designed the experiments: RIM EG. Analyzed the data: RIM EG. Wrote the paper: RIM EG.

References

1. IPCC (2007) Climate Change 2007: The Physical Science Basis; Change IPoC, editor. Cambridge, UK: Cambridge University Press.

2. Howden SM, Soussana JF, Tubiello FN, Chhetri N, Dunlop M, et al. (2007) Adapting agriculture to climate change. Proceedings of the National Academy of Sciences of the United States of America 104: 19691–19696.

3. Schmidhuber J, Tubiello FN (2007) Global food security under climate change. Proceedings of the National Academy of Sciences of the United States of America 104: 19703–19708.

4. Tubiello FN, Fischer GI (2007) Reducing climate change impacts on agriculture: Global and regional effects of mitigation, 2000–2080. Technological Forecasting and Social Change 74: 1030–1056.

5. Fischer G, Shah M, Tubiello FN, van Velhuizen H (2005) Socio-economic and climate change impacts on agriculture: an integrated assessment, 1990–2080. Philosophical Transactions of the Royal Society B-Biological Sciences 360: 2067–2083.

6. Schlenker W, Roberts MJ (2009) Nonlinear temperature effects indicate severe damages to US crop yields under climate change. Proceedings of the National Academy of Sciences of the United States of America 106: 15594–15598.

7. Tubiello FN, Soussana JF, Howden SM (2007) Crop and pasture response to climate change. Proceedings of the National Academy of Sciences of the United States of America 104: 19686–19690.

8. Luedeling E, Steinmann KP, Zhang MH, Brown PH, Grant J, et al. (2011) Climate change effects on walnut pests in California. Global Change Biology 17: 228–238.

9. Rosegrant MW, Ringler C, Zhu TJ (2009) Water for Agriculture: Maintaining Food Security under Growing Scarcity. Annual Review of Environment and Resources 34: 205–222.

10. Connor J, Schwabe K, King D, Kaczan D, Kirby M (2009) Impacts of climate change on lower Murray irrigation. Australian Journal of Agricultural and Resource Economics 53: 437–456.

11. Taylor R, Scanlon B, Doll P, Rodell M, van Beek R, et al. (2012) Ground water and climate change. Nature Climate Change 1744: http://dx.doi.org/10.1038/nclimate1744.

12. Sun G, McNulty SG, Myers JAM, Cohen EC (2008) Impacts of Multiple Stresses on Water Demand and Supply Across the Southeastern United States. Journal of the American Water Resources Association 44: 1441–1457.

13. Palmer MA, Liermann CAR, Nilsson C, Florke M, Alcamo J, et al. (2008) Climate change and the world's river basins: anticipating management options. Frontiers in Ecology and the Environment 6: 81–89.

14. Alcamo J, Florke M, Marker M (2007) Future long-term changes in global water resources driven by socio-economic and climatic changes. Hydrological Sciences Journal-Journal Des Sciences Hydrologiques 52: 247–275.

15. Elbakidze L (2006) Potential economic impacts of changes in water availability on agriculture in the Truckee and Carson River Basins, Nevada, USA. Journal of the American Water Resources Association 42: 841–849.

16. Bureau of Reclamation (2011) SECURE Water Act Section 9503(c)-Reclamation Climate Change and Water, Report to Congress. Washington, DC: US Department of the Interior, Bureau of Reclamation. 206 p.

17. Fischer G, Tubiello FN, Van Velthuizen H, Wiberg DA (2007) Climate change impacts on irrigation water requirements: Effects of mitigation, 1990–2080. Technological Forecasting and Social Change 74: 1083–1107.

18. Schlenker W, Hanemann WM, Fisher AC (2005) Will US agriculture really benefit from global warming? Accounting for irrigation in the hedonic approach. American Economic Review 95: 395–406.

19. Thomson AM, Rosenberg NJ, Izaurralde RC, Brown RA (2005) Climate change impacts for the conterminous USA: An integrated assessment - Part 5. Irrigated agriculture and national grain crop production. Climatic Change 69: 89–105.

20. Rosenzweig C, Strzepek KM, Major DC, Iglesias A, Yates DN, et al. (2004) Water resources for agriculture in a changing climate: international case studies. Global Environmental Change-Human and Policy Dimensions 14: 345–360.

21. Schlenker W, Hanemann WM, Fisher AC (2007) Water availability, degree days, and the potential impact of climate change on irrigated agriculture in California. Climatic Change 81: 19–38.

22. Kenny J, Barber N, Hutson S, Linsey K, Lovelace J, et al. (2009) Estimated Use of Water in the United States in 2005. Reston, VA: U.S. Geological Survey.

23. USDA (2008) Farm and Ranch Irrigation Survey. Washington, DC: U.S. Department of Agriculture.

24. USDA (2009) US Census of Agriculture:2007. Washington, DC: United States Department of Agriculture.

25. Daly C, Neilson RP, Phillips DL (1994) A statistical-topographic model for mapping climatological precipitation over mountainous terrain. Journal of applied Meteorology 33: 140–158.

26. Meehl G, Covey C, Delworth T, Latif M, McAvaney B, et al. (2007) The WCRP CMIP3 Multimodel Dataset: A new era in climate change research. Bulletin of the American Meterological Society September: 1383–1394.

27. Girvetz EH, Zganjar C, Raber GT, Maurer EP, Kareiva P, et al. (2009) Applied Climate-Change Analysis: The Climate Wizard Tool. Plos One 4.

28. IPCC (2000) Special Report on Emissions Scenarios. Geneva: Intergovernmental Panel on Climate Change.

29. Lu J, Sun G, McNulty SG, Amataya D (2005) A comparison of six potential evapotranspiration methods for regional use in the southeastern United States. Journal of the American Water Resources Association 41: 621–633.

30. Vörösmarty CJ, Federer C, Schloss AL (1998) Potential evapotranspiration functions compared on U.S. watersheds: Possible implications for global-scale water balance and terrestrial ecosystem modelling. Journal of Hydrology 207: 147–169.

31. Kingston D, Todd M, Taylor R, Thompson JN, Arnell NW (2009) Uncertainty in the estimation of potential evapotranspiration under climate change. Geophysical Research Letters 36: 1–6.

32. Penuelas J, Canadell JG, Ogaya R (2011) Increased water-use efficiency during the 20th century did not translate into enhanced tree growth. Global Ecology and Biogeography 20: 597–608.

33. Huang JG, Bergeron Y, Denneler B, Berninger F, Tardif J (2007) Response of forest trees to increased atmospheric CO_2. Critical Reviews in Plant Sciences 26: 265–283.

34. Morgan JA, Pataki DE, Korner C, Clark H, Del Grosso SJ, et al. (2004) Water relations in grassland and desert ecosystems exposed to elevated atmospheric CO_2. Oecologia 140: 11–25.

35. Wullschleger SD, Tschaplinski TJ, Norby RJ (2002) Plant water relations at elevated CO_2 - implications for water-limited environments. Plant Cell and Environment 25: 319–331.

36. Medlyn BE, Barton CVM, Broadmeadow MSJ, Ceulemans R, De Angelis P, et al. (2001) Stomatal conductance of forest species after long-term exposure to elevated CO_2 concentration: a synthesis. New Phytologist 149: 247–264.

37. Moore BD, Cheng SH, Sims D, Seemann JR (1999) The biochemical and molecular basis for photosynthetic acclimation to elevated atmospheric CO_2. Plant Cell and Environment 22: 567–582.

38. Drake BG, GonzalezMeler MA, Long SP (1997) More efficient plants: A consequence of rising atmospheric CO_2? Annual Review of Plant Physiology and Plant Molecular Biology 48: 609–639.

39. Manderscheid R, Weigel HJ (2007) Drought stress effects on wheat are mitigated by atmospheric CO_2 enrichment. Agronomy for Sustainable Development 27: 79–87.

40. Hunsaker DJ, Kimball BA, Pinter PJ, Wall GW, LaMorte RL, et al. (2000) CO_2 enrichment and soil nitrogen effects on wheat evapotranspiration and water use efficiency. Agricultural and Forest Meteorology 104: 85–105.

41. Easterling WE (1996) Adapting North American agriculture to climate change in review. Agricultural and Forest Meteorology 80: 1–53.

42. Hellegers P, Perry CJ (2006) Can irrigation water use be guided by market forces? Theory and practice. International Journal of Water Resources Development 22: 79–86.

43. Tardieu H, Prefol B (2002) Full cost or "sustainability cost" pricing in irrigated agriculture. Charging for water can be effective, but is it sufficient? Irrigation and Drainage 51: 97–107.

44. Savenije HHG, van der Zaag P (2002) Water as an economic good and demand management - Paradigms with pitfalls. Water International 27: 98–104.

45. Hooker MA, Alexander WE (1998) Estimating the demand for irrigation water in the Central Valley of California. Journal of the American Water Resources Association 34: 497–505.

20

Rivermouth Alteration of Agricultural Impacts on Consumer Tissue δ^{15}N

James H. Larson*, William B. Richardson, Jon M. Vallazza, John C. Nelson

Upper Midwest Environmental Sciences Center, United States Geological Survey, La Crosse, Wisconsin, United States of America

Abstract

Terrestrial agricultural activities strongly influence riverine nitrogen (N) dynamics, which is reflected in the δ^{15}N of riverine consumer tissues. However, processes within aquatic ecosystems also influence consumer tissue δ^{15}N. As aquatic processes become more important terrestrial inputs may become a weaker predictor of consumer tissue δ^{15}N. In a previous study, this terrestrial-consumer tissue δ^{15}N connection was very strong at river sites, but was disrupted by processes occurring in rivermouths (the 'rivermouth effect'). This suggested that watershed indicators of N loading might be accurate in riverine settings, but could be inaccurate when considering N loading to the nearshore of large lakes and oceans. In this study, the rivermouth effect was examined on twenty-five sites spread across the Laurentian Great Lakes. Relationships between agriculture and consumer tissue δ^{15}N occurred in both upstream rivers and at the outlets where rivermouths connect to the nearshore zone, but agriculture explained less variation and had a weaker effect at the outlet. These results suggest that rivermouths may sometimes be significant sources or sinks of N, which would cause N loading estimates to the nearshore zone that are typically made at discharge gages further upstream to be inaccurate. Identifying definitively the controls over the rivermouth effect on N loading (and other nutrients) will require integration of biogeochemical and hydrologic models.

Editor: Candida Savage, University of Otago, New Zealand

Funding: This project was funded by a Great Lakes Restoration Initiative Grant (Project #82). Any use of trade, product or firm names does not imply endorsement by the U.S. Government. The funders had no role in study design, data collection and analysis, decision to publish, or preparation of the manuscript.

Competing Interests: The authors have declared that no competing interests exist.

* E-mail: jhlarson@usgs.gov

Introduction

Terrestrial land cover is often strongly related to the supply of essential elements (nutrients) in nearby aquatic ecosystems (e.g., [1–3]). One consistently observed relationship is between agriculture and the nitrogen (N) isotopic composition of dissolved N, seston and consumers (e.g., [4–6]). This relationship is strong because agricultural N sources have a distinct isotopic ratio relative to other N sources [4]. However, aquatic processes can remove large quantities of N (e.g., denitrification) and in locations where these processes are prominent the movement of agricultural N downstream may be reduced and/or its isotopic signature altered [7]. Previous studies have suggested that depositional habitats (where waters slow) can significantly alter N dynamics [7].

Among aquatic ecosystem types, streams have the shortest water retention time, and thus it is unsurprising that watershed agriculture and the N isotopic composition of stream consumers is strongly correlated (e.g., [4,8]). Other ecosystems are less cleanly connected to their upstream watersheds [7,8]. The connection between terrestrial agriculture and nutrient loading into the nearshore zone of large lakes and oceans is particularly interesting, as these coastal areas are economically important. However, estimates of nutrient loading to nearshore areas are often made in association with monitoring stations located upstream of any direct lake or ocean influence [9]. These estimates do not include the effect of low-flow, depositional areas associated with the rivermouth itself, where water residence times are longer than in streams[10,11]. These wetlands and embayments associated with

rivermouths and estuaries may significantly alter nutrient delivery to the nearshore [12,13]. The gap in monitoring between the river and the nearshore corresponds to a gap in the understanding of nutrient delivery to nearshore zones.

In a recent manuscript [8], we reported a relationship between agriculture and N isotopic composition of consumers in tributary rivers of Lake Michigan. That study suggested the influence of agriculture was disrupted by significant N sources or processing occurring within the rivermouth [8]. In other words, loading estimates of N to the nearshore of Lake Michigan might be inaccurate if the effect of rivermouth processing is not included. The conclusions of that study were tentative for several reasons (e.g., small sample size and limited spatial extent), but the implications if accurate are significant. For example, if certain rivermouths provide a degree of buffering to nearshore zones from upstream agricultural inputs, then short-term and long-term nutrient reduction strategies may opt to incorporate this information into prioritization schemes. For this reason, a new and expanded sampling effort was necessary to determine whether these results are apparent at a larger spatial extent.

The primary question addressed in this study is: Are models relating landscape characteristics to indices of nutrient loading dependent on aquatic ecosystem type? To address this question, consumers were collected from rivers and rivermouth ecosystems throughout the Great Lakes. The isotopic composition of these primary consumers was used as a time-integrated indicator of N loading [4]. Previous work has established a strong relationship between agricultural activities and the δ^{15}N in tissues of aquatic consumers [4,5,8,14]. We predict agricultural land cover and consumer tissue $\delta^{15}N$ will have a positive linear relationship, as observed previously [8]. Some previous work has also suggested that low-flow aquatic habitats tend to be N sinks due to denitrification and sedimentation, which might also influence $\delta^{15}N$ in consumer tissues [7,8,15]. We predict that land cover indicators of depositional or low-flow habitats such as wetlands and open surface waters (i.e., lake area) will be negatively related to consumer tissue $\delta^{15}N$. Finally, we predict that the magnitude of land cover effects on consumer tissue $\delta^{15}N$ will vary with ecosystem type (river and rivermouth). Aquatic processes that remove or retain N are more effective in lower-flow environments [7], and by definition rivermouths are areas where lotic waters merge and become more lentic [10,11]. Removal of agricultural N in rivermouths would lessen the strength of relationships between watershed agriculture and the N present in consumers (as observed previously [8]).

Materials and Methods

Ethics statement

No permits were required for the sample collection described herein. All of the sites sampled here are publicly accessible and no threatened or endangered species were collected as a part of this study.

Study Sites

Twenty-five tributary systems of the Laurentian Great Lakes were sampled during June-August of 2011 (Figure 1, Table 1). Two sites in each of these tributary systems were sampled: 1) At an upstream location outside the influence of seiche-driven lake water inputs (River; R) and 2) at the outlet where the rivermouth entered the adjacent lake (RM). R and RM sites were separated by an average of 7.7 river km (range 0.9–17.8; Table 1). Site selection was constrained by logistical issues: Sampling in Canada was not permitted and many rivers and rivermouths were inaccessible with our field equipment. We also excluded any site with obvious surface water inputs between R and RM sampling locations. Of the remaining sites, all sites with long-term discharge monitoring were sampled and additional sites were randomly selected to reach at least 5 sites per Great Lake. We also sampled Oak Creek (WI) opportunistically. Aerial photographs were used to confirm that there were minimal surface water inputs between our R and RM sites. At each site conductivity, pH and temperature were measured using a YSI probe once at the time of consumer sampling (pH calibrated daily; Model no. 600XLM).

Consumers

At each location (R, RM), filter-feeding consumers were collected that would imply the isotopic composition of material entering the base of the food web. When possible, dreissenid mussels (either *Dreissena polymorpha* or *D. bugensis*) were collected, but dreissenids did not occur at many of the R sites, so Hydropsychidae caddisflies were also collected. All individuals were morphologically identified as *Dreissenna polymorpha*, although cryptic species variation cannot be ruled out [16]. No filter feeding basal consumers were available at a few locations, and these sites are excluded from the following analysis.

To evaluate the similarity of $\delta^{15}N$ in caddisflies and dreissenids, samples of both taxa were collected whenever possible (5 sites). To strengthen this cross-taxa comparison, both caddisflies and dreissenids were collected from 7 nearshore lake sites (off shorelines and the outside walls of harbor walls). These nearshore lake sites were close to the RM sites for the Ford, Manitowoc,

Cheboygan, Little Salmon, Cataragas, 12-Mile and Au Sable (sites and data in File S1).

Dreissenids and caddisflies were typically collected off of breakwalls and rocks within the outlet of the rivermouth, although some were found on woody debris. Consumers at R sites were collected from hard substrates in areas with flowing water. In smaller streams, these were collected from near the thalweg, but in larger streams where wading was not possible these individuals were taken from rocks and woody debris along the shoreline. All individuals were collected by hand. Target size for dreissenids was from ~2–3 cm (to insure enough sample material), but size was not recorded and a few individuals outside of this range may have been collected. All consumers were taken from less than 1 m depth. A minimum of 3 dreissenids or 5 caddisflies were collected and grouped into a single sample. Consumers were kept chilled in coolers with ice until they could be processed (~6 hours). Dreissenids had shells and byssal threads removed, and soft tissues were placed in cryovials prior to storage in liquid nitrogen. Caddisflies were placed whole in cryovials and then stored in liquid nitrogen.

Stable Isotope Analysis

Consumers were stored in the field in liquid nitrogen, and then returned to the lab. In the lab, samples were stored in an $-80°C$ freezer until they were lyophilized and shipped to the Colorado Plateau Stable Isotope Laboratory (http://www.mpcer.nau.edu/isotopelab/isotope.html) for analysis. Samples were analyzed using a CE Instruments NC2100 Elemental Analyzer interfaced to a Thermo-Electron Delta V gas-isotope ratio mass spectrometer. Analytical duplicates indicate small analytical error rates, with standard deviation $<\pm0.1$ $^0/_{00}$ for $\delta^{13}C$ and $\delta^{15}N$.

Land cover

The watershed properties for the 23 rivermouths (where consumers were found) are based on watershed boundaries from the USGS Watershed Boundary Dataset (WBD) and user defined boundaries created from 10-meter DEMs from the USGS National Elevation Dataset (NED). The entire watershed basin properties (above the RM sampling location) were calculated based on the WBD and the properties above the R locations were based on those created from 10-meter NED using Pour Point in ArcGIS 10.0 [17]. After the watershed was created, summaries of land cover based on the 2006 National Land Cover Database were created [18]. From the full summaries, two categories were created: Agriculture and Depositional. The Agriculture category was the Cultivated Crops class plus the Pasture/Hay class. The Depositional category was the combination of Surface Water, Woody Wetlands and Emergent Wetlands classes.

The depositional areas specifically associated with the rivermouths below the R were estimated by subtracting the area of depositional habitat of the whole watershed at the RM site by the area of depositional habitat occurring above the R site. This is referred to as the RM depositional.

The area of the rivermouth itself was also estimated. The lakeward boundary was across the outlet or harbor outlet for rivermouths with harbor walls that extend into the lake. For many tributary rivers, as the river approaches the lake, the river widens and backwater areas and islands become apparent in aerial photographs. The upstream boundary was placed at the most downstream point where the river still appeared to be in a constrained channel without these obvious backwater areas.

Figure 1. Map of sites sampled during this study.

Statistical analyses

All statistical analyses were conducted in R (version 2.11.1 [19]). Bayesian statistics were conducted using the BRugs package, which interfaces R to OpenBUGS [20]. Examples and descriptions of the code used for statistical analysis are provided in File S2. Mean values and 95% credible intervals of seston and consumer FAs at R, RM and L sites were made using the approach described in McCarthy, pp 66–67 [21] (see example code in File S2). Comparisons between R, RM and L sites in mean FA values were made by comparing the overlap of 95% credible intervals around the mean. In this approach, a 'significant' difference is inferred when 95% credible intervals do not overlap.

Although a single taxonomic group could be sampled at all R sites (the Hydropsychidae family), both RM sites had a mix of Hydropsychidae caddisflies and dreissenid mussels. To determine whether or not dreissenid and caddisfly consumers had similar tissue δ^{15}N, a simple linear regression (the lm() function in R) was performed. A total of 13 pairs of caddisflies and dreissenids were used to evaluate this model. Since the model fit was strong (see Results), the regression model was used to convert dreissenid tissue δ^{15}N from sites with only dreissenids into equivalent caddisfly tissue δ^{15}N values. Other possible sources of variation in this model could not be evaluated with the available data (e.g., whether this relationship varies by ecosystem type).

The support for models relating land cover data and consumer δ^{15}N was assessed using the deviance information criterion (DIC; [21]). The use of DIC is analogous to the use of the more common Akaike's information criterion (AIC; [22]). DIC differences (ΔDIC) are used to estimate the rank of the fit of the models to reality [21], with ΔDIC ≤2 indicating substantial support, ΔDIC from 4–7 indicating 'considerably less' support and ΔDIC >10 indicating essentially no support [21,23]. Mean and 95% credible intervals were estimated for model parameters of models with ΔDIC <5.0. The Bayesian correlation coefficient (R^2_B) and its 95% credible interval were also estimated [24]. Credible intervals are similar to confidence intervals (see discussions in [21,25,26]). Together, estimates of ΔDIC, R^2_B and model parameters indicate different

aspects of statistical significance in these models (best model, the amount of variation being explained and the effect size, respectively). Example code used for this analysis is included in File S2. Several models were evaluated for significance. These linear models related variation of either agriculture, depositional areas or a combination of these two effects on consumer tissue δ^{15}N. The agriculture data was included either untransformed or natural log-transformed (base e) agriculture (log$_e$[% agriculture+1]). Logarithmic transformation was evaluated after data were plotted and a non-linear relationship seemed apparent. Depositional area data was untransformed.

The use of a Bayesian approach has several advantages on theoretical grounds that have been described elsewhere [21]. Pragmatically, this approach is ideal for this study as it allows the explicit incorporation of previously collected data into a given analysis. The data from Larson et al. [8] was used to create the prior distributions for model parameters and precision (see File S2). To do this, models from Larson et al. [8] were re-estimated using the Bayesian approach described here. To be sure the statistical method used here would not alter the conclusions of Larson et al. [8], the previous data was completely re-analyzed using this statistical approach (File S3).

Results

Variation in δ^{15}N among sites, habitat types and taxa

There was considerable variation in the δ^{15}N of the consumers sampled in this study (see File S1). Caddisfly tissue δ^{15}N values ranged from 2.18 to 12.4 (R mean = 7.93±2.64 [standard deviation], RM mean = 6.87±2.12), while dreissenid mussel tissue δ^{15}N ranged from 3.74 to 10.89 (R mean = 6.53±3.28, RM mean = 7.98±1.80). The δ^{15}N of consumer tissues in rivers and rivermouths could not be directly compared because of taxonomic differences between these habitat types. Caddisflies were found at all R sites, but dreissenids were only found at 2 R sites (Table 1). Similarly, dreissenids were found at most RM sites, but at 5 sites only caddisflies could be found. Among-site variation in caddisfly and dreissenid tissue δ^{15}N was strongly correlated, but caddisfly

Table 1. Study sites and characteristics.

Site	Date	Ag	WDep	RMDep	RM consumers	Distance from R to RM (river km)
Ford (MI)	8/22–8/23/11	4.2	47.4	0.4	DM	14.8
Kewaunee (WI)	6/21–6/23/11	78.1	6.2	1.4	DM	8.9
Manitowoc (WI)	8/29–8/30/11	69.3	15.0	0.1	DM	8.9
Pere Marquette (MI)	8/9/11	12.2	15.8	1.7	DM	16.2
Betsie (MI)	7/6–7/7/11	7.7	24.7	1.2	DM	10.1
Tahquamenon	6/16/2011				–	
Cheboygan	6/14/11	6.8	26.0	0.2	CF, DM	3.2
Ocqueoc	6/15/11	6.2	33.0	1.4	CF	0.9
Little Salmon (NY)	6/27/11	13.9	16.7	1.3	CF	4.8
Salmon (NY)	6/26–6/28/11	3.8	19.6	0.6	CF	7.7
Knife River (MN)	7/12–7/13/11				–	
Bois Brule (WI)	7/12–7/13/11				–	
Ontonagon (MI)	7/13–7/14/11	3.8	21.9	0.3	CF	5.0
Conneaut (OH)	7/20/11	32.5	5.9	0.9	DM	7.1
Grand (OH)	7/21/11	33.8	8.9	0.2	DM	13.6
Cataraugas (NY)	7/26–7/27/11	35.4	2.7	0.5	CF, DM	17.8
Genesee (NY)	6/28–6/29/11	45.9	4.6	0.0	DM	9.6
Twelvemile (NY)	7/27–7/29/11	66.5	5.3	5.3	DM	4.8
Johnson (NY)	7/28–7/29/11	59.4	9.2	0.9	DM	8.6
Pigeon (MI)	8/1/11	79.4	7.7	4.2	DM	5.0
Black (MI)	8/3/11				–	3.3
Au Sable (MI)	8/2–8/4/11	3.2	14.9	0.6	DM	3.0
Crane (OH)	8/10–8/11/11	72.1	11.3	11.1	DM	9.9
Old Woman (OH)	7/19/11	68.8	1.4	1.2	DM	5.6
Firesteel (MI)	8/24/11				–	
Oak (WI)	8/30–8/31/11	13.0	4.4	4.4	DM	1.2

Land cover (in percentage) is presented for the watershed upstream of the rivermouth (RM).
Ag = % of watershed with agricultural land cover; WDep = % of watershed covered by low-flow aquatic habitats (lakes +.
wetlands); RMDep = % of watershed covered by low-flow aquatic habitats below the R site; CF = caddisflies; DM = dreissenid mussels.
CF consumers were collected at all R sites. DM were collected at Cheboygan and Genesee R sites. Watershed land cover is not reported for sites where filter-feeding consumers could not be found.

tissues had higher δ^{15}N values ($R^2 = 0.74$; Figure 2). The following regression model was used to convert dreissenid δ^{15}N values from sites with only dreissenids into equivalent caddisfly δ^{15}N values.

$$((Dresissenid \delta 15N) \times 0.8715) + 1.5612 = Caddisfly \delta 15N$$

This conversion places caddisflies and dreissenids in a common framework in regards to tissue δ^{15}N. For rivermouth consumers, the tissue δ^{15}N in this common framework was used as the consumer tissue δ^{15}N.

Relationships between land cover and consumer tissue δ^{15}N

At R sites, one model (R1) had substantial support (ΔDIC <2; Table 2) and a second model (R2) was also supported to a somewhat lesser degree (ΔDIC ~2.9; Table 2). These models both suggested agriculture had an association with consumer tissue δ^{15}N, and the nature of this relationship appeared to be logarithmic (Table 2, Figure 3). The proportion of depositional areas in the watershed is also included in the best model, but the addition of depositional areas only marginally increases the R^2_B from 0.86 to 0.87 (Table 2). These two models have similar support from ΔDIC and R^2_B indicates both explain about the same amount of variation. The magnitude of the agricultural effect in each is approximately the same (95% credible intervals for slope estimates overlap). There seems to be little substantial difference between the best two models at R sites. Equivalent models that did not use $\log_e$-transformed data had substantially less support (ΔDIC >5; Table 2, Figure 4), indicating the effect of agriculture is more likely to be non-linear.

At RM sites, the model selection procedure yielded largely similar results (Table 2). The two strongly supported models both included the logarithmic association with agriculture and barely differed in DIC or R^2_B (Table 2, Figure 3). The difference between these models was caused by the inclusion of whole watershed depositional areas. However, the slope estimate for depositional areas was negative (opposite expectations [4] and estimate in R1) and that parameter estimate was not statistically different from zero (95% credible interval −0.11 to 0.023; Table 2). Untrans-

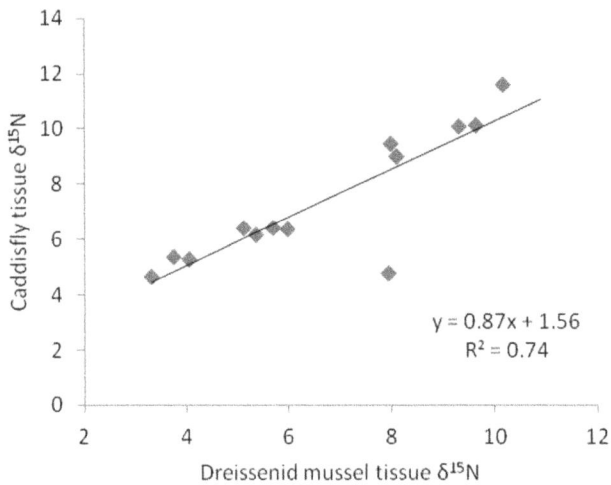

Figure 2. Relationship between caddisfly and dreissenid mussel tissue $\delta^{15}N$. Each point represents a location where both caddisflies and dreissenid mussels were collected.

formed agricultural data in the models also resulted in models with some support (ΔDIC <5; Table 2), although R^2_B were lower. Using just the depositional areas associated with the rivermouth instead of using the depositional areas of the entire watershed

created weaker models (models with RMDep instead of WDep were not strongly supported).

Comparing the models of association between agriculture and consumer tissue $\delta^{15}N$ at R and RM sites reveals some differences (Table 2). The most strongly supported model using R site data has a $\log_e$-transformed agriculture parameter coefficient (R1 $= 2.0_{[1.7-2.4]}$; 95% credible intervals in brackets) that is significantly higher than the either of the strongest models created using RM data (RM1 $= 1.4_{[0.73-1.54]}$ and RM2 $= 1.0_{[0.38-1.7]}$; Table 2). However, these models still have broad overlap in credible intervals (incorporating variation in both slope and intercept; Figure 3). Also differing between the best R and RM models is in the amount of variation being explained. Models derived from R data have a significantly higher R^2_B ($R^2_B = 0.87_{[0.81-0.92]}$) than models derived using RM data ($R^2_B = 0.47_{[0.10-0.69]}$) and a narrower credible interval on the agricultural parameter coefficient (R1 $= 0.58$; RM1 $= 0.81$; RM2 $= 1.32$). For these reasons, the RM data suggest a weaker relationship between agriculture and consumer tissue $\delta^{15}N$ than the R data.

Relationships from Larson et al. [8]

In Larson et al. [8] only untransformed agriculture data were used to relate agriculture and depositional areas to consumer $\delta^{15}N$. Those models were not among the most strongly supported in the current analysis (Table 2). Including $\log_e$-transformed agriculture data in the analysis of the earlier dataset would not qualitatively change the inferences of Larson et al. [8] because the earlier data

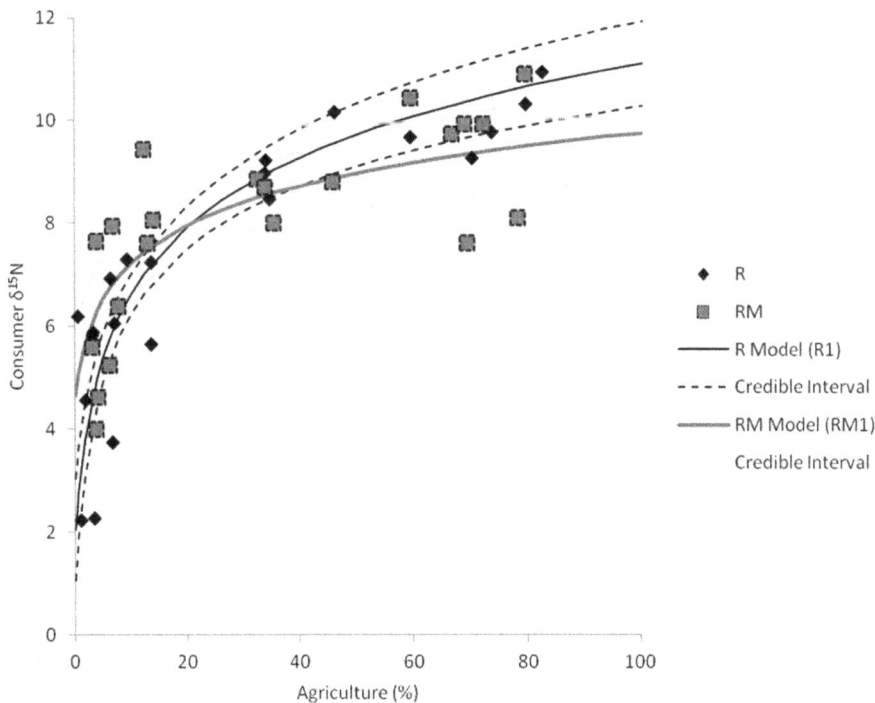

Figure 3. Modeled relationships between $\log_e$-transformed watershed agriculture and consumer tissue $\delta^{15}N$ as estimated by the most strongly supported models (models R1 and RM1 from Table 2). Squares denote data from rivermouth (RM) consumers, diamonds are river (R) consumers. Solid lines denote the relationships between agriculture) and consumer $\delta^{15}N$ for each of R and RM sites. The black line and black dashed lines are model and 95% credible intervals for the relationship derived at the R sites. For the purpose of this figure, the watershed depositional areas are held constant at 16.9% (the overall average for R sites). The grey dashed line and dotted lines are model and 95% credible intervals for the relationship derived at RM sites. Parameter values are listed in Table 2. 95% credible intervals in model estimates incorporate variation in both slope and intercept.

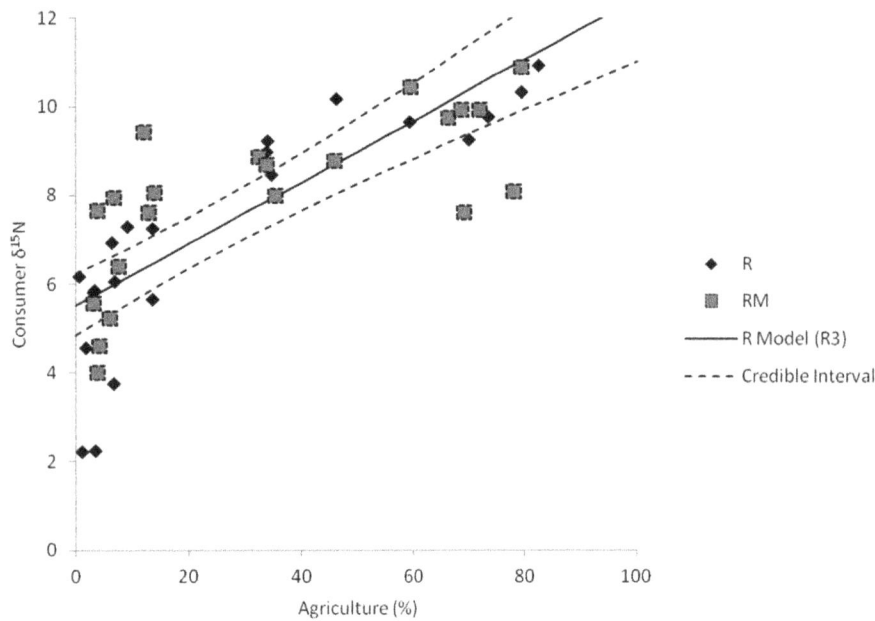

Figure 4. Relationship between watershed agriculture and consumer tissue δ^{15}N. Squares are rivermouth (RM) consumers, diamonds are river (R) consumers. The lines represent the linear relationship between agriculture and consumer δ^{15}N at the R sites (with 95% credible interval). The linear relationship for the RM sites was not statistically significant (parameters in Table 2).

included too few sites to adequately characterize a logarithmic relationship (see results of such a re-analysis in File S3).

The other relationship identified in Larson et al [8] was one between rivermouth depositional areas (RMDep) and consumer tissue δ^{15}N. Inclusion of rivermouth depositional areas did not improve model fit in the current dataset (even though that earlier data was explicitly incorporated into this analysis through the prior distributions). However, the relationship in Larson et al. [8] was

largely driven by 2 sites. With this expanded dataset, that relationship is no longer among most supported models, even if only the linear models are considered.

Discussion

Agricultural land cover and consumer tissue δ^{15}N

Agriculture and consumer tissue δ^{15}N were strongly, positively related in this study, as has been observed in a variety of locations

Table 2. Results of model selection for relationship between watershed properties and δ^{15}N in consumers and seston.

Location	Model No.	Model	ΔDIC	R^2_B
R	R1	$1.1_{(-0.36 \text{ to } 2.5)}$ + **WDep*0.041**$_{(0.0095 \text{ to } 0.072)}$ + **Ln(Ag)*2.0**$_{(1.7 \text{ to } 2.4)}$	0	**0.87 (0.81 to 0.92)**
	R2	$2.9_{(2.0 \text{ to } 3.7)}$ +**Ln(Ag)*1.6**$_{(1.3 \text{ to } 1.9)}$	2.87	**0.86 (0.78 to 0.91)**
	R3	$5.5_{(4.8 \text{ to } 6.2)}$+**Ag*0.069**$_{(0.052 \text{ to } 0.086)}$	5.02	**0.67 (0.44 to 0.81)**
	R4	WDep + Ag	6.01	–
	R5	WDep	68.4	–
RM	RM1	$4.5_{(3.2 \text{ to } 5.8)}$ + **Ln(Ag)*1.4**$_{(0.73 \text{ to } 1.54)}$	0	**0.47 (0.10 to 0.69)**
	RM2	$5.4_{(2.5 \text{ to } 8.2)}$ + WDep*$-0.043_{(-0.11 \text{ to } 0.023)}$ + **Ln(Ag)*1.0**$_{(0.38 \text{ to } 1.7)}$	0.1	**0.47 (0.09 to 0.70)**
	RM3	$8.1_{(6.6 \text{ to } 9.6)}$ + **WDep** *$-0.07_{(-0.13 \text{ to } -0.014)}$ + **Ag** *$0.027_{(0.0035 \text{ to } 0.051)}$	2.39	0.42 (−0.01 to 0.66)
	RM3	$6.5_{(5.8 \text{ to } 7.3)}$+**Ag*0.042**$_{(0.02 \text{ to } 0.06)}$	4.90	0.34 (−0.11 to 0.61)
	RM4	WDep	5.81	–
	RM5	RMDep + Ln(Ag)	8.71	–
	RM6	RMDep+Ag	14.99	–
	RM7	RMDep	25.89	–

Estimates of intercepts, coefficients and R^2_B values (with 95% credible intervals) are shown only for models with a ΔDIC of less than 5. R = river sites; RM = rivermouth sites; Ag = percentage of watershed that is agriculture; WDep = percentage of watershed that is depositional habitats (lakes plus wetlands); RMDep = depositional habitats below the R site as a percentage of the entire watershed; Ln(Ag) = the natural log of Ag plus 1.

previously [4,14,27,28]. This observation is consistent with the understanding that agricultural and urban sources of labile N either have a high $\delta^{15}N$ or become more enriched by processing in aquatic ecosystems [4,29]. The strong relationship between agriculture and the tissue $\delta^{15}N$ of primary consumers is well-established on both empirical and theoretical grounds.

Somewhat unexpectedly, the nature of the relationship between agriculture and consumer tissue $\delta^{15}N$ that was most strongly supported was not one that we initially considered. Indeed, the decision to look for a logarithmic model form (i.e., using log-transformed data) was driven by a visual inspection of the data rather than an a priori hypothesis. However, consumer tissue $\delta^{15}N$ reflects agricultural land cover because agricultural N has a distinct $\delta^{15}N$ composition [4,29]. As agricultural N becomes an increasing component of the total N consumed, the consumer tissue $\delta^{15}N$ more closely reflects that agricultural N isotopic composition. The relationship between agricultural land cover and N loading from agriculture is presumably linear, but this does not necessarily mean that biotic accumulation of this N is similarly linear [14]. If, for example, phosphorus (P) is limiting, then once the biotic demand from N is met, additional N inputs might not lead to further alterations in the isotopic N composition in biota. Similarly, if agricultural N is more labile than other N sources, then all biotic N demand may be met by agricultural N even if other N sources are available. For these reasons, a logarithmic model that approximates a saturation curve over the gradient of possible agricultural land cover values is mechanistically reasonable, although we are unaware of previous studies that have seen such a relationship.

Depositional habitats and consumer tissue $\delta^{15}N$

Depositional habitats (wetlands and other low-flow areas) are locations where denitrification rates can be high and for this reason depositional habitats are thought to influence the overall movement of N across landscapes [7,15]. Denitrification effectively removes N from aquatic systems by producing N_2 (which is inert) and increases the $\delta^{15}N$ of remaining N. Denitrification is such a widespread process that even fertilizer inputs (with an initial $\delta^{15}N$ of ~0) appear to have a positive association with consumer tissue $\delta^{15}N$ [4]. However, other than our previous work [8], there do not appear to have been studies that explicitly investigated the relationship between watershed depositional areas (e.g., wetlands) and consumer tissue $\delta^{15}N$. The positive effect of depositional habitats observed here in river sites is conceptually consistent with the literature [15,29], but this effect differs from that observed at rivermouth sites and in our previous work [8]. Other studies have used aggregated land cover variables (e.g., [5,27]) that typically include wetlands on the opposite end of the spectrum from agriculture, implying that wetlands cause lower consumer tissue $\delta^{15}N$ by not being a source of labile N. Obviously, for denitrification to have a significant effect on consumer tissue $\delta^{15}N$ requires there to be N available to be denitrified. For this reason, landscape context may be an important consideration when assessing the impact of depositional habitats on $\delta^{15}N$ in aquatic systems. Possibly, wetlands with "upstream" sources of N from agriculture will have a greater effect on food web $\delta^{15}N$ than wetlands that do not have significant labile N inputs. This study lacks any measure of that landscape context, potentially explaining both the difference in results between this and other studies [8], as well as the inconsistency between results from the river and rivermouth consumer data.

Are models relating landscape characteristics to indices of nutrient loading dependent on aquatic ecosystem type?

In contrast to our expectations [8], models relating land cover to consumer tissue $\delta^{15}N$ were strongly supported at both river and rivermouth sites. Both the direction and overall shape of these models were very similar in both ecosystem types. Further, river and rivermouth consumers sampled did not appear to differ significantly in average tissue $\delta^{15}N$ (after correction for taxonomic differences). At some level, this contradicts the results of our earlier work and suggests the answer to the primary question of this study is 'no.' More importantly, this result suggests that land cover does have a substantial influence over the $\delta^{15}N$ in rivermouth food webs, and that this influence is mechanistically similar to the influence land cover has over river food webs.

However, the details of these models vary considerably between these ecosystem types. This is most easily seen by the amount of variation explained by the models, which is considerably lower at rivermouth sites, indicating land cover models are less predictive of actual consumer tissue $\delta^{15}N$ in rivermouth ecosystems. The magnitude of the agricultural land cover coefficient is lower in rivermouths and both magnitude and direction of the depositional habitat coefficient differs between river and rivermouth models. The variation in predictive capability is consistent with other studies: Peterson et al. [5] saw the R^2 of simple linear models relating agriculture to benthic consumer tissue $\delta^{15}N$ was greater in coastal wetlands (0.82) than in adjacent nearshore waters (0.47). Whether the diminished strength of this relationship at the rivermouth is because of processing within these mixing zones, simple mixing with nearshore zone waters or because of some other factor is unclear.

The evidence presented here does suggest ecosystem-specific differences in the relationship between agriculture and consumer tissue $\delta^{15}N$. However, important caveats from this analysis should be kept in mind. For example, there is no way to effectively evaluate the possible effects of taxonomic variation as a cause of the reduced relationship at the rivermouth sites. For river sites, all consumers came from a single family (Hydropsychidae caddisflies), whereas rivermouth consumers were a mix of caddisflies and dreissenid mussels. Although tissue $\delta^{15}N$ varied similarly among sites for caddisflies and dreissenid mussels, species-specific differences may have had an effect on the ability of the best models to relate agriculture to consumer tissue $\delta^{15}N$.

Conclusions

Consumers can be considered time-integrated samplers of their available resources, making them excellent indicators of the abundance of distinct resources. In this context, previous work has suggested using primary consumers as indicators of agricultural N, which has a distinct isotopic signature (e.g., [4]). This concept makes the most sense conceptually in stream and river systems, where water residence times are low, and aquatic processes have little opportunity to alter nutrient loads and isotopic signatures. As rivers approach rivermouths, waters slow, suspended loads are deposited and water residence times increase [11], all factors that can alter N dynamics and isotopic signature [4,7]. Consistent with this reasoning, we found strong relationships between consumer N and agriculture in upstream watersheds in stream and river sites, but somewhat weaker relationships in those same tributaries at the outlet to the nearshore zone of large lakes. This result implicitly suggests that significant N processing (or sources) occur in rivermouths that do not occur in rivers.

The identity of these rivermouth-specific processes (or sources) is unclear. Earlier work suggested depositional habitats associated with the rivermouth were driving this additional variation in N dynamics in rivermouths [8], but this study offers no support for that conclusion. Rivermouths do receive regular inputs of nearshore zone water from seiche activities [30] and these inputs might transport significant quantities of N to the rivermouth. Further, these seiche events cause river water to slow, potentially increasing the likelihood of denitrification [7,31]. Few studies have documented the dynamics of water mixing in rivermouths, but those that have demonstrate that mixing regimes vary dramatically in response to season, storm events and differences in water density [11,30–33]. This mixed hydrology lacks a riverine analog and seems capable of significantly altering N dynamics (thus influencing $\delta^{15}N$ composition in the food web). Determining whether among-system variation in the frequency or extent of seiche-driven lake influences alter N dynamics in rivermouths should be a focus of future research efforts.

Many estimates of nutrient loading from rivers to coastal zones occur upstream of rivermouths (and estuaries; [9]). This study suggests processes or sources within rivermouths are altering N dynamics. This raises the possibility that estimates of nutrient loading to the nearshore zone are inaccurate. Determining the likelihood and magnitude of systematic inaccuracies in these loading estimates will require more detailed hydrologic and biogeochemical modeling of the rivermouth itself, or validated indices of these processes.

References

1. Larson JH, Frost PC, Zheng Z, Johnston CA, Bridgham SD, et al. (2007) Effects of upstream lakes on dissolved organic matter in streams. Limnology and Oceanography 52: 60–69.
2. Frost PC, Kinsman LE, Johnston CA, Larson JH (2009) Watershed discharge modulates relationships between landscape components and nutrient ratios in stream seston. Ecology 90: 1631–1640.
3. Renwick WH, Vanni MJ, Zhang Q, Patton J (2006) Water quality trends and changing agricultural practices in a midwest U.S. watershed, 1994–2006. Journal of Environmental Quality 37: 1862–1874.
4. Diebel MW, Vander Zanden MJ (2009) Nitrogen stable isotopes in streams: effects of agricultural sources and transformations. Ecological Applications 19: 1127–1134.
5. Peterson GS, Sierszen ME, Yurista PM, Kelly JR (2007) Stable nitrogen isotopes of plankton and benthos reflect a landscape-level influence on Great Lakes coastal ecosystems. Journal of Great Lakes Research 33: 27–41.
6. McClelland JW, Valiela I, Michener RH (1997) Nitrogen-stable isotope signatures in estuarine food webs: a record of increasing urbanization in coastal watersheds. Limnology and Oceanography 42: 930–937.
7. Saunders DL, Kalff J (2001) Nitrogen retention in wetlands, lakes and rivers. Hydrobiologia 443: 205–212.
8. Larson JH, Richardson WB, Vallazza JM, Nelson JC (2012) An exploratory investigation of the landscape-lake interface: Land cover controls over consumer N and C isotopic composition in Lake Michigan rivermouths. Journal of Great Lakes Research 38: 610–619.
9. Robertson DM, Saad DA (2011) Nutrient Inputs to the Laurentian Great Lakes by Source and Watershed Estimated Using SPARROW Watershed Models. Journal of the American Water Resources Association 47: 1011–1033.
10. Mikhailov VN, Gorin SL (2012) New definitions, regionalization, and typification of river mouth areas and estuaries as their parts. Water Resources 39: 247–260.
11. Larson JH, Trebitz AS, Steinman AD, Wiley MJ, Mazur MC, et al. (2013) Great Lakes Rivermouth Ecosystems: Scientific Synthesis and Management Implications. Journal of Great Lakes Research In press.
12. Krieger K (2003) Effectiveness of a coastal wetland in reducing pollution of a Laurentian Great Lake: hydrology, sediment, and nutrients. Wetlands 23: 778–791.
13. Stephens BM, Minor EC (2010) DOM characteristics along the continuum from river to receiving basin: a comparison of freshwater and saline transects. Aquatic Sciences 72: 403–417. 14.
14. Anderson C, Cabana G (2006) Does delta15N in river food webs reflect the intensity and origin of N loads from the watershed? The Science of the Total Environment 367: 968–978.
15. Bedard-Haughn A, van Groenigen JW, van Kessel C (2003) Tracing 15N through landscapes: potential uses and precautions. Journal of Hydrology 272: 175–190.
16. Grigorovich I, Kelly J, Darling J (2008) The quagga mussel invades the Lake Superior basin. Journal of Great Lakes Research 34: 342–350.
17. ESRI (2011) ArcGIS Desktop.
18. Fry J, Xian G, Jin S, Dewitz J, Homer C, et al. (2011) Completion of the 2006 national land cover database for the conterminous United States. Photogrammetric Engineering and Remote Sensing 77: 858–864.
19. R Development Core Team (2010) R: A language and environment for statistical computing.
20. Openbugs T, Best N, Lunn D (2007) The BRugs Package.
21. McCarthy M (2007) Bayesian methods for ecology. New York, New York, USA: Cambridge University Press. 296 p.
22. Burnham KP, Anderson DR (2001) Kullback-Leibler information as a basis for strong inference in ecological studies. Wildlife Research 28: 111–119.
23. Burnham KP, Anderson DR (1998) Model Selection and Inference A Practical Information-Theoretic Approach. New York, New York, USA: Springer-Verlag. 349 p.
24. Ntzoufras I (2009) Bayesian modeling in WinBugs. Hoboken, NJ, USA: John Wiley & Sons, Inc. 492 p.
25. Ellison AM (2004) Bayesian inference in ecology. Ecology Letters 7: 509–520.
26. Wade P (2000) Bayesian methods in conservation biology. Conservation Biology 14: 1308–1316.
27. Clapcott JE, Young RG, Goodwin EO, Leathwick JR (2010) Exploring the response of functional indicators of stream health to land-use gradients. Freshwater Biology 55: 2181–2199.
28. Hebert CE, Arts MT, Weseloh DVC (2006) Ecological tracers can quantify food web structure and change. Environmental Science & Technology 40: 5618–5623.
29. Nestler A, Berglund M, Accoe F, Duta S, Xue D, et al. (2011) Isotopes for improved management of nitrate pollution in aqueous resources: review of surface water field studies. Environmental Science and Pollution Research 18: 519–533.
30. Trebitz AS (2006) Characterizing seiche and tide-driven daily water level fluctuations affecting coastal ecosystems of the Great Lakes. Journal of Great Lakes Research 32: 102–116.
31. Trebitz AS, Morrice JA, Cotter AM (2002) Relative role of Lake and tributary in hydrology of Lake Superior coastal wetlands. Journal of Great Lakes Research 28: 212–227.
32. Baker DB (2011) The sources and transport of bioavailable phosphorus to Lake Erie final report: Part 2. U.S. EPA/GLNPO Assistance ID: GL 00E75401-1.
33. Kaur J, Jaligama G, Atkinson JF, DePinto JV, Nemura AD (2007) Modeling Dissolved Oxygen in a Dredged Lake Erie Tributary. Journal of Great Lakes Research 33: 62–82.

Supporting Information

File S1 Raw and transformed data used in this analysis. Six spreadsheets are included here, with the spreadsheet labeled "MetaData" explaining what the other spreadsheets contain.

File S2 Statistical appendix. This supplemental file includes a description of the model selection procedure, a description of the procedure used to generate Figures 4 and 5 and a description of the process used to generate the mean and 95% credible intervals presented in Figure 3.

File S3 Table showing results of the present data analysis performed on data previously published in Larson et al. [8].

Acknowledgments

Brian Gray reviewed an earlier version of this manuscript and provided many helpful comments.

Author Contributions

Conceived and designed the experiments: JHL WBR JMV JCN. Performed the experiments: JHL WBR JMV JCN. Analyzed the data: JHL WBR JMV JCN. Contributed reagents/materials/analysis tools: JHL WBR JMV JCN. Wrote the paper: JHL WBR JMV JCN.

21

Examining the 10-Year Rebuilding Dilemma for U.S. Fish Stocks

Wesley S. Patrick[1]*, **Jason Cope**[2]

1 Office of Sustainable Fisheries, National Marine Fisheries Service, Silver Spring, Maryland, United States of America, **2** Northwest Fisheries Science Center, National Marine Fisheries Service, Seattle, Washington, United States of America

Abstract

Worldwide, fishery managers strive to maintain fish stocks at or above levels that produce maximum sustainable yields, and to rebuild overexploited stocks that can no longer support such yields. In the United States, rebuilding overexploited stocks is a contentious issue, where most stocks are mandated to rebuild in as short a time as possible, and in a time period not to exceed 10 years. Opponents of such mandates and related guidance argue that rebuilding requirements are arbitrary, and create discontinuities in the time and fishing effort allowed for stocks to rebuild due to differences in productivity. Proponents, however, highlight how these mandates and guidance were needed to curtail the continued overexploitation of these stocks by setting firm deadlines on rebuilding. Here we evaluate the statements made by opponents and proponents of the 10-year rebuilding mandate and related guidance to determine whether such points are technically accurate using a simple population dynamics model and a database of U.S. fish stocks to parameterize the model. We also offer solutions to many of the issues surrounding this mandate and its implementation by recommending some fishing mortality based frameworks, which meet the intent of the 10-year rebuilding requirement while also providing more flexibility.

Editor: Brian R. MacKenzie, Technical University of Denmark, Denmark

Funding: These authors have no support or funding to report.

Competing Interests: The authors have declared that no competing interests exist.

* Email: Wesley.Patrick@noaa.gov

Introduction

Managing marine fisheries for sustainable yield has been a goal of fishery managers for centuries [1,2,3], yet today many of the world's fisheries still suffer from overexploitation [4,5,6]. The various consequences of depleting a fishery resource include economic (e.g., sub-optimal yields), social (e.g., reduced workforce), and ecological (e.g., reductions in the resiliency of the marine ecosystem) impacts [7,8]. Rebuilding overexploited fisheries to sustainable levels of catch can take several years to decades, depending on the productivity of the stocks (which may change due to environmental and biological conditions), the history and degree of depletion, and fishing mortality rate within those fisheries [9,10]. Thus, fishery managers must consider the ecological, social, and economic trade-offs of rebuilding immediately versus rebuilding more slowly over time.

In the United States, federally managed marine fisheries are mandated to rebuild the biomass (B) of overfished stocks (i.e., often defined as $B < \frac{1}{2} B_{msy}$) to levels that support maximum sustainable yield (B_{msy}) in as short a time as possible, accounting for the status and biology of the stock, the needs of the fishing communities, recommendations by international organizations in which the U.S. participates, and the interactions within the marine ecosystem (Section 304(e)(4) of the Magnuson-Stevens Fishery Conservation and Management Act (MSA), as amended by the Sustainable Fisheries Act (SFA) (11). Furthermore, overfished stocks must be rebuilt within 10 years, except in cases where the life history

characteristics of the stock, environmental conditions or management measures under an international agreement dictate otherwise [11].

The legislative history behind the 10-year requirement was not documented by Congress; however, Safina et al. [9] asserts that the 10-year requirement to rebuild was the result of several population dynamics experts stating, during the drafting of the SFA in 1996, that many overfished stocks were capable of rebuilding to maximum sustainable yield within 5 years if there was a moratorium on fishing. The drafters of the SFA then looked at balancing the short- and long-term trade-offs, and decided that 10 years (twice the time needed for most stocks to rebuild) was a reasonable timeframe to ensure stocks rebuild in a timely manner while accounting for socio-economic impacts [9].

In 1998, NOAA's National Marine Fisheries Service (NMFS; the federal agency responsible for managing marine fisheries) developed national guidance on rebuilding overfished stocks to operationalize the 1996 SFA mandate to rebuild in as short a time as possible [12]. The guidance provided managers with a framework to determine the targeted time to rebuild (T_{target}) by specifying a minimum (or quickest) time for rebuilding a stock (T_{min}) and a maximum time allowable for rebuilding a stock (T_{max}). T_{target} is then set somewhere between T_{min} and T_{max} based on an analysis of the factors listed previously (MSA Section 304(e)(4)). T_{min} is defined as the expected amount of time a stock needs to rebuild to B_{msy} in the absence of fishing mortality. In this context, the term "expected" means a 50 percent probability of attaining the B_{msy} given inherent uncertainty in projecting biomass. For stocks that have a T_{min} of 10 years or less, the T_{max} cannot exceed 10 years. If T_{min} exceeds 10 years, then T_{max} is calculated as T_{min} plus one generation time for that stock, where "generation time" is defined as the average age of spawning individuals within a population [13]. Once T_{target} has been chosen by fishery managers based on their 304(e)(4) analysis, a rebuilding plan is developed which often specifies a constant rebuilding fishing mortality rate ($F_{rebuild}$) that is some percentage of the rate associated with achieving maximum sustainable yield (F_{msy}) [14].

Since the implementation of the SFA in 1996 and NMFS's 1998 guidance, these rebuilding requirements have been both praised and criticized publicly [12,15]. More recently, issues with rebuilding led Congress to require that NMFS fund a study by the National Academy of Science's National Research Council (NRC) to evaluate the effectiveness of the current rebuilding requirements [8]. In general, proponents believe the requirements were needed to curtail practices of inaction by managers to prevent overfishing (i.e., $F > F_{msy}$) on rebuilding stocks and laissez-faire attempts to meet rebuilding targets [9,16,17]. For example, both Rosenberg et al. [18] and Milazzo [14] found overfishing occurring in 40 to 45% of the stocks under rebuilding plans, and in some cases overfishing had persisted for more than 5 years. Furthermore, at least 22% of plans had reset rebuilding deadlines back to year 1 when the plans were revised, instead of using the existing time frame. This practice allowed managers to extend the rebuilding time frames well beyond the plain language of the SFA [18].

However, opponents note that the 10-year requirement has limited the way in which managers can consider the socio-economic impacts of rebuilding plans. For example, stocks unlucky enough to have a T_{min} of 10 years would be subject to a 10-year moratorium [8,12,15]. Such a moratorium could wreak havoc on the infrastructure and markets of the fishing industry if the stock makes up a key component of the fishery [19,20], or could severely limit fishing opportunities for other stocks in the fishery due to bycatch issues [21]. Such a scenario is unlikely when $T_{min} > 10$ years. The discontinuity in the treatment of T_{max} is also viewed as unfair [8]. For example, if a stock could be rebuilt in 11 years in the absence of fishing pressure (instead of 10), the stock would not be subject to a moratorium and could have a longer T_{max} (i.e., 11 years plus one generation time of the stock). Opponents also point to discontinuities in the guidance that allows stocks that can rebuild in less than 10 years (using the T_{min} plus one generation calculation) to still have a 10-year T_{max}, because of the SFA mandate that specifies that stocks should rebuild in as short a time as possible [12,15].

We evaluate these statements made by proponents and opponents of the 10-year rebuilding mandate and guidance to determine whether such points are technically accurate, and offer a resolution to some of the issues surrounding this mandate and its implementation. Lastly, it is worth noting, that many of the statements evaluated below are based on the findings of Safina et al. [9]. Thus, for comparison sake, our modelling exercises replicate that of Safina et al. [9], rather than using more sophisticated modelling techniques that are more commonly used in fisheries management.

Can Most Stocks in the United States Rebuild in 5 Years under Moratorium Conditions?

As mentioned earlier, Safina et al. [9] is the primary source of information that explains why the 10-year rebuilding timeframe was chosen (i.e., twice the time needed to rebuild most stocks). In

that article, the authors relied on a Graham-Schaefer model to describe why most stocks can rebuild within 5 years. The model estimated rebuilding times (t) based on the intrinsic rate of population increase (r), fishing mortality (F) relative to the rate associated with MSY (F_{msy}, thus $F_{ratio} = F/F_{msy}$), and the biomass of the stock at the onset of rebuilding relative to the biomass needed to produce MSY (B_{ratio}).

$$t = \frac{1}{r - F_{ratio}} \ln \frac{2B_{ratio}^{-1}\left(1 - \frac{F_{ratio}}{r}\right) - 1}{2\left(1 - \frac{F_{ratio}}{r}\right) - 1}$$

Safina et al. [9] did not explicitly state what r values were used to describe "most stocks", but they did illustrate that r values ranged from 0.1 to 1.5 for 242 fish populations based on the work of Myers et al. [22,23], with the highest counts occurring between 0.4 and 0.6, forming a bell-shaped curve. Jensen et al. [24] recently reviewed 170 populations of fish and found a similar range of r values (0.1 to 1.3), but the distribution was highly skewed toward the left, with 0.1 having the highest counts. Given the disparities in r distributions between these two studies, we created r distributions specifically for U.S. fish populations, by reviewing 154 stock assessments conducted between 2000 and 2012 (Table S1). Intrinsic rate of increase estimates could be coarsely calculated for 62 of those stocks by doubling the reported F_{msy} (or the harvest rate at MSY; U_{msy}) value based on the logistic model relationship of $F_{msy} = r/2$ [25]. The other 92 stock assessments we considered only provided proxies of F_{msy} or U_{msy}, and in some cases a fishing mortality estimate was lacking. The resulting r distribution for U.S. fish populations was essentially a hybrid of the other two studies, which had a bimodal distribution with peaks at 0.05 and 0.35 (Figure 1).

Similarly, Safina et al. [9] did not discuss what B_{ratio} values were used to describe "most stocks" in a rebuilding plan. In the United States most stocks are declared overfished when B_{ratio}s fall below $\frac{1}{2} B_{msy}$, although several stocks have more conservative overfished thresholds (e.g., (1-M)*B_{msy}), where M is the natural mortality rate). Rather than use $\frac{1}{2} B_{msy}$ as B_{ratio} in our analysis, we reviewed 41 stocks that were in rebuilding plans and documented the biomass of the stock when it was declared overfished and related overfished threshold definition to determine the distribution of B_{ratio}s of U.S. rebuilding stocks (Table S2). The most current stock assessments were used, because they are considered the best available scientific information, which required eliminating seven stocks from consideration because biomasses never dropped below the overfished threshold according to the newest assessments, a result not uncommon when biomass uncertainty across assessments is large [8,26,27]. Of the remaining 34 stocks, the current B_{ratio} distribution for U.S. fish populations ranged from 0.01 to 0.82 (Figure 2).

Considering measures of central tendency across these stocks, the average U.S. fish stock had an r value of 0.40 and the average U.S. overfished stock had a B_{ratio} value of 0.32 at the onset of rebuilding (Table 1; median values were not that different 0.37 and 0.34, respectively). Taking these measures of central tendency and using the Graham-Schaefer model, rebuilding is predicted to occur for the average overfished stock within 4.1 years when $F = 0$. Thus, Safina et al. [9] statement that "most" stocks can rebuild within 5 years appears to be true.

It is also important to note that these rebuilding times are based on a Graham-Schaefer model that assumes constant conditions of productivity. In reality, marine environments are not constant and

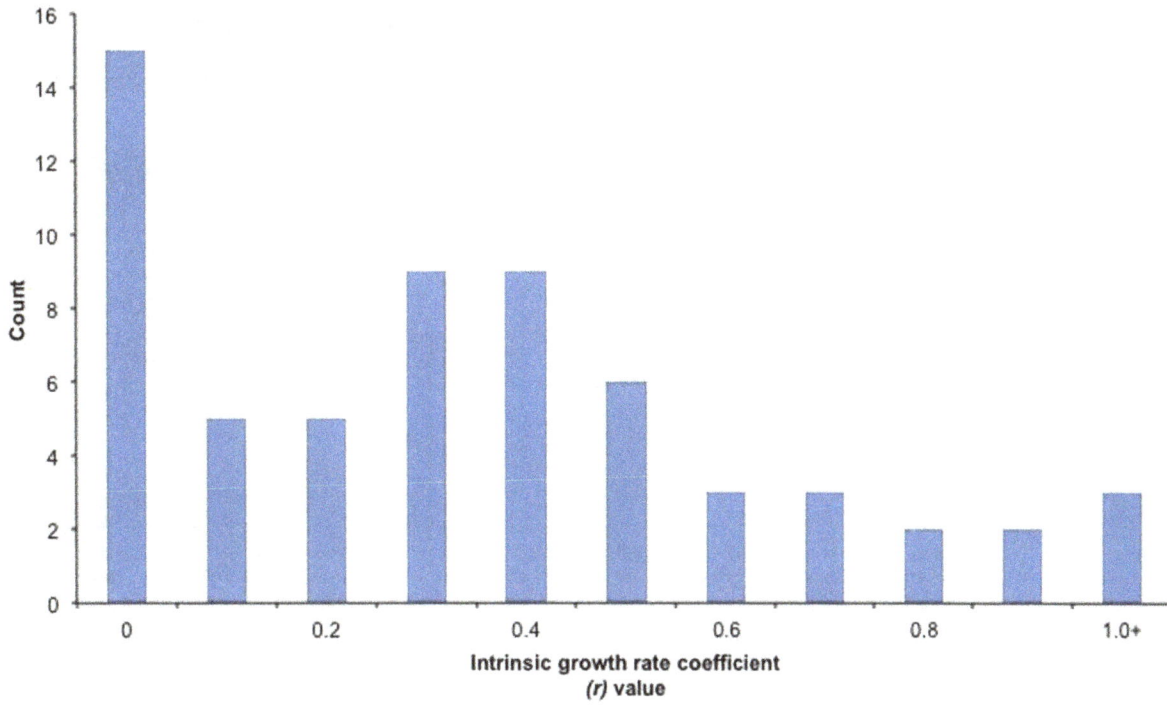

Figure 1. The distribution of intrinsic growth rate coefficients (_r_) for U.S. fish stocks, based on 62 stocks for which F_{msy} or U_{msy} values were available. Note that _r_ values labeled 0 on the _x_-axis actually represent 0.01 to 0.09 values.

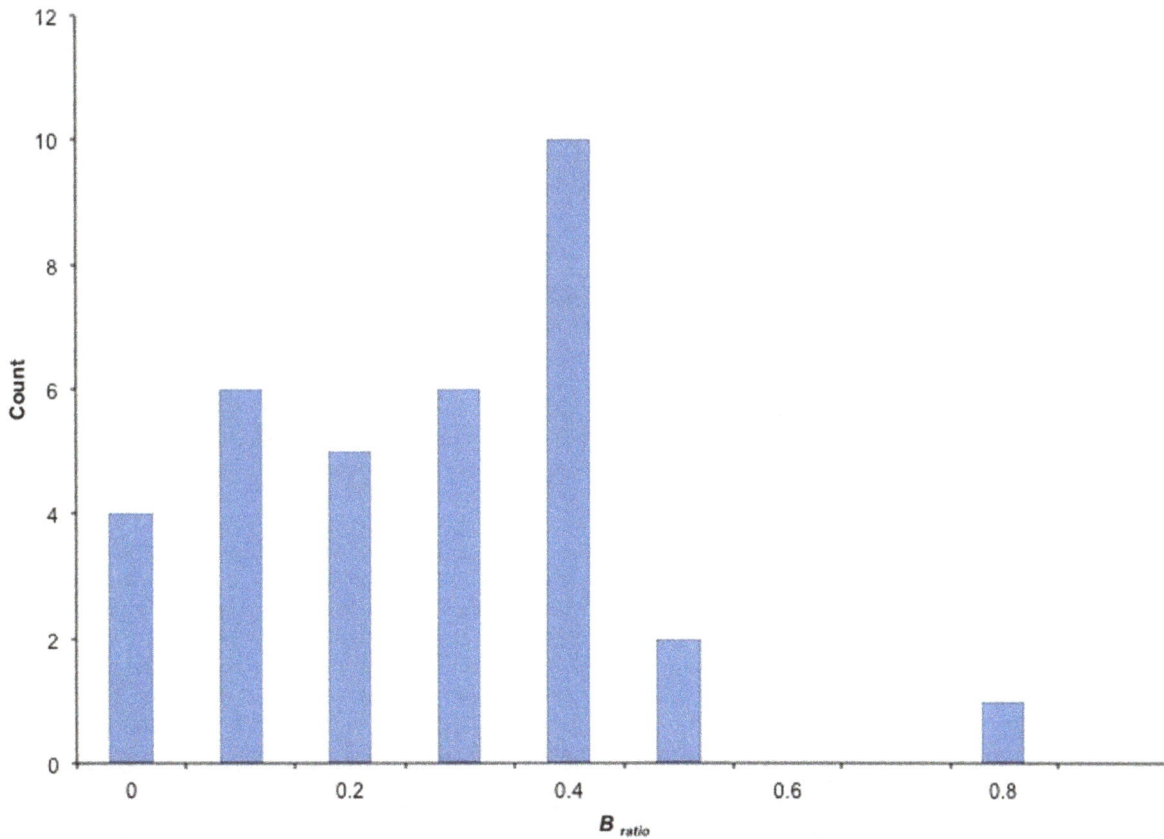

Figure 2. The ratio distribution of biomass at the onset of rebuilding relative to the biomass associated with maximum sustainable yield (_B_{ratio}_) for U.S. fish stocks, based on 34 stocks that were determined to be overfished based on the most current stock assessment. Note that B_{ratio} values labeled 0 on the _x_-axis actually represent 0.01 to 0.09 values.

Table 1. A comparison of F_{ratios} (maximum allowed $F_{rebuild}/F_{msy}$), given various intrinsic growth rate coefficients (r) values, an initial biomass ratio (B_{ratio}) of 0.32, and a rebuilding time (t) of 10 years.

r	B_{ratio}	t (yrs)	F_{ratio}	10 year exemption
0.050	0.32	10	0.00	Yes
0.100	0.32	10	0.00	Yes
0.200	0.32	10	0.00	Yes
0.300	0.32	10	0.23	No
0.400	0.32	10	0.76	No
0.500	0.32	10	0.85	No
0.600	0.32	10	0.91	No
0.700	0.32	10	0.94	No
0.800	0.32	10	0.97	No
0.900	0.32	10	0.98	No
1.000	0.32	10	0.99	No

Stocks with an r value of 0.200 or less are exempt from the mandated 10 year rebuilding requirement, while other stocks are subject to F_{ratios} ranging from 0.23 to 0.99. Under normal circumstances, the harvest policy for healthy U.S. fish stocks (i.e., not overfished) often ranges between 75% and 95% of F_{msy} (Carmichael and Fenske 2010).

the productivity of stocks can be sporadic. Thus, in many cases rebuilding timelines are based on more complex models that use age structured stochastic rebuilding dynamics that better reflect the highly variable nature of recruitment events and uncertainty in the marine environment [28,29,30]. Stochastic models generally provide more precautious estimates of population growth compared to the deterministic biomass models used here [31,32]. Therefore, the deterministic rebuilding timelines presented here are likely to be more optimistic, resulting in shorter rebuilding times.

How Many Stocks Are Susceptible to a 10-Year Moratorium?

Although no stocks have been subject to a 10-year moratorium, the threat of such a scenario is a major talking point for opponents of the 10-year rebuilding requirement. Sewell et al. [17] recently summarized the rebuilding timelines of 44 U.S. fish stocks, and showed that 23 (52%) of the stocks had 10-year rebuilding timelines. However, none of these stocks were subject to a 10-year moratorium. Instead, the high percentage of 10-year rebuilding plans is the result of managers choosing to set their T_{target} to the maximum allowed under the SFA, because the calculation of T_{min} was something less than 10 years. However, this is not to say that some of these stocks may have dramatically reduced fishing mortality rates in order to achieve the 10-year rebuilding timeline (Table 1). For example, the Southern New England/Mid-Atlantic winter flounder stock is in a 10-year rebuilding plan and has a F_{ratio} (i.e., maximum $F_{rebuild}/F_{msy}$) that is ~25% of F_{msy} [8], whereas the normal harvest policy for this groundfish stock is 75% of F_{msy} [33].

To determine how many U.S. stocks could qualify for a 10-year moratorium, we used the Graham-Schaefer model to identify combinations of r and B_{ratio} values that trigger 10-year moratoriums, where F was equal to zero. Our analysis revealed stocks with r values ranging from 0.11 to 0.53 and B_{ratio} values ranging from 0.01 to 0.49 would trigger the 10-year moratorium. Of the 62 stocks for which we have r value estimates, half fell in this range, though none of those stocks to date have been subjected to a 10-year moratorium. The lack of 10-year fishing moratoriums

suggests that the likelihood of the right conditions occurring (i.e., r and B_{ratio} values that result in a 10-year moratorium) is either not high or that managers are capable of avoiding the moratorium via rebuilding scenarios where T_{min} is slightly higher than 10 years and thus set T_{max} higher.

Does the Use of Generation Time Result in Equitable F_{ratios} among Rebuilding Stocks?

Historically, the use of the generation time in calculating T_{max} has not typically been a point of contention in terms of its applicability to rebuilding guidance, although how it is defined can vary. For example, Safina et al. [9] note that the mean generation time for an unfished stock may be much longer than for an overfished stock that has a highly truncated age distribution.

However, we found it interesting that in the past, stakeholders have not questioned why this particular life history characteristic was used in the 1998 guidance. NMFS guidance only notes that it places a reasonable, species-specific cap on the maximum time to rebuild [12,15,34]. Although the logarithmic inverse relationship between r and generation time is well documented in the scientific literature [35], its utility for scaling T_{max} with the productivity of the stock has not been empirically investigated to our knowledge.

To evaluate the relationship between r values and generation time, we used the database of 154 U.S. stocks that have stock assessments (described earlier), of which 62 stocks had r values calculated. Generation times for these 62 corresponding stocks were produced using the "life history tool" found within FishBase [13]. In FishBase, generation time is derived from relationships in optimum age (t_{opt}), age at length zero (t_0), optimum length, infinity length (L_{inf}), and the von Bertlanffy growth function (K) [13]. Although the accuracy of generation time outputs in FishBase has not been investigated, Thorson et al. [36] recently estimated biases of age at maturity (t_{mat}) outputs in FishBase, which was found to be relatively accurate and relies on some of the same input parameters as generation time (i.e., L_{inf}, K, and t_0). Additionally, the generation times reported in 21 U.S. rebuilding plans and their related r values were available for comparison (Table S1); how these generation times were calculated were not provided but we understand that proxies are often used.

The logarithmic inverse relationship between r values and generation times of stocks is shown in Figure 3. In general, the fit of data is relatively poor (FishBase $R^2 = 0.31$ – dashed line; Reported values $R^2 = 0.52$ – dotted line), but more importantly the contrast between r and generation time is lost when r values are greater than 0.20. This lack of contrast between r and generation time means that stocks with an r value of 0.20 or greater will have very similar generation times (i.e., ~5 years), which disproportionally affects the calculation of T_{max} and related F_{ratio} of the stock. For example, using the Graham-Schaefer model and NMFS rebuilding guidance, the average overfished stock ($r = 0.40$ and $B_{ratio} = 0.32$) that has a generation time of 5.0 years, a T_{min} of 4.1 years, a T_{max} of 9.1 years, and an allowable F_{ratio} that is 71% of F_{MSY}. Whereas the F_{ratio} for a stock with the same generation time (5.0 years) and B_{ratio} (0.32), but a lower r value (0.10) is only 31% of F_{MSY} ($T_{min} = 16.6$ years and $T_{max} = 21.6$ years), a 56% reduction compared to the example above.

Although our observations are consistent with the expected inverse relationships between r and generation time, these results are based on non-validated data from FishBase and reported generation times for which the methodology used to calculate the values is unknown. Therefore the poor relationships observed here could be the result of various proxies being used to calculated generation time, as opposed to more reliable methods [34]. Regardless, it appears that the generation time lacks the contrasts to be useful scalar of productivity for the majority of U.S. fish stocks and its use likely results in disproportional estimates of T_{max} and F_{ratio}.

Does Overfishing Still Threaten the Success of Rebuilding Plans?

In the past, several researchers have shown that overfishing during the initial phases of a rebuilding plan or chronic overfishing of a stock throughout the rebuilding plan were the primary causes for a stock not to rebuild [8,14,16,17,18]. For stocks that could rebuild in 10 years, the 10-year rebuilding mandate was effective in that it placed a backstop on the time allowed to rebuild, and made preventing overfishing a priority because otherwise drastic cuts to fishing effort near the end of the rebuilding timeline may be needed to meet the 10-year maximum [9]. However, much has changed since 1996 in the way U.S. fisheries are managed. In 2006, the MSA was reauthorized (MSRA) and required, among other things, the use of annual catch limits and accountability measures to prevent overfishing of all stocks within a fishery management plan [14,15]. These new requirements were implemented through national guidance in 2009 [5], and created an annual catch limit framework that accounts for the scientific and management uncertainty in fisheries management by setting precautionary catch limits (i.e., annual catch limits) below the amount corresponding to maximum sustainable yield (i.e., overfishing limit) (Figure 4). When annual catch limits are exceeded, accountability measures are triggered to correct for the overage and to help prevent chronic overfishing [37]. Annual catch limits and accountability measures are also applied to rebuilding plans, to ensure rebuilding catch limits are not exceeded.

Given the reauthorized MSA, we evaluated the performance of fisheries management in preventing overfishing by reviewing the

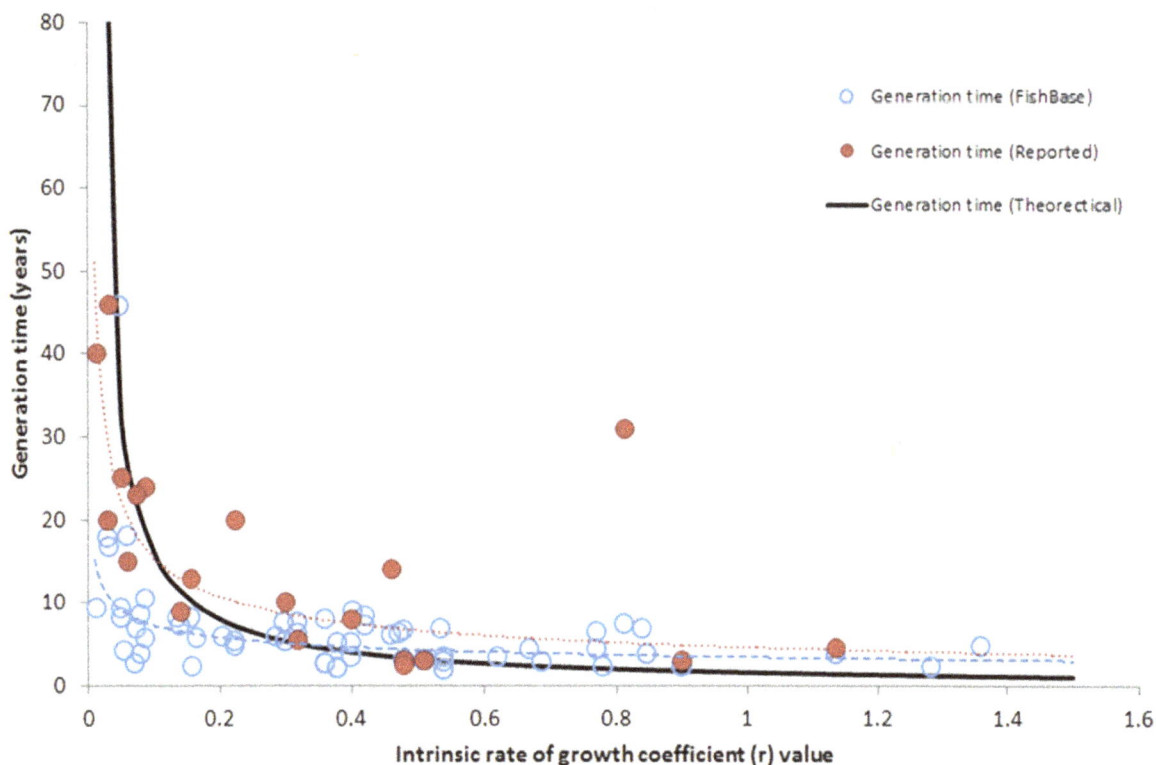

Figure 3. The empirical relationship between intrinsic rate of growth coefficient (r) and generation time, for which r values (based on F_{msy} and U_{msy}) and generation time (based on FishBase and reported values from rebuilding plans) are available. The theoretical relationship between r and generation time (G) is provided by the equation $G = \ln R_0 / r$, where R_0 (the net reproductive rate per generation) is assumed to be 5, which is roughly the median value of R_0 for the r and either FishBase or reported generation times.

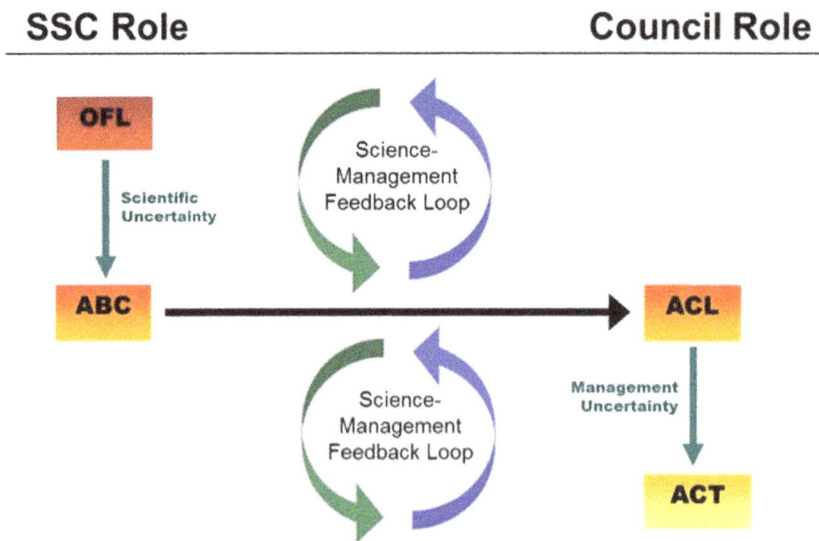

Figure 4. The annual catch limit (ACL) framework, describing how the acceptable biological catch level (ABC) is reduced from the overfishing limit (OFL) based on scientific uncertainty in the estimate of OFL, how the ACL can be set at or below ABC, and how an annual catch target (ACT) can be set below the ACL to account for management uncertainty. The Scientific and Statistical Committee of a Regional Fishery Management Council sets an OFL and ABC, while the Regional Fishery Management Council sets the ACL and ACT; each decision is based on a science-management feedback loop.

Status of U.S. Fisheries Report to Congress [38]. These reports summarize the number of stocks determined to be undergoing overfishing or in an overfished state between 2000 and 2013 (at time of writing 2013 data are based on second quarter reports) [38]. During this period, the number of stocks undergoing overfishing declined from 48 to 28 (a 41% decrease) while stocks in an overfished state declined from 52 to 40 (a 23% decrease) (Figure 5). Of the stocks with a known status in 2013, only 13% (26 of 194) were undergoing overfishing and 21% (37 of 175) were overfished. The percentage of stocks undergoing overfishing and stocks in an overfished condition are likely to continue to decrease given that annual catch limits and accountability measures were not fully implemented until 2012 by the U.S. regional fishery management councils. Preliminary data on annual catch limit performance in 2011 and 2012 suggest that approximately 10% of the stocks exceeded their annual catch limit and 7% exceeded the overfishing limit [39]. Therefore it appears that the reauthorized Magnuson-Stevens Act and annual catch limit framework have been successful at limiting (with the intention of ending) overfishing and have added effective provisions that were not in place when the 10-year rebuilding requirement was initially implemented.

The rebuilding framework developed to implement the 10-year requirement, however, has been a very useful concept. Prior to the SFA and related 1998 rebuilding guidance, the only guidance provided for rebuilding plans was that a program must be established for rebuilding the stock over a period of time specified by the Regional Fishery Management Council and acceptable to the Secretary of Commerce [40]. As a result, the rationales for how timelines for rebuilding were chosen were not as transparent as they have been under the original 1998 or revised 2009 guidelines. With the 1998 rebuilding guidance, the T_{min}, T_{target}, and T_{max} framework was created and essentially established lower and upper limits on the allowable time to rebuild, which set the stage for discussing the socio-economic trade-offs in setting the T_{target} somewhere between T_{min} and T_{max}.

The rebuilding framework is also good for developing a roadmap to recovery, because it necessitates the development of long-term projection models to predict how the stock will respond to the fishing mortality rate associated with T_{target} ($F_{rebuild}$). Managers can then use updated data and assessments and subsequent projection models to determine whether they are making adequate progress (i.e., more or less on schedule). If adequate progress is not being made, managers can then evaluate whether the lack of progress is due to inadequate management control (i.e., $F > F_{rebuild}$), unfounded assumptions or other needed updates to the projection model, or identifiable environmental conditions that changed the expectation of recruitment to the fishery [40,41,42,43,44]. Depending on the factors identified, managers have an array of management tools they can consider to resolve the underlying issue.

Alternative Approaches to Calculating T_{max} Using Fishing Mortality

Given the pros and cons of current rebuilding mandates and related guidance, we recommend two alternative approaches to developing rebuilding timelines. Both approaches rely on constant fishing mortality rates to calculate T_{max} and avoid discrepancies in how T_{max} is calculated between short- and long-lived species in terms of F_{ratios} and the 10-year time limit. Both approaches also fit within the existing rebuilding guidance framework, where T_{min} and T_{max} are calculated to define the minimum and maximum times to rebuild, while T_{target} is still set somewhere in between the two reference points to rebuild in as short a time as possible while taking into account the needs of fishing communities and interactions within the marine ecosystem. However, given the uncertainty in stock assessment projections, the rebuilding framework would only be used for planning purposes, and less for delineating or defining hard deadlines to rebuild. Instead, strong accountability measures (e.g., in-season closure authority, payback provisions, annual catch targets, etc.) could be used to reduce the effects of implementation error and ensure that

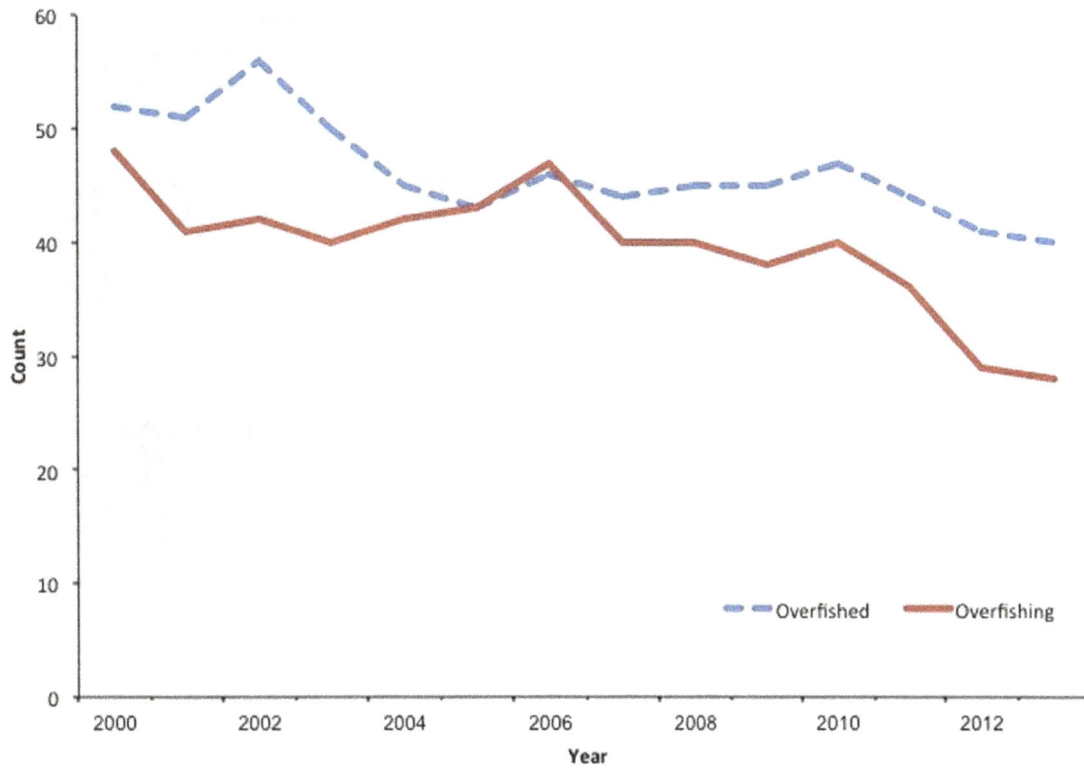

Figure 5. Between 2000 and 2013, overfishing determinations declined from 48 to 28 (a 41% reduction) while overfished determinations declined from 52 to 40 (a 23% reduction).

rebuilding occurs in a timely manner. Such an approach is also recommended in the recent NRC report on U.S. rebuilding plans [8], noting that rebuilding plans that focus more on meeting selected fishing mortality targets than on exact schedules for attaining biomass targets may be more robust to assessment uncertainties, natural variability, and ecosystem consideration, and may have lower social and economic impacts.

The key to these alternative approaches is identifying an acceptable F that both meets the existing 10-year rebuilding mandate for the average overfished stock, and is acceptably high to support the fishery during rebuilding. As noted earlier, the average overfished stock ($r = 0.40$ and $B_{ratio} = 0.36$) can rebuild in 4.1 years (roughly 5 years) when the fishing mortality is set to zero and productivity is constant, and this was presumably the rationale used for creating the 10-year rebuilding timeframe (i.e., twice the time to rebuild for "most" stocks). Thus, one alternative to calculating T_{max} is to simply multiply T_{min} by two ($T_{max} = 2 * T_{min}$) to allow the stock twice the time to rebuild. Using this alternative approach means the average overfished stock would be expected to rebuild in 8.2 years, and the maximum allowable F_{ratio} for the stock would be 66% of F_{msy}. Furthermore, the F_{ratio} of 0.66 is constant among different r types of stocks, because F_{msy} is a function of r ($F_{msy} = r/2$; 25). Lastly, it is worth noting that New Zealand's Ministry of Primary Industry uses this approach for calculating T_{max} and it was highlighted by the NRC report on U.S. rebuilding plans as a model to consider [8,45].

The second alternative tries to rectify the difference between a T_{max} that is 8.2 years (based on $T_{min} * 2$) and the 10 years allowed for the average overfished stock. We ran the Graham-Schaefer model with different F_{ratios} (assuming T_{target} equals T_{max}) until the

time to rebuild (t) equaled 10 years. Our analysis revealed that the average overfished stock could rebuild in 10 years (T_{max}) using a F_{ratio} that was 76% of F_{msy} (Figure 6). Though the use of surplus production models admittedly may produce more optimistic results than stochastic age structured models, the F_{ratio} of 76% of F_{msy} coincidentally aligns with the 75% F_{msy} harvest control rule that is commonly used in U.S. fisheries [33,46]. The 75% F_{msy} harvest control rule gained popularity in the 1990s as a precautionary approach to fisheries management, when studies revealed that such a rule reduces the chances of overfishing, results in equilibrium yields of 94% of MSY or higher, and equilibrium biomass levels between 125% and 131% B_{msy}—a relatively small sacrifice in yield for a relatively large gain in biomass [34,47,48]. Additionally, NMFS national guidance on rebuilding overfished stocks notes that if a stock has not rebuilt by T_{max}, then the fishing mortality rate should be maintained at $F_{rebuild}$ or 75% F_{msy}, whichever is less [15]. Given the similarities to the 75% harvest control rule and current rebuilding guidance, we re-ran the analysis using 75% as the F_{ratio} and found that the average overfished stock could rebuild in 9.8 years. Since there is essentially no difference in the rebuilding time (i.e., 9.8 vs. 10.0), we will refer to this alternative approach as the 75% F_{msy} rebuilding approach.

The intent of our analyses was to demonstrate that a T_{max} approach based on a constant F could meet the 10-year rebuilding requirement, and result in a more consistent and simplified rebuilding strategy. Assuming the history behind the 10 year requirement reported in Safina et al. [9] was accurate, our two approaches to defining T_{max} (i.e., $T_{min} * 2$ and 75%F_{msy}), along with ACL and AM management and using T_{max} as a planning tool

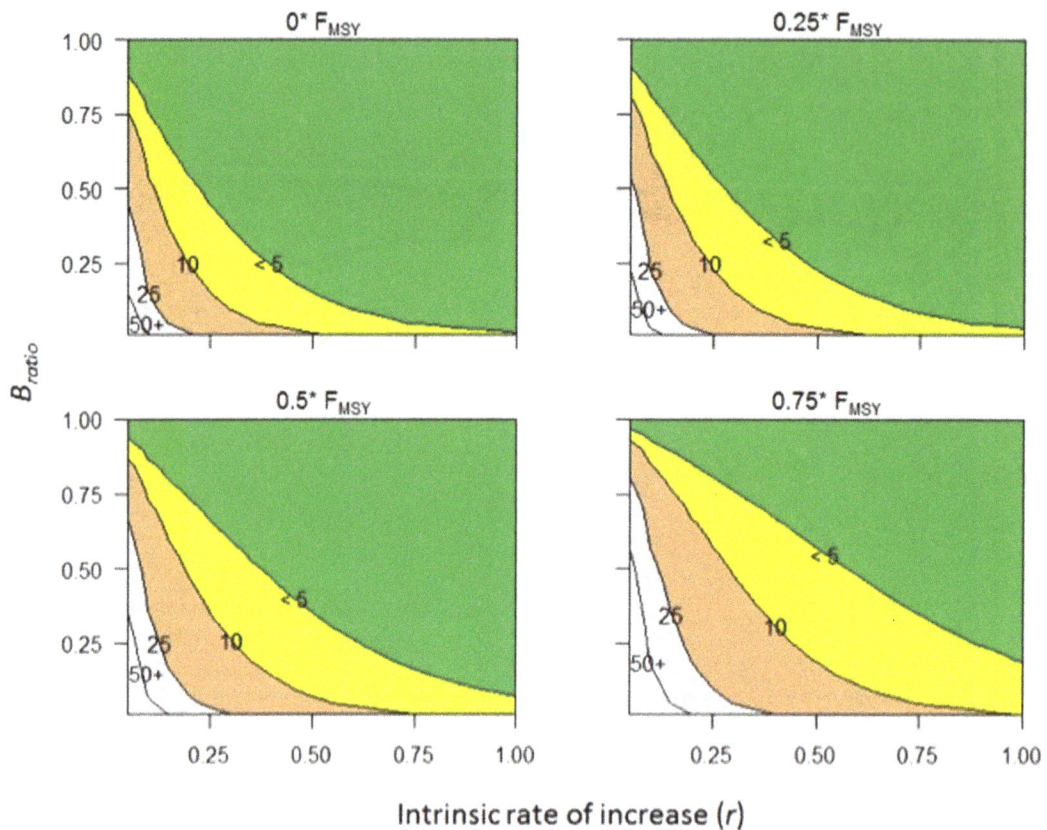

Figure 6. Rebuilding times (t) with various F_{ratios} (0%, 25%, 50%, and 75%), using the Graham-Schaefer model. Rebuilding time depends only on the intrinsic rate of increase (r), fishing mortality relative to F_{msy} (F_{ratio}), and the biomass at the onset of rebuilding relative to B_{msy} (B_{ratio}). The average overfished stock ($r = 0.40$ and $B_{ratio} = 0.32$) can rebuild in 4.1 years with $F_{ratio} = 0.0$, or 10.0 years when $F_{ratio} = 0.76$.

rather than a hard deadline, address the majority of stakeholders concerns with the current rebuilding framework including: (1) most stocks (i.e., over 50%) can still rebuild within 10 years, (2) rebuilding stocks would no longer be susceptible to a10-year moratorium; (3) there would be no discontinuities in allowable F_{ratios} among short- and long-lived species due to the use of 1+ generation time calculations; and (4) managers could focus on controlling $F_{rebuild}$, as opposed to trying to control the biomass of the stock which is harder to estimate and can be driven by environmental conditions. Regarding this later point, evaluating the biomass of the stock is still needed to determine whether the rebuilding target (i.e., B_{msy}) has been reached or not.

Our suggested approaches, however, are limited by the current T_{min}, T_{target}, T_{max} framework recommended in NMFS national guidance on rebuilding overfished stocks that operationalizes the rebuilding mandates of the MSA. Alternatives to this framework could be considered, if these mandates were revised. For example, the NRC report on U.S. rebuilding encouraged the use of harvest control rules that reduce fishing mortality as the biomass of the stock declines [8]. Harvest control rules constructed in such a manner would likely prevent stocks from ever being declared overfished in the first place, as more conservative fishing mortality rates are applied as the stock declines [8,49,50]. These types of harvest control rules can also be constructed in such a way that there are tiered categories of management action, where severely depleted stocks may be subject moratorium until the biomass of the stock reaches a minimum threshold (e.g., ¼ B_{msy}), whereby a

more structured F-based approach is applied until the stock rebuilds to its target level of biomass [45].

Lastly, while we have only focused on trying to resolve the 10-year rebuilding dilemma of the current rebuilding mandates and guidelines, there is much more analysis that goes into minimizing the economic, social, and ecological impacts of rebuilding a fishery. There is also a myriad of other management tools that managers can use to rebuild fish stocks, such as the use of marine protected areas [8,51,52], buy-back programs to reduce fishing capacity [53,54], stock enhancement [55], and habitat restoration [56,57]. The most effective rebuilding plans will likely take a portfolio approach to rebuilding stocks, by deploying a wide array of management tools, rather than solely relying on controlling the fishing mortality or the biomass of the stock.

Supporting Information

Table S1 A list of 154 assessed stocks used in the analysis related to estimates of intrinsic rate of growth (r), generation time, and age at first maturity.

Table S2 A list of the 34 U.S. fish stocks in rebuilding plans for which the biomass at the onset of rebuilding (B_{ratio}) was reported.

Acknowledgments

Disclaimer: The views expressed in this article are the authors' own and do not necessarily represent the view of NMFS.

We would like to thank T. Quinn II for supplying us with the Graham-Schaefer model used in this analysis. We are also grateful for the editorial contributions provided by S. Balwani, J. Field, O. Hamel, J. Hare, M. McClure, F. Pflieger, J. Stein, A. Risenhoover, G. Tromble, and our anonymous reviewers.

Author Contributions

Conceived and designed the experiments: WSP JMC. Performed the experiments: WSP JMC. Analyzed the data: WSP JMC. Contributed reagents/materials/analysis tools: JMC. Wrote the paper: WSP JMC.

References

1. Smith TD (1994) Scaling Fisheries: The Science of Measuring the Effects of Fishing, 1855–1955. Cambridge: Cambridge University Press. 392 p.

2. Caddy JF, Cochrane KL (2001) A review of fisheries management past and present and some future perspectives for the third millennium. Ocean Coast Manage 44: 653–682.

3. Longhurst A (2010) Mismanagement of Marine Fisheries. Cambridge: Cambridge University Press. 315p.

4. Pauly D (2007) The Sea Around Us project: documenting and communicating global fisheries impacts on marine ecosystems. AMBIO 36(4): 290–245.

5. Worm B, Hilborn R, Baum JK, Branch TA, Collie JS, et al. (2009) Rebuilding global fisheries. Science 325(5940): 578–585.

6. Costello C, Ovando D, Hilborn R, Gaines SD, Deschenes O, et al. (2012) Status and solutions for the world's unassessed fisheries. Science 338(6106): 517–520.

7. Hilborn R, Stewart IJ, Branch TA, Jensen OP (2012) Defining trade-offs among conservation, profitability, and food security in the California Current Bottom-Trawl fishery. Conserv Biol 26(2): 257–266.

8. National Research Council (NRC) (2013) Evaluating the effectiveness of Fish Stock Rebuilding Plans in the United States. Washington: The National Academies Press. 292 p.

9. Safina C, Rosenberg AA, Myers RA, II Quinn TJ, Collie JS (2005) U.S. ocean fish recovery: staying the course. Science 309: 707–708.

10. Neubauer P, Jensen OP, Hutchings JA, Baum JK (2013) Resilience and recovery of overexploited marine populations. Science 340: 347–349.

11. Sustainable Fisheries Act of the Magnuson-Stevens Fishery Conservation and Management Act (SFA) (1996) Silver Spring: U.S. Department of Commerce, National Oceanic and Atmospheric Administration, National Marine Fisheries Service. 170 p.

12. Federal Register (1998) Magnuson-Stevens Act Provision; National Standard Guidelines – Final Rule. Federal Register 63(84): 24212–24237.

13. FishBase (2013) Estimation of life-history key facts. Available: http://www.fishbase.org/MANUAL/key%20facts.htm. Accessed 2013 Aug 28.

14. Milazzo MJ (2011) Progress and problems in U.S. marine fisheries rebuilding plans. Rev Fish Biol Fisheries 22 (1): 273–296.

15. Federal Register (2009) Magnuson-Stevens Act Provisions; Annual Catch Limits; National Standard Guidelines – Final Rule. Federal Register 74(11): 3178–3213.

16. Murawski SA (2010) Rebuilding depleted fish stocks: the good, the bad, and, mostly, the ugly. ICES J Mar Sci 67(9): 1830–1840.

17. Sewell B, Atkinson S, Newman D, Suatoni L (2013) Bringing back the fish: an evaluation of U.S. fisheries rebuilding under the Magnuson-Stevens Fishery Conservation and Management Act. Washington: Natural Resource Defense Council Report R: 13-01-A. 26 p.

18. Rosenberg AA, Swasey JH, Bowman M (2006) US fisheries: progress and problems. Front Ecol Environ 4(6): 303–308.

19. Jacob S, Jepson M (2009) Creating a community context for the fishery stock sustainability index. Fisheries 34: 228–231.

20. Kasperski S, Holland DS (2013) Income diversification and risk for fishermen. Proc Natl Acad Sci USA 110(6): 2076–2081.

21. Patrick WS, Benaka LR (2013) Estimating the economic impacts of bycatch in U.S. commercial fisheries. Mar Policy 38: 470–475.

22. Myers RA, Mertz G, Fowlow PS (1997) Maximum population growth rates and recovery times for Atlantic cod, *Gadus morhua*. Fish Bull 95(4): 762–772.

23. Myers RA, Bowen KG, Barrowman NJ (1999) Maximum reproductive rate of fish at low population sizes. Can J Fish Aquat Sci 56(12): 2404–2419.

24. Jensen OP, Branch TA, Hilborn R (2012) Marine fisheries as ecological experiments. Theoretical Ecology 5: 3–22.

25. Jacobson LD, Cadrin SX (2002) Stock-rebuilding time isopleths and constant-*F* stock-rebuilding plans for overfished stocks. Fish Bull 100: 519–536.

26. Cadrin SX, Vaughan DS (1997) Retrospective analysis of virtual population estimates for Atlantic menhaden stock assessment. Fish Bull 95(3): 445–455.

27. Ralston S, Punt AE, Hamel OS, DeVore JD, Conser RJ (2011) A meta-analytic approach to quantifying scientific uncertainty in stock assessments. Fish Bull 109: 217–231.

28. Brodziak J, Legault CM (2005) Modeling averaging to estimate rebuilding targets for overfished stocks. Can J Fish Aquat Sci 62: 544–562.

29. Stewart IJ (2009) Rebuilding analysis for canary rockfish based on the 2009 updated stock assessment. Portland: Pacific Fishery Management Council. 254 p.

30. Pacific Fishery Management Council (2012) Terms of Reference for the Groundfish and Coastal Pelagic Species Stock Assessment and Review Process for 2013-2014. Portland: Pacific Fishery Management Council. 29 p.

31. Lande R, Saether B, Engen S (2001) Sustainable exploitation of fluctuating populations. In: Reynolds JD, Mace GM, Redford KH, Robinson JG, editors. Conservation of Exploited Species. Cambridge: Cambridge University Press. Pp. 67–86.

32. Punt AE, Szuwalski C (2012) How well can F_{MSY} and B_{MSY} be estimated using empirical measures of surplus production? Fish Res 134–136: 113–124.

33. Carmichael J, Fenske K (2011) Report of a National SSC Workshop on ABC Control Rules Implementation and Peer Review Procedures. Charleston: South Atlantic Fishery Management Council. 95 p.

34. Restrepo VR, Thompson GG, Mace PM, Gabriel WL, Low LL, et al. (1998) Technical guidance on the use of precautionary approaches to implementing National Standard 1 of the Magnuson-Stevens Fishery Conservation and Management Act. Silver Spring: National Marine Fisheries Service, NOAA Tech. Memo. NMFS-F/SPO-31. 54 p.

35. Caswell H (2001) Matrix population models: construction, analysis and interpretation, 2nd edition. Sinauer, Sunderland, MA. 722 p.

36. Thorson JT, Cope JM, Patrick WS (2014) Assessing the quality of life history information in publicly available databases. Ecol Appl 24(1): 217–226.

37. Methot RD, Tromble GR, Lambert DM, Greene KE (2013) Implementing a science-based system for preventing overfishing and guiding sustainable fisheries in the U.S. ICES J Mar Sci 71(2): 183–194.

38. NMFS (National Marine Fisheries Service) (2013) Status of U.S. fisheries. Silver Spring: National Marine Fisheries Service. Available: http://www.nmfs.noaa.gov/sfa/fisheries_eco/status_of_fisheries/archive/stock_status_archive.html. Accessed 2014 Jul 10.

39. Rauch S (2013) Written testimony by Samuel D. Rauch III Acting Assistant Administrator for the National Marine Fisheries Service National Oceanic and Atmospheric Administration US Department of Commerce – Hearing on Magnuson-Stevens Fisheries Conservation and Management Act – Before the Committee on Natural Resources United States House of Representatives, September 11, 2013. Available: http://www.legislative.noaa.gov/Testimony/Rauch091113.pdf. Accessed 2013 Aug 28.

40. Federal Register (1989) Guidelines for fishery management plans – Final Rule. Federal Register 54(140): 30826–30844.

41. Patrick WS, Morrison W, Nelson M, González Marrero RL (2013) Factors affecting management uncertainty in U.S. fisheries and methodological solutions. Ocean & Coastal Management 71: 64–72.

42. Hsieh CH, Glaser SM, Lucas AJ, Sugihara G (2005) Distinguishing random environmental fluctuations from ecological catastrophes for the North Pacific Ocean. Nature 435: 336.

43. Hollowed AB, Barange M, Ito S, Kim S, Loeng H, et al. (2011) Effects of climate change on fish and fisheries: forecasting impacts, assessing ecosystem responses, and evaluating management strategies. ICES J Mar Sci 68: 984–985.

44. Punt AE (2011) The impact of climate change on the performance of rebuilding strategies for overfished groundfish species of the U.S. West Coast. Fish Res 109: 320–329.

45. New Zealand Ministry of Fisheries (2014) Operational guidelines for New Zealand's Harvest Strategy Standard. Wellington: Ministry of Fisheries. 67 p.

46. Berkson J, Barbieri L, Cadrin S, Cass-Calay S, Crone P, et al. (2011) Calculating acceptable biological catch for stocks that have reliable catch data only (Only Reliable Catch Stocks – ORCS). Miami: National Marine Fisheries Service, NOAA Tech. Memo. NMFS-SEFSC-616. 56 p.

47. Thompson GG (1993) A proposal for a threshold stock size and maximum fishing mortality rate. In: Smith J, Hunt JJ, Rivard D, editors. Risk evaluation and biological reference points for fisheries management. Ottawa: National Research Council of Canada, Can Spec Publ Fish Aquat Sci. pp 303–320.

48. Mace PM (1994) Relationships between common biological reference points used as thresholds and targets of fisheries management strategies. Can J Fish Aquat Sci 51: 110–122.

49. Restrepo VR, Powers JE (1999) Precautionary control rules in US fisheries management: specification and performance. ICES J Mar Sci 56: 846–852.

50. Punt AE (2003) Evaluating the efficacy of managing West Coast groundfish resources through simulations. Fish Bull 101: 860–873.

51. Guenette S, Pitcher TJ, Walters CJ (2000) The potential of marine reserves for the management of northern cod in Newfoundland. Bull Mar Sci 66(3): 831–852.

52. Collie J, Rochet MJ, Bell R (2013) Rebuilding fish communities: the ghost of fisheries past and the virtue of patience. Ecol App23(2): 374–391.

53. Clark CW, Munro GR, Sumaila UR (2005) Subsidies, buybacks, and sustainable fisheries. J Env Econ Management 50(1): 47–58.

54. Curtis R, Squires D (2007) Fisheries Buybacks. Hoboken: Blackwell Publishing. 288 p.

55. Bell JD, Bartley DM, Lorenzen K, Loneragan NR (2006) Restocking and stock enhancement of coastal fisheries: potential, problems and progress. Fish Res 80(1): 1–8.

56. Williams RN, Bisson PA, Bottom DL, Calvin LD, Coutant CC, et al. (1999) Scientific issues in the restoration of salmonid fishes in the Columbia River. Fisheries 24(3): 10–19.

57. Hasselman DJ, Limburg KE (2012) Alosine restoration in the 21st century: challenging the status quo. Mar Coast Fish 4(1): 174–187.

22

Effects of Management Tactics on Meeting Conservation Objectives for Western North American Groundfish Fisheries

Michael C. Melnychuk*, Jeannette A. Banobi, Ray Hilborn

University of Washington, School of Aquatic and Fishery Sciences, Seattle, Washington, United States of America

Abstract

There is considerable variability in the status of fish populations around the world and a poor understanding of how specific management characteristics affect populations. Overfishing is a major problem in many fisheries, but in some regions the recent tendency has been to exploit stocks at levels *below* their maximum sustainable yield. In Western North American groundfish fisheries, the status of individual stocks and management systems among regions are highly variable. In this paper, we show the current status of groundfish stocks from Alaska, British Columbia, and the U.S. West Coast, and quantify the influence on stock status of six management tactics often hypothesized to affect groundfish. These tactics are: the use of harvest control rules with estimated biological reference points; seasonal closures; marine reserves; bycatch constraints; individual quotas (i.e., 'catch shares'); and gear type. Despite the high commercial value of many groundfish and consequent incentives for maintaining stocks at their most productive levels, most stocks were managed extremely conservatively, with current exploitation rates at only 40% of management targets and biomass 33% above target biomass on average. Catches rarely exceeded TACs but on occasion were far below TACs (mean catch:TAC ratio of 57%); approximately $150 million of potential landed value was foregone annually by underutilizing TACs. The use of individual quotas, marine reserves, and harvest control rules with estimated limit reference points had little overall effect on stock status. More valuable fisheries were maintained closer to management targets and were less variable over time than stocks with lower catches or ex-vessel prices. Together these results suggest there is no single effective management measure for meeting conservation objectives; if scientifically established quotas are set and enforced, a variety of means can be used to ensure that exploitation rates and biomass levels are near to or more conservative than management targets.

Editor: Howard Browman, Institute of Marine Research, Norway

Funding: Funding for this study was provided by a CAMEO grant through National Science Foundation and National Oceanic and Atmospheric Administration (cameo.noaa.gov; grant number 1041570), by the Walton Family Foundation (www.waltonfamilyfoundation.org/), and by a Natural Sciences and Engineering Research Council of Canada fellowship (www.nserc-crsng.gc.ca; number BPDF-424586-2012). The funders had no role in study design, data collection and analysis, decision to publish, or preparation of the manuscript.

Competing Interests: The authors have declared that no competing interests exist.

* E-mail: mmel@u.washington.edu

Introduction

Marine fish populations around the world show a tremendous diversity of exploitation status. Some populations are severely overfished while others appear to be managed sustainably, with strong differences emerging at the regional level [1]. This regional variation likely involves differences in historical and current fishery management regimes, but it is unclear what particular aspects of management systems tend to lead to successful biological outcomes for some populations and overfishing for others. We can turn to regions in which stocks tend to be managed sustainably and ask what specific characteristics of those management systems tend to make them successful [2,3].

Groundfish are fish species associated with the seafloor during their adult lives. Groundfish stocks, including rockfish, flatfish and roundfish such as cod, hake, and pollock, support important commercial fisheries around the world. The groundfish fisheries of western North America have a diverse history of exploitation. No Alaskan groundfish stocks have been overfished since the exclusion of foreign fleets in the late 1970s [1,4]. In contrast, eight groundfish stocks off the U.S. West Coast (USWC) of Washington, Oregon and California have been classified as overfished [5–7]. In British Columbia, Canada (B.C.), four rockfish species have been designated as 'Threatened' by scientists (www.cosewic.gc.ca), but no recovery plans have been implemented at the political level (www.sararegistry.gc.ca). There is some debate regarding the appropriateness of IUCN population decline criteria (which are used by COSEWIC) to determine stock status for fisheries management [8,9]. However, regardless of differences among regions in the specific criteria used to characterize population depletions, these criteria all form the basis for deciding whether to implement rebuilding plans for depleted stocks, and hence whether the targeting of a stock is reduced or prohibited. In addition to differences among regions, there is considerable diversity in the current status of individual groundfish stocks within each of these regions.

Fishery management systems are also diverse among these regions. For several decades Alaska has conducted extensive stock assessments and used harvest control rules for most stocks to limit overexploitation [4,10] (harvest control rules are explicit rules for adjusting catches in response to observed changes in stock biomass). A wide range of methods are used to allocate quota to different fishing fleets and sectors. Stock assessments and harvest control rules were generally adopted somewhat later for the U.S. West Coast, after some of the overfished stocks had already been depleted. Severe catch restrictions for overfished stocks have resulted in many of these beginning to rebuild [1,11]. In British Columbia, a system using trip limits transitioned into one using individual transferable quotas beginning in the mid-1990s [12]. The current system operates across several fleets and allows for a full accounting of catch mortality including discard mortality and bycatch [13–15]. Precautionary harvest control rules are used in British Columbia when estimates of stock status are available, but several stocks have not recently been assessed, thereby limiting the widespread use of harvest control rules. Individual quota systems have in the last few years been adopted for increasingly more stocks in Alaska and the U.S. West Coast. All regions conduct comprehensive fishery-independent surveys. In the context of global fisheries, the management systems in these regions may generally be regarded as successful, but the considerable variability in the particular aspects of how these management systems operate has likely influenced the variability in the biological status of individual stocks.

The conservation status of fish stocks is often framed in terms of recent biomass and exploitation rate (the proportion of the stock caught each year) relative to the levels that would in the long term produce maximum sustainable yield (MSY). In addition to reconstructed time series of biomass (B) and exploitation rate (F), stock assessments typically generate estimates of (or estimates of proxies for) the biomass, B_{MSY}, and fishing mortality rate, F_{MSY}, that would produce MSY. The total allowable catch (TAC) for a given stock is typically set based on the current stock status relative to these MSY-based biological reference points. For a variety of reasons stocks are sometimes overfished, including uncertain or biased stock status estimates and unreported or illegal catches. In other cases, stocks are 'underfished', or exploited at levels below the most productive yield possible [16–18]. Fish stocks undergo abundance fluctuations due to biological reasons beyond the control of management actions so we can not expect stocks to produce MSY every year, but maintaining stocks near their MSY-based targets on average is often considered to be a good management strategy.

Several hypotheses have been proposed about specific measures deemed necessary, sufficient, or contributory to successful fisheries management. Proponents of marine reserves suggest they are necessary for the protection of some stocks from fishing mortality [19,20]. Spatial and/or temporal closures are used for western North American groundfish stocks to limit the extent of fishing effort, though the proportion of a stock's area of distribution or proportion of the year closed to fishing varies considerably. It is commonly suggested that a key ingredient to fisheries management is setting appropriate science-based catch limits and properly enforcing them [21]. Further, proponents of individual quota systems (i.e., 'catch shares') suggest that economic incentives for fishermen may promote resource stewardship and thus desirable conservation outcomes [22,23]. Almost all western North American groundfish stocks are managed using annual TACs, though for some stocks these TACs are allocated to individuals while for others they are fished in an Olympic 'race-to-fish'. The implementation of harvest control rules relies on stock assessment outputs, and these together are widely suggested as key measures for safeguarding against overexploitation [24,25]. Assessments for most major groundfish stocks and several secondary target stocks are conducted in western North American regions, but the

number of stocks within a region that are regularly assessed may limit the use of harvest control rules. Bycatch constraints are often adopted to limit the catch of sensitive or depleted fish stocks under rebuilding plans. In a multispecies fishery where several stocks are caught together, such bycatch limits may limit the quota allocated to more productive stocks in order to avoid further depleting the sensitive stocks; this certainly applies to western North American groundfish stocks [26–28].

We present an analysis of western North American groundfish stocks to quantify the influence on stock status of six management attributes often hypothesized to affect the status of groundfish. These attributes are: the use of harvest control rules with estimated biological reference points; the use of temporal closures; the extent of permanent or semi-permanent spatial closures; the level of catch constraints on target stocks from bycatch limits of sensitive species; the use of individual quotas; and the proportion of catch taken with bottom trawls. We consider as performance measures commonly-reported quantities representing stock status relative to management targets: the ratio of biomass to target biomass, the ratio of exploitation rate to target exploitation rate, the ratio of catch to TAC, as well as the proportion of catch discarded at sea. These performance measures have been increasingly used in recent years to compare the status of fish stocks [1,29,30], but few attempts have been made to attribute their variability to specific management measures. We recognize that predictor and response variables were not selected as part of a carefully controlled experiment and thus we are not able to demonstrate perfect causal relationships. However, since the predictor variables are generally employed independently of the response variables and cause-effect relationships between these variables are commonly hypothesized, we use terminology such as "influence of predictor variable on stock status" when describing the associations between variables.

Results

Current status of groundfish stocks

Groundfish stocks on the west coast of North America are, for the most part, managed very conservatively. These stocks include benthopelagic-oriented species such as walleye pollock, Pacific whiting, and spiny dogfish (see Supporting Information Table S1 for a full list of stocks). Stocks are generally defined separately for each of the three regions, with the exception of Pacific halibut and Pacific whiting, for which coastwide stock assessments are conducted. Over 70% of assessed groundfish stocks had recent biomass (5-year average) above the management target (Fig. 1); many of the depleted stocks, especially rockfish, are under formal rebuilding plans (Fig. 1a). Only 11% of targeted stocks had recent exploitation rates above the management target. The stock with the highest estimated exploitation rate in the last 5 years, USWC petrale sole, just entered into a rebuilding plan (Fig. 1a). Five stocks were previously under a rebuilding plan during their history, but recovered to near management targets for biomass and exploitation rate (Fig. 1b). A few stocks are currently co-caught with the USWC stocks under rebuilding plans (Fig. 1c), which limits exploitation rates of these co-caught stocks in order to reduce bycatch of rebuilding stocks. The majority of groundfish stocks have never been under a formal rebuilding plan, but are still fished conservatively, with exploitation rates of most stocks below levels (and biomass above levels) that are predicted to result in long term optimal yield (Fig. 1d).

Regional differences are apparent in terms of rebuilding plans for overfished stocks. No Alaskan groundfish stocks are currently or have recently been classified as overfished or under rebuilding plans. Several stocks from B.C. have been identified as possibly

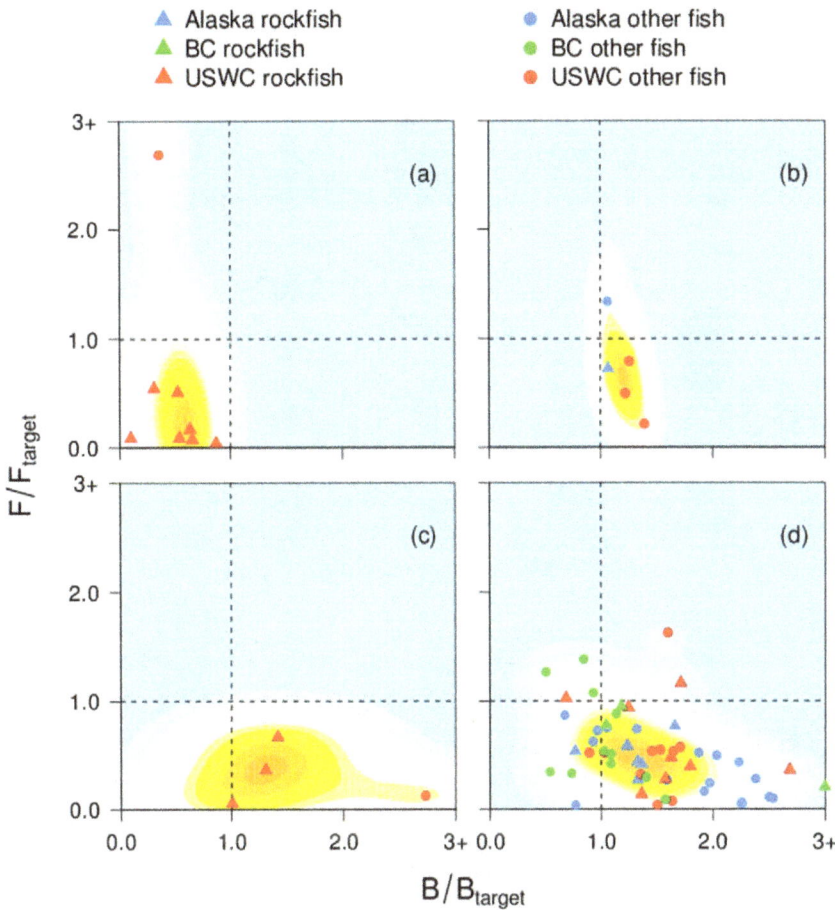

Figure 1. Current groundfish stock status in terms of recent exploitation rates to target exploitation rates and recent biomass to target biomass levels. Stocks are categorized as: (a) currently under a rebuilding plan; (b) previously under a rebuilding plan; (c) currently co-caught with rebuilding stocks; and (d) neither under a rebuilding plan nor co-caught with stocks under rebuilding. Stocks with little or no commercial value are excluded. Data points show averages of the most recent 5-year period available for each stock, and are separated by region and rockfish or other groundfish. Kernel density contour lines are shown, calculated over all data points assuming a bandwidth (smoothing parameter) of 2. Dashed lines show management targets for biomass and exploitation rate.

being 'Threatened' or of 'Special Concern', but no groundfish stocks are under recovery plans. Stocks from the USWC under rebuilding plans (Fig. 1a) or co-caught with rebuilding stocks (Fig. 1c) are not heavily targeted. Since we restrict the remainder of our analysis to targeted stocks, we exclude the stocks shown in Fig. 1a and c, and also exclude stocks with little or no commercial value.

Diverse management tactics across regions

Tactics for managing groundfish fisheries are very diverse across stocks and regions (Fig. 2). The proportion of a stock's area under permanent spatial closures is similar for stocks from Alaska and B.C., typically 10–20% of the total area. Rockfish conservation areas on the USWC result in spatial closures of >20% for some stocks, but many stocks are unaffected by these closures so on average area closures are <10%. The estimated proportion of the hypothetical total catch that is constrained by bycatch limits (i.e., an estimate of foregone catch due to bycatch limits) typically ranged up to 15% for Alaskan and USWC stocks, but was >50% for Aleutian Islands walleye pollock, whose catch is constrained by protection measures for Steller sea lions. (Other forms of catch constraints, such as the 2 million t cap for Bering Sea groundfish, are not included in this measure.) This foregone catch was nearly

20% for most B.C. stocks, whose catch was constrained by bycatch limits for bocaccio and canary rockfish [31]. Strong regional differences were observed in the use of catch share programs. Nearly all groundfish stocks from B.C. had 100% of the total catch under individual transferable quotas. Although a few stocks from Alaska and the USWC were under partial catch share management including fishing co-operatives, most were competitive – still under limited entry systems, but without quota allocations to individuals (catch share programs have recently become more common in these regions, but the period of data availability is before these programs were implemented). The proportion of a stock's total catch caught by bottom trawls ranged widely from near 0% to 100%, but on average stocks from the USWC were more frequently caught with fixed gear – longlines or pots (59% of total catch by bottom trawl) – compared with stocks from Alaska (75%) and B.C. (69%). Harvest control rules with estimated limit biomass reference points are commonly employed for Alaskan and USWC stocks, but are employed for less than one third of B.C. stocks owing to a smaller proportion of stocks with assessments in which reference points are estimated. Finally, seasonal closures are used for essentially all groundfish stocks in B.C.; they are implemented mainly for lingcod and halibut spawning closures, but affect the fisheries for other stocks. In contrast, about 50% of

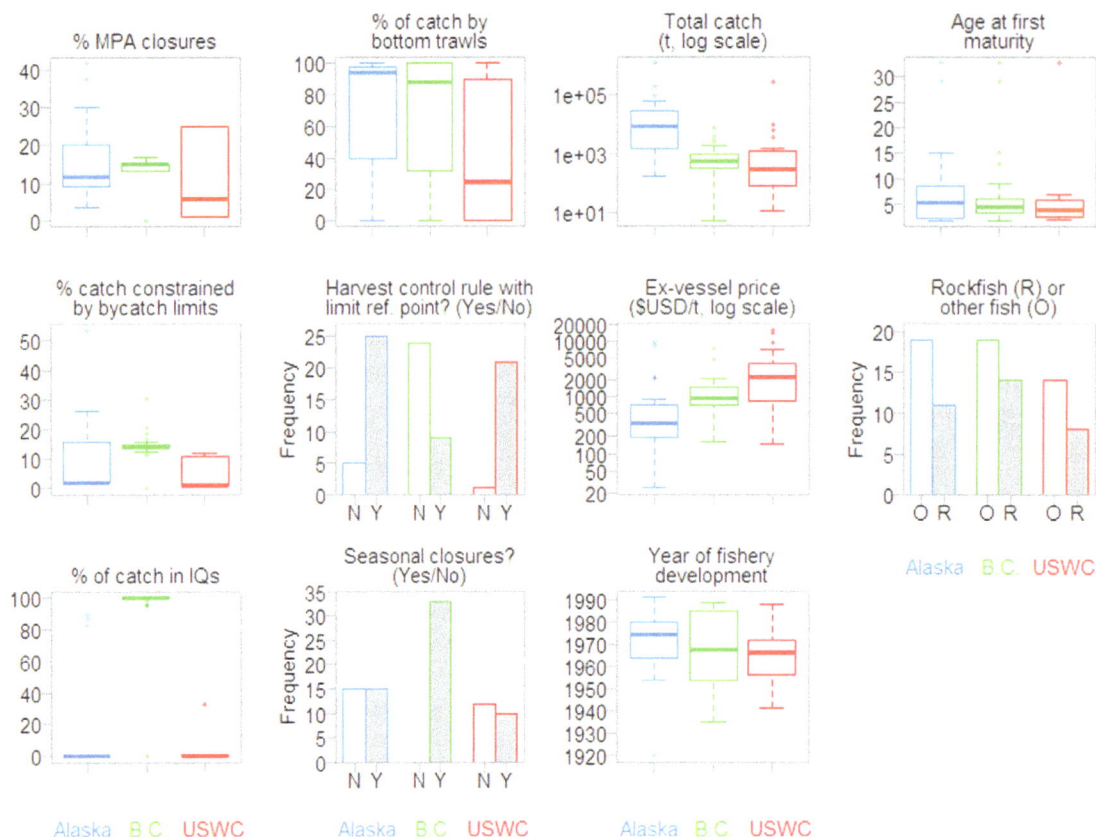

Figure 2. Boxplots of continuous predictor variables and barplots of categorical predictor variables used in analyses. Plots in two leftmost columns show management tactics and plots in two rightmost columns show other variables accounted for. Data are separated by region. See Table 3 for variable descriptions.

the stocks from Alaska and the USWC were affected by seasonal closures.

We accounted for several other factors in our analyses that may influence groundfish stock status, and these also varied widely across stocks (Fig. 2). The median total catch of Alaskan stocks was about 10 times greater than that of B.C. or USWC stocks. Ex-vessel prices were typically lower for Alaskan stocks, and greatest for USWC stocks. Fisheries developed almost a decade later in Alaska, in the 1970s, than they did in B.C. or the USWC. The age at maturity was similar across the regions, as many species are found in two or all three of these regions, and because there was a similar proportion of rockfish stocks (compared to flatfish and roundfish) across regions. There is considerable variability among stocks in both management tactics (Fig. 2) and stock status (Fig. 1). While some stocks are larger than others in terms of biomass, catch, or landed value, all stocks were equally weighted, so are equally informative and influential for analyses. We now turn to look at the effects these management tactics have on biological performance measures describing stock status.

Effects of management tactics on current stock status

Total catches of western North American groundfish stocks were considerably less than annual TACs (mean catch:TAC ratio of 0.57; 5th, 50th and 95th percentiles of 0.16, 0.75, and 1.00). Stocks from B.C. tended to have recent catch:TAC ratios closest to the target (average, 0.69) than those from Alaska (0.57) or the USWC (0.39; Fig. 3). Stocks with seasonal closures had catch:TAC ratios that were considerably higher (closer to

management targets) and less variable than stocks without seasonal closures (Table 1). Within the regions of greatest data density (i.e., focusing on the white regions in Fig. 3 instead of grey-shaded regions), we observed little influence on catch:TAC variables from the proportion of catch constrained by either spatial closures or bycatch limits, the proportion of catch under individual quota systems, or the proportion caught by bottom trawls. In contrast, we observed a strong effect of catch volume on the mean and variability of catch:TAC ratios. Stocks with greater total catch volume showed reduced variability in the ratio of catch:TAC as might be expected, but also had high catch:TAC ratios (closer to management targets), while smaller fisheries had ratios of catch:TAC much further from management targets (Fig. 3).

The mean ratio of current exploitation rate to target exploitation rates was only 0.40 (5th, 50th and 95th percentiles of 0.06, 0.52, and 1.27). Compared to the ratio of catch:TAC, there was less variation among regions in the mean ratio of $F:F_{target}$ (Alaska, 0.34; B.C., 0.51; USWC, 0.41; Fig. 4). Stocks with seasonal closures had higher mean levels of $F:F_{target}$, closer to management targets, compared to stocks without seasonal closures (Table 1). Groundfish stocks with areas of permanent spatial closures greater than 5–10% of the stock's distributional area had relatively low interannual variability in $F:F_{target}$, but as %MPA (marine protected area) levels dropped below 5–10%, variability increased (Fig. 4). This was not, however, associated with a similar increase in the semi-deviation of $F:F_{target}$ as %MPA levels decreased (semi-deviation is a measure of 'downside risk', or the asymmetrical variability of exceeding F_{target}; see Methods), suggesting the high

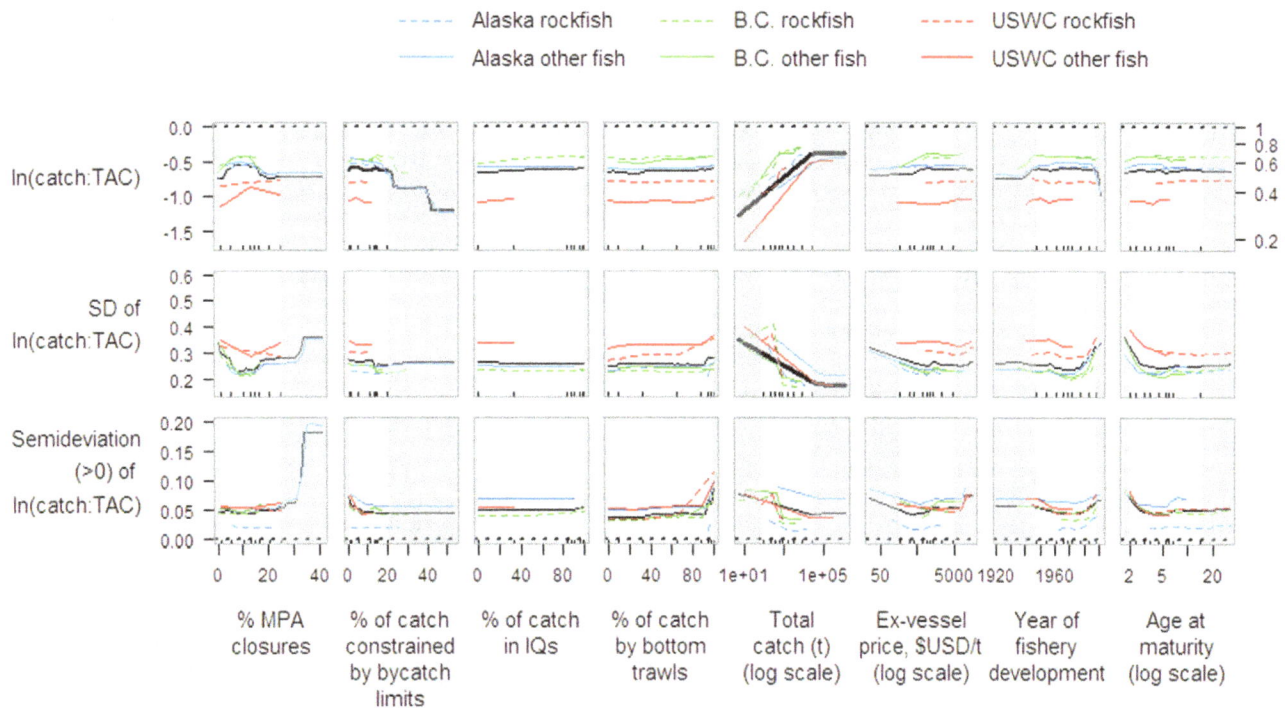

Figure 3. Partial dependence plots of three catch:TAC-related variables on eight numerical stock-level covariates of North American Pacific groundfish. The mean, standard deviation, and semideviation of the log-ratio of catch to TAC were calculated for each stock from the most recent 5-year period available. The three variables were analyzed independently using random forests (10,000 trees, 5 covariates randomly sampled at each split). The solid black line shows the marginal effect of a covariate across all stocks; thickness of the line is proportional to the covariate's relative importance score. Horizontal dotted lines at y = 0 represent general management objectives. Tick marks in each plot show deciles of covariate values; the region between grey-shaded areas contain 80% of the covariate values while grey-shaded areas contain the upper and lower 10% of covariate values. Right hand axis values show catch:TAC values on linear scale.

Table 1. Marginal mean responses of performance measures for each level of categorical management covariates.

| Performance measure | Categorical covariate | | | | | | | | |
| | HCR with limit RP? | | | Seasonal closures? | | | Taxa | | |
	No	Yes	% diff	No	Yes	% diff	Rockfish	non-RF	% diff
ln(catch:TAC)									
Mean	−0.63	−0.61	6.8%	−0.76	−0.57	61.3%	−0.63	−0.60	7.0%
Standard deviation	0.27	0.26	11.6%	0.28	0.25	25.1%	0.27	0.26	7.0%
Semi-deviation	0.051	0.051	0.2%	0.051	0.051	0.0%	0.052	0.050	4.3%
ln(F:F$_{target}$)									
Mean	−0.90	−0.87	6.7%	−1.00	−0.87	27.4%	−0.94	−0.89	10.0%
Standard deviation	0.233	0.232	1.0%	0.235	0.231	4.6%	0.233	0.232	0.7%
Semi-deviation	0.076	0.060	24.1%	0.066	0.065	1.2%	0.064	0.067	5.4%
ln(B:B$_{target}$)									
Mean	0.32	0.26	31.1%	0.31	0.27	17.9%	0.29	0.28	5.1%
Standard deviation	0.103	0.104	2.2%	0.098	0.105	13.1%	0.104	0.100	8.2%
Semi-deviation	0.110	0.095	16.4%	0.098	0.099	1.5%	0.098	0.099	0.1%
Discard proportion	15.1%	14.4%	8.6%	17.1%	13.7%	41.7%	15.2%	14.2%	11.7%

Ten performance measures are shown, and for each random forest analysis (10,000 trees, 5 covariates randomly sampled at each split) the three categorical covariates were included along with the numerical covariates shown in Figs. 3–6. For each analysis, marginal means are given for both levels of each covariate. '% diff' indicates the relative difference between the two levels, calculated as the absolute difference in means between the two levels as a percentage of the span of the 95% confidence interval of all data for the performance measure.

interannual variability observed did not involve exploitation rates exceeding management targets. Instead, stocks managed under harvest control rules which used limit reference points (implying also that stock assessments were conducted in which limit reference points were estimated) had lower semi-deviations of $F:F_{target}$, while stocks managed without these harvest control rules and limit reference points had a greater tendency for exploitation rates to exceed target levels (Table 1). Although the interannual variability in $F:F_{target}$ was similar across the entire range of the proportion of catch caught by bottom trawls, the semi-deviation increased as the % catch by bottom trawls dropped below about 20%, suggesting that stocks fished primarily with fixed gear or mid-water trawls had a greater tendency for exploitation rates to exceed F management targets (Fig. 4). Similar to the patterns observed for catch:TAC ratios, larger fisheries had low interannual variability and higher levels overall of $F:F_{target}$, while smaller fisheries had higher variability and lower overall levels of $F:F_{target}$, further away from management targets (Fig. 4). Stocks with relatively high ex-vessel prices (and also those that developed earlier) had higher $F:F_{target}$ ratios, closer to management targets, than did lower-priced or later-developed stocks.

The mean ratio of current biomass to target biomass was 1.33 (5th, 50th and 95th percentiles of 0.68, 1.36, and 2.49). The biomasses of B.C. stocks were on average at target levels (1.04), while average biomasses of stocks from Alaska (1.45) and the USWC (1.42) were much higher (Fig. 5; with stocks under rebuilding plans included, the average biomass ratio for USWC stocks drops to 1.08). Flatfish and roundfish stocks from B.C. also showed greater interannual variability in $B:B_{target}$ and, as average levels were so close to the management target, had a greater tendency for biomass levels to drop below target levels compared to stocks from other regions (Fig. 5). There was a slight tendency for $B:B_{target}$ levels of Alaskan flatfish and roundfish stocks to be lower, closer to management targets, for stocks with a greater proportion of their catch under individual quotas. (There were insufficient data to evaluate biological responses after the establishment of recent Alaskan catch share programs —the Gulf of Alaska Rockfish Pilot Program and the Bering Sea Amendment 80 Program—so the only Alaskan stocks with IQs in this analysis were sablefish, halibut, and Bering Sea pollock.) Mean $B:B_{target}$ levels were also lower, closer to management targets, for stocks that had harvest control rules with limit reference points (Table 1). Over most of the range of the proportion of total catch by bottom trawling, there was little observable effect of this covariate on any $B:B_{target}$ variable. However, when total catches taken by bottom trawling were nearly 100%, the mean $B:B_{target}$ was surprisingly much lower, closer to management target levels (Fig. 5). Interannual variation and especially the semi-deviation in $B:B_{target}$ increased at this level of near-100% bottom trawling, suggesting that biomass had a greater tendency to fall below management targets for stocks caught exclusively by bottom trawling. Mean $B:B_{target}$ levels were slightly lower, closer to target levels, for stocks with relatively early year of development and early year at maturity (Fig. 5).

Estimated discard rates ranged from 8.3% on average for B.C. rockfish stocks to 21.1% on average for Alaskan flatfish and roundfish stocks. Ex-vessel prices had the strongest influence by far on discard rates, with discard rates dropping from >20% at low prices to 10% at medium and high prices (Fig. 6). Effects of the other numerical covariates were negligible, although stocks with seasonal closures had lower discard rates on average than those without seasonal closures (Table 1).

Exploratory and sensitivity analyses

We found limited reason for concern about the independence of categorical predictor variables (see Supporting Information Fig.

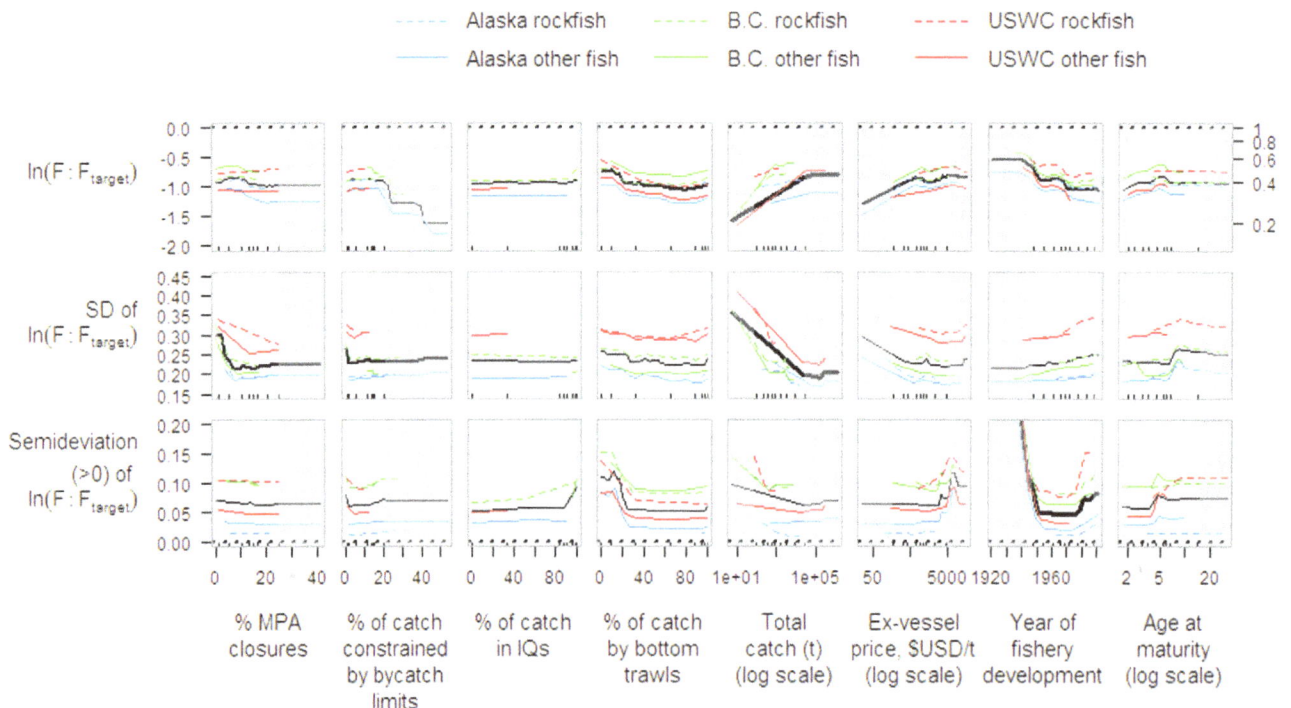

Figure 4. Partial dependence plots of three exploitation rate-related variables on eight numerical stock-level covariates of North American Pacific groundfish. The mean, standard deviation, and semideviation of the log-ratio of current exploitation rate to the target exploitation rate were calculated for each stock from the most recent 5-year period available. See Fig. 3 caption for further details.

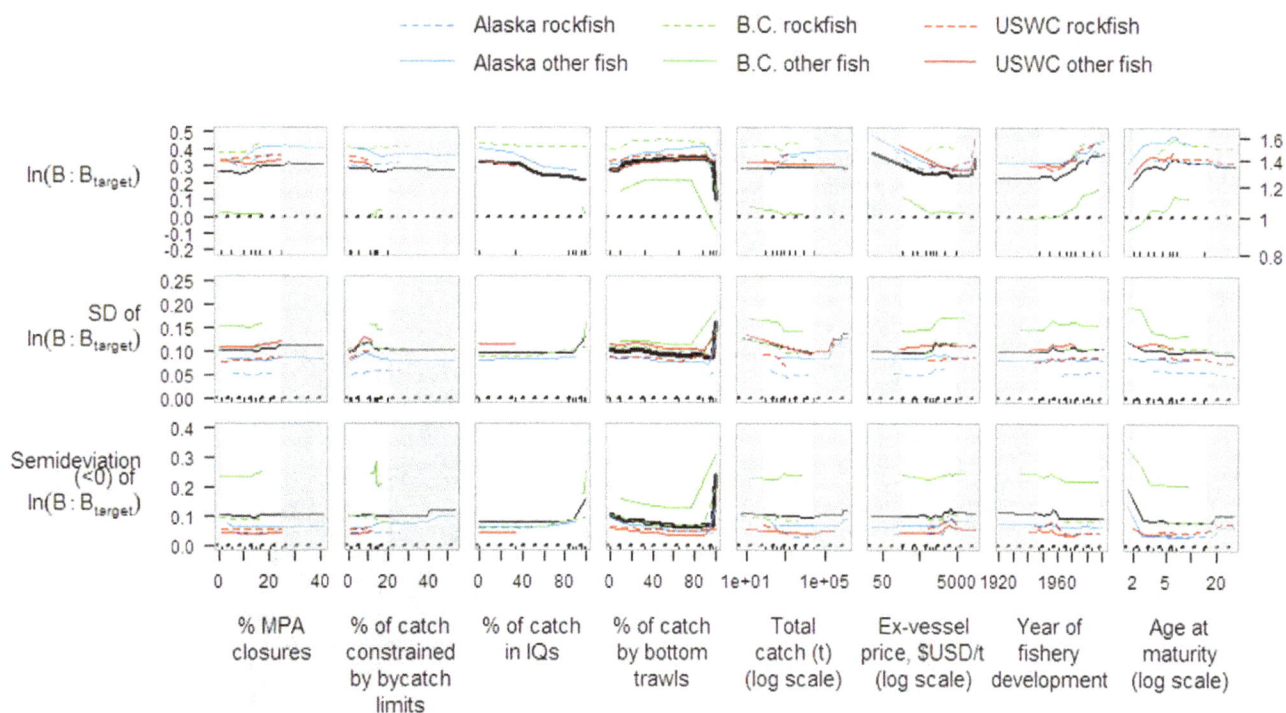

Figure 5. Partial dependence plots of three biomass-related variables on eight numerical stock-level covariates of North American Pacific groundfish. The mean, standard deviation, and semideviation of the log-ratio of current biomass to target biomass were calculated for each stock from the most recent 5-year period available. See Fig. 3 caption for further details.

S1). Similarly, we found little evidence of colinearity among the eight numerical predictor variables (all pairwise correlation coefficients were <0.5; Fig. S2). Scatterplots of observed response variables versus numerical predictor variables (Figs. S3, S4, S5) and scatterplots of observed versus predicted response variable values (Fig. S6) are also shown in the Supporting Information section. Partial dependence plots similar to Figs. 3, Fig. 4, Fig. 5,Fig. 6 are shown in the Supporting Information section for the five sensitivity analyses conducted (Fig. S7, Fig S8, Fig S9, Fig S10, Fig S11,Fig S12, Fig S13, Fig S14, Fig S15, Fig S16). In general, there were few noteworthy changes in the observed results from excluding stocks that are predominantly caught in recrea-

tional fisheries, filtering out less valuable stocks from the dataset (secondary target stocks or stocks with catch:TAC ratios <50%), or adding an extra predictor variable into random forest analyses (either region or maximum body length).

Discussion

Conservation objectives of western North American groundfish fisheries are clearly being met for the vast majority of stocks, with catches, exploitation rates, and biomass levels adhering to management targets. There are few stocks whose recent catches exceed TACs or whose recent exploitation rates exceed targets. Of

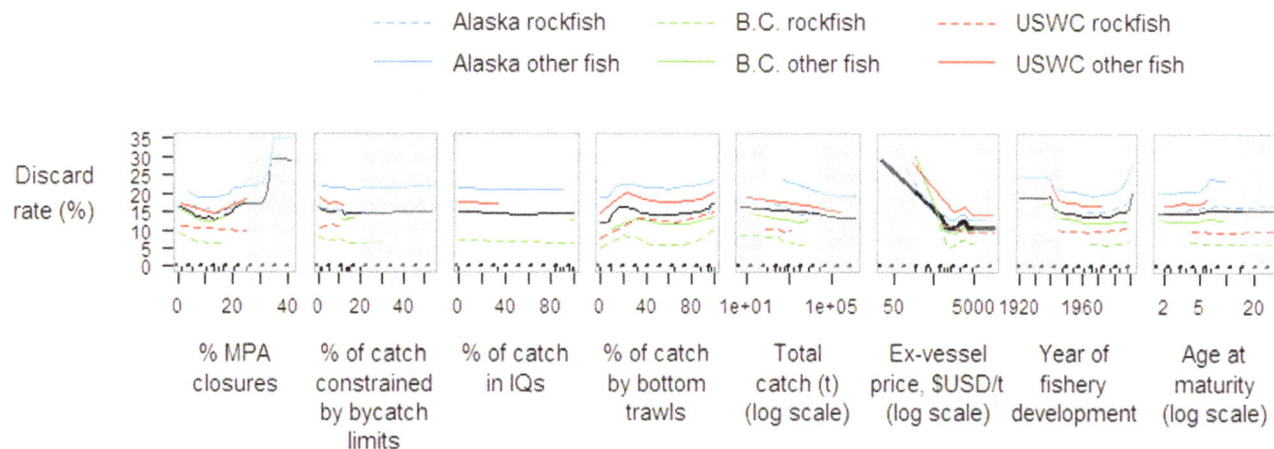

Figure 6. Partial dependence plots of discard proportion on eight numerical stock-level covariates of North American Pacific groundfish. Discard rates were calculated for each stock from the most recent 5-year period available. See Fig. 3 caption for further details.

the stocks with recent biomass below target levels, most have had exploitation rates reduced to below MSY-based target levels and thus are expected to rebuild. Although overfishing is a problem in many regions of the world [1,32,33], in these three regions the general pattern across stocks appears to be quite different in recent years. Catches and exploitation rates are on average far below the levels that would in the long term provide maximum sustainable yield, greatest revenue, and maximum benefits of food security.

Overexploitation and underexploitation are both associated with loss of potential yield and revenue from a fished stock [16,17]. If catch targets (TACs) are set adhering to sustainable yield recommendations from stock assessments, the loss of potential revenue from catching less than the TAC may be substantial. Based on simple calculations of stock-specific discrepancies between recent TAC and catch multiplied by ex-vessel prices (i.e., not accounting for long-term population dynamics impacts from set harvest policies), the total lost revenue from catching less than TAC in recent years was approximately 9% in Alaska (USD$ 100.7 million), 13% in B.C. ($14.7 million), and 43% on the U.S. West Coast ($40.8 million; Fig. 7). Catches generally adhere closely to TACs for the most valuable stocks in each region, but the cumulative percent loss of potential revenue increases as more of the lower value stocks are accounted for (Fig. 7). Allowable biological catches (ABC) are routinely estimated in assessments from Alaska and the USWC, and there is further revenue foregone from setting TACs lower than these ABC's (Fig. 7; TAC's are set lower than ABC's for a wide variety of ecological, social, and economic reasons). For some stocks, typically those of lower value, catches are below TACs due to market limitations. For others, even those with high ex-vessel prices, catches are below TACs because bycatch limits for weaker stocks are encountered before TACs for more targeted stocks are fulfilled [18]. This is especially the case for USWC stocks, where the percent loss of potential revenue for even the 10 most valuable stocks is >20% because TACs are not fully utilized (Fig. 7). Throughout the history of the USWC groundfish fishery, the loss in potential yield due to underexploitation (15–33% since 1990) has been considerably greater than the loss due to overexploitation (up to 3%) [18].

The total catch volume and ex-vessel price of stocks, both contributing to total value, were often stronger determinants of stock status than were any of the management tactics considered. The value of these groundfish fisheries was highly skewed towards a small number of species: in all three regions, the five most valuable stocks contributed 80–90% of the total catch value across all stocks (Fig. 7). More valuable stocks had lower discard rates and were kept closer to TAC, exploitation rate, and biomass targets than were less valuable stocks that were typically 'underfished'. Similarly, fisheries that developed earlier were closer to exploitation rate and biomass targets than were later-developing fisheries. These findings are consistent with results from a global analysis at the species level, as the best business opportunities (higher prices, greater catch potential) are often developed first [34]. For more valuable fisheries, there is greater economic incentive to maintain catches closer to MSY levels each year. The interannual variability and tendency to exceed TAC and exploitation rate targets were consequently greater for stocks with lower average catch volume. This has important implications from a food security perspective. Stocks providing the greatest contributions towards food security (via magnitude of catch volume) are also the ones with a more reliable provision of food (via lower interannual variability), not by top-down design but rather because they're more valuable. This observed pattern for groundfish is unlikely to hold across all taxa. Small pelagic fish stocks, for example, often provide large catch volumes but are also highly variable interannually.

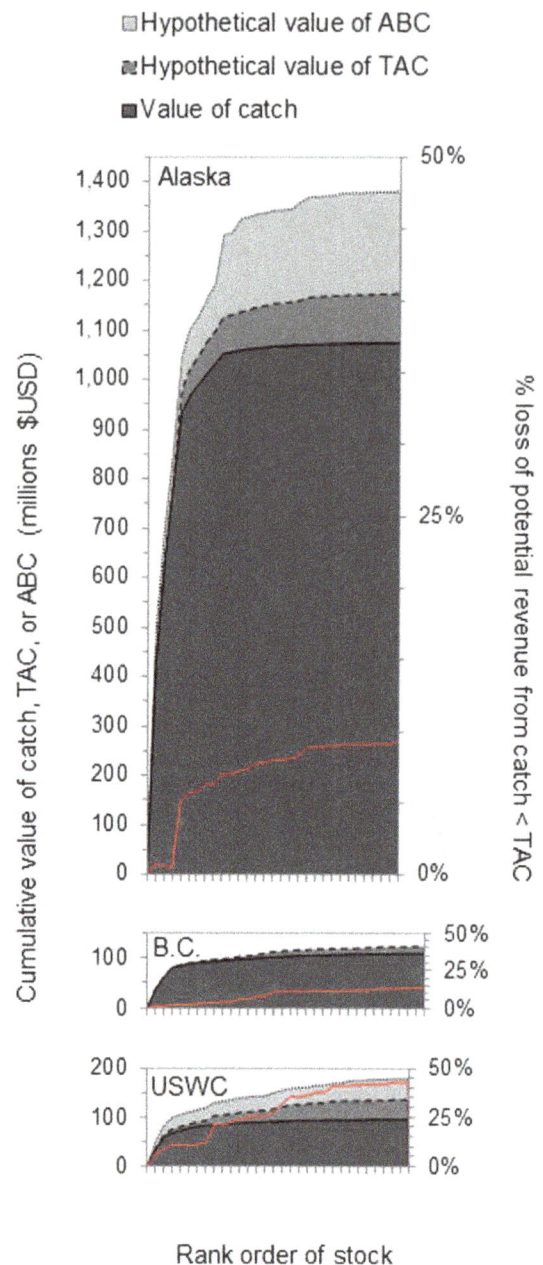

Figure 7. Cumulative distribution of groundfish value in Alaska, British Columbia, and the U.S. West Coast. The ex-vessel price of each stock is multiplied by the stock's total catch, total allowable catch (TAC), or allowable biological catch (ABC). Values represent recent 5-year means, and stocks are ranked within each region by catch value. % loss of potential revenue is calculated from the cumulative TAC–catch discrepancy as cumulative TAC/cumulative catch − 100%.

A harvest control rule which relies on estimated biological reference points is a management strategy that, if properly adhered to, should prevent overexploitation in the long term regardless of which management tactics are utilized to limit exploitation. Such rules [24,35] are a transparent framework for adjusting exploitation rates in response to biomass changes. Stocks managed under harvest control rules with lower limit reference points had less tendency for exploitation rates and biomass to exceed management targets despite mean exploitation rate and biomass levels being closer to biomass targets. In other words, harvest control rules maintained stocks closer to management targets without exceeding them. This observation is confounded with regional differences, however, as assessments with estimated limit reference points are more common for the two U.S. regions (Fig. 2). On the USWC, several stocks were previously overfished, but after strong measures to cut exploitation rates were implemented, stocks have now rebuilt or are recovering [1,11,36]. Adherence to control rules is also stronger in the two U.S. regions (Fig. 8). Written into fisheries legislation, the targeting of a stock is automatically ceased if estimated biomass falls below the lower limit [24]. Conversely, in B.C. the response of managers and the federal fisheries Minister to decreases in estimated abundance is more discretionary. In B.C., managers and the fisheries Minister thus have more flexibility to also consider social and economic factors when deciding on management actions to rebuild depleted stocks. The observed effect of harvest control rules is therefore unclear in this study, as other management differences between B.C. and the two U.S. regions may confound effects on stock status.

Several management tactics for limiting the catch of fish stocks are employed around the world, consisting of output controls (e.g. individual quotas, trip limits, fleet-wide TAC caps), input controls (e.g. effort limits, gear restrictions, time and area closures), or a combination of these [1,37–39]. We evaluated three of the most common tactics for constraining catch: temporal closures, permanent spatial closures, and bycatch limits of stocks co-caught with target stocks. In some ways, effects of these tactics on stock status may be moot as exploitation rates rarely exceeded management targets, so they are perhaps less useful as a catch-constraining tool compared to regions where stocks are more commonly overexploited. Still, some unexpected patterns emerged. Stocks with low proportions of spatial closures had increased variance in $F{:}F_{target}$, but exploitation rates rarely exceeded targets even at these low levels of spatial closure. Stocks with seasonal closures had lower discard rates and considerably higher catch:TAC and $F{:}F_{target}$ ratios, closer to management targets, but these are not necessarily causal relationships. Use of seasonal closures may be more likely for more valuable stocks in order to protect them during sensitive times like spawning periods, and these stocks may be maintained closer to their most productive levels and discarded less frequently because of their high value.

Bottom trawling has received considerable criticism in recent decades, not only for destructive habitat effects but also for the non-selective nature of the fishing gear [40]. Multispecies bottom trawl fisheries are generally thought to be less selective than mid-water trawls, pot, or longline fisheries [41], and bycatch limitations may thus be severe [26,27]. For example, stocks from the USWC multispecies fisheries had the lowest catch:TAC ratios (furthest from management targets) and the greatest interannual variability in catch:TAC, largely as a result of such bycatch limits. Interestingly, the tendency to exceed management targets across all regions occurred at very low and high levels of bottom trawling, but not at intermediate levels. High semi-deviation of $F{:}F_{target}$ occurred for stocks whose proportion of the total catch by bottom trawling was <20% and high semi-deviation of $B{:}B_{target}$ occurred for stocks with >90% catch by bottom trawling (most apparent for stocks from B.C.). In the latter case, the frequency of falling below biomass targets was relatively high because biomass levels of stocks caught entirely by bottom trawls were lower, *closer* to management targets on average, than were biomass levels of stocks caught with a mixture of gear types or without bottom trawling. This suggests that bottom trawl fisheries can target individual stocks [28] well

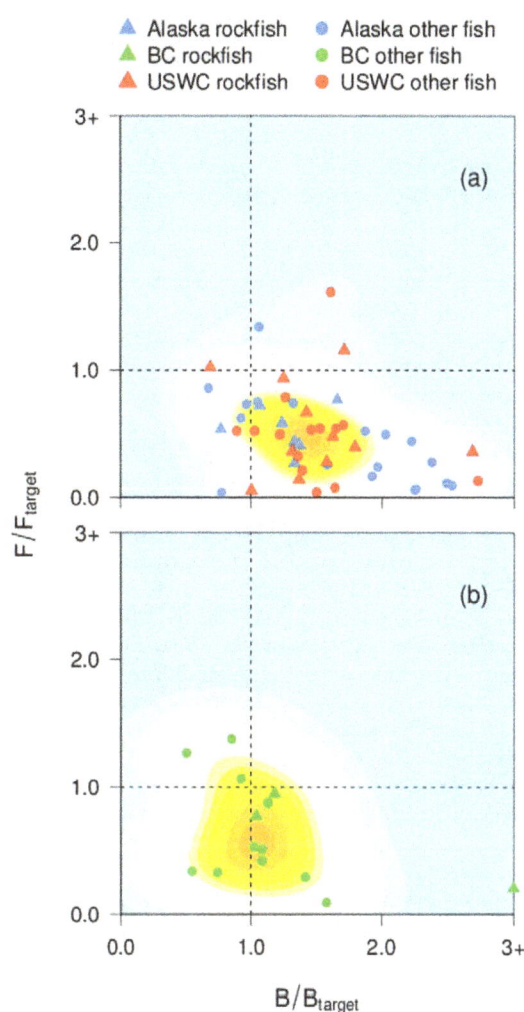

Figure 8. Influence of region and strength of adherence to harvest control rules on current groundfish stock status. Stock status is shown in terms of recent exploitation rates to target exploitation rates and recent biomass to target biomass levels. Stocks are categorized by the strength of adherence to harvest control rules: (a) automatic adherence; (b) discretionary adherence. Stocks with little or no commercial value, currently under a rebuilding plan, or co-caught with other stocks under rebuilding plans are excluded. See Fig. 1 caption for further details.

enough that on average stock biomasses are maintained near target levels even in multispecies fisheries.

Individual quota management has garnered much attention recently as a possible approach to end the race-to-fish. Catch-share systems are thought to provide incentives for fishermen to align their behavior with conservation objectives [22,42,43], and they have been shown to allow fisheries to better adhere to management targets and reduce interannual variation in catches and exploitation rates [30,44]. Most groundfish stocks from B.C. are under full catch share management, but under the 5-year period of data availability only Pacific whiting from the USWC and Pacific halibut, sablefish, and walleye pollock from Alaska were under partial catch share systems. These are three of the four most valuable Alaskan stocks, and the economic incentive to maintain stocks near their most productive levels likely explains why biomass levels were lower, closer to management targets, for these stocks under catch shares. The non-catch share stocks, i.e.

those managed with fleet-wide TAC caps but without allocation of quota to individuals, were also generally maintained within management targets. Thus it appears that as long as established TACs adhere to scientific recommendations and are adequately enforced, there are several possible allocation approaches for meeting groundfish conservation objectives. As a caveat, it is likely that catch share effects on response variables were underestimated for B.C. stocks because of the lack of contrast, i.e., there were very few non-catch share stocks (see Fig. 2) to contribute to observed relationships with response variables. The importance of catch shares to meeting management targets is likely especially important for B.C. stocks given that there are fewer stock assessments, less comprehensive survey programs, and less stringent enforcement of management plans compared to stocks in the U.S. regions.

Rockfish may be particularly susceptible to overfishing because they are often long lived and have relatively low fecundity, as has been shown for the California Current ecosystem [5,7]. Although we did not detect any strong differences in recent stock status between rockfish and other groundfish after accounting for other factors, we restricted our analysis to targeted stocks, so we did not include the stocks under rebuilding plans from the USWC, most of which are rockfish (Fig. 1a). Taxonomic effects may also arise through the age of maturity covariate, as rockfish tend to mature relatively late; stocks of younger age at maturity were on average associated with lower biomass, closer to management targets, and a greater tendency to fall below target biomass.

While it is common practice to evaluate exploitation rates and biomass relative to targets, there is generally less consideration given to their interannual variability even though effects of management strategies and tactics may have a greater impact on variances than on mean responses [45]. Interannual variability may be as great a concern as the current stock status for fishermen who must make a living every year including the poor years [46]. Further, it is often interannual variability in stock status that leads to ratcheting effects, with fleets building up during good years to capacities beyond what can be sustained in poorer years [47]. Identifying management actions and strategies that reduce the interannual variability in stock status may therefore be an important goal along with identifying actions that lead to favorable stock status on average.

We found that management of groundfish stocks is highly conservative in western North American regions; these three regions tend to be considered successfully managed in the context of global fisheries. Overfishing of individual stocks was rare, and across all stocks the tendency was to catch less than science-based TACs. However, the relatively low exploitation rates and high biomass in these regions are not representative of other regions around the world where overexploitation is a more serious problem [1,48]. These commercial fisheries also have a more recent history and fewer fishermen involved compared to many other regions like Europe and eastern North America, and this has possibly allowed for a greater conservation-oriented focus than other regions where social considerations are more prominent. No single management attribute stood out as being critical for sustainable management in these relatively successful regions. In the two U.S. regions, this success is likely attributable to strong science programs of surveys and assessments, legal requirements for conservation, and adequate enforcement. In B.C., catch shares are likely important for meeting stock management targets, despite not detecting this in analyses, as discussed above. Further studies at the global scale would provide more contrast in stock status, and in regions more characterized by overexploitation the influence of particular tactics on stock status may be stronger.

Methods

Data collection

We take a 'snapshot' approach and consider the most recent 5-year period of available data for each stock. Data used to calculate biological status were drawn from stock assessments. Data describing management attributes were compiled by reviewing assessments, fishery management plans, government databases or reports, the peer-reviewed literature, and especially by interviewing fisheries scientists, managers, and representatives of industry associations familiar with particular stocks. Values for management attributes corresponded to the same 5-year period of data availability for biological status. If a stock underwent a major management change during the most recent 5-year period (e.g., establishment of individual quota systems), we used instead the 5-year period immediately preceding the change.

All assessed stocks and many unassessed stocks from the three focal regions were included in our compiled databases. Time series and target reference point estimates for biomass and exploitation rates are routinely published in assessments. These are publicly available from the RAM Legacy Stock Assessment Database (ramlegacy.marinebiodiversity.ca) [29]. Occasionally time series estimates were available but target reference points were not estimated or published. In these few cases (4 of 60 stocks), a Schaefer [49] surplus production model was fit to time series data of total catch and total biomass estimates in order to estimate MSY-based reference points. The estimated ratios of $B:B_{MSY}$ and $F:F_{MSY}$ from a Schaefer model are a reasonable approximation to estimated ratios from stock assessments which typically use age-structured models [1,30].

Although many of the stocks from these regions are assessed annually, for others assessments are conducted less frequently. If time series estimates from assessments were only available for years prior to 2000, we did not include the stock in our analysis. Because we consider measures of biological status relative to management targets, we include in our analysis only targeted stocks. We therefore also exclude stocks that had little or no commercial value, were under a rebuilding plan during the 5-year period, or were co-caught with stocks under rebuilding plans (Table S2; we show the current status of these stocks, but exclude them from random forest data analyses). After these exclusions a total of 85 stocks were included in our data set, and for any given performance measure data were available for 60–77 stocks (Table 2).

Biological status as performance measures

We considered performance measures that related to the biological status of groundfish stocks: biomass, exploitation rate, adherence of catch to TAC, and discard rate. For the first three of these, we explicitly accounted for management targets; performance measures consisted of the ratio of current biomass to target biomass, the ratio of current exploitation rate to target exploitation rate, and the ratio of current catch to current TAC. These targets are generally more conservative than values associated with maximum sustainable yield (e.g., $B_{40\%}$ is a conservative proxy for B_{MSY}, the biomass level estimated to return long-term optimal yield). There is uncertainty as to which reference points are optimal [50,51], and occasionally other targets are used by managers; we used as the denominator whatever was the explicitly stated management target. The ratio of catch:TAC represents a measure of implementation error [52] (more recently termed outcome error). To calculate this ratio, time series values were carefully screened to ensure that catches and TACs represented the same quantities in terms of spatial area, fishing gears, inclusion

Table 2. Sample sizes for types of biological performance measures used in analyses.

Region Taxa	Catch and TAC data	F and F_{target} estimates	B and B_{target} estimates	% Discard estimates
Alaska, U.S.				
Rockfish	11	8	7	11
Other groundfish	19	18	18	19
U.S. West Coast (continental)				
Rockfish	7	8	8	4
Other groundfish	9	13	13	10
British Columbia, Canada (B.C.)				
Rockfish	12	3	3	14
Other groundfish	18	11	11	10
Total	76	61	60	77

Sampling units are individual groundfish stocks as defined in stock assessments. Stocks with little commercial value, under rebuilding plans in the last 5 years of data availability, or co-caught with stocks under rebuilding plans in the last 5 years are excluded.

or exclusion of discards, and inclusion or exclusion of recreational catches. In some cases this involved using only a subset of the total catch of a stock so that it corresponded appropriately with the TAC value for the same year.

For each of these three types of performance measures, we calculated three metrics: the mean response over the 5-year period, the 5-year standard deviation around this mean, and the 5-year semi-deviation. Semi-deviation is often used as a measure of asymmetric risk around a target value: it represents the variability only on the undesirable side of the management target (i.e., catches exceeding TACs, exploitation rates exceeding target exploitation rate, and biomass below target biomass) [46,53]. The 5-year semi-deviation around a target, $\delta(5, target)$, is zero if current values (x_i) are at or more conservative than management targets throughout the 5-year period, and increases as current values become increasingly undesirable with respect to target values:

$$\delta(5,\text{target}) = \sqrt{\frac{1}{5} \sum_{i=1}^{5} \begin{cases} (x_i - \text{target})^2 & \text{if } x_i > \text{target} \\ 0 & \text{if } x_i < \text{target} \end{cases}} . \quad (1)$$

Equation 1 is expressed for the cases of 0 if $catch_i < TAC_i$ or $F_i < F_{target}$; for biomass, the condition would instead switch to 0 if $B_i > B_{target}$. For all nine of these performance measures, the ratios were treated in log space. A tenth performance measure consisted of the proportion of a stock's total catch discarded at sea.

Management tactics as predictor variables

We considered six management attributes that are commonly hypothesized to affect the biological status of fish stocks (Table 3):

(i) A harvest control rule is a management strategy for reducing exploitation rates in response to declines of stock biomass below a target reference point. Some harvest control rules require that fishing stops if estimated biomass falls below a lower limit reference point, which implies that this limit reference point is estimated in stock assessments. Stocks were classified by whether or not the latter type of harvest control rule was in place.

(ii) Stocks were classified by whether or not seasonal closures were employed, which include either a period during the fishing season or the part of the year outside the fishing season. Closures may be designed for either the target stock or other stocks. We did not include cases where a fishery is closed for the remainder of a year simply after reaching the TAC.

(iii) The proportion of a stock's distributional area that is under spatial closure to fishing or extractive use (e.g., marine reserves; rockfish conservation areas). If distributional areas were not available, the area over which fishing typically occurs was used instead as the denominator. If a spatial closure did not impact the gear(s) with which a stock is normally caught (e.g. mid-water trawls), it was not considered a closure for that stock. Rotational or temporary closures were not counted.

(iv) The level of catch constraints due to bycatch limits for other species was estimated by fishery managers and/or industry association representatives. This is a subjective measure of the foregone catch of the target stock, expressed as a percentage of the hypothetical total catch in the absence of any bycatch limits. Bycatch limits may be for other exploited stocks, or for threatened species such as Steller sea lions in the Aleutian Islands. Note that four stocks were excluded from analyses because they were frequently co-caught with rebuilding stocks (Table S2), but catch constraints may also apply to the remaining 85 stocks.

(v) The proportion of a stock's total catch under catch share programs. These programs include any quota system that allocates harvesting rights to individuals, such as individual transferable, vessel, or fishing quotas (ITQ, IVQ, IFQ) and industry co-operatives with quota allocations.

(vi) The proportion of a stock's total catch that was taken by bottom trawls.

Other covariates that may potentially influence performance measures were also considered (Table 3). The total catch and ex-vessel price contribute to the value of a fishery, and there may be greater attention given to more valuable fisheries to ensure that management objectives are met [34]. The year of fishery development was determined for each stock, defined as the first

Table 3. Fishery management covariates and other factors accounted for in analyses.

Covariate	Description
HCR with limit RP	Yes/no. Is there a harvest control rule used which involves a limit reference point? I.e., is there a stock size below which all directed fishing should stop? This implies that stock assessments are conducted in which limit reference points are estimated.
Seasonal closures	Yes/no. Are seasonal closures used?
% MPA closures	Numerical. The proportion of the stock's area of distribution under permanent spatial closure to fishing or extractive use (e.g., marine protected areas).
% catch constrained by bycatch limits	Numerical. The (hypothetical) proportion of the stock's total possible catch that is foregone as a result of bycatch limits for other exploited stocks or for threatened species (e.g., marine mammals).
% of catch in IQs	Numerical. The proportion of the total catch that is caught under catch share programs (e.g. individual quotas, co-operatives with allocated harvesting rights).
% of catch by bottom trawling	Numerical. The proportion of the total catch that is caught by bottom trawls.
Total catch	Numerical. Includes recreational catches and discards if data are available.
Ex-vessel price	Numerical. Average price for commercial landings.
Year of fishery development	Numerical. The year in which total landings first reached 25% of the historic maximum annual landings in the entire time series.
Age at maturity	Numerical. The average age at reproductive maturity.
Taxa	Boolean. Broad taxonomic division between rockfish and other groundfish, which include flatfish and roundfish.

All covariates were calculated or determined at the stock level for the same recent 5-year period considered for response variables.

year in which the total landings reached 25% of the maximum historic landings in the full time series [34]. The average age at maturity represents an important life history trait affecting a stock's potential to rebuild. Values were drawn at the species level from FishBase. Finally, a taxonomic division was considered between rockfish and other groundfish, as the biological status and influence of management tactics may differ between these groups.

Data analyses

We used random forests to assess the influence of management tactics and other covariates (i.e., predictor variables) on the biological performance indicators (response variables). Random forests [54] are an ensemble of regression trees, which are a non-parametric recursive data splitting method for identifying covariates with relatively strong influence on a numerical response variable. At a given node of a regression tree, values of the response variable and one predictor variable are split into two groups based on whichever predictor variable's split results in the greatest sum of squares reduction of the response variable. The procedure is repeated such that within a single tree, multiple predictor variables can be shown to influence the response variable. Although single regression trees are unstable in terms of the order of variable importance among covariates, random forests involve bootstrapping the dataset (each component regression tree is constructed from one resampled dataset) and only allowing a random subset of covariates to be included at any given node of a component tree, with the result being a more robust measure of variable importance across the aggregated set of trees [55]. Random forests have been used increasingly in ecology and fisheries research [46,56,57]. They are attractive for cases like ours in that they allow for non-linear relationships between a predictor and response variable, they do not make any parametric assumptions about the distribution of a response variable, interactions among predictor variables are accounted for implicitly, they can handle missing values of predictor variables, and they are less susceptible to over-fitting compared with parametric

methods such as generalized linear models because the number of predictor variables available for selection at any given node of a tree is limited to a specified number.

For each of the ten performance measures, we conducted a separate random forest analysis using the 'randomForest' package (version 4.6-2) [58] in R (version 2.14.1) [59]. Stocks were weighted equally. Forests of 10,000 trees were used, which were more than adequate from visual inspection of model diagnostics. The cross-validation prediction accuracy represented by the mean square error of model fit is sensitive to $mtry$, a tuning parameter which limits the number of predictor variables allowed for selection at any one node of a component regression tree. Larger values of $mtry$ are often less susceptible to overfitting large models and allow for higher order interactions among predictors, while smaller values of $mtry$ often have greater cross-validation prediction accuracy. The mean square error of model fit was plotted over a range of $mtry$ values to determine an appropriate value. A value of $mtry = 5$ was selected based on these diagnostics, which is reasonably close to rules of thumb of $mtry \approx 1/3p$ for continuous predictors or $\sqrt{p}$ for categorical predictors, where p is the number of predictor variables (in our case, $p = 11$).

We show effects of management tactics and other covariates on response variables using partial dependence plots for the eight continuous predictors and marginal means of performance measures for the three categorical predictors listed in Table 3. Partial dependence plots show the effect of a predictor variable of interest on a response variable after accounting for the average effects of the other predictor variables in the model. At a given value of predictor variable x, a value of the response variable is predicted from all the combinations of observed values of the other predictor variables in the random forest dataset, and the average predicted response variable is determined. This process is repeated for many values of x to construct a dependence plot (see [58] for further details). Partial dependence plots are not constrained by linear relationships through the range of covariate values. We show marginal relationships for six separate groups (3 regions ×

categories of rockfish or other groundfish) as well as the overall marginal relationship. Although 'region' was not included as a predictor variable in the main analysis (only in a sensitivity analysis), the three regions differ in their values of other predictor variables, which permits separate partial dependence functions to be calculated for each region. We show a measure of relative predictor variable importance for each of the ten random forest analyses. The importance score is determined with cross-validation and reflects the loss of prediction accuracy associated with removing each predictor variable in turn (for further details see the 'importance {randomForest}' function in [58]). As the modest sample sizes available sometimes resulted in sparse data in the tails of predictor variable values, we draw the reader's attention in these partial dependence plots to the middle 80% of the values of each predictor variable. For categorical predictor variables, we show the difference in marginal means of response variables between the different levels of the predictor. We express this difference as a proportion of the span of the 95% confidence interval of response variable values, which provides a relative index of the effect of categorical predictors on a response variable.

Prior to analyses, we tested for colinearity among predictor variables using generalized variance inflation factors (GVIF) [60]. All GVIF values were <2.5 suggesting little possibility of confounding among the 11 predictor variables [60]. (We had originally considered 'region' as a twelfth predictor, categorical with 3 levels, but this was highly confounded with other predictors on the basis of high GVIF scores so was dropped as a predictor.) We confirmed this visually, plotting all pair-wise combinations of the eight continuous predictor variables and finding little evidence of colinearity. Similarly, we inspected mosaic plots of pair-wise combinations of the three categorical predictors and found limited reason for concern about the independence of these predictors. We plotted response variables versus predictor variables to visually assess relationships and plotted observed versus predicted response variable values to visually assess model fit. Finally, we conducted five sensitivity tests for random forest analyses to assess the influence on observed results of subsetting the dataset or adding predictor variables into the models:

(a) We excluded stocks if >50% of their total catch was taken by the recreational sector.

(b) We excluded stocks if their catch:TAC ratio was <50%, as this likely represents stocks that are not heavily targeted.

(c) We excluded stocks identified by managers or in fishery management plans as being secondary targets. Note these exclusions a–c are in addition to the exclusions mentioned previously (stocks with little or no commercial value, under rebuilding plans, or co-caught with rebuilding stocks), which were common across all analyses.

(d) We added a 3-level 'region' predictor variable to the model even though it was highly confounded with other predictor variables in order to assess whether observed effects of other predictors would change once region was accounted for explicitly (rather than implicitly, through differences among regions in the ranges of covariate values).

(e) We added maximum length (drawn at the species level from FishBase) as a continuous predictor variable in the model to allow for a second life history variable to possibly explain variation in performance measures.

Supporting Information

Figure S1 Values of management-related categorical covariates used in the analysis. Mosaic plots are shown for each of the three pair-wise combinations of categorical covariates.

Figure S2 Values of management-related numerical covariates used in the analysis. Lower panels show pair-wise scatterplots between covariates. Upper panels show correlation coefficients for the same pairs. Histograms of covariate values are shown on the diagonal. A Lowess fit with smoothing parameter = 2 is shown on each scatterplot. Data points show values for individual stocks, separated by color: Alaska—blue, U.S. west coast—red, B.C.—green; and by symbol: rockfish—triangles, other groundfish—circles.

Figure S3 Scatterplots of three catch:TAC response variables versus eight numerical stock-level covariates. The mean, standard deviation, and semideviation of the log-ratio of catch to TAC were calculated for each stock from the most recent 5-year period available. Data points are shown by region and rockfish/other groundfish groupings. Horizontal dotted lines at y = 0 represent general management objectives. Right hand axis values show catch:TAC values on linear scale.

Figure S4 Scatterplots of three $F:F_{target}$ response variables versus eight numerical stock-level covariates. See Fig. 3 caption for details.

Figure S5 Scatterplots of three $B:B_{target}$ response variables versus eight numerical stock-level covariates. See Fig. 3 caption for details.

Figure S6 Scatterplots of observed versus predicted response variable values. Three metrics (mean, standard deviation, and semideviation of the most recent 5-year period of data available) for each of three variables (catch:TAC, $F:F_{target}$, and $B:B_{target}$) were calculated for each stock. Predicted values are from the key run of random forest analyses. Dotted line shows the 1:1 relationship.

Figure S7 Partial dependence plots for sensitivity analyses showing the influence of numerical covariates on the mean catch:TAC ratio. The key run and five sensitivity scenarios are labelled in the right margin. See Fig. 3 caption in main text for further details.

Figure S8 Partial dependence plots for sensitivity analyses showing the influence of numerical covariates on the interannual variability of the catch:TAC ratio. The key run and five sensitivity scenarios are labelled in the right margin. See Fig. 3 caption in main text for further details.

Figure S9 Partial dependence plots for sensitivity analyses showing the influence of numerical covariates on the semi-deviation of the catch:TAC ratio. The key run and five sensitivity scenarios are labelled in the right margin. See Fig. 3 caption in main text for further details.

Figure S10 Partial dependence plots for sensitivity analyses showing the influence of numerical covariates on the mean F:F$_{target}$ ratio. The key run and five sensitivity scenarios are labelled in the right margin. See Fig. 3 caption in main text for further details.

Figure S11 Partial dependence plots for sensitivity analyses showing the influence of numerical covariates on the interannual variability of the F:F$_{target}$ ratio. The key run and five sensitivity scenarios are labelled in the right margin. See Fig. 3 caption in main text for further details.

Figure S12 Partial dependence plots for sensitivity analyses showing the influence of numerical covariates on the semi-deviation of the F:F$_{target}$ ratio. The key run and five sensitivity scenarios are labelled in the right margin. See Fig. 3 caption in main text for further details.

Figure S13 Partial dependence plots for sensitivity analyses showing the influence of numerical covariates on the mean B:B$_{target}$ ratio. The key run and five sensitivity scenarios are labelled in the right margin. See Fig. 3 caption in main text for further details.

Figure S14 Partial dependence plots for sensitivity analyses showing the influence of numerical covariates on the interannual variability of the B:B$_{target}$ ratio. The key run and five sensitivity scenarios are labelled in the right margin. See Fig. 3 caption in main text for further details.

Figure S15 Partial dependence plots for sensitivity analyses showing the influence of numerical covariates on the semi-deviation of the B:B$_{target}$ ratio. The key run and five sensitivity scenarios are labelled in the right margin. See Fig. 3 caption in main text for further details.

Figure S16 Partial dependence plots for sensitivity analyses showing the influence of numerical covariates on the proportion of catch discarded. The key run and five

sensitivity scenarios are labelled in the right margin. See Fig. 3 caption in main text for further details.

Table S1 Stocks included in random forest data analyses (n = 85).

Table S2 Stocks excluded from stock status presentation and from random forest data analyses.

Text S1 This section contains the results of exploratory data analyses and sensitivity tests as mentioned in the main text. Also listed are the groundfish stocks that were included in analyses (Table S1), and stocks excluded from analyses (Table S2).

Acknowledgments

We thank Trevor Branch for helpful comments on the manuscript. We thank many people for patiently answering our questions about management attributes of these groundfish stocks. For Alaska, this includes: Jane DiCosimo, Diana Stram, Mark Fina, Jeannie Heltzel (North Pacific Fishery Management Council); Mary Furuness, Jessica Gharrett, Josh Keaton, Ingrid Spies, Jim Ianelli (NOAA); and Cathy Tide (Alaska Fish & Game). For British Columbia, this includes Bruce Turris (Pacific Fisheries Management Inc.); Tameezan Karim, Rob Tadey, Barry Ackerman, Adam Keizer, John Davidson (Fisheries and Oceans Canada, Management); Greg Workman, Rick Stanley, Jackie King, Andy Edwards, Rob Kronlund, Lynne Yamanaka, Robyn Forrest, Nathan Taylor, Max Stocker (Fisheries and Oceans Canada, Science); Dennis Chalmers, and Carmen Matthews (B.C. Fisheries). For the U.S. West Coast, this includes John Devore, Jason Cope (Pacific Fishery Management Council); Bruce Leaman, Heather Gilroy, and Steven Hare (International Pacific Halibut Commission). We thank Trevor Branch and Coilin Minto for code used to create Fig. 1. We thank Trevor Branch and Olaf Jensen for insightful discussions.

Author Contributions

Conceived and designed the experiments: MCM RH. Performed the experiments: MCM JAB. Analyzed the data: MCM. Contributed reagents/materials/analysis tools: MCM RH. Wrote the paper: MCM. Edited manuscript for intellectual content: JAB RH.

References

1. Worm B, Hilborn R, Baum JK, Branch TA, Collie JS, et al. (2009) Rebuilding global fisheries. Science 325: 578–585.

2. Parma AM, Hilborn R, Orensanz JM (2006) The good, the bad, and the ugly: Learning from experience to achieve sustainable fisheries. Bulletin of Marine Science 78: 411–427.

3. Hilborn R (2007) Moving to sustainability by learning from successful fisheries. Ambio 36: 296–303.

4. DiCosimo J, Methot RD, Ormseth OA (2010) Use of annual catch limits to avoid stock depletion in the Bering Sea and Aleutian Islands management area (Northeast Pacific). ICES Journal of Marine Science 67: 1861–1865.

5. Parker SJ, Berkeley SA, Golden JT, Gunderson DR, Heifetz J, et al. (2000) Management of Pacific rockfish. Fisheries 25: 22–30.

6. Ralston S (2002) West Coast groundfish harvest policy. North American Journal of Fisheries Management 22: 249–250.

7. Levin PS, Holmes EE, Piner KR, Harvey CJ (2006) Shifts in a Pacific Ocean fish assemblage: The potential influence of exploitation. Conservation Biology 20: 1181–1190.

8. Davies TD, Baum JK (2012) Extinction Risk and Overfishing: Reconciling Conservation and Fisheries Perspectives on the Status of Marine Fishes. Sci Rep 2.

9. Rice JC, Legacè È (2007) When control rules collide: a comparison of fisheries management reference points and IUCN criteria for assessing risk of extinction. ICES Journal of Marine Science: Journal du Conseil 64: 718–722.

10. Ianelli J (2005) Assessment and fisheries management of Eastern Bering Sea walleye pollock: Is sustainability luck? Bulletin of Marine Science 76: 321–335.

11. Murawski SA (2010) Rebuilding depleted fish stocks: the good, the bad, and, mostly, the ugly. ICES Journal of Marine Science: Journal du Conseil 67: 1830–1840.

12. Turris BR (2000) A comparison of British Columbia's ITQ fisheries for groundfish trawl and sablefish: similar results from programmes with differing objectives, designs and processes. In: Shotton R, editor. Use of Property Rights in Fisheries Management FAO Fisheries Tech Paper 404/1 FishRights 99 Conference. Fremantle: FAO. pp. 254–261.

13. Branch TA (2006) Discards and revenues in multispecies groundfish trawl fisheries managed by trip limits on the US West Coast and by ITQs in British Columbia. Bulletin of Marine Science 78: 669–689.

14. Mawani T (2009) Evaluation of the Commercial Groundfish Integration Pilot Program in British Columbia: Royal Roads University. 124 p.

15. Yamanaka KL, Logan G (2010) Developing British Columbia's inshore rockfish conservation strategy. Marine and Coastal Fisheries 2: 28–46.

16. Hilborn R (2007) Managing fisheries is managing people: what has been learned? Fish and Fisheries 8: 285–296.

17. Hilborn R (2010) Pretty Good Yield and exploited fishes. Marine Policy 34: 193–196.

18. Hilborn R, Stewart IJ, Branch TA, Jensen OP (2012) Defining trade-offs among conservation, profitability, and food security in the California current bottom-trawl fishery. Conserv Biol 26: 257–266.

19. Allison GW, Lubchenco J, Carr MH (1998) Marine reserves are necessary but not sufficient for marine conservation. Ecological Applications 8: S79–S92.

20. Pauly D, Christensen V, Guenette S, Pitcher TJ, Sumaila UR, et al. (2002) Towards sustainability in world fisheries. Nature 418: 689–695.

21. Beddington JR, Agnew DJ, Clark CW (2007) Current problems in the management of marine fisheries. Science 316: 1713–1716.

22. Grafton RQ, Arnason R, Bjorndal T, Campbell D, Campbell HF, et al. (2006) Incentive-based approaches to sustainable fisheries. Canadian Journal of Fisheries and Aquatic Sciences 63: 699–710.

23. Costello C, Gaines SD, Lynham J (2008) Can catch shares prevent fisheries collapse? Science 321: 1678–1681.

24. Restrepo VR, Thompson GG, Mace PM, Gabriel WL, Low LL, et al. (1998) Technical guidance on the use of precautionary approaches to implementing National Standard 1 of the Magnuson–Stevens Fishery Conservation and Management Act. National Oceanic and Atmospheric Administration (US) Technical Memorandum NMFS-F/SPO-31. 54 p.

25. Mace (2001) A new role for MSY in single-species and ecosystem approaches to fisheries stock assessment and management. Fish and Fisheries 2: 2–32.

26. Adlerstein SA, Trumble RJ (1998) Pacific halibut bycatch in Pacific cod fisheries in the Bering Sea: an analysis to evaluate area-time management. Journal of Sea Research 39: 153–166.

27. Crowder LB, Murawski SA (1998) Fisheries bycatch: Implications for management. Fisheries 23: 8–17.

28. Branch TA, Hilborn R (2008) Matching catches to quotas in a multispecies trawl fishery: targeting and avoidance behavior under individual transferable quotas. Canadian Journal of Fisheries and Aquatic Sciences 65: 1435–1446.

29. Ricard D, Minto C, Jensen OP, Baum JK (2012) Examining the knowledge base and status of commercially exploited marine species with the RAM Legacy Stock Assessment Database. Fish and Fisheries 13: 380–398.

30. Melnychuk MC, Essington TE, Branch TA, Heppell SS, Jensen OP, et al. (2012) Can catch share fisheries better track management targets? Fish and Fisheries 13: 267–290.

31. Fraser and Associates (2008) Linkages between the primary fish production and fish processing sectors in British Columbia. Final Phase 3 Report. Prepared for the Department of Fisheries and Oceans Canada, Pacific Region, Policy Branch. 29 p.

32. Froese R, Proelß A (2010) Rebuilding fish stocks no later than 2015: will Europe meet the deadline? Fish and Fisheries 11: 194–202.

33. Piet GJ, van Overzee HMJ, Pastoors MA (2010) The necessity for response indicators in fisheries management. ICES Journal of Marine Science 67: 559–566.

34. Sethi SA, Branch TA, Watson R (2010) Global fishery development patterns are driven by profit but not trophic level. Proceedings of the National Academy of Sciences of the United States of America 107: 12163–12167.

35. Deroba JJ, Bence JR (2008) A review of harvest policies: Understanding relative performance of control rules. Fisheries Research 94: 210–223.

36. Rosenberg AA, Swasey JH, Bowman M (2006) Rebuilding US fisheries: progress and problems. Frontiers in Ecology and the Environment 4: 303–308.

37. Sutinen JG (1999) What works well and why: evidence from fishery-management experiences in OECD countries. ICES Journal of Marine Science 56: 1051–1058.

38. Cochrane KL (2002) A fishery manager's guidebook. Management measures and their application. FAO Fisheries Technical Paper 424. Rome. 231 p.

39. Degnbol P, Gislason H, Hanna S, Jentoft S, Raakjarnielsen J, et al. (2006) Painting the floor with a hammer: Technical fixes in fisheries management. Marine Policy 30: 534–543.

40. Hiddink JG, Jennings S, Kaiser MJ, Queiros AM, Duplisea DE, et al. (2006) Cumulative impacts of seabed trawl disturbance on benthic biomass, production, and species richness in different habitats. Canadian Journal of Fisheries and Aquatic Sciences 63: 721–736.

41. Chuenpagdee R, Morgan LE, Maxwell SM, Norse EA, Pauly D (2003) Shifting gears: assessing collateral impacts of fishing methods in US waters. Frontiers in Ecology and the Environment 1: 517–524.

42. Fujita RM, Foran T, Zevos I (1998) Innovative approaches for fostering conservation in marine fisheries. Ecological Applications 8: S139–S150.

43. Hilborn R, Orensanz JM, Parma AM (2005) Institutions, incentives and the future of fisheries. Philosophical Transactions of the Royal Society B-Biological Sciences 360: 47–57.

44. Essington TE (2010) Ecological indicators display reduced variation in North American catch share fisheries. Proceedings of the National Academy of Sciences of the United States of America 107: 754–759.

45. Essington TE, Melnychuk MC, Branch TA, Heppell SS, Jensen OP, et al. (2012) Catch shares, fisheries, and ecological stewardship: a comparative analysis of resource responses to a rights-based policy instrument. Conservation Letters 5: 186–195.

46. Sethi SA, Dalton M, Hilborn R (2012) Quantitative risk measures applied to Alaskan commercial fisheries. Canadian Journal of Fisheries and Aquatic Sciences 69: 487–498.

47. Ludwig D, Hilborn R, Walters C (1993) Uncertainty, resource exploitation, and conservation: Lessons from history. Science 260: 17–18.

48. FAO (2012) State of World Fisheries and Aquaculture. Rome: Food and Agriculture Organization of the United Nations. Fisheries and Aquaculture Department.

49. Schaefer MB (1954) Some aspects of the dynamics of populations, important for the management of the commercial fisheries. 56 p p.

50. Brodziak J (2002) In search of optimal harvest rates for West Coast groundfish. North American Journal of Fisheries Management 22: 258–271.

51. Hilborn R, Parma A, Maunder M (2002) Exploitation rate reference points for West Coast rockfish: Are they robust and are there better alternatives? North American Journal of Fisheries Management 22: 365–375.

52. Rosenberg AA, Restrepo VR (1994) Uncertainty and risk evaluation in stock assessment advice for U.S. maine fisheries. Canadian Journal of Fisheries and Aquatic Sciences 51: 2715–2720.

53. Porter RB (1974) Semivariance and stochastic dominance: A comparison. American Economic Review 64: 200–204.

54. Breiman L (2001) Random forests. Machine Learning 45: 5–32.

55. Breiman L (2001) Statistical modeling: The two cultures. Statistical Science 16: 199–231.

56. Cutler DR, Edwards TC, Beard KH, Cutler A, Hess KT, et al. (2007) Random forests for classification in ecology. Ecology 88: 2783–2792.

57. Gutierrez NL, Hilborn R, Defeo O (2011) Leadership, social capital and incentives promote successful fisheries. Nature 470: 386–389

58. Liaw A, Wiener M (2002) Classification and regression by randomForest. R News. pp. 18–22.

59. R Development Core Team (2012) R: A language and environment for statistical computing. 2.14.1 ed. Vienna, Austria: R Foundation for Statistical Computing.

60. Zuur AF, Ieno EN, Elphick CS (2009) A protocol for data exploration to avoid common statistical problems. Methods in Ecology and Evolution 1: 3–14.

PERMISSIONS

LIST OF CONTRIBUTORS

Sandrine Petit, Nicolas M. Munier-Jolain and Martin Lechenet
Institut National de la Recherche Agronomique, Unité Mixte de Recherche 1347 Agroécologie, Dijon, Côte d'Or, France

Vincent Bretagnolle
Centre d'Etudes Biologiques de Chizé - Centre National de Recherche Scientifique, Beauvoir sur Niort, Deux-Sévres, France

Christian Bockstaller
Institut National de la Recherche Agronomique, Unité de Recherche 1121 Agronomie et Environnement, Colmar, Haut-Rhin, France
Université de Lorraine, Vandoeuvre-lés-Nancy, Meurthe-et-Moselle, France

François Boissinot
Chambre d'Agriculture des Pays de la Loire, Angers, Maine-et-Loire, France

Marie-Sophie Petit
Chambre Régionale d'Agriculture de Bourgogne, Quetigny, Côte d'Or, France

Florian Zabel, Birgitta Putzenlechner and Wolfram Mauser
Department of Geography, Ludwig Maximilians University, Munich, Germany

Luis Fernando Chaves
Graduate School of Environmental Sciences, Hokkaido University, Sapporo, Japan
Programa de Investigació'n en Enfermedades Tropicales (PIET), Escuela de Medicina Veterinaria, Universidad Nacional, Heredia, Costa Rica
Institute of Tropical Medicine (NEKKEN), Nagasaki University, Nagasaki, Japan

Nathan G. Taylor, Murdoch K. McAllister and Tom Carruthers
Fisheries Center, University of British Columbia, Vancouver, British Columbia, Canada

Gareth L. Lawson
Department of Biology, Woods Hole Oceanographic Institution, Woods Hole, Massachusetts, United States of America

Barbara A. Block
Hopkins Marine Station, Stanford University, Pacific Grove, California, United States of America

Ussif Rashid Sumaila, Andrew Dyck, Vicky Lam and Wilf Swartz
Fisheries Economics Research Unit, Fisheries Centre, University of British Columbia, Vancouver, British Columbia, Canada

William Cheung, Daniel Pauly, Reginald Watson and Dirk Zeller
Sea Around Us Project, Fisheries Centre, University of British Columbia, Vancouver, British Columbia, Canada

Kamal Gueye
The United Nations Environment Programme, Geneva, Switzerland

Ling Huang
Department of Economics, University of Connecticut, Storrs, Connecticut, United States of America

Thara Srinivasan
Pacific Ecoinformatics and Computational Ecology Lab, Berkeley, California, United States of America

Belén Carbonetto, Nicolás Rascovan and Martin P. Vázquez
Instituto de Agrobiotecnología de Rosario (INDEAR), Predio CCT Rosario, Santa Fe, Argentina

Roberto Àlvarez
Facultad de Agronomía, Universidad de Buenos Aires, Buenos Aires, Argentina

Alejandro Mentaberry
Departamento de Fisiología y Biología Molecular y Celular, Facultad de Ciencias Exactas y Naturales, Universidad de Buenos Aires, Buenos Aires, Argentina

Catherine J. Payne, Euan G. Ritchie and Dale G. Nimmo
Centre for Integrative Ecology, School of Life and Environmental Sciences, Deakin University, Melbourne, Victoria, Australia

Luke T. Kelly
Australian Research Council Centre of Excellence for Environmental Decisions, School of Botany, University of Melbourne, Melbourne, Victoria, Australia

Masami Fujiwara
Department of Wildlife and Fisheries Sciences, Texas A&M University, College Station, Texas, United States of America

Xuelin Zhang, Qun Wang, Yilun Wang and Chaohai Li
The Incubation Base of the National Key Laboratory for Physiological Ecology and Genetic Improvement of Food Crops in Henan Province, Zhengzhou, China; Agronomy College of Henan Agricultural University, Zhengzhou, China

Frank S. Gilliam
Department of Biological Sciences, Marshall University, Huntington, West Virginia, United States of America

Feina Cha
Meteorological Bureau of Zhengzhou, Zhengzhou, China

Kosuke Takemura
Graduate School of Management, Kyoto University, Kyoto, Japan

Yukiko Uchida and Sakiko Yoshikawa
Kokoro Research Center, Kyoto University, Kyoto, Japan

Sarah J. Helyar, Hywel ap D Lloyd, Mark de Bruyn and Gary R. Carvalho
Molecular Ecology and Fisheries Genetics Laboratory, Bangor University, Bangor, Wales, United Kingdom

Jonathan Leake
Sunday Times, London, United Kingdom

Niall Bennett
Greenpeace UK, London, United Kingdom

Noél Michael André Holmgren and Niclas Norrström
Systems Biology Research Centre, School of Bioscience, University of Skövde, Skövde, Sweden

Robert Aps
University of Tartu, Estonian Marine Institute, Tallinn, Estonia

Sakari Kuikka
Fisheries and Environmental Management Group, Department of Environmental Sciences, University of Helsinki, Helsinki, Finland

Colin D. Buxton, Klaas Hartmann and Caleb Gardner
Fisheries Aquaculture and Coasts Centre, Institute for Marine and Antarctic Studies, University of Tasmania, Hobart, Tasmania, Australia

Robert Kearney
Institute for Applied Ecology, University of Canberra, Canberra, Australian Capital Territory, Australia

Martin Patenaude-Monette and Jean-François Giroux
Groupe de recherche enécologie comportementale et animale, Département des sciences biologiques, Université du Québec á Montréal, Montréal, Québec, Canada

Marc Bélisle
Département de biologie, Université de Sherbrooke, Sherbrooke, Québec, Canada

Nicolás L. Gutiérrez, David J. Agnew, Patricia L. Bianchi, Daniel D. Hoggarth, Chris Ninnes and Amanda Stern-Pirlot
Marine Stewardship Council, London, United Kingdom

Sarah R. Valencia and Christopher Costello
Bren School of Environmental Science and Management, University of California Santa Barbara, Santa Barbara, California, United States of America

Trevor A. Branch, Timothy E. Essington, Ray Hilborn and James T. Thorson
School of Aquatic and Fishery Sciences, University of Washington, Seattle, Washington, United States of America

Julia K. Baum
Department of Biology, University of Victoria, Victoria, British Columbia, Canada

Jorge Cornejo-Donoso
Interdepartmental Graduate Program in Marine Science, Marine Science Institute, University of California Santa Barbara, Santa Barbara, California, United States of America
Universidad Austral de Chile, Centro Trapananda, Coyhaique, Chile

Omar Defeo
UNDECIMAR, Facultad de Ciencias, Montevideo, Uruguay

Ashley E. Larsen, Seeta Sistla, Rebecca L. Selden and Sarah J. Teck
Department of Ecology, Evolution and Marine Biology, University of California Santa Barbara, Santa Barbara, California, United States of America

Keith Sainsbury
University of Tasmania, Tasmanian Aquaculture & Fisheries Inst, Taroona, Tasmania, Australia

Anthony D. M. Smith
Commonwealth Scientific and Industrial Research Organization, Wealth from Oceans Flagship, Hobart, Tasmania, Australia

Nicholas E. Williams
Department of Anthropology, University of California Santa Barbara, Santa Barbara, California, United States of America

Chris Pagnutti
School of Environmental Sciences, University of Guelph, Guelph, Ontario, Canada
Department of Mathematics and Statistics, University of Guelph, Guelph, Ontario, Canada

Chris T. Bauch
Department of Mathematics and Statistics, University of Guelph, Guelph, Ontario, Canada
Department of Ecology and Evolutionary Biology, Princeton University, Princeton, New Jersey, United States of America

Madhur Anand
School of Environmental Sciences, University of Guelph, Guelph, Ontario, Canada
Department of Ecology and Evolutionary Biology, Princeton University, Princeton, New Jersey, United States of America

Gema Escribano-Avila, Emilio Virgós and Adrián Escudero
Departamento de Biología y Geología, Universidad Rey Juan Carlos, Madrid, Spain

Virginia Sanz-Pérez
Facultad de Ciencias Ambientales, Universidad de Castilla-La Mancha, Toledo, Spain

Beatriz Pías
Departamento de Biología Vegetal I, Universidad Complutense de Madrid, Madrid, Spain

Fernando Valladares
Departamento de Biología y Geología, Universidad Rey Juan Carlos, Madrid, Spain
Museo Nacional de Ciencias Naturales, Consejo Superior de Investigaciones Cientı́ficas, Madrid, Spain

Esperanza Huerta and Salvador Hernandez-Daumas
El Colegio de la Frontera Sur, Unidad Campeche, Dpto. Agroecología, Campeche, México

Christian Kampichler
Universidad Juárez Autónoma de Tabasco, División de Ciencias Biológicas, Villahermosa, Tabasco, México
Sovon Dutch Centre for Field Ornithology, Natuurplaza (Mercator 3), Nijmegen, The Netherlands

Susana Ochoa-Gaona and Ben De Jong
El Colegio de la Frontera Sur, Unidad Campeche, Dpto. Sustainability Sciences, Campeche, México

Violette Geissen
University of Bonn - INRES, Bonn, Germany
Wageningen University and Research Center – Alterra, Wageningen, Gelderland, Netherlands

Robert I. McDonald
Central Science Program, The Nature Conservancy, Arlington, Virginia, United States of America

Evan H. Girvetz
Global Climate Change Program, The Nature Conservancy, Seattle, Washington, United States of America

James H. Larson, William B. Richardson, Jon M. Vallazza and John C. Nelson
Upper Midwest Environmental Sciences Center, United States Geological Survey, La Crosse, Wisconsin, United States of America

Wesley S. Patrick
Office of Sustainable Fisheries, National Marine Fisheries Service, Silver Spring, Maryland, United States of America

Jason Cope
Northwest Fisheries Science Center, National Marine Fisheries Service, Seattle, Washington, United States of America

Michael C. Melnychuk and Jeannette A. Banobi, Ray Hilborn
University of Washington, School of Aquatic and Fishery Sciences, Seattle, Washington, United States of America

Index